Emergence and Control of African Swine Fever

Emergence and Control of African Swine Fever

Guest Editors

Lauro Velazquez-Salinas
Gisselle N. Medina
Elizabeth Ramirez-Medina

MDPI Basel • Beijing • Wuhan • Barcelona • Belgrade • Novi Sad • Cluj • Manchester

Guest Editors

Lauro Velazquez-Salinas
ARS
Plum Island Animal Disease Center
New York
United States

Gisselle N. Medina
ARS
Plum Island Animal Disease Center
New York
United States

Elizabeth Ramirez-Medina
ARS
Plum Island Animal Disease Center
New York
United States

Editorial Office
MDPI AG
Grosspeteranlage 5
4052 Basel, Switzerland

This is a reprint of the Special Issue, published open access by the journal *Pathogens* (ISSN 2076-0817), freely accessible at: www.mdpi.com/journal/pathogens/special_issues/emergence_control_african_swine_fever.

For citation purposes, cite each article independently as indicated on the article page online and using the guide below:

Lastname, A.A.; Lastname, B.B. Article Title. *Journal Name* **Year**, *Volume Number*, Page Range.

ISBN 978-3-7258-4038-0 (Hbk)
ISBN 978-3-7258-4037-3 (PDF)
https://doi.org/10.3390/books978-3-7258-4037-3

Contents

pathogens

Editorial

Emergence and Control of African Swine Fever

Lauro Velazquez-Salinas

National Bio and Agro-Defense Facility (NBAF), Agricultural Research Service, U.S. Department of Agriculture, Manhattan, KS 66502, USA; lauro.velazquez@usda.gov

African swine fever (ASF) is a highly lethal and contagious viral disease found in domestic pigs, wild boars, and wild suids, and it has significant economic consequences [1]. This disease is currently endemic in many parts of the world, but the lack of an effective commercial vaccine for prevention and control makes the threat of ASF spreading to ASF-free regions more prominent [1]. The identification of ASF in the Americas for the first time in almost 40 years demonstrates the urgent need to develop novel countermeasures to control this disease [2]. This editorial aims to announce the completion (31/01/2025) of the Special Issue "Emergence and Control of African Swine Fever" (https://www.mdpi.com/journal/pathogens/special_issues/emergence_control_african_swine_fever) (accessed on 13 March 2025). A total of 20 publications are included in this Special Issue. These publications can be classified into the categories of control (vaccination, drugs, disinfectants, and immunology) ($n = 6$), epidemiology ($n = 6$), pathogenesis ($n = 5$), and diagnostics ($n = 3$). Below, I highlight some of the main findings published in this Special Issue.

Currently, the use of live attenuated vaccines (LAVs) based on the deletion of specific viral genes appears to be one of the most promising strategies for controlling ASF [3]. However, multiple challenges are associated with the development of LAVs, such as the lack of adapted cell lines that support the replication of ASFV and the limited knowledge about the safety profile of LAV candidates in the long term. In light of this, Ramirez-Medina et al. (2024) reported the ability of IPKM (Immortalized Porcine Kidney Macrophage) cells to be used in the production of stocks for LAVs, which only replicate in primary swine macrophages. The authors showed the efficacy of these cells in producing the LAV candidate ASFV-G-Δ9GL/ΔUKp10, which was able to protect pigs via the virulent ASFV-G (parental strain). On the other hand, Borca et al. (2023) documented the absence of the long-term (180 days) residual virulence of ASFV-G-ΔI177L, highlighting the safety profile of this LAV candidate.

In terms of the protective mechanisms induced by the LAV candidates of ASFV, Silva et al. (2022) contributed a study using ASFV-G-Δ9GL/ΔUK and ASFV-G-ΔI177L, showing the relevance of neutralizing antibodies during protection induced by these LAV candidates. Also, Wu et al. (2024) published an interesting review on the implications of T/B-cell epitopes and vaccine adjuvants in the immune response and vaccine development of ASFV.

As an alternative mechanism of control of ASF, Jackman et al. (2023) explored the significance of glycerol monolaurate (GML) in inhibiting the infection of a highly virulent strain of ASFV (isolate Armenia/07) in porcine macrophages, showing a > 99% decrease in viral infectivity. Furthermore, Frost et al. (2023) evaluated the efficacy of 24 commercial disinfectants, demonstrating the importance of disinfectants containing formic acid, phenolic compounds, and oxidizing agents in decreasing the viral titers of ASFV.

The infectious cycle of ASFV in nature involves complex epidemiological dynamics that complicate the control and eradication of ASF [4]. Different epidemiologic studies

Received: 5 March 2025
Revised: 11 March 2025
Accepted: 13 March 2025
Published: 14 March 2025

Citation: Velazquez-Salinas, L. Emergence and Control of African Swine Fever. *Pathogens* **2025**, *14*, 283. https://doi.org/10.3390/pathogens14030283

were published in this Special Issue, giving more insights into the factors predisposing the circulation of ASFV worldwide. Schambow et al. (2023) conducted an epidemiologic assessment in the Dominican Republic, showing the correlation of low biosecurity practices with the presentation of outbreaks in backyard farms. Spatiotemporal analysis suggested the existence of an endemic pattern of ASF located in the central region of this country. In the Philippines, Hsu et al. (2023) carried out an epidemiological study to identify the factors influencing the diagnosis, spread, and control of ASF in this country. This study highlights the relevance of timely reporting and enhanced biosecurity measures to manage ASF outbreaks in the Philippines. Additionally, Moskalenko et al. (2024) evaluated the awareness and perceptions of pig keepers about ASF in Ukraine, demonstrating the necessity of improving communication strategies about ASF in this country.

Two interesting studies involving molecular epidemiology were published in this special edition. On the one hand, Okwasiimire et al. (2023) conducted a study in Uganda, confirming the circulation of the ASFV genotype IX in domestic pigs in this country. Whole genome sequencing analysis revealed details about the evolution of this genotype, indicating the necessity of characterizing current viral strains circulating in Uganda to identify potentially new clinical phenotypes. On the other hand, Mazloum et al. (2022) conducted the genetic characterization of the ASFV isolates associated with genotype II in the Russian Federation between 2013 and 2017. Using the central variable region (CVR) of the ASFV gene B602L, it was shown that viral strains that circulate in twenty-three different regions within the Russian Federation have diverged into six different phylogenetic groups. The authors propose using the CVR of B602L to conduct future epidemiology studies in Europe and Asia.

Finally, in this category of epidemiology, Richter et al. (2023) presented an epidemiologic analysis comparing the circulation patterns of ASF in wild boars in Germany (Saxony) and Latvia. The results of this comparison indicated that the rapid implementation of new control strategies based on previous experiences from other countries might have improved Saxony's response to ASF in wild boars.

Understanding the pathogenesis of ASFV is a key aspect in controlling the disease. Indeed, identifying reliable experimental models that replicate the clinical outcomes observed in the field is an important aspect to be determined by pathogenesis studies [5]. In this sense, Olesen et al. (2025) contributed an interesting study to this Special Issue, in which they evaluated the dose of the ASFV strain (ASFV POL/2015/Podlaskie strain) required to establish infections in pigs following oral uptake (natural route of infection). The authors concluded that compared with the intranasal route, higher doses are needed to establish an infection via the oral route. These results highlight the implications of control strategies for ASF using baited vaccines containing LAVs. In another study using the oral route of inoculation, Nguyen et al. (2023) conducted a pathogenies study in pigs to characterize the ASFV strain (VNUA/HY/ASF1) associated with the first reported outbreak of ASF in Vietnam. Pigs infected with VNUA/HY/ASF1 produced disparate clinical outcomes (acute and subacute), producing clinical signs between 4 and 14 days post infection (dpi) and causing the death of the pigs between 10 and 27 dpi. This study provides relevant information about the pathogenesis of ASFV using a natural exposure model.

Some years after the emergence of the highly virulent ASFV genotype II in 2007, clinical outcomes observed in the field started varying from fatal acute to subclinical, indicating the existence of low-virulence variants [6]. In this sense, Avagyan et al. (2024) conducted a study to characterize ASFV variants producing chronic and persistent infections in pigs, concluding that chronic forms of ASF are associated with a decreased immune response and lower infectious titers found in the blood and tissues of the infected pigs.

Similarly, Sehl-Ewert et al. (2022) conducted a study characterizing the viral variants emerging from the ASFV lineage genotype II circulating in wild boars in Germany. The pathological analysis of carcasses obtained from naturally infected wild boars showed potential disparate levels of virulence among natural ASFV variants circulating in Germany. This study highlights the relevance of variants producing extended disease outcomes and their potential impact on shedding and transmission.

Furthermore, in this category of pathogenesis, Li et al. (2022) published a transcriptomic analysis of macrophages infected with ASFV. The results of this study emphasized the potential role of the NF-κB signaling pathway in the early stage of infection with ASFV in macrophages, providing more insights into the pathogenesis of ASFV.

Undeniably, diagnostics play a key role in the control of ASF [7]. In this sense, research associated with the development of new diagnostic methodologies that improve the detection of ASFV constitutes an important milestone in the rapid and accurate detection of this disease. As part of this Special Issue, Friedrichs et al. (2024) evaluated qPCR protocols for ASFV diagnosis in wild boars using semen as a target sample. As a result, the authors published a workflow that detected the genome of ASFV in the semen of wild boars infected as early as two days post infection. Another qPCR protocol was published by Shi et al. (2023). These authors developed a triplex crystal digital PCR able to detect and rule out the presence of the ASFV genes B646L, MGF505-2R, and I177L. Notably, the last two genes are implicated in the development of LAVs. This condition highlights this design as a diagnostic tool capable of differentiating vaccinated from infected animals with ASFV.

Lastly, Watanabe et al. (2023) developed an indirect enzyme-linked immunosorbent assay (ELISA) using recombinant p11.5 protein to detect antibodies against ASFV. By using a collection of serum samples from pigs and wild boars experimentally infected with multiple ASFV strains, the authors demonstrated the value of this serological method in detecting antibodies early during infection.

In conclusion, the studies published in this Special Issue on the emergence and control of African swine fever contribute to improving the knowledge of some critical aspects of this complex disease. I encourage the scientific community of ASFV to review these published studies.

Funding: This research received no external funding.

Data Availability Statement: No primary data are included in this article.

Acknowledgments: The author thanks Ninnet Gomez-Romero for her critical review of this editorial.

Conflicts of Interest: The author declares no conflicts of interest.

List of Contributions:

1. Ramirez-Medina, E.; Rai, A.; Espinoza, N.; Spinard, E.; Silva, E.; Burton, L.; Clark, J.; Meyers, A.; Valladares, A.; Velazquez-Salinas, L.; et al. Recombinant Vaccine Strain ASFV-G-Δ9GL/ΔUK Produced in the IPKM Cell Line Is Genetically Stable and Efficacious in Inducing Protection in Pigs Challenged with the Virulent African Swine Fever Virus Field Isolate Georgia 2010. *Pathogens* **2024**, *13*, 319.
2. Borca, M.V.; Ramirez-Medina, E.; Silva, E.; Rai, A.; Espinoza, N.; Velazquez-Salinas, L.; Gladue, D.P. ASF Vaccine Candidate ASFV-G-ΔI177L Does Not Exhibit Residual Virulence in Long-Term Clinical Studies. *Pathogens* **2023**, *12*, 805.
3. Silva, E.B.; Krug, P.W.; Ramirez-Medina, E.; Valladares, A.; Rai, A.; Espinoza, N.; Gladue, D.P.; Borca, M.V. The Presence of Virus Neutralizing Antibodies Is Highly Associated with Protection against Virulent Challenge in Domestic Pigs Immunized with ASFV live Attenuated Vaccine Candidates. *Pathogens* **2022**, *11*, 1311.

4. Wu, Q.; Li, C.; Zhu, B.; Zhu, J.; Yang, K.; Liu, Z.; Liu, W.; Gao, T.; Yuan, F.; Guo, R.; et al. Advancement in the Antigenic Epitopes and Vaccine Adjuvants of African Swine Fever Virus. *Pathogens* **2024**, *13*, 706.
5. Jackman, J.A.; Arabyan, E.; Zakaryan, H.; Elrod, C.C. Glycerol Monolaurate Inhibits Wild-Type African Swine Fever Virus Infection in Porcine Macrophages. *Pathogens* **2023**, *12*, 1193.
6. Frost, L.; Tully, M.; Dixon, L.; Hicks, H.M.; Bennett, J.; Stokes, I.; Marsella, L.; Gubbins, S.; Batten, C. Evaluation of the Efficacy of Commercial Disinfectants against African Swine Fever Virus. *Pathogens* **2023**, *12*, 855.
7. Schambow, R.A.; Hussain, S.; Antognoli, M.C.; Kreindel, S.; Reyes, R.; Perez, A.M. Epidemiological Assessment of African Swine Fever Spread in the Dominican Republic. *Pathogens* **2023**, *12*, 1414.
8. Hsu, C.H.; Schambow, R.; Montenegro, M.; Miclat-Sonaco, R.; Perez, A. Factors Affecting the Spread, Diagnosis, and Control of African Swine Fever in the Philippines. *Pathogens* **2023**, *12*, 1068.
9. Moskalenko, L.; Schulz, K.; Nedosekov, V.; Motus, K.; Viltrop, A. Understanding Smallholder Pigkeepers' Awareness and Perceptions of African Swine Fever and Its Control Measures in Ukraine. *Pathogens* **2024**, *13*, 139.
10. Okwasiimire, R.; Flint, J.F.; Kayaga, E.B.; Lakin, S.; Pierce, J.; Barrette, R.W.; Faburay, B.; Ndoboli, D.; Ekakoro, J.E.; Wampande, E.M.; et al. Whole Genome Sequencing Shows that African Swine Fever Virus Genotype IX Is Still Circulating in Domestic Pigs in All Regions of Uganda. *Pathogens* **2023**, *12*, 912.
11. Mazloum, A.; Van Schalkwyk, A.; Chernyshev, R.; Shotin, A.; Korennoy, F.I.; Igolkin, A.; Sprygin, A. Genetic Characterization of the Central Variable Region in African Swine Fever Virus Isolates in the Russian Federation from 2013 to 2017. *Pathogens* **2022**, *11*, 919.
12. Richter, M.; Schulz, K.; Elflein, T.; Achterberg, J.; Olsevskis, E.; Serzants, M.; Lamberga, K.; Conraths, F.J.; Sauter-Louis, C. The First Eighteen Months of African Swine Fever in Wild Boar in Saxony, Germany and Latvia—A Comparison. *Pathogens* **2023**, *12*, 87.
13. Olesen, A.S.; Lazov, C.M.; Accensi, F.; Johnston, C.M.; Rasmussen, T.B.; Botner, A.; Lohse, L.; Belsham, G.J. Evaluation of the Dose of African Swine Fever Virus Required to Establish Infection in Pigs Following Oral Uptake. *Pathogens* **2025**, *14*, 119.
14. Nguyen, T.T.H.; Nguyen, V.T.; Le, P.N.; Mai, N.T.A.; Dong, V.H.; Bui, T.A.D.; Nguyen, T.L.; Ambagala, A.; Le, V.P. Pathological Characteristics of Domestic Pigs Orally Infected with the Virus Strain Causing the First Reported African Swine Fever Outbreaks in Vietnam. *Pathogens* **2023**, *12*, 393.
15. Avagyan, H.; Hakobyan, S.; Baghdasaryan, B.; Arzumanyan, H.; Poghosyan, A.; Bayramyan, N.; Semerjyan, A.; Sargsyan, M.; Voskanyan, H.; Vardanyan, T.; et al. Pathology and Clinics of Naturally Occurring Low-Virulence Variants of African Swine Fever Emerged in Domestic Pigs in the South Caucasus. *Pathogens* **2024**, *13*, 130.
16. Sehl-Ewert, J.; Deutschmann, P.; Breithaupt, A.; Blome, S. Pathology of African Swine Fever in Wild Boar Carcasses Naturally Infected with German Virus Variants. *Pathogens* **2022**, *11*, 1386.
17. Li, Z.; Chen, W.; Li, X.; Wu, K.; Wang, X.; Wang, W.; Li, Y.; Yi, L.; Zhao, M.; Ding, H.; et al. Transcriptome Profiling in Swine Macrophages Infected with African Swine Fever Virus (ASFV) Uncovers the Complex and Close Relationship with Host. *Pathogens* **2022**, *11*, 1411.
18. Friedrichs, V.; Reicks, D.; Zimmerman, J.J.; Nelson, E.A.; Sauter-Louis, C.; Beer, M.; Christopher-Hennings, J.; Blome, S. Establishment of a Suitable Diagnostic Workflow to Ensure Sensitive Detection of African Swine Fever Virus Genome in Porcine Semen. *Pathogens* **2024**, *13*, 537.
19. Shi, K.; Zhao, K.; Wei, H.; Zhou, Q.; Shi, Y.; Mo, S.; Long, F.; Hu, L.; Feng, S.; Mo, M. Triplex Crystal Digital PCR for the Detection and Differentiation of the Wild-Type Strain and the MGF505-2R and I177L Gene-Deleted Strain of African Swine Fever Virus. *Pathogens* **2023**, *12*, 1092.
20. Watanabe, M.; Kitamura, T.; Nagata, K.; Ikezawa, M.; Kameyama, K.I.; Masujin, K.; Kokuho, T. Development of a Novel Indirect ELISA for the Serological Diagnosis of African Swine Fever Using p11.5 Protein as a Target Antigen. *Pathogens* **2023**, *12*, 774.

References

1. Li, Z.; Chen, W.; Qiu, Z.; Li, Y.; Fan, J.; Wu, K.; Li, X.; Zhao, M.; Ding, H.; Fan, S.; et al. African Swine Fever Virus: A Review. *Life* **2022**, *12*, 1255. [CrossRef] [PubMed]
2. Schambow, R.A.; Carrasquillo, N.; Kreindel, S.; Perez, A.M. An update on active and passive surveillance for African swine fever in the Dominican Republic. *Sci. Rep.* **2025**, *15*, 2244. [CrossRef] [PubMed]
3. Vu, H.L.X.; McVey, D.S. Recent progress on gene-deleted live-attenuated African swine fever virus vaccines. *NPJ Vaccines* **2024**, *9*, 60. [CrossRef] [PubMed]
4. Bergmann, H.; Dups-Bergmann, J.; Schulz, K.; Probst, C.; Zani, L.; Fischer, M.; Gethmann, J.; Denzin, N.; Blome, S.; Conraths, F.J.; et al. Identification of Risk Factors for African Swine Fever: A Systematic Review. *Viruses* **2022**, *14*, 2107. [CrossRef] [PubMed]
5. Ruedas-Torres, I.; Thi To Nga, B.; Salguero, F.J. Pathogenicity and virulence of African swine fever virus. *Virulence* **2024**, *15*, 2375550. [CrossRef] [PubMed]
6. Chen, S.; Wang, T.; Luo, R.; Lu, Z.; Lan, J.; Sun, Y.; Fu, Q.; Qiu, H.J. Genetic Variations of African Swine Fever Virus: Major Challenges and Prospects. *Viruses* **2024**, *16*, 913. [CrossRef] [PubMed]
7. Pikalo, J.; Deutschmann, P.; Fischer, M.; Roszyk, H.; Beer, M.; Blome, S. African Swine Fever Laboratory Diagnosis-Lessons Learned from Recent Animal Trials. *Pathogens* **2021**, *10*, 177. [CrossRef] [PubMed]

pathogens

MDPI

Communication

Recombinant Vaccine Strain ASFV-G-Δ9GL/ΔUK Produced in the IPKM Cell Line Is Genetically Stable and Efficacious in Inducing Protection in Pigs Challenged with the Virulent African Swine Fever Virus Field Isolate Georgia 2010

Elizabeth Ramirez-Medina [1,2,†], Ayushi Rai [1,3,†], Nallely Espinoza [1,2,†], Edward Spinard [1,2], Ediane Silva [2], Leeanna Burton [2], Jason Clark [2], Amanda Meyers [1,3], Alyssa Valladares [1,3], Lauro Velazquez-Salinas [1,2], Cyril G. Gay [4], Douglas P. Gladue [1,2,*] and Manuel V. Borca [1,2,*]

1 U.S. Department of Agriculture, Agricultural Research Service, Foreign Animal Disease Research Unit, Plum Island Animal Disease Center, Orient, NY 11957, USA; elizabeth.ramirez@usda.gov (E.R.-M.); ayushi.rai@usda.gov (A.R.); nallely.espinoza@usda.gov (N.E.); edward.spinard@usda.gov (E.S.); amanda.meyers@usda.gov (A.M.); alyssa.valladares@usda.gov (A.V.); lauro.velazquez@usda.gov (L.V.-S.)

2 U.S. Department of Agriculture, Agricultural Research Service, Foreign Animal Disease Research Unit, National Bio and Agro-Defense Facility, Manhattan, KS 66502, USA; ediane.silva@usda.gov (E.S.); leeanna.burton@usda.gov (L.B.); jason.clark@usda.gov (J.C.)

3 Oak Ridge Institute for Science and Education (ORISE), Oak Ridge, TN 37830, USA

4 U.S. Department of Agriculture, Agricultural Research Service, Beltsville, MD 20705, USA; cyril.gay@usda.gov

* Correspondence: douglas.gladue@usda.gov (D.P.G.); manuel.borca@usda.gov (M.V.B.)

† These authors contributed equally to this work.

Citation: Ramirez-Medina, E.; Rai, A.; Espinoza, N.; Spinard, E.; Silva, E.; Burton, L.; Clark, J.; Meyers, A.; Valladares, A.; Velazquez-Salinas, L.; et al. Recombinant Vaccine Strain ASFV-G-Δ9GL/ΔUK Produced in the IPKM Cell Line Is Genetically Stable and Efficacious in Inducing Protection in Pigs Challenged with the Virulent African Swine Fever Virus Field Isolate Georgia 2010. *Pathogens* **2024**, *13*, 319. https://doi.org/10.3390/pathogens13040319

Academic Editors: Yuying Liang and Fun-In Wang

Received: 28 February 2024
Revised: 10 April 2024
Accepted: 11 April 2024
Published: 13 April 2024

Abstract: We have previously reported that the recombinant African Swine Fever (ASF) vaccine candidate ASFV-G-Δ9GL/ΔUK efficiently induces protection in domestic pigs challenged with the virulent strain Georgia 2010 (ASFV-G). As reported, ASFV-G-Δ9GL/ΔUK induces protection, while intramuscularly (IM), administered at doses of 10^4 HAD_{50} or higher, prevents ASF clinical disease in animals infected with the homologous ASFV g strain. Like other recombinant vaccine candidates obtained from ASFV field isolates, ASFV-G-Δ9GL/ΔUK stocks need to be produced in primary cultures of swine macrophages, which constitutes an important limitation in the production of large virus stocks at the industrial level. Here, we describe the development of ASFV-G-Δ9GL/ΔUK stocks using IPKM (Immortalized Porcine Kidney Macrophage) cells, which are derived from swine macrophages. We show that ten successive passages of ASFV-G-Δ9GL/ΔUK in IPKM cells induced small changes in the virus genome. The produced virus, ASFV-G-Δ9GL/ΔUKp10, presented a similar level of replication in swine macrophages cultures to that of the original ASFV-G-Δ9GL/ΔUK (ASFV-G-Δ9GL/ΔUKp0). The protective efficacy of ASFV-G-Δ9GL/ΔUKp10 was evaluated in pigs that were IM-inoculated with either 10^4 or 10^6 HAD_{50} of ASFV-G-Δ9GL/ΔUKp10. While animals inoculated with 10^4 HAD_{50} present a partial protection against the experimental infection with the virulent parental virus ASFV-G, those inoculated with 10^6 HAD_{50} were completely protected. Therefore, as was just recently reported for another ASF vaccine candidate, ASFV-G-ΔI177L, IPKM cells are an effective alternative to produce stocks for vaccine strains which only grow in swine macrophages.

Keywords: ASFV; ASF; ASFV vaccine; ASFV-G-Δ9GL/ΔUK

1. Introduction

African Swine Fever virus (ASFV) produces a lethal disease in domestic swine. African Swine Fever (ASF), which severely affects the production of domestic pigs worldwide, is currently widely distributed, being present in Africa, Asia, Europe, and, recently, in the Caribbean region [1–3]. ASFV is a large and structurally complex virus. Its genome is composed of double-stranded DNA of approximately 180–190 kilobase pairs which encode more than 160 genes [4–6].

Although the development of vaccines to prevent ASF was attempted for years [7], commercial vaccines were not available until very recently. Therefore, the control of the disease was based on the elimination of infected and susceptible animals, as well as the restricted mobility of infected herds [8,9].

The use of recombinant attenuated virus strains as potential vaccine candidates has recently increased [10–12]. Novel vaccine candidates were developed by deleting ASFV genes involved in the production of the disease in pigs. In general, these recombinant vaccine strains were efficacious in protecting pigs against the challenge with the homologous virulent field isolate [13–22].

Among those recombinant live attenuated viruses is the ASFV-G-Δ9GL/ΔUK [23]. ASFV-G-Δ9GL/ΔUK was developed by a double gene deletion in the genome of the virulent isolate Georgia 2010 (ASFV-G) of the 9GL and the UK genes. This recombinant virus was shown to have no residual virulence in domestic pigs even when parenterally inoculated at doses as high as 10^6 HAD_{50} [23]. Importantly, ASFV-G-Δ9GL/ΔUK efficaciously induces protection in vaccinated pigs against the experimental infection using the highly virulent homologous ASFV g [23]. ASFV-G-Δ9GL/ΔUK was reported to produce protection as early as 2 weeks after a single vaccine dose of at least 10^4 HAD_{50} [23]. ASFV-G-Δ9GL/ΔUK was developed, purified, and stock was generated using primary cell cultures of swine macrophages. The use of primary cell cultures is a difficulty in the necessary scaling up of vaccine production with commercial purposes. As an alternative, the adaptation of ASFV to grow in a cell line is usually accompanied by significant and undesirable modifications in the virus genome that appear during the process of the virus adaptation [24]. Here, we report the production of ASFV-G-Δ9GL/ΔUK stocks utilizing the cell line IPKM as growing substrate [25]. It is demonstrated that ASFV-G-Δ9GL/ΔUK replicates efficiently in IPKM cells incorporating minimal genomic changes. In addition, pigs were partially and fully protected against the challenge with the virulent parental virus when they received a single vaccine dose containing 10^6 or 10^4 HAD_{50} of ASFV-G-Δ9GL/ΔUK, respectively. Consequently, IPKM cells can be an option for the massive production of ASFV-G-Δ9GL/ΔUK stocks.

2. Materials and Methods

2.1. Viruses and Cells

The ASFV-G-Δ9GL/ΔUK vaccine was produced in the Plum Island Animal Disease Center, as previously reported [13]. Primary cultures of peripheral blood swine macrophages were produced, as reported earlier [22]. Macrophages were used at a final concentration of 5×10^6 cells/mL to coat 96 or 6 well plates, as needed. Titrations of ASFV-G-Δ9GL/ΔUK and its parental field isolate ASFV g were implemented in primary swine macrophage cultures as earlier reported [22] and summarized below. The immortalized porcine kidney macrophage derived cell line IPKM [25] was kindly provided by Dr Kokuho Takehiro from the National Institute of Animal Health of Japan. IPKM cells were grown under culture conditions, as previously described [25]. ASFV-G-Δ9GL/ΔUK (ASFV-G-Δ9GL/ΔUKp0) and the parental ASFV g (ASFV-Gp0) were sequentially passed 10 times in IPKM cultures (MOI = 1), producing ASFV-G-Δ9GL/ΔUKp10 and ASFV-Gp10, respectively. Every passage proceeded until cytopathic effect reached approximately 80% of the cells. Intermediate virus stocks were prepared by one freezing and thawing step, the clarification of the obtained cell suspension by centrifugation, and their titration on primary swine macrophage cultures, as described below. Growth kinetics studies comparing different ASFV-G-Δ9GL/ΔUKp0 and p10 stocks and those of ASFV-Gp0 and p10 were performed at a MOI of 0.01 HAD_{50} using previously published protocols [22]. Virus yields were titrated at different times post-infection on swine macrophages. The presence of virus-infected cells was evaluated by hemadsorption (HA), as previously reported, and determined by the Reed and Muench method [26].

2.2. Sequencing and Analysis of the ASFV-G-ΔI177L Genome

The procedures to obtain the full genomic sequence of ASFV-G-Δ9GL/ΔUKp0 and ASFV-G-Δ9GL/ΔUKp10 were exactly those recently reported [27].

2.3. Evaluation of ASFV-G-Δ9GL/ΔUKp10 Efficacy in Domestic Pigs

The ability of ASFV-G-Δ9GL/ΔUKp10 to protect domestic pigs against infection with the highly virulent ASFV g was assessed in 35–40 kg crossbreed pigs. Groups of 5 animals were inoculated intramuscularly (IM), receiving either 10^4 or 10^6 HAD_{50} of ASFV-G-Δ9GL/ΔUKp10. Another group of pigs was mock-inoculated with the culture medium. The presence of the ASF clinical disease as well as changes in body temperature reads were recorded daily for 28 days. The harvesting of blood samples (in heparinized blood collection tubes) to quantify the presence of the virus was scheduled to be performed at days 0, 4, 7, 11, 14, 21, and 28 post-infection (pi). By day 28 pi, all groups of pigs were IM-challenged with 10^2 HAD_{50} of ASFV-G. Animals were monitored and sampled, as described above, until day 21 post challenge (pc). The experiments with animals were performed under biosafety level 3 conditions in the animal facilities at Plum Island Animal Disease Center, strictly following a protocol approved by the Institutional Animal Care and Use Committee (225.06-19-R_090716, approved on 9 June 2019).

3. Results and Discussion

3.1. Effect of Sequential Passages of ASFV-G-Δ9GL/ΔUK in IPKM Cells

In most cases, the growth of an ASFV strain in a cell line requires a process of adaptation, usually accompanied by important genomic changes. To evaluate the capability of IPMK cells to allow the growth of ASFV-G-Δ9GL/ΔUK without a preliminary period of adaptation, ASFV-G-Δ9GL/ΔUK was subjected to a set of 10 successive passages in IPMK cells. In these passages, infection was performed using an MOI of 1 (the initial and all intermediate stocks' titers were calculated based on titrations implemented in swine macrophages).

All virus passages in IPKM cells were halted when the cytopathic effect reached approximately 80% of the cells. At that point, cultures were frozen and each of the intermediate stocks prepared, as described in Materials and Methods. A similar experiment was also performed in parallel with the parental virus ASFV-G.

The results showed that the virus yields of the recombinant ASFV-G-Δ9GL/ΔUK as well as ASFV g remained without large fluctuations during the 10 passages in the IPKM cells (Figure 1). Virus yield along the passages remained within a range, with titer values of $10^{5.4–7.3}$ HAD_{50}/mL for ASFV-G-Δ9GL/ΔUK and between $10^{5.8}$ to 10^8 HAD_{50}/mL for ASFV-G. Therefore, the two viruses effectively replicate in the IPKM cells without needing a clear phase of adaptation. These data results support results already reported, showing that several ASFV isolates (Armenia07, Kenya05/Tk-1, Espana75 and Lisbon60) easily replicate in IPKM cells [28], reaching virus yields similar to those obtained in the primary cultures of swine macrophages.

3.2. Genomic Stability of ASFV-G-Δ9GL/ΔUK in IPKM Cells

To assess the genomic stability of ASFV-G-Δ9GL/ΔUK during the 10 successive passages in IPKM cells (ASFV-G-Δ9GL/ΔUK p10), ASFV-G-Δ9GL/ΔUKp10 was sequenced and the result was compared to that of the ASFV-G-Δ9GL/ΔUKp0. Just one mutation of high confidence (over 70% of the reads at that position contained the SNP) was observed, at nucleotide position 127,166 within the CP530R gene, where a G-to-A mutation produced a Gly-to-Ser substitution. This gene encodes for the virus polyprotein pp62, described as being essential to the processes of virus core development [29]. It is not clear if the glycine-to-serine amino acid substitution detected in the ASFV-G-Δ9GL/ΔUKp10 may affect the function of pp62L protein. Similar results in terms of the genome stability of ASFV strains sequentially passed in IPKM cells have been reported for Armenia 2007, ASFV g isolates and the vaccine candidate ASFV-G-ΔI177L [27,28].

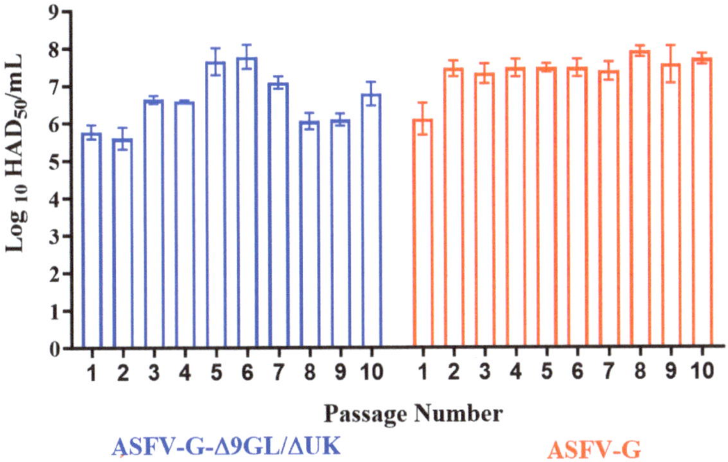

Figure 1. Virus yields of ASFV-G-Δ9GL/ΔUK in sequential passages in IPKM cells. ASFV-G-Δ9GL/ΔUK and ASFV g were sequentially passed 10 times (MOI = 1) in IPKM cell cultures. Viral titers in each passage were evaluated in primary swine macrophages and values expressed as HAD_{50}/mL. Data represent averages and SD of two experiments.

3.3. Evaluation of the Virus Replication of the Recombinant ASFV-G-Δ9GL/UK in IPKM Cells

The growth kinetics of ASFV-G-Δ9GL/ΔUKp10 on IPKM cells was assessed by comparison with that of the original parental virus stock, ASFV-G-Δ9GL/ΔUKp0. In addition, the replication ability of both ASFV-Gp0 and ASFV-Gp10 was assessed. The study was conducted as a multistep growth assay on primary swine macrophage cultures. Infections were performed (MOI = 0.01) and virus yields assessed at 2, 24, 48, 72, and 96 h pi by titration performed in swine macrophages.

As previously reported, ASFV-Gp10 presented (at MOI = 0.01) growth kinetics very similar to the original virus stock (ASFV-Gp0). (Figure 2). On the other hand, ASFV-G-Δ9GL/ΔUKp0 showed lower virus yields than ASFV-G-Δ9GL/ΔUKp10. In fact, ASFV-G-Δ9GL/ΔUKp10 exhibited growth kinetics similar to those of ASFV-Gp0 and ASFV-Gp10 viruses. Virus yield differences between ASFV-G-Δ9GL/ΔUKp0 and ASFV-G-Δ9GL/ΔUKp10 ranged between $10^{1.5}$ to $10^{2.5}$ HAD_{50}/mL, regarding the evaluated sample time point post-infection. It is not clear if these differences could be completely justified by the small genomic changes acquired during the successive passages in IPKM cells.

3.4. ASFV-G-Δ9GL/ΔUKp10 Replication in Experimentally Infected Domestic Pigs

To assess the ability of ASFV-G-Δ9GL/ΔUKp10 to replicate in domestic pigs and efficaciously protect them against infection with the virulent ASFV g isolate, two groups (n = 5) of 35–40 kg pigs were IM-inoculated with either 10^4 or 10^6 HAD_{50}. These doses of the vaccine were chosen since, when originally reported [23], ASFV-G-Δ9GL/ΔUK demonstrated the ability to protect against ASFV challenge. A group of similar characteristics was used as a control. The potential presence of clinical signs related to ASF was checked daily for 28 days after inoculation. Results demonstrated that all animal groups remained clinically normal for the whole observational period (Figure 3), indicating that ASFV-G-Δ9GL/ΔUKp10 continues to be fully attenuated when inoculated in domestic pigs, even at doses as high as 10^6 HAD_{50}.

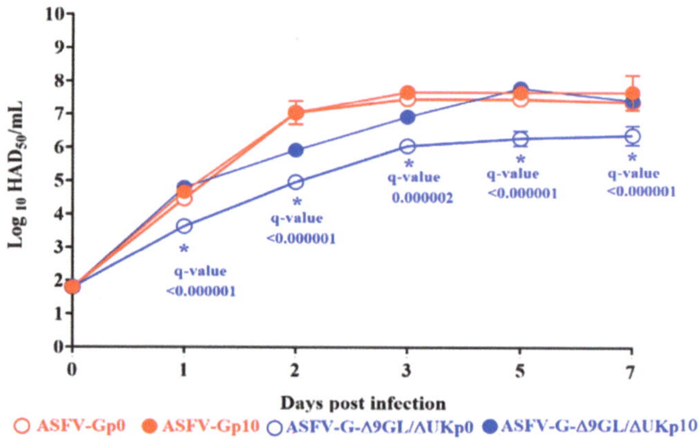

Figure 2. In vitro growth kinetics of ASFV-G-Δ9GL/ΔUkp10 in primary swine macrophages (MOI = 0.01). Samples were taken at the indicated time points and titrated in swine macrophages. Titrations were performed in swine macrophages. Data represent means and standard deviations from two independent experiments. The sensitivity of virus detection is ≥log10 1.8 HAD_{50}/mL. The symbol (*) indicates significant differences between ASFV-G-Δ9GL/ΔUKp0 and ASFV-G-Δ9GL/ΔUKp10 at specific time points. Differences were inferred by the unpaired *t* test using the two-stage set up (Benjamine, Krieger and Yekutieli) method. The reliability of multiple comparisons was evaluated by the false discovery rate method (FDR), considering a q-value < 0.05.

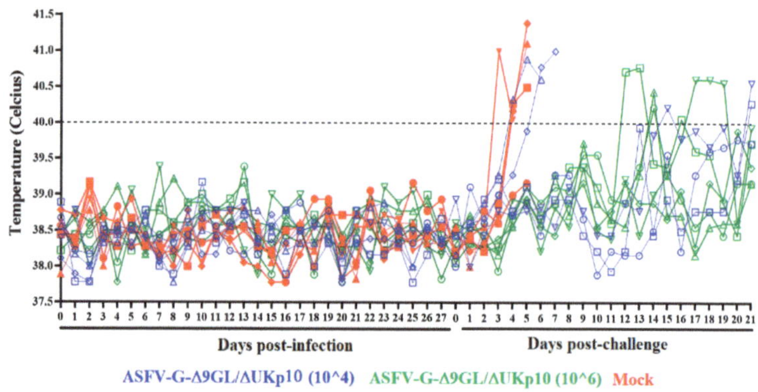

Figure 3. Body temperature in pigs (n = 5) IM inoculated (or Mock inoculated) with either 10^4 or 10^6 HAD_{50} of ASFV-G-Δ9GL/ΔUKp10 and challenged 28 days later with 10^2 HAD_{50} of parental virulent ASFV-G. Data represent individual animals.

The ability of ASFV-G-Δ9GL/ΔUKp10 to replicate after its inoculation was assessed by quantifying its viremia titers. The pattern of viremia in the inoculated animals was heterogeneous in both groups (Figure 4). In the group inoculated with 10^4 HAD_{50}/mL, two of the animals showed undetectable levels of viremia throughout the observational period of 28 days pi. Another animal remained negative until day 21 pi, when it showed low titers ($10^{2.3}$ HAD_{50}/mL) that remained at that level until the day of challenge (28 days pi). Viremia in the fourth animal inoculated with 10^4 HAD_{50}/mL remained undetectable until day 11 pi, when high titers were detected ($10^{5.05}$ HAD_{50}/mL), and it showed variable titers (ranging between $10^{5.8–3.8}$ HAD_{50}/mL) until day 28 pi. The fifth animal in the group showed relatively high viremia titers (ranging between $10^{4.3–6.3}$ HAD50/mL) from day 7 pi until the challenge day.

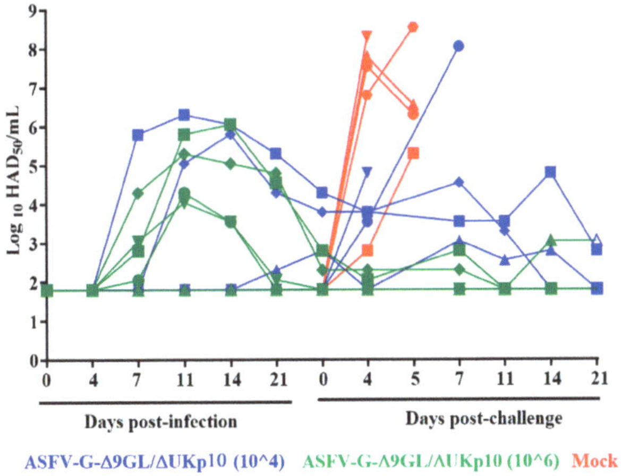

Figure 4. Viremias observed in pigs (n = 5) IM-inoculated (or Mock inoculated) with either 10^4 or 10^6 HAD_{50} of ASFV-G-Δ9GL/ΔUKp10 or mock-inoculated and challenged 28 days later with 10^2 HAD_{50} of ASFV-G. Data represent individual animals. Sensitivity of virus detection: $\geq 10^{1.8}$ $TCID_{50}$/mL.

With the exception of one pig, which presented undetectable viremia titers during the 28 days following inoculation, all animals receiving 10^6 HAD_{50}/mL of ASFV-G-Δ9GL/ΔUKp10 showed low-to-medium titers by day 7 pi ($10^{2.05–4.3}$ HAD50/mL). Two of these animals evolved, showing medium-to-high viremia titers (ranging between $10^{4.55–5.8}$ HAD50/mL) by days 11 to 21 pi, and lower titers by the day of challenge ($10^{2.3–2.8}$ HAD50/mL). The remaining two pigs of the group presented medium titers ($10^{2.05–4.3}$ HAD50/mL) on days 11 and 21 pi and undetectable levels by day 28 pi.

These levels of replication are clearly lower than viremia titers detected in pigs IM-inoculated with similar doses of the original stock ASFV-G-Δ9GL/ΔUK (ASFV-G-Δ9GL/ΔUKp0) [23].

3.5. Evaluation of the Protective Effect of ASFV-G-Δ9GL/ΔUKp10 on Experimental Inoculation with the Parental ASFV-G

The ability of ASFV-G-Δ9GL/ΔUK to protect animals against the experimental infection with the parental virus ASFV g was evaluated by IM challenging the animals previously inoculated with ASFV-G-Δ9GL/ΔUK 28 days later with 10^2 HAD_{50} of ASFV-G. The mock vaccinated control pigs were similarly infected.

The mock-inoculated pigs presented initial signs of ASF by days 3–4 post challenge (dpc), quickly getting worse with all of them euthanized due to the severity of the clinical signs by day 4, one of them early, and the remaining four on day 5 pc (Figures 3 and 5). Conversely, all the animals inoculated with 10^6 HAD_{50} of ASFV-G-Δ9GL/ΔUKp10 remained clinically normal during the observation period, except for two, presenting one period of transient mild body temperature elevation without any other additional sign associated with ASF (Figures 3 and 5). Animals in the group receiving 10^4 HAD_{50} of ASFV-G-Δ9GL/ΔUKp10 presented a heterogenous response after the challenge. Two of the animals presented initial signs of disease by days 4–5 post challenge, followed by a quick evolution to the severe clinical form of the disease and the need to be euthanized by days 6–7 pc. The remaining three animals did not show any clinical sign associated with ASF during the 21-day observational period, with the exception of very mild and transitory elevation in body temperature (Figures 3 and 5).

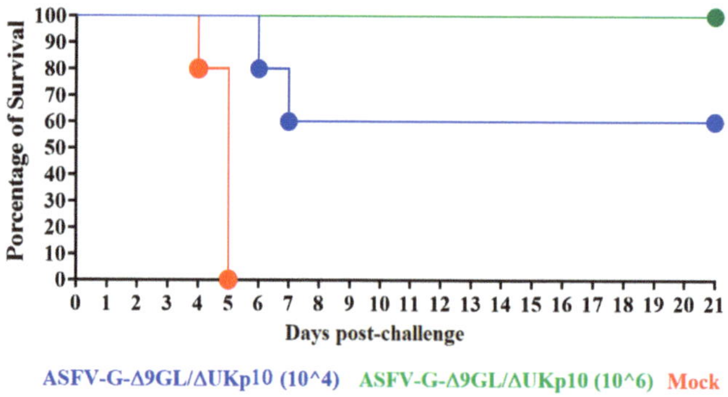

Figure 5. Mortality in animals (n = 5) IM-inoculated (or mock-inoculated) with either 10^4 or 10^6 HAD_{50} of ASFV-G-Δ9GL/ΔUKp10 and challenged 28 days later with 10^2 HAD_{50} of ASFV-G.

Viremias in the mock-vaccinated animals increased after the challenge. In all but one animal, titers ranged between $10^{6.5}$ and $10^{8.3}$ HAD_{50}/mL by day 4 pc and remained at that level until all the animals were euthanized. The remaining animal in this group presented a low ($10^{2.5}$ HAD_{50}) viremia by day 4 pc, increasing to 10^5 HAD_{50} at the time of the euthanasia (Figure 4). Animals inoculated with 10^6 HAD_{50} of ASFV-G-Δ9GL/ΔUKp10 presented low viremias after the challenge, ranging between $10^{1.8}$ and 10^3 HAD_{50}/mL. Viremia in most of these animals, except for one, decreased until it hit undetectable levels ($\leq 10^{1.8}$ HAD_{50}/mL) before the end of the 21-day pc observational period. Animals receiving 10^4 HAD_{50} of ASFV-G-Δ9GL/ΔUKp10 had a viremia pattern in accordance with their clinical status. The three animals surviving the challenge presented low viremias after the challenge, ranging between undetectable ($\leq 10^{1.8}$ HAD_{50}/mL) and $10^{2.8}$ HAD_{50}/mL. The two animals that succumbed to the challenge quickly presented medium-to-high viremias by the day they were euthanized because of the severity of the disease ($10^{4.5}$ and $10^{7.8}$ HAD_{50}/mL, respectively). Viremia in two of the remaining three animals decreased until it reached undetectable levels.

It is shown that the recombinant vaccine candidate ASFV-G-Δ9GL/ΔUK may be produced in the cell line IPKM. A stock virus produced after 10 sequential passages in these cells is still safe and efficacious in protecting animals against the infection with the virulent ASFV g when used at a dose of 10^6 HAD_{50}/mL. It is not clear why, though no main genetic changes are observed in the ASFV-G-Δ9GL/ΔUKp10 genome; this strain is slightly less efficacious than the parental ASFV-G-Δ9GL/ΔUK, which is able to protect pigs at doses of 10^4 HAD_{50}/mL.

As it was previously described for the ASFV-G-ΔI177L vaccine strain [27], these results demonstrate the feasibility of employing the IPKM cell line to produce ASFV vaccine strains, which were initially developed and grown in primary swine macrophage cell cultures, signifying a restriction in the manufacture of a vaccine with commercial purposes.

Author Contributions: Conceptualization, M.V.B. and D.P.G.; Data curation, A.R., L.V.-S., A.V., A.M., J.C., E.S. (Ediane Silva) and L.B.; Formal analysis, M.V.B., A.R., E.R.-M., E.S. (Edward Spinard), L.V.-S., E.S. (Ediane Silva), C.G.G. and D.P.G.; Funding acquisition, M.V.B. and D.P.G.; Investigation, N.E. and D.P.G.; Methodology, A.R., N.E. and E.R.-M.; Writing—original draft, M.V.B., L.V.-S., C.G.G. and D.P.G.; Writing—review and editing, M.V.B., A.R., N.E., E.R.-M., E.S. (Edward Spinard), L.V.-S., A.V., E.S. (Ediane Silva), L.B., C.G.G., A.M., J.C. and D.P.G. All authors have read and agreed to the published version of the manuscript.

Funding: This project was partially and by the National Pork Board Project #21-137 and partially funded through the Foundation for Food and Agriculture Rapid Outcomes from Agriculture Research (ROAR) grant # 21-000064. This research was supported in part by an appointment to the Plum Island Animal Disease Center (PIADC) Research Participation Program administered by the Oak Ridge Institute for Science and Education (ORISE) through an interagency agreement between the U.S.

Department of Energy (DOE) and the U.S. Department of Agriculture (USDA). ORISE is managed by ORAU under DOE contract number DE-SC0014664.

Institutional Review Board Statement: Not applicable.

Informed Consent Statement: Not applicable.

Data Availability Statement: All data are included in the manuscript.

Acknowledgments: IPKM cells were kindly provided by Kokuho Takehiro from the National Institute of Animal Health of Japan. We thank the Plum Island Animal Disease Center Animal Care Unit staff for their excellent technical assistance. We wish to particularly thank Carmen V. Borca-Carrillo for editing the manuscript. All opinions expressed in this paper are the author's and do not necessarily reflect the policies and views of USDA, ARS, APHIS, DOE, or ORAU/ORISE.

Conflicts of Interest: The authors declare no conflict of interest.

References

1. Gonzales, W.; Moreno, C.; Duran, U.; Henao, N.; Bencosme, M.; Lora, P.; Reyes, R.; Nunez, R.; De Gracia, A.; Perez, A.M. African swine fever in the Dominican Republic. *Transbound. Emerg. Dis.* **2021**, *68*, 3018–3019. [CrossRef] [PubMed]
2. Schambow, R.A.; Hussain, S.; Antognoli, M.C.; Kreindel, S.; Reyes, R.; Perez, A.M. Epidemiological Assessment of African Swine Fever Spread in the Dominican Republic. *Pathogens* **2023**, *12*, 1414. [CrossRef] [PubMed]
3. Teklue, T.; Sun, Y.; Abid, M.; Luo, Y.; Qiu, H.J. Current status and evolving approaches to African swine fever vaccine development. *Transbound. Emerg. Di.s* **2020**, *67*, 529–542. [CrossRef] [PubMed]
4. Dixon, L.K.; Chapman, D.A.; Netherton, C.L.; Upton, C. African swine fever virus replication and genomics. *Virus Res.* **2013**, *173*, 3–14. [CrossRef] [PubMed]
5. Tulman, E.R.; Delhon, G.A.; Ku, B.K.; Rock, D.L. African Swine Fever Virus. In *Lesser Known Large dsDNA Viruses*; Etten, V., Ed.; Current Topics in Microbiology and Immunology; Springer: Berlin/Heidelberg, Germany, 2009; Volume 328, pp. 43–87.
6. Spinard, E.; Azzinaro, P.; Rai, A.; Espinoza, N.; Ramirez-Medina, E.; Valladares, A.; Borca, M.V.; Gladue, D.P. Complete Structural Predictions of the Proteome of African Swine Fever Virus Strain Georgia 2007. *Microbiol. Resour. Announc.* **2022**, *11*, e0088122. [CrossRef] [PubMed]
7. Revilla, Y.; Perez-Nunez, D.; Richt, J.A. African Swine Fever Virus Biology and Vaccine Approaches. *Adv. Virus Res.* **2018**, *100*, 41–74.
8. Dixon, L.K.; Stahl, K.; Jori, F.; Vial, L.; Pfeiffer, D.U. African Swine Fever Epidemiology and Control. *Annu. Rev. Anim. Biosci.* **2020**, *8*, 221–246. [CrossRef]
9. Simulundu, E.; Lubaba, C.H.; van Heerden, J.; Kajihara, M.; Mataa, L.; Chambaro, H.M.; Sinkala, Y.; Munjita, S.M.; Munang'andu, H.M.; Nalubamba, K.S.; et al. The Epidemiology of African Swine Fever in "Nonendemic" Regions of Zambia (1989–2015): Implications for Disease Prevention and Control. *Viruses* **2017**, *9*, 236. [CrossRef]
10. Deutschmann, P.; Carrau, T.; Sehl-Ewert, J.; Forth, J.H.; Viaplana, E.; Mancera, J.C.; Urniza, A.; Beer, M.; Blome, S. Taking a Promising Vaccine Candidate Further: Efficacy of ASFV-G-DeltaMGF after Intramuscular Vaccination of Domestic Pigs and Oral Vaccination of Wild Boar. *Pathogens* **2022**, *11*, 996. [CrossRef]
11. Vu, H.L.X.; McVey, D.S. Recent progress on gene-deleted live-attenuated African swine fever virus vaccines. *NPJ Vaccines* **2024**, *9*, 60. [CrossRef]
12. Zhang, H.; Zhao, S.; Zhang, H.; Qin, Z.; Shan, H.; Cai, X. Vaccines for African swine fever: An update. *Front. Microbiol.* **2023**, *14*, 1139494. [CrossRef] [PubMed]
13. Borca, M.V.; Ramirez-Medina, E.; Silva, E.; Vuono, E.; Rai, A.; Pruitt, S.; Holinka, L.G.; Velazquez-Salinas, L.; Zhu, J.; Gladue, D.P. Development of a Highly Effective African Swine Fever Virus Vaccine by Deletion of the I177L Gene Results in Sterile Immunity against the Current Epidemic Eurasia Strain. *J. Virol.* **2020**, *94*, e02017-19. [CrossRef] [PubMed]
14. Tran, X.H.; Le, T.T.P.; Nguyen, Q.H.; Do, T.T.; Nguyen, V.D.; Gay, C.G.; Borca, M.V.; Gladue, D.P. African swine fever virus vaccine candidate ASFV-G-DeltaI177L efficiently protects European and native pig breeds against circulating Vietnamese field strain. *Transbound. Emerg. Dis.* **2022**, *69*, e497–e504. [CrossRef] [PubMed]
15. Lewis, T.; Zsak, L.; Burrage, T.G.; Lu, Z.; Kutish, G.F.; Neilan, J.G.; Rock, D.L. An African swine fever virus ERV1-ALR homologue, 9GL, affects virion maturation and viral growth in macrophages and viral virulence in swine. *J. Virol.* **2000**, *74*, 1275–1285. [CrossRef]
16. Zhang, Y.; Ke, J.; Zhang, J.; Yang, J.; Yue, H.; Zhou, X.; Qi, Y.; Zhu, R.; Miao, F.; Li, Q.; et al. African Swine Fever Virus Bearing an I226R Gene Deletion Elicits Robust Immunity in Pigs to African Swine Fever. *J. Virol.* **2021**, *95*, e0119921. [CrossRef]
17. Gladue, D.P.; Ramirez-Medina, E.; Vuono, E.; Silva, E.; Rai, A.; Pruitt, S.; Espinoza, N.; Velazquez-Salinas, L.; Borca, M.V. Deletion of the A137R Gene from the Pandemic Strain of African Swine Fever Virus Attenuates the Strain and Offers Protection against the Virulent Pandemic Virus. *J. Virol.* **2021**, *95*, e0113921. [CrossRef] [PubMed]

18. Ran, Y.; Li, D.; Xiong, M.G.; Liu, H.N.; Feng, T.; Shi, Z.W.; Li, Y.H.; Wu, H.N.; Wang, S.Y.; Zheng, H.X.; et al. African swine fever virus I267L acts as an important virulence factor by inhibiting RNA polymerase III-RIG-I-mediated innate immunity. *PLoS Pathog.* **2022**, *18*, e1010270. [CrossRef] [PubMed]
19. Li, D.; Zhang, J.; Yang, W.; Li, P.; Ru, Y.; Kang, W.; Li, L.; Ran, Y.; Zheng, H. African swine fever virus protein MGF-505-7R promotes virulence and pathogenesis by inhibiting JAK1- and JAK2-mediated signaling. *J. Biol. Chem.* **2021**, *297*, 101190. [CrossRef] [PubMed]
20. Li, D.; Liu, Y.; Qi, X.; Wen, Y.; Li, P.; Ma, Z.; Liu, Y.; Zheng, H.; Liu, Z. African Swine Fever Virus MGF-110-9L-deficient Mutant Has Attenuated Virulence in Pigs. *Virol. Sin.* **2021**, *36*, 187–195. [CrossRef]
21. O'Donnell, V.; Holinka, L.G.; Krug, P.W.; Gladue, D.P.; Carlson, J.; Sanford, B.; Alfano, M.; Kramer, E.; Lu, Z.; Arzt, J.; et al. African swine fever virus Georgia 2007 with a deletion of virulence-associated gene 9GL (B119L), when administered at low doses, leads to virus attenuation in swine and induces an effective protection against homologous challenge. *J. Virol.* **2015**, *89*, 8556–8566. [CrossRef]
22. O'Donnell, V.; Holinka, L.G.; Gladue, D.P.; Sanford, B.; Krug, P.W.; Lu, X.; Arzt, J.; Reese, B.; Carrillo, C.; Risatti, G.R.; et al. African Swine Fever Virus Georgia Isolate Harboring Deletions of MGF360 and MGF505 Genes Is Attenuated in Swine and Confers Protection against Challenge with Virulent Parental Virus. *J. Virol.* **2015**, *89*, 6048–6056. [CrossRef]
23. O'Donnell, V.; Risatti, G.R.; Holinka, L.G.; Krug, P.W.; Carlson, J.; Velazquez-Salinas, L.; Azzinaro, P.A.; Gladue, D.P.; Borca, M.V. Simultaneous Deletion of the 9GL and UK Genes from the African Swine Fever Virus Georgia 2007 Isolate Offers Increased Safety and Protection against Homologous Challenge. *J. Virol.* **2016**, *91*, e01760-16. [CrossRef] [PubMed]
24. Krug, P.; Holinka, L.; O'Donnell, V.; Rees, B.; Sanford, B.; Fernandez Sainz, I.; Gladue, D.P.; Arzt, J.; Rodriguez, R.L.; Risatti, G.R.; et al. The progressive adaptation of a Georgian isolate of African swine fever virus to Vero cells leads to a gradual attenuation of virulence in swine corresponding with major modifications of the viral genome. *J. Virol.* **2015**, *89*, 2324–2332. [CrossRef] [PubMed]
25. Takenouchi, T.; Masujin, K.; Suzuki, S.; Haraguchi, S.; Hiramatsu, K.; Kokuho, T.; Uenishi, H. Establishment and characterization of the immortalized porcine lung-derived mononuclear phagocyte cell line. *Front. Vet. Sci.* **2022**, *9*, 1058124. [CrossRef] [PubMed]
26. Reed, L. A simple method of estimating fifty percent endpoints. *Am. J. Hyg.* **1938**, *27*, 493–497.
27. Borca, M.V.; Rai, A.; Espinoza, N.; Ramirez-Medina, E.; Spinard, E.; Velazquez-Salinas, L.; Valladares, A.; Silva, E.; Burton, L.; Meyers, A.; et al. African Swine Fever Vaccine Candidate ASFV-G-ΔI177L Produced in the Swine Macrophage-Derived Cell Line IPKM Remains Genetically Stable and Protective against Homologous Virulent Challenge. *Viruses* **2023**, *15*, 2064. [CrossRef] [PubMed]
28. Masujin, K.; Kitamura, T.; Kameyama, K.; Okadera, K.; Nishi, T.; Takenouchi, T.; Kitani, H.; Kokuho, T. An immor-talized porcine macrophage cell line competent for the isolation of African swine fever virus. *Sci. Rep.* **2021**, *11*, 4759. [CrossRef]
29. Suárez, C.; Salas, M.L.; Rodríguez, J.M. African swine fever virus polyprotein pp62 is essential for viral core development. *J. Virol.* **2010**, *84*, 176–187. [CrossRef]

pathogens

MDPI

Communication

ASF Vaccine Candidate ASFV-G-ΔI177L Does Not Exhibit Residual Virulence in Long-Term Clinical Studies

Manuel V. Borca [1,*], Elizabeth Ramirez-Medina [1], Ediane Silva [1], Ayushi Rai [1,2], Nallely Espinoza [1], Lauro Velazquez-Salinas [1] and Douglas P. Gladue [1,*]

1 Plum Island Animal Disease Center, ARS, USDA, Greenport, NY 11944, USA; elizabeth.ramirez@usda.gov (E.R.-M.); ediane.silva@usda.gov (E.S.); ayushi.rai@usda.gov (A.R.); nallely.espinoza@usda.gov (N.E.); lauro.velazquez@usda.gov (L.V.-S.)

2 Oak Ridge Institute for Science and Education (ORISE), Oak Ridge, TN 37830, USA

* Correspondence: manuel.borca@usda.gov (M.V.B.); douglas.gladue@usda.gov (D.P.G.)

Abstract: African swine fever (ASF) is an important disease in swine currently producing a pandemic affecting pig production worldwide. Except in Vietnam, where two vaccines were recently approved for controlled use in the field, no vaccine is commercially available for disease control. Up to now, the most effective vaccines developed are based on the use of live-attenuated viruses. Most of these promising vaccine candidates were developed by deleting virus genes involved in the process of viral pathogenesis and disease production. Therefore, these vaccine candidates were developed via the genomic modification of parental virus field strains, producing recombinant viruses and reducing or eliminating their residual virulence. In this scenario, it is critical to confirm the absence of any residual virulence in the vaccine candidate. This report describes the assessment of the presence of residual virulence in the ASFV vaccine candidate ASFV-G-ΔI177L in clinical studies conducted under high virus loads and long-term observation periods. The results demonstrated that domestic pigs intramuscularly inoculated with 10^6 HAD_{50} of ASFV-G-ΔI177L did not show the presence of any clinical sign associated with ASF when observed daily either 90 or 180 days after vaccination. In addition, necropsies conducted at the end of the experiment confirmed the absence of macroscopic internal lesions associated with the disease. These results corroborate the safety of using ASFV-G-ΔI177L as a vaccine candidate.

Keywords: ASFV; virus virulence; long-term clinical studies; live-attenuated vaccines; ASFV-G-ΔI177L

Citation: Borca, M.V.; Ramirez-Medina, E.; Silva, E.; Rai, A.; Espinoza, N.; Velazquez-Salinas, L.; Gladue, D.P. ASF Vaccine Candidate ASFV-G-ΔI177L Does Not Exhibit Residual Virulence in Long-Term Clinical Studies. *Pathogens* **2023**, *12*, 805. https://doi.org/10.3390/pathogens12060805

Academic Editor: Fun-In Wang

Received: 15 May 2023
Revised: 2 June 2023
Accepted: 4 June 2023
Published: 6 June 2023

Correction Statement: This article has been republished with a minor change. The change does not affect the scientific content of the article and further details are available within the backmatter of the website version of this article.

1. Introduction

African swine fever virus (ASFV), the causative agent of ASF, is a structurally complex virus with a large DNA double-stranded genome encoding more than 150–160 genes [1]. The disease is currently present as a pandemic, affecting swine production worldwide. Historically secluded in Africa, after an initial outbreak in the Republic of Georgia in 2007, the disease quickly spread to central and eastern Europe, reaching China in 2018 and later spreading throughout southeast Asia [2–4]. In addition, the disease was reported in Malaysia and the Hispaniola Islands in 2021, constituting the first report of ASF in the western hemisphere since the 1970s [5]. Continued outbreaks of ASF in affected countries are causing devastating economic losses to swine production as well as a shortage in worldwide protein availability [6].

Live-attenuated vaccine candidates have been developed via genetic manipulation, inducing protection against the current pandemic strain [7–11]. Recombinant viruses, developed by removing viral genes associated with virulence, efficiently protect pigs against infection with the homologous virulent parental virus [7–15]. With the exception of ASFV-G-ΔI177L (available in Vietnam), there is no commercially available vaccine to prevent ASF; therefore, disease control is based on the elimination of affected animals and following strict quarantine regulations. Vietnam licensed the first commercial genetically modified live vaccine in 2022, the vaccine strain ASFV-G-ΔI177L [9]. ASFV-G-ΔI177L was developed by partially removing the I177L gene from the genome of the highly virulent strain Georgia (ASFV-G). ASFV-G-ΔI177L has been shown to effectively protect pigs against infection with ASFV-G even when used in doses as low as 10^2 HAD_{50}. In addition, ASFV-G-ΔI177L induces sterile protection when used in doses of 10^4 HAD_{50} or higher and, importantly, lacks residual virulence even in doses as high as 10^6 HAD_{50} when tested in short-term studies [9,16]. The efficacy profile of commercially produced stocks of ASFV-G-ΔI177L were recently reported on regarding their use against Vietnamese field strains using animals of different genetic backgrounds [16].

A critical issue in the use of live-attenuated vaccine strains is the evaluation of the absence of residual virulence in the vaccine virus. This is particularly important in ASFV, where residual virulence is a recurrent issue in natural as well as genetically modified attenuated strains [17–19]. In this regard, the appearance of chronic clinical forms of the disease has been reported several times as the main example of residual virulence. We report here an evaluation of the potential presence of residual virulence in the ASFV-G-ΔI177L vaccine in long-term clinical studies performed using high doses of the virus. The results showed the absence of clinical ASF disease in vaccinated animals evaluated for up to 180 days, confirming the safety of ASFV-G-ΔI177L as a vaccine candidate.

2. Materials and Methods

2.1. *Viruses and Cells*

The vaccine candidate ASFV-G-ΔI177L was produced via recombination, as previously described [8]. The recombinant virus contains a partial gene deletion that interrupts the expression of the I77L protein in the virulent ASFV strain Georgia 2007 [8]. Cell cultures of primary swine macrophages were used to assess the presence of ASFV-G-ΔI177L in blood samples collected during this experiment. These cell cultures were prepared as previously described [8]. The assessment was conducted on 96-well plates containing a total of 1×10^6 cells per plate, with the presence of infected cells evaluated using hemadsorption (HA). Final viral titers were determined as previously described [8].

2.2. *Evaluation of ASFV-G-ΔI177L Virulence in Long-Term Experiments on Domestic Pigs*

The virulence of ASFV-G-ΔI177L was evaluated in 15–20 kg commercial Yorkshire cross-breed swine. Two independent experiments were performed where animals intramuscularly (IM) inoculated with 10^6 HAD_{50} of ASFV-G-ΔI177L were daily monitored for clinical signs associated with ASF (fever, anorexia, depression, diarrhea, purple skin discoloration, staggering gait, and cough). In the first experiment, 6 pigs were inoculated with ASFV-G-ΔI177L while another 5 animals were mock-inoculated with a macrophage culture medium. In the second experiment, 10 animals were inoculated with ASFV-G-ΔI177L, while 5 animals were mock-inoculated. All animals were observed for 90 and 180 days, respectively. Blood samples were collected at various days post-inoculation as depicted in the results section. All experiments were conducted under biosafety level 3 conditions at the Plum Animal Disease Center. Animal procedures were approved by the Institutional Animal Care and Use Committee under protocol 225.06-19-R_090716 (approved on 9 June 19).

2.3. *Detection of ASFV-Specific Antibodies*

The presence of antibodies against ASFV was determined using an in-house ELISA, as previously described [8]. Briefly, 96-well plates were coated with 1 µg per well of infected or uninfected cell extracts produced in Vero cells. Plates were blocked with 10% skim milk and 5% normal goat serum. Multiple dilutions of each serum were evaluated against both infected and uninfected cell antigens. The detection of specific antibodies against ASFV was conducted using an anti-swine IgG-horseradish peroxidase conjugate and SureBlue Reserve peroxidase substrate. Finally, plates were read at OD630 nm using a plate reader (ELX800).

The titer of different serums represents the inverse $\log_{10}$ of the highest serum dilution where the readings of the tested sera at least duplicated the readings of the mock-infected sera.

3. Results and Discussion

Assessment of the Presence of the Residual Virulence of ASFV-G-ΔI177L in Long-Term Experiments on Domestic Pigs

To assess the potential presence of the residual virulence of ASFV-G-ΔI177L in domestic pigs, the effect of inoculation using high doses of the virus was evaluated in two long-term experiments, evaluating the appearance of the clinical signs of the disease in animals observed for 3 or 6 months after inoculation with the vaccine virus.

In the first experiment, a group of five pigs was intramuscularly (IM) inoculated with 10^6 HAD_{50} of ASFV-G-ΔI177L. The presence of clinical signs associated with ASF was evaluated daily over an observational period of 90 days. A control group of five animals with the same characteristics as the ones inoculated with ASFV-G-ΔI177L was mock-inoculated with a macrophage culture medium.

The clinical evaluation performed on a daily basis demonstrated the complete absence of any significant abnormality or clinical signs that may be associated with the disease. The animals remained clinically normal (bright, alert, and responsive) during the three months of observation. No differences were observed in this group of animals in terms of clinical observation related to that of the mock-inoculated group of animals. The evolution of the daily body temperatures of the animals inoculated with ASFV-G-ΔI177L demonstrated that, with the exception of a few isolated cases, all animals presented values lower than 40 °C for the three months after vaccination, showing body temperature readings similar to the mock-inoculated control pigs (Figure 1A,B).

At the end of the experiment, all animals from both groups were euthanized, and necropsies were conducted in order to detect the presence of internal macroscopical lesions usually found in ASF-infected animals. No macroscopical lesions that could be associated with ASF were detected in the ASFV-G-ΔI177L-inoculated animals, whose internal organ conditions were indistinguishable from those observed in animals from the mock-inoculated group.

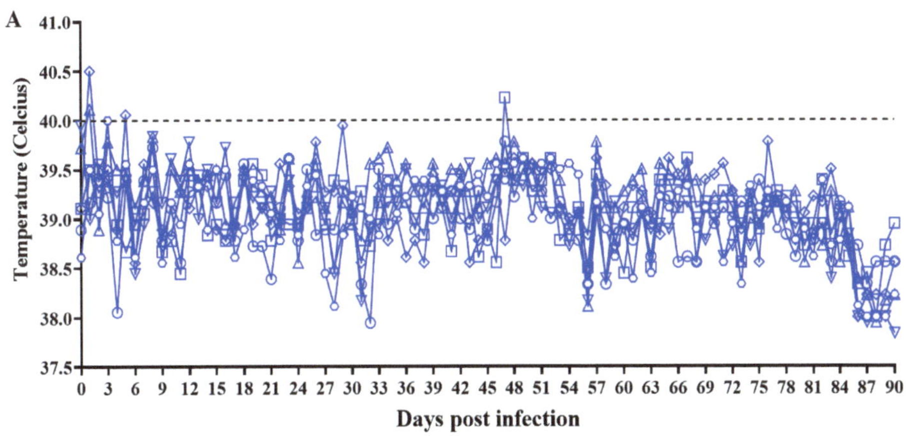

Figure 1. *Cont.*

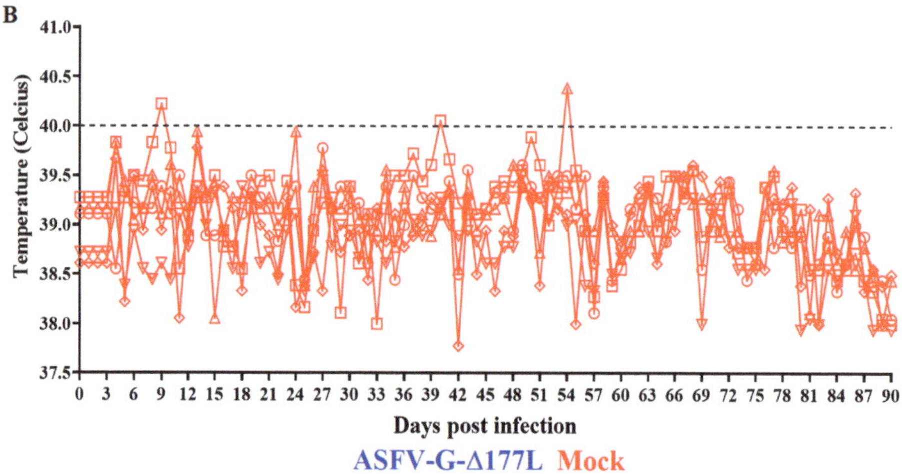

Figure 1. Evolution of body temperature in animals IM-inoculated with 10^6 HAD_{50} of ASFV-G-ΔI177L (**A**) or mock-inoculated (**B**) and observed for 90 days post-inoculation. Symbols represent specific animals at each experimental group.

The evolution of viremia in the animals inoculated with ASFV-G-ΔI177L indicates the expected kinetics based on previous results [9,16,20]. All animals presented extended viremias within the first month of inoculation, showing titers ranging from undetectable to $10^{6.05}$ HAD_{50}/mL by day 30 post-inoculation (PI) (Figure 2A). Viremias tested on day 60 PI demonstrated that all animals remained negative except one, which had a titer of $10^{3.55}$ HAD_{50}/mL. At 90 days PI, the virus in all the animals in the group was undetectable in their blood (titers $\leq 10^{1.8}$ HAD_{50}/mL).

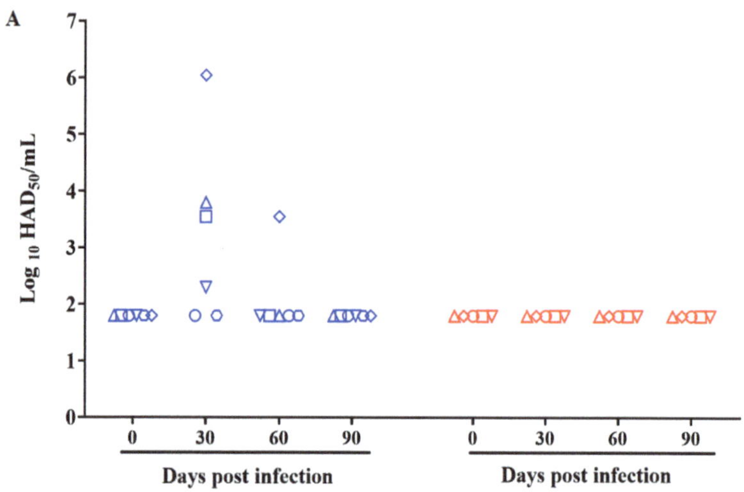

Figure 2. *Cont.*

B

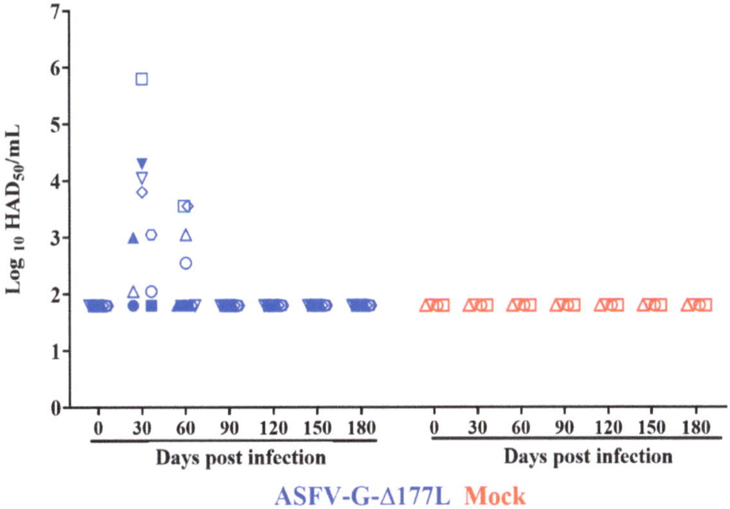

Figure 2. Viremia titers in animals IM-inoculated with 10^6 HAD_{50} of ASFV-G-ΔI177L and observed for either (**A**) 90 or (**B**) 180 days post-inoculation. Virus titers are expressed as log10 HAD_{50}/mL of blood. Sensitivity of detection: $\leq$1.8 HAD_{50}/mL. Symbols represent specific animals at each experimental group.

The detection of ASFV-specific antibody responses was performed with an in-house-developed direct ELISA using extracts of cell cultures infected with ASFV as an antigen. At 30 days PI, all animals presented high antibody titers, ranging from 10^3 to 10^4/mL (Figure 3A). Antibody titers increased when measured at 60 days PI, reaching titers of 10^3–10^5/mL that remained at those values until the end of the experimental period, at 90 days PI.

The second experiment performed was aimed at evaluating the potential presence of the residual virulence of ASFV-G-ΔI177L over a 180-day observational period. For that purpose, a group of ten pigs was IM-inoculated (also with 10^6 HAD_{50} of ASFV-G-ΔI177L), while a control group of five animals was mock-inoculated with a macrophage culture medium. Both groups were monitored daily for 180 days, watching for the presence of clinical signs associated with ASF.

No clinical signs were detected in any of the animals inoculated with ASFV-G-ΔI177L during the daily evaluation performed for 6 months after inoculation. No differences compared with the clinical data obtained from the group of mock-inoculated animals were observed. The body temperature kinetics of animals inoculated with ASFV-G-ΔI177L indicate that all animals showed values within the normal range (always below 40 °C) during the 180 days PI, with a few exceptions, in particular, at very early days after inoculation in animals inoculated with ASFV-G-ΔI177L. Animals in the mock-inoculated group presented similar body temperature values to the ASFV-G-ΔI177L-inoculated ones (Figure 4).

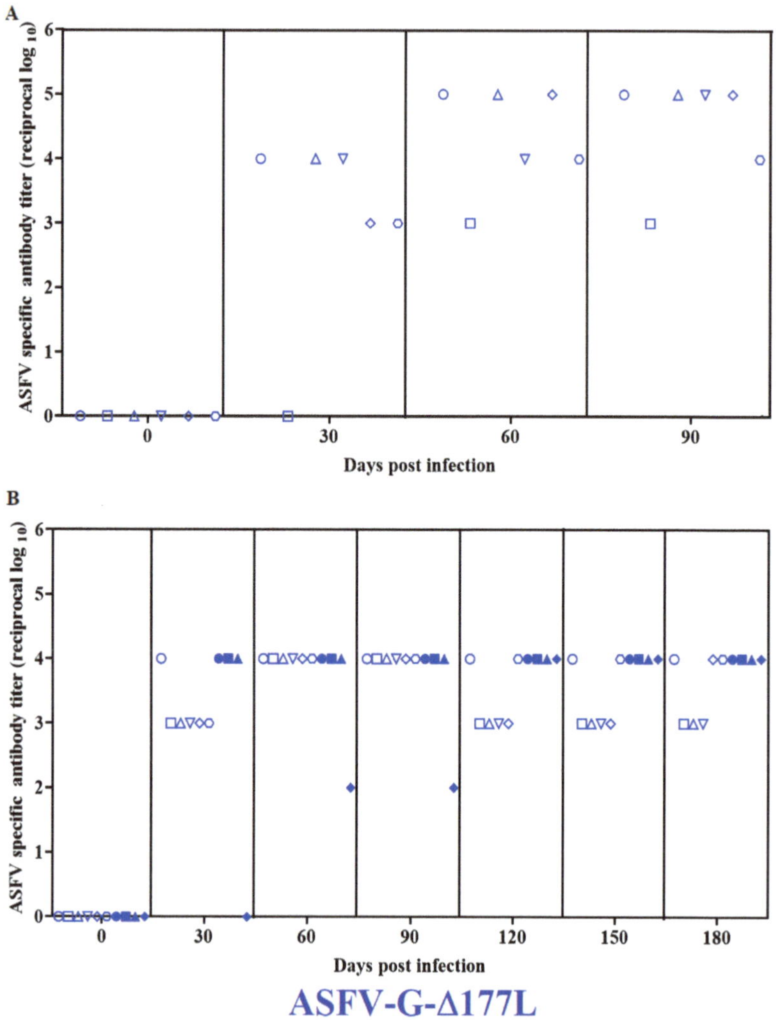

Figure 3. ASFV-specific antibody titers detected in pigs IM-inoculated with 10^6 HAD_{50} of ASFV-G-ΔI177L and observed for either (**A**) 90 or (**B**) 180 days post-inoculation. Symbols represent different animals in the experimental group.

As in the 90-day experiment, all animals from both groups were euthanized and necropsied, looking for the presence of macroscopical lesions associated with ASF-infected animals. No alterations were found in the animals inoculated with ASFV-G-ΔI177L or in the mock-inoculated animals.

The viremia kinetics of ASFV-G-ΔI177L-inoculated animals presented, as in the previous experiment, variable virus titers during the first month PI, with values ranging between $10^{2.05}$ and $10^{6.8}$ HAD_{50}/mL by day 30 PI (Figure 2B). At day 60 PI, viremia titers showed low titers ranging between undetectable and $10^{3.55}$ HAD_{50}/mL. The virus became undetectable in the blood of all the animals by day 90 PI and remained at that level in all animals when tested 120-, 150-, and 180-days PI.

The presence of virus-specific antibodies showed, as in the first experiment, serum titers ranging from 10^3 to 10^4/mL (Figure 2B) at day 30 pi. They remained at similar levels in all animals when tested at time points 60-, 90-, 120-, 150-, and 180-days PI. Antibody titers remained at background levels in all mock-inoculated animals at all time points tested during the 180-day experiment.

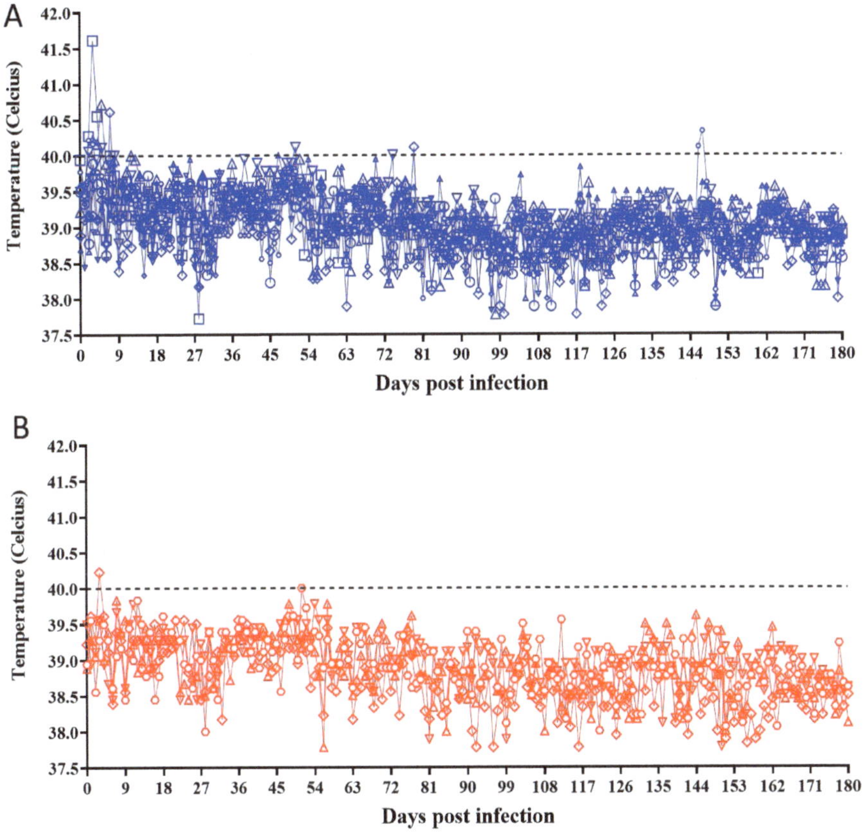

Figure 4. Evolution of body temperature in animals IM-inoculated with 10^6 HAD_{50} of ASFV-G-ΔI177L (**A**) or mock-inoculated (**B**) and observed for 180 days post-inoculation. Symbols represent specific animals at each experimental group.

In domestic swine, ASFV produces a clinical form of the disease that ranges from a subclinical to a highly lethal presentation. The disease outcome heavily depends on the active virus strain [1,2]. Therefore, residual virus virulence can present a variety of clinical forms of the disease. ASFV-G-ΔI177L has been tested in the absence of residual virulence in terms of causing an acute form of the disease [8,9,16,20], but data demonstrating complete attenuation in long-term experiments were not assessed until now. Natural attenuated strains usually present residual virulence in a chronic form of the disease, characterized by joint inflammation and skin lesions that appear several weeks after infection [1,2,17–19].

Taken together, the results presented here indicate that the ASFV-G-ΔI177L vaccine candidate lacks residual virulence even when tested in long-term studies on animals that have been inoculated with relatively high doses of the virus. It should be noted that the dose tested in this study is 10,000 times higher than the minimal protective doses used in experimental conditions [8,16,20]. Currently, ASFV-G-ΔI177L is being actively tested in field conditions at doses of $10^{2.6}$ HAD_{50} via the IM route. Therefore, the circumstances used in this report to test the residual virulence of ASFV-G-ΔI177L far exceed the conditions that the vaccine virus will be used in during field vaccinations. However, we should be cautious in directly extrapolating the results reported here to animals of different ages, reproductive statuses, or genetic backgrounds.

4. Conclusions

The data presented here, along with those from previously published studies [7,8,16,20], indicate that ASFV-G-ΔI177L lacks residual virulence even when used at relatively high doses (approximately 10,000 times the minimal protective dose) and evaluated during long observational periods. These results confirm the safety of the use of ASFV-G-ΔI177L as a vaccine candidate.

Author Contributions: Conceptualization, D.P.G. and M.V.B.; data curation, E.R.-M., A.R., N.E., E.S. and L.V.-S.; formal analysis, E.R.-M., D.P.G. and M.V.B.; funding acquisition, D.P.G. and M.V.B.; visualization, L.V.-S.; writing—original draft, D.P.G. and M.V.B.; writing—review and editing, E.R.-M., A.R., N.E., E.S., D.P.G. and M.V.B. All authors have read and agreed to the published version of the manuscript.

Funding: This project was partially funded by National Pork Board Project #21-137 and partially through the Foundation for Food and the Agriculture Rapid Outcomes from Agriculture Research (ROAR). Grant number 202-624-0700.

Institutional Review Board Statement: Animal experiments to collect blood for swine macrophages were performed under biosafety level 3AG conditions at the animal facilities of the Plum Island Animal Disease Center (PIADC). All experimental procedures were carried out in compliance with the Animal Welfare Act (AWA), the 2011 Guide for the Care and Use of Laboratory Animals, the 2002PHS Policy for the Humane Care and Use of Laboratory Animals, U.S. Government Principles for the Utilization and Care of Vertebrate Animals Used in Testing, Research, and Training (IRAC 1985), as well as specific animal protocols reviewed and approved by the PIADC Institutional Animal Care and Use Committee of the U.S. Departments of Agriculture and Homeland Security (protocol numbers 205.03-17-R, 205.03-19-R, and 225.02-19-R, approved on 28 September 2017, 10 September 2019, and 10 September 2019, respectively).

Informed Consent Statement: No applicable.

Data Availability Statement: The data are contained within the article.

Acknowledgments: We thank the Plum Island Animal Disease Center Animal Care Unit staff for their excellent technical assistance. We wish to particularly thank Carmen V. Borca Carrillo for editing the manuscript. This research was supported in part by an appointment to the Plum Island Animal Disease Center (PIADC) Research Participation Program administered by the Oak Ridge Institute for Science and Education (ORISE) through an interagency agreement between the U.S. Department of Energy (DOE) and the U.S. Department of Agriculture (USDA). ORISE is managed by ORAU under DOE contract number DE-SC0014664. All opinions expressed in this paper are the authors' and do not necessarily reflect the policies and views of the USDA, ARS, APHIS, DOE, or ORAU/ORISE.

Conflicts of Interest: The authors declare no conflict of interest.

References

1. Tulman, E.R.; Delhon, G.A.; Ku, B.K.; Rock, D.L. African swine fever virus. *Curr. Top. Microbiol. Immunol.* **2009**, *328*, 43–87. [PubMed]
2. Costard, S.; Wieland, B.; de Glanville, W.; Jori, F.; Rowlands, R.; Vosloo, W.; Roger, F.; Pfeiffer, D.U.; Dixon, L.K. African swine fever: How can global spread be prevented? *Philos. Trans. R. Soc. Lond. B Biol. Sci.* **2009**, *364*, 2683–2696. [CrossRef] [PubMed]
3. Rajukumar, K.; Senthilkumar, D.; Venkatesh, G.; Singh, F.; Patil, V.P.; Kombiah, S.; Tosh, C.; Dubey, C.K.; Sen, A.; Barman, N.N.; et al. Genetic characterization of African swine fever virus from domestic pigs in India. *Transbound. Emerg. Dis.* **2021**, *68*, 2687–2692. [CrossRef] [PubMed]
4. Sauter-Louis, C.; Forth, J.H.; Probst, C.; Staubach, C.; Hlinak, A.; Rudovsky, A.; Holland, D.; Schlieben, P.; Göldner, M.; Schatz, J.; et al. Joining the club: First detection of African swine fever in wild boar in Germany. *Transbound. Emerg. Dis.* **2021**, *68*, 1744–1752. [CrossRef] [PubMed]
5. Ramirez-Medina, E.; O'donnell, V.; Silva, E.; Espinoza, N.; Velazquez-Salinas, L.; Moran, K.; Daite, D.A.; Barrette, R.; Faburay, B.; Holland, R.; et al. Experimental Infection of Domestic Pigs with an African Swine Fever Virus Field Strain Isolated in 2021 from the Dominican Republic. *Viruses* **2022**, *14*, 1090. [CrossRef] [PubMed]

6. Malogolovkin, A.; Kolbasov, D. Genetic and antigenic diversity of African swine fever virus. *Virus Res.* **2019**, *271*, 197673. [CrossRef] [PubMed]
7. Borca, M.V.; Ramirez-Medina, E.; Silva, E.; Vuono, E.; Rai, A.; Pruitt, S.; Espinoza, N.; Velazquez-Salinas, L.; Gay, C.G.; Gladue, D.P. ASFV-G-ΔI177L as an Effective Oral Nasal Vaccine against the Eurasia Strain of Africa Swine Fever. *Viruses* **2021**, *13*, 765. [CrossRef] [PubMed]
8. Borca, M.V.; Ramirez-Medina, E.; Silva, E.; Vuono, E.; Rai, A.; Pruitt, S.; Holinka, L.G.; Velazquez-Salinas, L.; Zhu, J.; Gladue, D.P. Development of a Highly Effective African Swine Fever Virus Vaccine by Deletion of the I177L Gene Results in Sterile Immunity against the Current Epidemic Eurasia Strain. *J. Virol.* **2020**, *94*, e02017-19. [CrossRef] [PubMed]
9. Gladue, D.P.; Ramirez-Medina, E.; Vuono, E.; Silva, E.; Rai, A.; Pruitt, S.; Espinoza, N.; Velazquez-Salinas, L.; Borca, M.V. Deletion of the A137R Gene from the Pandemic Strain of African Swine Fever Virus Attenuates the Strain and Offers Protection against the Virulent Pandemic Virus. *J. Virol.* **2021**, *95*, e0113921. [CrossRef] [PubMed]
10. O'Donnell, V.; Holinka, L.G.; Gladue, D.P.; Sanford, B.; Krug, P.W.; Lu, X.; Arzt, J.; Reese, B.; Carrillo, C.; Risatti, G.R.; et al. African Swine Fever Virus Georgia Isolate Harboring Deletions of MGF360 and MGF505 Genes Is Attenuated in Swine and Confers Protection against Challenge with Virulent Parental Virus. *J. Virol.* **2015**, *89*, 6048–6056. [CrossRef] [PubMed]
11. O'Donnell, V.; Risatti, G.R.; Holinka, L.G.; Krug, P.W.; Carlson, J.; Velazquez-Salinas, L.; Azzinaro, P.A.; Gladue, D.P.; Borca, M.V. Simultaneous Deletion of the 9GL and UK Genes from the African Swine Fever Virus Georgia 2007 Isolate Offers Increased Safety and Protection against Homologous Challenge. *J. Virol.* **2017**, *91*, e01760-16. [CrossRef] [PubMed]
12. Chen, W.; Zhao, D.; He, X.; Liu, R.; Wang, Z.; Zhang, X.; Li, F.; Shan, D.; Chen, H.; Zhang, J.; et al. A seven-gene-deleted African swine fever virus is safe and effective as a live attenuated vaccine in pigs. *Sci. China Life Sci.* **2020**, *63*, 623–634. [CrossRef] [PubMed]
13. Teklue, T.; Wang, T.; Luo, Y.; Hu, R.; Sun, Y.; Qiu, H.-J. Generation and Evaluation of an African Swine Fever Virus Mutant with Deletion of the *CD2v* and *UK* Genes. *Vaccines* **2020**, *8*, 763. [CrossRef]
14. Zhang, J.; Zhang, Y.; Chen, T.; Yang, J.; Yue, H.; Wang, L.; Zhou, X.; Qi, Y.; Han, X.; Ke, J.; et al. Deletion of the L7L-L11L Genes Attenuates ASFV and Induces Protection against Homologous Challenge. *Viruses* **2021**, *13*, 255. [CrossRef] [PubMed]
15. Zhang, Y.; Ke, J.; Zhang, J.; Yang, J.; Yue, H.; Zhou, X.; Qi, Y.; Zhu, R.; Miao, F.; Li, Q.; et al. African Swine Fever Virus Bearing an I226R Gene Deletion Elicits Robust Immunity in Pigs to African Swine Fever. *J. Virol.* **2021**, *95*, e01199-21. [CrossRef] [PubMed]
16. Tran, X.H.; Le, T.T.P.; Nguyen, Q.H.; Do, T.T.; Nguyen, V.D.; Gay, C.G.; Borca, M.V.; Gladue, D.P. African swine fever virus vaccine candidate ASFV-G- Δ I177L efficiently protects European and native pig breeds against circulating Vietnamese field strain. *Transbound. Emerg. Dis.* **2021**, *69*, e497–e504. [CrossRef] [PubMed]
17. Brake, D.A. African Swine Fever Modified Live Vaccine Candidates: Transitioning from Discovery to Product Development through Harmonized Standards and Guidelines. *Viruses* **2022**, *14*, 2619. [CrossRef] [PubMed]
18. Gladue, D.P.; Borca, M.V. Recombinant ASF Live Attenuated Virus Strains as Experimental Vaccine Candidates. *Viruses* **2022**, *14*, 878. [CrossRef] [PubMed]
19. Rock, D.L. Thoughts on African Swine Fever Vaccines. *Viruses* **2021**, *13*, 943. [CrossRef] [PubMed]
20. Tran, X.H.; Phuong, L.T.T.; Huy, N.Q.; Thuy, D.T.; Nguyen, V.D.; Quang, P.H.; Ngôn, Q.V.; Rai, A.; Gay, C.G.; Gladue, D.P.; et al. Evaluation of the Safety Profile of the ASFV Vaccine Candidate ASFV-G-DI177L. *Viruses* **2022**, *14*, 896. [CrossRef] [PubMed]

pathogens

MDPI

Brief Report

The Presence of Virus Neutralizing Antibodies Is Highly Associated with Protection against Virulent Challenge in Domestic Pigs Immunized with ASFV live Attenuated Vaccine Candidates

Ediane B. Silva [1,†], Peter W. Krug [1,†], Elizabeth Ramirez-Medina [1], Alyssa Valladares [1,2], Ayushi Rai [1,3], Nallely Espinoza [1], Douglas P. Gladue [1,*] and Manuel V. Borca [1,*]

1 Plum Island Animal Disease Center, ARS, USDA, Greenport, NY 11944, USA
2 College of Veterinary Medicine, Kansas State University, Manhattan, KS 66506, USA
3 Oak Ridge Institute for Science and Education (ORISE), Oak Ridge, TN 37830, USA
* Correspondence: douglas.gladue@usda.gov (D.P.G.); manuel.borca@usda.gov (M.V.B.)
† Contributed equally to this work.

Abstract: African swine fever virus (ASFV) is currently producing a pandemic affecting a large area of Eurasia, and more recently, the Dominican Republic in the Western Hemisphere. ASFV is a large and structurally complex virus with a large dsDNA genome encoding for more than 150 genes. Live attenuated virus strains can induce protection in domestic swine against disease produced by homologous virulent parental viruses. The roles of the different immune mechanisms induced by the attenuated strains in protection still need to be understood. In particular, the role of ASFV neutralizing antibody in protection still is an important controversial issue to be elucidated. Here we present the development of a novel methodology to detect virus neutralizing antibodies based on the reduction of virus infectivity in a Vero cell adapted ASFV strain. The described method was used to assess levels of virus neutralizing antibodies in domestic swine inoculated with live attenuated ASFV. Results demonstrated a high association between the presence of virus neutralizing antibodies and protection in 84 animals immunized with the recombinant vaccine candidates ASFV-G-Δ9GL/ΔUK or ASFV-G-ΔI177L. To our knowledge, this is the first report demonstrating an association between virus neutralizing antibodies and protection against virulent challenge in such a large number of experimental individuals.

Keywords: ASFV; ASF; African swine fever virus; virus neutralizing antibodies; protective immunity

Citation: Silva, E.B.; Krug, P.W.; Ramirez-Medina, E.; Valladares, A.; Rai, A.; Espinoza, N.; Gladue, D.P.; Borca, M.V. The Presence of Virus Neutralizing Antibodies Is Highly Associated with Protection against Virulent Challenge in Domestic Pigs Immunized with ASFV live Attenuated Vaccine Candidates. *Pathogens* **2022**, *11*, 1311. https://doi.org/10.3390/pathogens11111311

Academic Editor: Natasha N. Gaudreault

Received: 29 September 2022
Accepted: 5 November 2022
Published: 8 November 2022

1. Introduction

African swine fever (ASF) is a usually lethal disease of domestic pigs which is currently producing a pandemic affecting the swine production industry across Eurasia and just recently, the Hispaniola Island, after more than 40 years of being absent in the Western Hemisphere [1]. Since the first available commercial vaccine is still restricted in its use [2,3], the current control of the disease is based on culling all infected animals along with strict measures to avoid the mobilization of susceptible and potentially infected animals.

Protection against the disease has been experimentally demonstrated by immunizing animals with live attenuated strains of viruses. Live attenuated strains, particularly those produced by the deletion of specific genes involved in virulence from the genome of virulent field ASFV isolates, are generally effective at inducing protection against the challenge with the homologous virulent strains [4–8]. Nevertheless, the host immune mechanism mediating the protection induced by those live attenuated strains is far from being identified. Although several immune mechanisms have been described as associated with ASF infection [9,10], their role in the induction of protection has not been conclusively demonstrated. There is evidence that the T-cell immunity is important in the protective immune response elicited by live attenuated vaccine candidates [11,12]. Similarly, passive transfer of immunoglobulin from ASF convalescent animals protected a high proportion of recipient naïve animals against challenge with virulent virus [13]. Nevertheless, the immune effector mechanism mediated by the ASF specific antibodies remains to be elucidated. Several laboratories have developed different methodologies to detect the presence of ASFV neutralizing antibodies [14–18]. However, the potential role of virus neutralizing antibodies in protection against infection and disease after challenge with virulent virus strains remains uncertain. In this report, we attempt to associate the presence of virus neutralizing antibodies and protection using a reliable experimental model involving domestic pigs immunized with protective ASFV recombinant vaccine candidates followed by virulent challenge. By assessing the sera from over 80 experimentally vaccinated animals, our results demonstrated a close association between the presence of virus neutralizing antibodies and protection against lethal ASFV challenge

2. Materials and Methods

2.1. Preparation of Virus Neutralization Stock

Vero cells (CCL-81, ATCC) were passaged using growth media consisting of DMEM supplemented with 7% fetal bovine sera (FBS) and 1X antibiotics. High-titer Vero cell-adapted ASFV from the 30th passage (ASFV/VP30) stocks [19] were diluted tenfold in maintenance media (DMEM supplemented with 1% FBS and 1X antibiotics), then Tween 80 was added to a final concentration of 0.05%. After extensive mixing, the stock was sonicated on ice 3 times for 15 s each with a microtip sonicator (Branson Ultrasonics™ Sonifier™ Cell Disruptor) set at 30% power to prevent virion aggregates. The virus stock was then filtered on 0.45μm syringe filters, aliquoted and stored at −70 °C until use. To determine virus titer, serial dilutions of this stock were used to infect Vero cells on 12-well cell culture plates. One hour after infection, the inoculum was removed, and the wells were overlayed with methylcellulose media (0.5% methylcellulose in DMEM supplemented with 1% FBS and 1X Penicillin-streptomycin). Ten days post infection, the methylcellulose media was aspirated, the cells were washed with PBS, fixed with a 1:1 mixture of methanol:acetone and the plaques were visualized by immunostaining using a monoclonal antibody recognizing ASFV protein p30 as previously described [20]. Virus stocks that contained at least 1×10^7 PFU/mL were used for the neutralization assay.

2.2. Neutralization Assay

This assay is a modification of the procedure originally described by Zsak et al. [18]. Heat-inactivated swine sera was serially diluted tenfold with maintenance media in non-tissue culture-treated 96-well dilution plates. An equal volume containing 100 PFU of ASFV/VP30 neutralization stock was added to each well of diluted serum, and the plates were incubated at 37 °C in a cell culture incubator with 5% CO_2 for 18–24 h. Wells containing 100 PFU of virus mixed with naïve swine serum were included as controls. Twelve-well cell culture plates containing 90% confluent Vero cell monolayers were washed once with phosphate-buffered saline (GIBCO™) to remove growth media and then infected with the serum:virus neutralization mixtures at 37 °C in a 5% CO_2 cell culture incubator. After one hour incubation, the inoculum was removed, and methylcellulose media was added to each well. Plates were then placed in a 37 °C cell culture incubator with 5% CO_2 for up to 10 days. After confirming the naïve serum control wells exhibited visible virus plaques by light microscopy, the methylcellulose overlay was removed from all the wells by aspiration and the wells were washed once with PBS to remove the remaining methylcellulose media. The wells were fixed with a 1:1 mixture of ice-cold methanol:acetone for 15 min and after the fixative was removed, plates were allowed to dry. Presence of plaques were visualized by immunostaining performed as described elsewhere [20]. The number of plaques on each well was determined by counting under a dissecting microscope. The number of plaques for each serum dilution was plotted against each serum concentration and the trendline equation was generated using Microsoft Excel. The neutralization index (NI) is defined as the dilution representing the serum concentration for each sample that inibits 50% of the viral plaques counted in the naïve serum control wells.

2.3. Detection of ASFV Specific Antibody Response by ELISA

Detection of ASFV specific antibody was performed using an in-house ELISA as described previously [20]. Briefly, ELISA antigen was prepared from Vero cells infected with a Vero adapted Georgia strain ASFV. Maxisorb ELISA plates (Nunc, St. Louis, MO, USA) were coated with 1 µg per well of infected or uninfected cell extract. The plates were blocked with phosphate buffered saline containing 10% skim milk (Merck, Kenilworth, NJ, USA) and 5% normal goat serum (Sigma, Saint Louis, MO, USA). Each swine serum sample was tested at multiple dilutions against both infected and uninfected cellular antigen. ASFV-specific antibodies in the swine sera were detected using an anti-swine IgG-horseradish peroxidase conjugate (KPL, Gaithersburg, MD, USA) and SureBlue Reserve peroxidase substrate (KPL). Plates were read at OD630 nm in an ELx808 plate reader (BioTek, Shoreline, WA, USA). Antibody titers were expressed as the $\log_{10}$ of the inverse highest dilution where the OD630 nm reading of the tested sera at least duplicated the reading of the mock infected (obtained at day 0 post infection) sera.

2.4. Animal Study Design

In order to analyze the potential association of the presence of serum neutralizing antibodies and protection against disease, a set of sera from animals that were vaccinated with experimental recombinant live attenuated vaccines and further challenged with the virulent parental field strain were selected. A total of 84 sera obtained at the time of challenge from pigs vaccinated with experimental vaccine candidates ASFV-G-Δ9GL/ΔUK [4] and ASFV-G-ΔI177L [8] were analyzed for neutralizing antibodies. The serum samples used in this study had been collected in previous studies [4,8,21] run under protocol approved by the Plum Island Animal Disease Center Institutional Animal Care and Use Committee (225.01-16-R_090716). These animals were vaccinated using different vaccines; doses, routes of inoculation and the challenge were conducted at different times post vaccination as described in each of the corresponding figures and it is summarized in Table 1. This set of sera contains sample form animals selected to represent a variety of scenarios regarding the protective response induced by intramuscular (IM) or oronasal (ON) vaccination and considering a different level of maturation of the antibody response.

Table 1. Detail of the vaccine type, route of administration, dose, time of challenge and protection status of pigs' sera used in this report.

Vaccine Type, Dose, and Route of Administration *	Challenge at dpv †	Number of Protected/ Total at Challenge ††
ASFV-G-Δ9GL/ΔUK (IM) 10^2 HAD	28	3/9
ASFV-G-Δ9GL/ΔUK (IM) 10^4 HAD	28	13/15
ASFV-G-Δ9GL/ΔUK (IM) 10^6 HAD	28	15/15
ASFV-G-Δ9GL/ΔUK (IM) 10^4 HAD	7	1/5
ASFV-G-Δ9GL/ΔUK (IM) 10^4 HAD	14	5/5
ASFV-G-Δ9GL/ΔUK (IM) 10^4 HAD	21	4/5
ASFV-G-ΔI177L (IM) 10^2 HAD	28	20/20
ASFV-G-ΔI177L (IM) 10^6 HAD	28	5/5
ASFV-G-ΔI177L (O/N) 10^6 HAD	28	5/5

* IM: intramuscular; ON: oronasal. HAD: Hemadsorption. † dpv: days post vaccination. †† Clinical observation was performed for 21 days after challenge.

2.5. Statistical Analysis

Data were analyzed using Microsoft Excel 2010 software and GraphPad prism 7 software. Comparison of the treated vaccine groups was analyzed by ANOVA with Dunnett's multiple comparison test. Correlations between animal survival and sera virus neutralization capability and antibody specificity and virus neutralization were analyzed using Pearson correlation coefficients. The receiver operating characteristic (ROC) curve was analyzed to select optimal cutoff for virus neutralization value.

3. Results and Discussion

3.1. Association between the Presence of ASFV Specific Antibodies Detected by ELISA and Protection against the Virulent Challenge in Pigs Immunized with Live Attenuated Vaccine Candidates

First, the presence of ASFV specific antibodies in the set of the 84 sera was evaluated by the ELISA as described in Materials and Methods. Results indicated that most vaccinated animals possess levels of virus specific binding antibodies at the time of challenge (Figure 1). The exceptions were a subset of animals vaccinated with ASFV-G-Δ9GL/ΔUK either receiving 10^2 HAD and challenged at 28 dpv or those receiving 10^4 HAD and challenged at 7 dpv. As expected, titer values increased in groups of animals challenged at 28 dpv compared to earlier challenge at 7, 14 and 21 dpv ($p < 0.0001$, <0.0001, 0.0003). Detectable antibody titer over 10^1 is generally associated with protection against clinical disease after challenge (Figure 1) with the exception of five animals immunized with ASFV-G-Δ9GL/ΔUK (one vaccinated with 10^2 HAD and challenged at 28 dpv, and four vaccinated with 10^4 HAD that were challenged at 21 and 28 dpv). Conversely, only four animals inoculated with ASFV-G-Δ9GL/ΔUK were protected at challenge without showing virus specific antibody titers, both vaccinated with 10^4 HAD and challenged at 7 and 28 dpv, respectively. Therefore, out of the 84 cases reviewed, only 5 animals were not protected in the presence of antibodies and only 2 were protected in the absence of them. These results are in agreement with our previous observation that the presence of medium to high ASFV specific antibody titers at the time of challenge closely associates with protection against clinical disease [20].

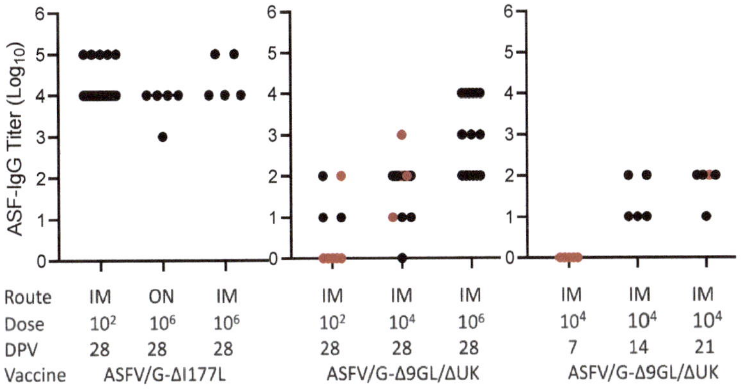

Figure 1. ELISA titers in sera obtained at the time of challenge in animals immunized with either ASFV-G-Δ9GL/ΔUK or ASFV-G-ΔI177L vaccine stains. Black dots and red dots indicate presence and absence of protection against the challenge, respectively.

3.2. Detection of Virus Neutralizing Antibodies in Pigs Immunized with Live Attenuated Vaccine Candidates

Next, to evaluate the relationship of virus neutralizing antibodies with ASF protection, the virus neutralization activity of the 84 sera from the pigs vaccinated with the two different recombinant virus strains was assessed (Figure 2). All animals vaccinated with ASFV-G-ΔI177L survived the challenge and exhibited NI values over 1.5, with the exception of two animals that were immunized with 10^2 HAD_{50} that did not show any neutralizing activity. As observed with the antibodies detected by ELISA, overall, animals vaccinated with the ASFV-G-ΔI177L strain had a higher neutralization index (NI) compared with those vaccinated with ASFV-G-Δ9GL/ΔUK ($p < 0.05$). This is particularly evident ($p = 0.0004$) when groups immunized with low doses of vaccine (10^2 HAD_{50}) and challenged at 28 dpv are compared: all animals receiving ASFV-G-ΔI177L developed NI over 2 while those immunized with group ASFV-G-Δ9GL/ΔUK barely reach that value. Considering the animals vaccinated with ASFV-G-Δ9GL/ΔUK, all of them demonstrated NI values over 1.5 and were protected against challenge with the exception of one receiving 10^2 HAD_{50} and six receiving 10^4 HAD_{50}, challenged at 7, 21 and 28 dpv. Therefore, out of the 84 cases considered in the study, only 7 pigs with NI values over 1.5 were not protected against the challenge. Interesting, only one animal, vaccinated with 10^4 HAD_{50} of ASFV-G-Δ9GL/ΔUK and challenged at 28 dpv, was protected without generating detectable neutralizing antibodies. Thus, in over 90% of the animals there existed an association between the presence of neutralizing antibody and protection against challenge.

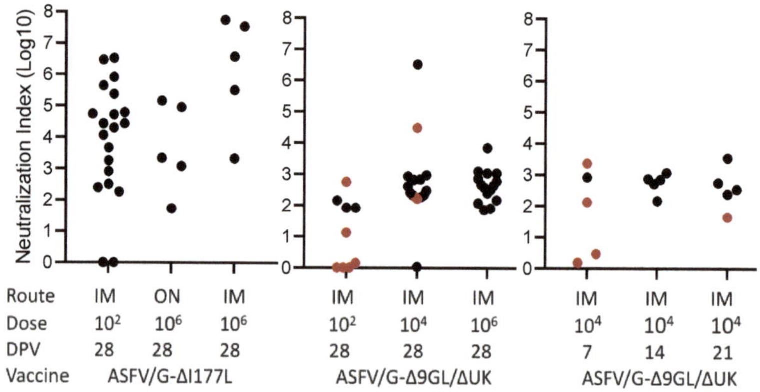

Figure 2. Neutralization index in sera obtained at the time of challenge in animals immunized with either ASFV-G-Δ9GL/ΔUK or ASFV-G-ΔI177L vaccine stains. The ROC curve was calculated with AUC = 0.7745 and Cut-off < 1.690 with specificity of 95.71% and Sensitivity of 50%. Black dots and red dots indicate presence and absence of protection against the challenge, respectively.

In the group of animals surviving the challenge, 95.7% of them demonstrated NI activity and only 3 out 70 (4.3%) failed to induce neutralizing antibodies against ASFV. Therefore, there was a strong association between survival after the challenge and NI activity. Our study, confirmed by ELISA (97.6%) and NI activity (95.7%) (Figure 3), suggests that survival from ASFV challenge is associated with the presence of antibodies.

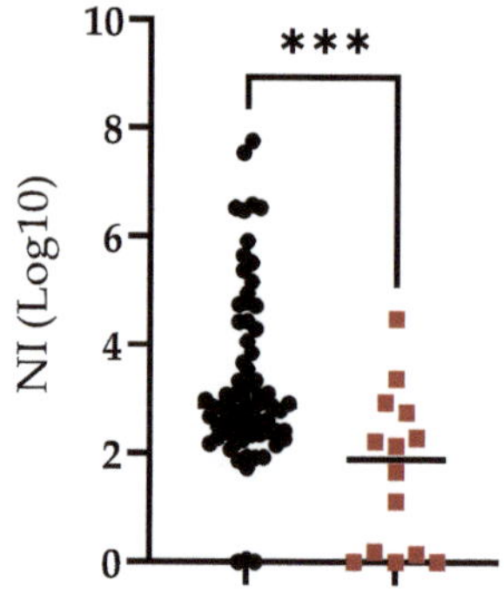

Figure 3. Neutralization index (NI) in sera of animals vaccinated with ASFV-G-Δ9GL/ΔUK or ASFV-G-ΔI177L vaccine strains and challenged with virulent field strain ASFV-G. Animals were grouped based on survival (black symbols) or absence of protection (red symbols) after the challenge. Asterisks indicate differences among groups ($p < 0.05$).

In addition, the correlation between the presence of ASFV specific antibodies detected by ELISA and those detected by virus neutralization was analyzed. A Pearson correlation coefficient was calculated to establish this potential association and results indicate the presence of a significant positive correlation (r = 0.65; p = 0.0001) between the two variables considered (Figure 4). Regardless of the characteristics and inherent limitations of both techniques, the correlation related to the presence of antibodies detected by either methodology is clear. This correlation is not unexpected since it is reasonable to assume that the majority of antibody mediating virus neutralization will also be detected by the ELISA test used here. Consequently, the presence of antibodies detected by both techniques is consistently associated with protection against the challenge.

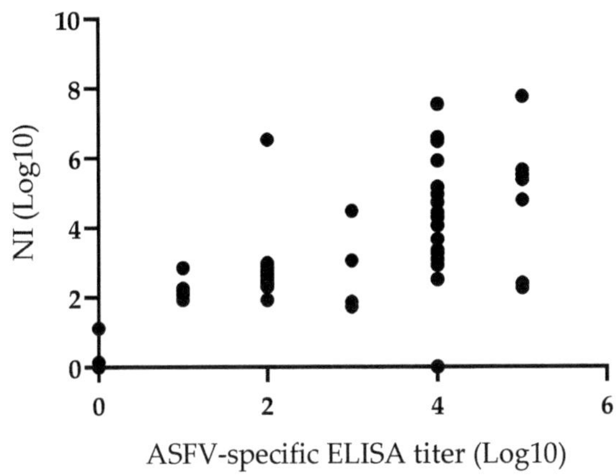

Figure 4. Comparison of neutralization index (NI) and anti-ASFV ELISA titers in sera of animals vaccinated with ASFV-G-Δ9GL/ΔUK or ASFV-G-ΔI177L vaccine stains. Pearson correlation coefficient (r) 0.65; 95% of interval 0.5004 to 0.7898; P (two-tailed) 0.0001.

In summary, out of the 84 animals considered in this study 70 of them were protected against the challenge while the other 14 of them succumbed to lethal infection. In the group of animals surviving the challenge, 95.7% of them presented virus neutralizing antibody while, on the other hand, only 3 out 70 (4.3%) of those animals failed to produce detectable neutralizing antibodies against ASFV. Therefore, this suggests an association between animal survival at the time of challenge and VN activity. The Pearson correlation between survival and antibody neutralizing activity further demonstrates this relationship (Figure 5). There was a positive correlation between the two variables (r = 0.68; p = 0.0421).

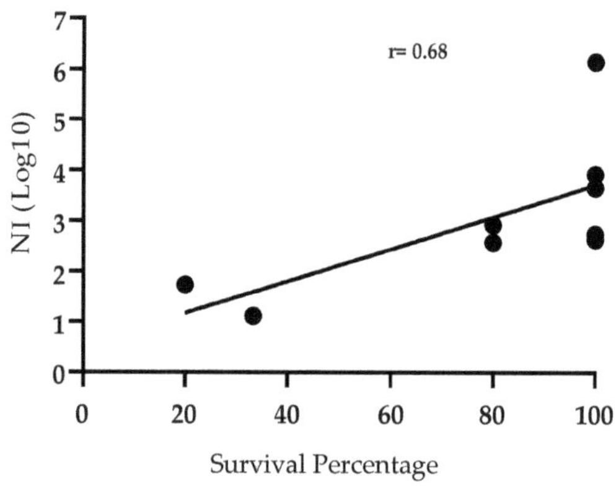

Figure 5. Correlation between protection after challenge and Neutralizing Index (NI) in sera of animals vaccinated with ASFV-G-Δ9GL/ΔUK or ASFV-G-ΔI177L vaccine stains. Survival percentage is represented by number of survivals groups described in Table 1. Pearson correlation coefficient r = 0.6841, $p < 0.05$; 95% confidence interval 0.0368 to 0.9271; P (two-tailed) 0.0421.

The neutralization assay developed here was based on a protocol designed to test the neutralizing capacity of monoclonal antibodies and swine serum to block ASFV infection as described in Zsak et al. [18]. The major difference between the assays is the use of a Vero cell-adapted Georgia strain virus at the 30th passage, which is representative of the virus circulating in the current Eurasian pandemic [19]. Another difference is based on the readout of the assay. While the previous work [18] determined a percentage reduction for each serum or antibody tested at a predetermined dilution, the assay described here uses multiple dilutions for each serum to determine a 50% inhibition titer. Since there is always a fraction of the virus that is not neutralizable, the calculations were carried out with multiple serum dilution controls for this fraction. Using the average neutralization index among each group of animals given various vaccines at different doses, we were able to correlate high survival percentage after virulent challenge with high neutralizing antibody levels using a non-linear curve fit analysis, suggesting that neutralizing antibodies are a correlate of protection. Potential modifications of this assay would include high-throughput sample processing to analyze neutralizing antibodies using fluorescently labeled ASFV, enabling neutralization index analysis to become part of the pre-challenge quality control for vaccine candidates prior to challenge.

The presence of neutralizing antibodies in ASFV has been an issue of historical controversy among research groups working in the identification of host immune mechanisms mediating protection against virulent challenge. Although early findings supported the absence of virus neutralizing antibodies in ASFV [22,23] (perhaps due to assay limitations), numerous reports have shown the existence and characterization of neutralizing antibodies in the sera of animals surviving the infection with attenuated viruses (reviewed in [14]). It is now widely accepted that antibodies in animals surviving infection or those that have been treated with vaccine candidates have activity-blocking virus infectivity. The mechanism of this inhibition may not necessarily be virus neutralization per se, as protective antibodies can function in other ways (e.g., antibody-dependent cellular cytotoxicity, opsonization, complement-mediated lysis). Additionally, antibodies may work in cooperation with T-cell-mediated mechanisms of protection [24]. The results presented here indicate that the presence of ASFV neutralizing antibodies is associated with the protection against clinical disease and death produced by virulent challenge. In this report, we show that virus neutralization activity was present in almost 100% of animals vaccinated with live attenuated vaccine candidates that survived to the ASV challenge. This study attempted, for the first time, to associate the presence of ASFV neutralizing antibodies with protection from challenge utilizing a large number of experimentally vaccinated animals. The presented results suggest that antibody plays an important role in ASF protection.

Author Contributions: Conceptualization, M.V.B. and D.P.G.; methodology, P.W.K. and E.B.S.; investigation, P.W.K., E.B.S., E.R.-M., A.R., A.V. and N.E.; data curation, P.W.K., E.B.S., E.R.-M., A.R., A.V. and N.E.; writing—original draft preparation, M.V.B. and D.P.G.; writing—review and editing, P.W.K., M.V.B., E.B.S., D.P.G., E.R.-M., A.R., A.V. and N.E.; funding acquisition, M.V.B. and D.P.G. All authors have read and agreed to the published version of the manuscript.

Funding: This project was funded through an interagency agreement with the Science and Technology Directorate of the U.S. Department of Homeland Security under Award Number 70RSAT19KPM000056.

Institutional Review Board Statement: This study run under protocol approved by the Plum Island Animal Disease Center Institutional Animal Care and Use Committee (225.01-16-R_090716).

Informed Consent Statement: Not applicable.

Data Availability Statement: Not applicable.

Acknowledgments: We thank the Plum Island Animal Disease Center Animal Care Unit staff for their excellent technical assistance. We wish to particularly thank Carmen V. Borca-Carrillo for editing the manuscript. This research was supported in part by an appointment to the Plum Island Animal Disease Center (PIADC) Research Participation Program administered by the Oak Ridge Institute for Science and Education (ORISE) through an interagency agreement between the U.S. Department of Energy (DOE) and the U.S. Department of Agriculture (USDA). ORISE is managed by ORAU under DOE contract number DE-SC0014664. All opinions expressed in this paper are the author's and do not necessarily reflect the policies and views of USDA, ARS, APHIS, DOE, or ORAU/ORISE.

Conflicts of Interest: The authors declare no conflict of interest.

References

1. Costard, S.; Wieland, B.; de Glanville, W.; Jori, F.; Rowlands, R.; Vosloo, W.; Roger, F.; Pfeiffer, D.U.; Dixon, L.K. African swine fever: How can global spread be prevented? *Philos. Trans. R. Soc. Lond. B Biol. Sci.* **2009**, *364*, 2683–2696. [CrossRef] [PubMed]
2. Tran, X.H.; Phuong, L.; Huy, N.Q.; Thuy, D.T.; Nguyen, V.D.; Quang, P.H.; Ngôn, Q.V.; Rai, A.; Gay, C.G.; Gladue, D.P.; et al. Evaluation of the Safety Profile of the ASFV Vaccine Candidate ASFV-G-ΔI177L. *Viruses* **2022**, *14*, 896. [CrossRef] [PubMed]
3. Tran, X.H.; Le, T.; Nguyen, Q.H.; Do, T.T.; Nguyen, V.D.; Gay, C.G.; Borca, M.V.; Gladue, D.P. African swine fever virus vaccine candidate ASFV-G-ΔI177L efficiently protects European and native pig breeds against circulating Vietnamese field strain. *Transbound Emerg. Dis.* **2022**, *69*, e497–e504. [CrossRef] [PubMed]
4. O'Donnell, V.; Risatti, G.R.; Holinka, L.G.; Krug, P.W.; Carlson, J.; Velazquez-Salinas, L.; Azzinaro, P.A.; Gladue, D.P.; Borca, M.V. Simultaneous Deletion of the 9GL and UK Genes from the African Swine Fever Virus Georgia 2007 Isolate Offers Increased Safety and Protection against Homologous Challenge. *J. Virol.* **2016**, *91*, e01760-16. [CrossRef]
5. O'Donnell, V.; Holinka, L.G.; Krug, P.W.; Gladue, D.P.; Carlson, J.; Sanford, B.; Alfano, M.; Kramer, E.; Lu, Z.; Arzt, J.; et al. African Swine Fever Virus Georgia 2007 with a Deletion of Virulence-Associated Gene 9GL (B119L), when Administered at Low Doses, Leads to Virus Attenuation in Swine and Induces an Effective Protection against Homologous Challenge. *J. Virol.* **2015**, *89*, 8556–8566. [CrossRef]
6. O'Donnell, V.; Holinka, L.G.; Gladue, D.P.; Sanford, B.; Krug, P.W.; Lu, X.; Arzt, J.; Reese, B.; Carrillo, C.; Risatti, G.R.; et al. African Swine Fever Virus Georgia Isolate Harboring Deletions of MGF360 and MGF505 Genes Is Attenuated in Swine and Confers Protection against Challenge with Virulent Parental Virus. *J. Virol.* **2015**, *89*, 6048–6056. [CrossRef]
7. Chen, W.; Zhao, D.; He, X.; Liu, R.; Wang, Z.; Zhang, X.; Li, F.; Shan, D.; Chen, H.; Zhang, J.; et al. A seven-gene-deleted African swine fever virus is safe and effective as a live attenuated vaccine in pigs. *Sci. China Life Sci.* **2020**, *63*, 623–634. [CrossRef]
8. Borca, M.V.; Ramirez-Medina, E.; Silva, E.; Vuono, E.; Rai, A.; Pruitt, S.; Holinka, L.G.; Velazquez-Salinas, L.; Zhu, J.; Gladue, D.P. Development of a Highly Effective African Swine Fever Virus Vaccine by Deletion of the I177L Gene Results in Sterile Immunity against the Current Epidemic Eurasia Strain. *J. Virol.* **2020**, *94*, e02017-19. [CrossRef]
9. Montoya, M.; Franzoni, G.; Pérez-Nuñez, D.; Revilla, Y.; Galindo, I.; Alonso, C.; Netherton, C.L.; Blohm, U. 3. Immune Responses against African Swine Fever Virus Infection. In *Understanding and Combatting African Swine Fever*; Wageningen Academic Publishers: Wageningen, The Netherlands, 2021; pp. 63–85. [CrossRef]
10. Martins, C.L.; Leitão, A.C. Porcine immune responses to African swine fever virus (ASFV) infection. *Vet. Immunol. Immunopathol.* **1994**, *43*, 99–106. [CrossRef]
11. Sánchez-Cordón, P.J.; Jabbar, T.; Chapman, D.; Dixon, L.K.; Montoya, M. Absence of Long-Term Protection in Domestic Pigs Immunized with Attenuated African Swine Fever Virus Isolate OURT88/3 or BeninΔMGF Correlates with Increased Levels of Regulatory T Cells and Interleukin-10. *J. Virol.* **2020**, *94*, e00350-20. [CrossRef]

12. Oura, C.; Denyer, M.S.; Takamatsu, H.; Parkhouse, R. In vivo depletion of CD8+ T lymphocytes abrogates protective immunity to African swine fever virus. *J. Gen. Virol.* **2005**, *86*, 2445–2450. [CrossRef] [PubMed]
13. Onisk, D.V.; Borca, M.V.; Kutish, G.; Kramer, E.; Irusta, P.; Rock, D.L. Passively transferred African swine fever virus antibodies protect swine against lethal infection. *Virology* **1994**, *198*, 350–354. [CrossRef] [PubMed]
14. Escribano, J.M.; Galindo, I.; Alonso, C. Antibody-mediated neutralization of African swine fever virus: Myths and facts. *Virus Res.* **2013**, *173*, 101–109. [CrossRef] [PubMed]
15. Gómez-Puertas, P.; Rodríguez, F.; Ortega, A.; Oviedo, J.M.; Alonso, C.; Escribano, J.M. Improvement of African swine fever virus neutralization assay using recombinant viruses expressing chromogenic marker genes. *J. Virol. Methods* **1995**, *55*, 271–279. [CrossRef]
16. Fernandez, A.; Perez, J.; Martin de las Mulas, J.; Carrasco, L.; Dominguez, J.; Sierra, M.A. Localization of African swine fever viral antigen, swine IgM, IgG and C1q in lung and liver tissues of experimentally infected pigs. *J. Comp. Pathol.* **1992**, *107*, 81–90. [CrossRef]
17. Neilan, J.G.; Zsak, L.; Lu, Z.; Burrage, T.G.; Kutish, G.F.; Rock, D.L. Neutralizing antibodies to African swine fever virus proteins p30, p54, and p72 are not sufficient for antibody-mediated protection. *Virology* **2004**, *319*, 337–342. [CrossRef]
18. Zsak, L.; Onisk, D.V.; Afonso, C.L.; Rock, D.L. Virulent African swine fever virus isolates are neutralized by swine immune serum and by monoclonal antibodies recognizing a 72-kDa viral protein. *Virology* **1993**, *196*, 596–602. [CrossRef]
19. Krug, P.W.; Holinka, L.G.; O'Donnell, V.; Reese, B.; Sanford, B.; Fernandez-Sainz, I.; Gladue, D.P.; Arzt, J.; Rodriguez, L.; Risatti, G.R.; et al. The progressive adaptation of a georgian isolate of African swine fever virus to vero cells leads to a gradual attenuation of virulence in swine corresponding to major modifications of the viral genome. *J. Virol.* **2015**, *89*, 2324–2332. [CrossRef]
20. Carlson, J.; O'Donnell, V.; Alfano, M.; Velazquez Salinas, L.; Holinka, L.G.; Krug, P.W.; Gladue, D.P.; Higgs, S.; Borca, M.V. Association of the Host Immune Response with Protection Using a Live Attenuated African Swine Fever Virus Model. *Viruses* **2016**, *8*, 291. [CrossRef]
21. Borca, M.V.; Ramirez-Medina, E.; Silva, E.; Vuono, E.; Rai, A.; Pruitt, S.; Espinoza, N.; Velazquez-Salinas, L.; Gay, C.G.; Gladue, D.P. ASFV-G-ΔI177L as an Effective Oral Nasal Vaccine against the Eurasia Strain of Africa Swine Fever. *Viruses* **2021**, *13*, 765. [CrossRef]
22. Viñuela, E. African swine fever virus. *Curr. Top. Microbiol. Immunol.* **1985**, *116*, 151–170. [CrossRef] [PubMed]
23. Hess, W.R. African swine fever: A reassessment. *Adv. Vet. Sci. Comp. Med.* **1981**, *25*, 39–69. [PubMed]
24. Goatley, L.C.; Nash, R.H.; Andrews, C.; Hargreaves, Z.; Tng, P.; Reis, A.L.; Graham, S.P.; Netherton, C.L. Cellular and Humoral Immune Responses after Immunisation with Low Virulent African Swine Fever Virus in the Large White Inbred Babraham Line and Outbred Domestic Pigs. *Viruses* **2022**, *14*, 1487. [CrossRef] [PubMed]

pathogens

MDPI

Review

Advancement in the Antigenic Epitopes and Vaccine Adjuvants of African Swine Fever Virus

Qiong Wu †, Chang Li †, Bo Zhu, Jiajia Zhu, Keli Yang, Zewen Liu, Wei Liu, Ting Gao, Fangyan Yuan, Rui Guo, Yongxiang Tian * and Danna Zhou *

Key Laboratory of Prevention and Control Agents for Animal Bacteriosis (Ministry of Agriculture and Rural Affairs), Hubei Provincial Key Laboratory of Animal Pathogenic Microbiology, Institute of Animal Husbandry and Veterinary, Hubei Academy of Agricultural Sciences, Wuhan 430064, China; wuqiong302@hbaas.com (Q.W.); lichang1113@hbaas.com (C.L.); 15972742735@163.com (B.Z.); xmszjj@hbaas.com (J.Z.); keliy6@hbaas.com (K.Y.); liuzwen2004@hbaas.com (Z.L.); liuwei@hbaas.com (W.L.); gaoting2017@hbaas.com (T.G.); fyyuan@hbaas.com (F.Y.); guorui@hbaas.com (R.G.)

* Correspondence: tyxanbit@hbaas.com (Y.T.); zdn66@hbaas.com (D.Z.)

† These authors contributed equally to this work.

Abstract: African swine fever virus (ASFV), a highly virulent double-stranded DNA virus, poses a significant threat to global pig farming, with mortality rates in domestic pigs reaching up to 100%. Originating in Kenya in 1921, ASFV has since proliferated to Western Europe, Latin America, Eastern Europe, and most recently China in 2018, resulting in substantial global agricultural losses. Antigenic epitopes, recognized by the immune system's T cells and B cells, are pivotal in antiviral immune responses. The identification and characterization of these antigenic epitopes can offer invaluable insights into the immune response against ASFV and aid in the development of innovative immunotherapeutic strategies. Vaccine adjuvants, substances that amplify the body's specific immune response to antigens, also play a crucial role. This review provides an overview of the progress in studying T/B-cell epitopes in ASFV proteins and ASFV vaccine adjuvants, highlighting their role in the immune response and potential use in new vaccine development.

Keywords: African swine fever virus; antigenic epitope; T-cell epitope; B-cell epitope; Vaccine adjuvants

Citation: Wu, Q.; Li, C.; Zhu, B.; Zhu, J.; Yang, K.; Liu, Z.; Liu, W.; Gao, T.; Yuan, F.; Guo, R.; et al. Advancement in the Antigenic Epitopes and Vaccine Adjuvants of African Swine Fever Virus. *Pathogens* **2024**, *13*, 706. https://doi.org/10.3390/pathogens13080706

Academic Editors: Gisselle N. Medina, Lauro Velazquez-Salinas and Elizabeth Ramirez-Medina

Received: 31 July 2024
Revised: 17 August 2024
Accepted: 19 August 2024
Published: 21 August 2024

1. Introduction

African Swine Fever (ASF) is an acute hemorrhagic viral disease, caused by African Swine Fever Virus (ASFV), that affects domestic pigs and various species of wild boars [1,2]. ASFV is a member of the Asfarviridae family and is classified as a large icosahedral double-stranded DNA virus. Its linear genome spans approximately 189 kilobases and encodes for more than 180 genes. Since the 1960s, concerted efforts have been made to develop effective ASF vaccines, including ASF gene deletion vaccines, inactivated vaccines, multiepitope vaccines, attenuated live vaccines, and subunit vaccines [3–5]. Despite advancements, the absence of a clear understanding of immune correlates of protection and the precise identification of protective antigens has resulted in the current absence of an effective vaccine to halt the virus's spread in pig farming. Consequently, it is essential to conduct comprehensive research into ASFV's protective antigens, antigenic epitopes, and innovative vaccine adjuvants.

Much like most viral infections, both humoral immunity and cellular immune responses play crucial roles in ASFV protective immunity [6]. The serum from pigs in the recovery phase of ASFV infection possessed the ability to neutralize the infection of homologous and some heterologous strains both in vivo and in vitro, potentially by inhibiting virus attachment and internalization [7–9]. Anti-ASFV antibodies typically became detectable around six days post-infection, and could persist for an extended period once they had survived. However, despite the presence of these antibodies, their ability to cross-neutralize in vitro did not correlate with ASFV cross-protection in pigs [10]. Thus, the specific role of antibodies in ASFV protection remains unclear. ASFV serogroup classification, based on erythrocyte adsorption inhibition tests, suggests that ASF protective immunity may be serotype-specific, as ASFV within the same serogroup can cross-protect each other, while viruses outside the serotype cannot. Furthermore, a wealth of data supports the pivotal role of T-cell immune responses in ASFV control [6,11–13]. Depletion of pig lymphocytes indicated that cytotoxic CD8 lymphocytes were vital for ASFV clearance and protection against the virus. Moreover, pigs vaccinated with DNA vaccines exhibited partial protection during an ASFV attack, even though no anti-ASFV antibodies were detected in the protected pigs. Additionally, the failure of adjuvant-formulated inactivated ASFV and recombinant vaccines to offer strong protection highlighted the critical importance of cytotoxic T lymphocytes (CTLs) in the protective immune response against the virus [14].

Antigenic epitopes, the unique structural features on antigen molecules, exhibit specific antigenic functions. They are differentiated into B-cell and T-cell epitopes based on their interactions with corresponding antigen receptors on immune cells [15]. The investigation of antigenic epitopes is crucial for elucidating virus-induced immune responses and forms the foundation for the development of antiviral strategies, thereby representing a dynamic field of research in virology. According to the Immune Epitope Database (IEDB), several ASFV epitopes have been identified [16]. Despite the significant progress in ASFV antigenic epitope research, which has advanced the development of antigenic epitope diagnostic methods and vaccines, the identification and application of ASFV antigenic epitopes continue to pose challenges.

Vaccine adjuvants are substances that can enhance the body's specific immune response to antigens [17–19]. When formulated with vaccines, they can effectively enhance the immune response to vaccine antigens. Adjuvants were first discovered in 1920 by French scientist Gaston Ramon. They usually do not have immunogenicity, but can guide humoral and cellular immune responses to produce specific immunity against pathogens [19]. Currently, adjuvants are used to enhance the immune response of vaccines and effectively reduce the dose of vaccine antigens, which is crucial in the production process of veterinary vaccines.

In this review, we mainly focus on the recent research progress of ASFV's T-cell epitopes, B-cell epitopes, and vaccine adjuvants, aiming to provide basic information for the formulation of ASFV vaccine development and control strategies.

2. Advancement in the Development of Antigenic Epitopes of ASFV

2.1. Antigenic Epitope-Identified Proteins in ASFV

The ASFV virion primarily consists of five components, including the viral genome, core shell, inner envelope, capsid, and outer envelope. The ASFV genome encodes approximately 150 to 200 proteins, of which about 60 are structural proteins. A large body of literature has confirmed the presence of B-cell and T-cell epitopes in ASFV. These antigenic epitope-identified proteins include p72, p37, p30, and CD2v (Figure 1).

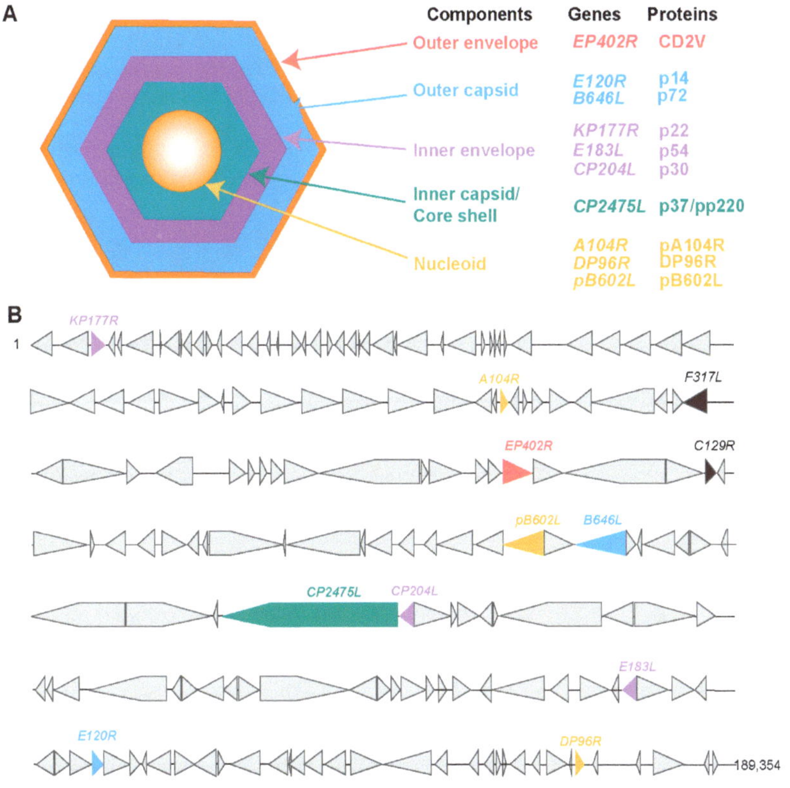

Figure 1. Schematic representation of ASFV structure and location of antigenic epitope-identified proteins in ASFV genomes. (**A**), Schematic diagram of the ASFV structure. ASFV is composed of an Outer envelope, outer capsid, inner envelope, inner capsid, and nucleoid. (**B**), Location of antigenic epitope-identified proteins in ASFV genomes.

2.2. Current Methods for the Identification of Antigenic Epitopes of the ASFV

2.2.1. Methods for the Identification of B-Cell Epitopes

ASFV protein B-cell epitope identification is a systematic procedure involving protein expression, mouse immunization, monoclonal antibody production, and bioinformatics-guided epitope analysis. Epitope validation is further accomplished via peptide synthesis, alanine scanning in conjunction with Enzyme-Linked Immunosorbent Assay (ELISA), and dot-ELISA characterization. These methodologies have led to the discovery of numerous B-cell epitopes, including the extracellular domain of CD2v and p37 [20,21]. These findings are pivotal for elucidating immune response mechanisms against ASFV and facilitating effective vaccine development.

Given the multitude of ASFV proteins, predicting B-cell epitopes using bioinformatics tools remains an important and challenging task [22]. The complexity inherent to antigen recognition complicates epitope prediction. For linear B-cell epitopes, a prevalent approach involves the use of the ABCpred server, trained on the Bcipep database via a recurrent neural network (RNN), for analysis and selection. Epitopes with higher scores indicate a greater likelihood of B-cell epitope prediction. Subsequent secondary structure prediction is performed using DNAStar Protean software 11. The protein's secondary structure significantly impacts the epitope. The robust chemical bonds of α-helices and β-sheets allow for the maintenance of protein structure, making such epitopes less likely to be antigenic due to their internal location and difficulty in antibody fitting. Conversely, the flexibility of β-turns and random coils in proteins allows for a looser structure, easily twisted and displayed on the protein surface. Such structures are more likely to serve as antigenic epitopes due to their prominence and ease of antibody fitting. For instance, Song et al. utilized ABCpred and DNAStar to predict the B-cell epitopes of pB602L, p30, p72, and CD2v proteins [12]. They evaluated the affinity of ASFV positive serum for synthetic peptides through peptide synthesis and dot blot analysis, successfully identifying immunodominant B-cell epitopes and incorporating them into nanoparticle vaccine antigen design [12].

2.2.2. Methods for the Identification of T-Cell Epitopes

The most common identification process for T-cell epitopes in ASFV proteins includes ASFV protein expression, mouse immunization, IFN-γ ELISpot identification of cellular immune responses in splenic cells, bioinformatic analysis of T-cell epitopes, peptide synthesis combined with IFN-γ ELISpot, flow cytometry, and T-cell proliferation for further confirmation of T-cell epitopes. For instance, one T-cell epitope, ^{246} SRRSLVNPWT ^{255}, has been identified on F317L using this method [23].

The use of bioinformatics tools for predicting ASFV T-cell epitopes is gaining traction. The most common approach for T-cell epitope prediction involves the use of software such as NetMHC 4.0 to screen potential T-cell epitopes. This software can predict peptides that bind with a multitude of SLA-I class molecules, from which high-affinity binding peptides are selected for biological synthesis and further epitope confirmation via IFN-γ ELISpot. For instance, Song et al. utilized NetMHC 4.0 to screen potential T-cell epitopes from CD2v and p72, and evaluated the cellular immune response of mice immunized with recombinant CD2v protein through peptide synthesis and IFN-γ ELISpot [12]. This validated the immunodominant T-cell epitopes, leading to the successful selection of immunodominant T-cell epitopes that were subsequently incorporated into the antigen design of nanoparticle vaccines.

2.3. *Research Progress on B-Cell Epitopes in ASFV Proteins*

2.3.1. p72-Derived B-Cell Epitopes

Given the complexity of the p72 trimer structure, information about specific p72 epitopes remains limited. Previous studies have identified a linear B-cell epitope located at ^{280}FPENSHNIQTAGKQD^{294} [24] and ^{281}PENSHNIQTA^{290} [25]. Recently, Tesfagaber et al. prepared anti-p72 mAb and identified a new linear B-cell epitope region between ^{283}NSHNIQ^{288} [26]. The identification of B-cell epitopes of p72 not only plays an important role in ASF serological diagnosis but may also lay a solid foundation for further research on the antigenic function of the p72 protein.

2.3.2. CD2v-Derived B-Cell Epitopes

Multiple linear B-cell epitopes in the extracellular domain of ASFV's CD2v have been identified through monoclonal antibodies (mAbs). For instance, a linear epitope formed by amino acids ^{28}LDSNITNDNNINGVSWNFFNNSF^{51} was identified by mAbs produced by truncated CD2v protein fused with Norovirus (NoV) P particles [27]. Using mAbs produced by eukaryotic cells expressing CD2v, two epitopes, ^{38}DINGVSWN^{45} and ^{134}GTNTNIY^{140}, were identified through peptide scanning [28]. Two linear B-cell epitopes, ^{128}TCKKNNGTNT^{137} and ^{148}VKYTNESILE^{157}, were recognized by mAbs produced by truncated CD2v expressed in CHO-S [29]. The B-cell epitope (^{154}SILE^{157}) of the CD2v extracellular domain was further ensured by dot-Blot, ELISA and IFA tests [20]. Jia et al. identified three linear B-cell epitopes, ^{147}FVKYT^{151}, ^{157}EYNWN^{161}, and ^{195}SSNY^{198}, by screening five types of mAbs produced by truncated CD2v protein expressed in baculovirus [30]. Song et al. recently predicted and identified a major linear B-cell epitope, ^{160}WNNSNINNFT^{169}, which induced humoral and cellular immune responses in a mouse model, strongly suggesting that linear B-cell epitopes may promote the design and development of ASFV subunit vaccines [31]. Lu et al. identified a novel epitope, ^{264}EPSPREP^{270}, located in the CD2v-IR structural domain, which could be used for the design and development of subunit vaccines [32]. It is noteworthy that some of the linear B-cell epitopes mentioned above overlap; therefore, the immunogenicity of these epitopes needs to be evaluated.

2.3.3. p30-Derived B-Cell Epitopes

p30 is a protein capable of generating neutralizing antibodies [33], with its C-terminus confirmed as an immunodominant region (aa 111–130) [34]. Petrovan et al. preliminarily identified a large polypeptide fragment (120–204 aa) at the C-terminus of the p30 protein by mAb 62-35 and 142-4 [35]. Recently, a study expressed overlapping truncated fragments in E. coli, determining that the linear epitope sequence recognized by monoclonal antibody 1B4G2–4 of p30 was within the range of ^{157}FNKVIRAHNFIQTIYGTPLK^{177}. After conducting amino acid-to-amino acid transfer identification, the smallest linear epitope was finally identified as ^{164}HNFIQTI^{170}. This epitope exhibited a strong antigen index and partial alpha-helical angles and coiled regions. It also showed high conservation across different strains, making it suitable for subsequent vaccine development [36].

2.3.4. p54-Derived B-Cell Epitopes

Envelope protein p54, similar to B646L, can induce neutralizing antibodies in pigs, although these antibodies cannot provide protection against potent ASFV attacks [37]. Utilizing high-throughput analysis technology based on gene chips, Desmet et al. identified two B-cell epitopes of p54 (IVLIYLFSSRKKKAA and AA 149–161) [38]. Zheng et al. identified a novel linear B-cell epitope (^{110}TMSAIENLR^{118}) on the ASFV P54 protein using monoclonal antibodies, which was conserved in all reference ASFV strains from different regions in China [39]. Nanobodies were used as a new tool to identify linear B-cell epitopes on the ASFV p54 and a novel minimal linear B-cell epitope, ^{76}QQWVEV^{81}, was identified with core binding site as ^{76}QQWV^{79} [40].

2.3.5. DP96R-Derived B-Cell Epitopes

DP96R, also known as uridine kinase (UK), encodes a protein associated with virulence, which can be utilized in the development of attenuated live vaccines [41–43]. Two B-cell epitopes (^{03}THDCSLKEK^{11} and ^{55}YWKGIKRGND^{64}) were found on ASFV's DP96R protein [44].

2.3.6. E120R-Derived B-Cell Epitopes

E120R is highly conserved among ASFV strains and serves as a target for the development of attenuated live vaccines against ASF [45]. Through incubation with a gene chip based on high-density peptide microarrays, it was discovered that the peptide sequence EEFEPIPDYDTTST of ASFV envelope protein E120R could react with ASF positive serum samples [38].

2.3.7. pA104R-Derived B-Cell Epitopes

The A104R gene is responsible for the synthesis of a protein, hypothesized to resemble a histone, which is strategically positioned at the sites of viral DNA replication and gene expression [46]. The dominant IgM epitope PEP23 (KAVKIRALK) and PEP15 (KFTVVTVKA) were identified by epitope modification [47]. Through the confirmatory analysis of the pA104R epitope using monoclonal antibodies, an immunodominant B-cell epitope (KPTITKOELYSI) was identified. This finding could potentially aid in the development of sensitive diagnostic tools and serve as a target for candidate vaccine development [48].

2.3.8. E184L-Derived B-Cell Epitopes

E184L serves as a crucial antagonist of the IFN signal and an immunogenic ASFV protein, capable of evading the host's innate antiviral immune response [49,50]. Through meticulous localization, the antigenic epitopes for the E184L mAbs have been identified as

^{119} IQRQGFL ^{125} and ^{153} DPTEFF ^{158}. These findings lay the groundwork for serological diagnosis and the development of epitope-based marker vaccines [51].

Taken together, numerous B-cell epitopes have been identified in ASFV, which guide the design of immunogenic peptides and novel vaccine molecules, and also facilitate the development of diagnostic reagents and clinical disease diagnosis (Table 1).

Table 1. Identification and application of B-cell epitopes of ASFV antigens.

ASFV (Protein)	Sequence	Application		References
		Vaccine	Diagnosis	
p72	^{156}TLVDPFGRPI^{165} ^{265}QRTCSHTNPKFLSQHF^{280} ^{280}FPENSHNIQTAGKQD^{294} ^{290}AGKQDITPITDATY^{303}	−	+	[24]
	^{281}PENSHNIQTA^{290}	−	+	[25]
	^{283}NSHNIQ^{288}	−	+	[26]
CD2v	^{154}SILE^{157}	+	+	[20]
	^{28}LDSNITNDNNINGVSWNFFNNSF^{51}	−	+	[27]
	^{38}DINGVSWN^{45} ^{134}GTNTNIY^{140}	+	−	[28]
	^{128}TCKKNNGTNT^{137} ^{148}VKYTNESILE^{157}	−	+	[29]
	^{147}FVKYT^{151} ^{157}EYNWN^{161} ^{195}SSNY^{198}	+	+	[30]
	^{160}WNNSNINNFT^{169}	+		[31]
	^{264}EPSPREP^{270}	+	+	[32]
	^{157}FNKVIRAHNFIQTIYGTPLK^{177}	+	−	[36]
p54	^{149}IVLIYLFSSRKKKAA^{161}	+	−	[38]
	^{110}TMSAIENLR^{118}	+	−	[39]
	^{76}QQWVEV^{81}	+	−	[40]
p37	^{58}LGDAIAGRLMQLD^{70} ^{100}DWKATVSAIELEY^{112} ^{163}TTGDTLAQVFESFPT^{177} ^{63}AGRLMQLDTEKNARI^{77}	+	−	[21]
E120R	EEFEPIPDYDTTST	+	−	[38]
DP96R	^{03}THDCSLKEK^{11} ^{55}YWKGIKRGND^{64}	−	+	[44]
pA104R	KPTITKOELYSI	+	+	[48]
	KFTVVTVKA KAVKIRALK	+	−	[52]
E184L	^{119}IQRQGFL^{125} ^{153}DPTEFF^{158}	+	+	[51]
pB602L	^{474}SKENLTPDE^{482}	+	+	[12,53]
p22	^{37}KVCKVDKDCGSGEHC^{51} ^{157}VYNNPHHPVLKYGKDHIIALP^{171}	−	+	[54]
MGF_110-13L	^{48}WDCQDGICKNKITESRFIDS^{67} ^{122}GDHQQLSIKQ^{131}	−	+	[55]

2.4. Research Progress on ASFV-Specific T-Cell Epitopes

2.4.1. CD2v-Derived T-Cell Epitopes

The intracellular epitope in the CD2v protein of ASFV has been demonstrated to induce both humoral and cellular immune responses [32,56]. The CD2v protein's proline-rich cytoplasmic domain demonstrated a high degree of conservation, with 79% to 100% amino acid identity, across various genotypes of ASFV [56]. The discovery of monoclonal antibody 1F3 (^{264}EPSPREP^{270}) has identified it as a T-cell epitope that was not only specific to ASFV but also remarkably conserved across different genotypes [32].

2.4.2. p72-Derived T-Cell Epitopes

The p72 protein is the major capsid protein and also one of the most immunogenic proteins of ASFV, making it an important target for detection and vaccine development [57]. Using the IEDB MHC-I binding prediction algorithm coupled with ELISPOT assay detection, Sun et al. have delineated the core peptides P351 (SRISNIKNNKY), P334 (SDYTL), and P366 (SSYGGAK) derived from the p72 protein as exhibiting elevated immunogenicity in pigs that have survived infection, offering a pivotal reference for the subsequent development of tetrameric constructs in immunological research [58].

2.4.3. F317L-Derived T-Cell Epitopes

The late F317L protein of ASFV induced immunosuppression by dampening the activation of the NF-κB pathway, making it a potential immunogenic antigen [59,60]. Through T-cell epitopes prediction and validation by the IFN-γ ELISpot assay, the peptide F25 (^{246}SRRSLVNPWT^{255}) was identified as a presumed T-cell epitope, capable of inducing a robust immune response [23].

2.4.4. C129R-Derived T-Cell Epitopes

The C129R protein of ASFV, known for its strong immunogenicity and ability to target the cyclic GMP-AMP pathway, was utilized in the development of recombinant adenovirus vaccines [10,61]. Recently, Zhai and colleagues successfully pinpointed T-cell epitopes within the C129R protein, specifically in peptides C11 (^{53}LQNPYEAVI^{61}), C14 (^{81}GHVTWAVPY^{89}), C16 (^{97}AKPDAIMLT^{105}), and C18 (^{116}ALNQNVLTL^{124}) [62].

Collectively, the validated ASFV-specific T-cell epitopes summarized in this review may play a crucial role in the design and development of novel ASFV vaccines (Table 2).

Table 2. Identification and application of T-cell epitopes of ASFV antigens.

ASFV (Protein)	Sequence	Application		Reference
		Vaccine	Diagnosis	
F317L	^{246}SRRSLVNPWT^{255}	+	–	[23]
CD2v Intracellular Epitope	^{264}EPSPREP^{270}	+	–	[32]
p72	P351(SRISNIKNNKY) P334(SDYTL) P366(SSYGGAK)	+	–	[58]
C129R	^{53}LQNPYEAVI^{61} ^{81}GHVTWAVPY^{89} ^{97}AKPDAIMLT^{105} ^{116}ALNQNVLTL^{124}	+	–	[62]

3. Advancement in the Development of Vaccine Adjuvants of ASFV

3.1. The Functions and Mechanism of Adjuvants in ASF Vaccines

In the development of ASFV vaccines, the selection and use of adjuvants is a critical component. Adjuvants can enhance the immunogenicity of vaccines, improve the durability of immunity, reduce vaccine dosage, and thus improve the cost-effectiveness of vaccines [17–19]. Currently, a variety of adjuvants have been used in the development of ASFV vaccines, including aluminum salts, oil emulsions, and polymer microspheres [63,64]. These adjuvants were shown to enhance the immune response through various mechanisms, such as stimulating immune cells, promoting antigen uptake by antigen-presenting cells (APCs), stimulating the secretion of various cytokines and chemokines, inducing the differentiation of CD4 + T cells into different types (Th1, Th2, Th17, etc.), prolonging antigen presentation time, and enhancing antigen presentation efficiency [63,64].

3.2. *Types of Adjuvants for ASF Vaccines*

3.2.1. Montanide™ ISA

Montanide™ ISA, the veterinary vaccine adjuvant, is a common choice in animal vaccines due to its ability to induce strong immune responses with non-toxicity, good tolerability, and the simplicity of mixing the antigen and adjuvant [65]. Zajac et al. used a recombinant adenovirus mixture containing ASFV antigens (Ad5-ASFV) with 42 multi-epitopes, combined with Montanide ISA-201™ adjuvant, to immunize pigs three times. This could induce a humoral immune response, but there was no protective effect after the immune challenge [64]. A recent study utilized an inactivated vaccine produced by gamma-irradiated ASFV in combination with Montanide™ ISA 201 VG adjuvant. This formulation was administered to five weaned piglets at three-week intervals for two immunizations. Despite all animals developing antibodies against ASFV p72, the response was insufficient to confer protection from ASFV strain attacks [66].

3.2.2. Polygen™

Polygen™ is a low molecular weight copolymer adjuvant that can cross-link in solution to form a high molecular weight gel. It has been demonstrated in studies to elicit significant interferon-gamma (IFN-γ) and interleukin-12 (IL-12) responses when used in bovine parasitic vaccines [67]. Researchers administered an inactivated vaccine, prepared by combining ASFV inactivated by gamma irradiation or binary ethylenimine (BEI) with the Polygen adjuvant, via intramuscular injection to weaned piglets. Despite the presence of ASFV-specific antibodies in all vaccinated subjects, no protective effect was observed, even after stringent homologous challenge [66,68].

3.2.3. Zoetis

Vaccine adjuvants produced by Zoetis have been used in vaccine immunization research for various pathogens, including *paratyphoid salmonella* and porcine reproductive and respiratory syndrome virus [69–71]. Pigs immunized with a cocktail of 12 ASFV antigens combined with the Zoetis adjuvant showed that the cocktail-ii- Zoetis vaccine recipients had a higher survival rate, but it did not prevent clinical disease [63]. Compared to the ENABL adjuvant, pigs vaccinated with Ad5-ASFV 4-way cocktail vaccine combined with Zoetis adjuvant exhibited higher humoral immune response [72].

3.2.4. BioMize®

BioMize® is an innovative, ready-to-use, and fully customizable adjuvant developed by VaxLiant company, offering great flexibility for vaccine development and commercialization [63,64,73]. Compared to the use of the Montanide ISA-201™ adjuvant, the Ad5-ASFV BioMize® immunogen elicited a relatively lower anti-pp62 specific IgG response [64]. Immunizing pigs with a 35 adenovirus-vectored ASFV cocktail, along with BioMize® adjuvant, generated robust ASFV-specific antibodies, IFN-γ cells, and CTL responses, yet failed to confer protection against the virulent Arm07 isolate in Eurasian wild boar [73].

3.2.5. MF59®

MF59® is an oil-in-water emulsion adjuvant that has been included in influenza vaccines approved in Europe since 1997 and has been administered to over 100 million people in more than 30 countries [74]. The BEI-inactivated ASFV vaccine, combined with the MF59 adjuvant and administered intradermally and intramuscularly in pigs, elicited a positive antibody response to ASFV. However, it did not provide effective protection against a lethal attack [75].

3.2.6. Adjuvants from Bacterial Component

The major outer membrane lipoprotein I (OprI) of *Pseudomonas aeruginosa* is a ligand for Toll-like receptor (TLR)-2 [76]. It can trigger dendritic cells (DC) to secrete pro-inflammatory cytokines in vivo, thereby indirectly regulating adaptive immune responses. OprI can serve as a natural adjuvant, and after fusion, it can induce a strong humoral and cytotoxic T-cell response against peptides/proteins. The application of OprI in fusion proteins has been extended to antigens encoded by ASFV's B646L and G1340L, and the resulting proteins can induce ASFV-specific CTL activity [76,77]. The different immune functions of OprI, including promoting Th1/Th2 responses, are all attributed to the activation of TLR-2 signaling. The immunoregulatory activity of OprI fusion proteins has also been used in vaccine development. A recent study has shown that the mixture of OprI fusion proteins formulated with ISA206 adjuvant can induce strong ASFV-specific humoral and cellular immune responses in pigs, providing valuable information for the further development of subunit vaccines against ASF [11]. Heat-labile enterotoxin B (LTB), when used as an adjuvant carrying antigens, can enhance the immunogenicity of vaccines and play a role in T-cell activation and differentiation [78]. Following oral immunization with recombinant Lactobacillus lactis expressing ASFV protein-LTB fusion protein, the local mucosal immunity, humoral immunity, and Th2 cell immunity were enhanced compared with no LTB adjuvant group [79]. These findings provide new insights into the design and development of ASFV subunit vaccines.

3.2.7. Nano-Adjuvants

Compared to traditional vaccines, nano-vaccines demonstrate enhanced efficacy due to their ability to accumulate, self-assemble in lymph nodes, and be readily uptaken by APC cells [80]. Recently, a self-assembling nano ASFV candidate vaccine (NanoFVax) targeting dendritic cells has been demonstrated to elicit a potent T-cell response, with high-level antibody responses against ASFV persisting for over 231 days [12].

In summary, the selection and optimization of adjuvants for ASFV vaccines remains an important area of research that requires further exploration and study (Table 3).

Table 3. Research progress on adjuvants in ASFV vaccines.

Adjuvant Name	The Components of ASFV Vaccine	Immune Rout	Reference
Montanide ISA	OPMT + OPET + OCET + MontanideTM ISA206™	i.m.	[11]
	Ad5-ASFV Mix + Montanide ISA-201™	i.m.	[64]
	30 kGy Irradiated ASFV + Montanide ISA-201™	i.m.	[66]
	Inactivated ASFV + MontanideTM ISA201™	i.m.	[75]
Polygen™	30 kGy Irradiated ASFV + Polygen™	i.m.	[66]
	BEI-inactivated ASFV + Polygen™	i.m.	[68]
Zoetis	Ad-ASFV cocktail-II + Zoetis	i.m.	[63]
	Ad5-ASFV 4-way cocktail + Zoetis	i.m.	[72]
ENABL	Ad5-ASFV 4-way cocktail + ENABL	i.m.	[72]

Table 3. *Cont.*

Adjuvant Name	The Components of ASFV Vaccine	Immune Rout	Reference
BioMize®	Ad-ASFV cocktail-I + BioMize	i.m.	[63]
	Ad5-ASFV + BioMize	i.m.	[73]
MF59®	Inactivated ASFV + MF59®	i.m.	[75]
Emulsigen D	BEI-inactivated ASFV + Emulsigen D	i.m.	[68]
Silica oil	Inactivated ASFV + Silica oil	i.m.	[75]
mGNE	Inactivated ASFV + mGNE	i.m.	[75]
LTB	MG1363/pMG36e-p30 + p54 + p72-LTB-His	Oral	[79]
OprI	OPMT + OPET + OCET + OprI	i.m.	[11]
FlaB	rAd-ASFV CD2v-p30-p54-FlaB	i.n.	[81]
Hsp70	rAd-ASFV CD2v-p30-p54-Hsp70	i.n.	[81]
IL-33	NC8-pLP-S-p14.5-IL-33-Mus	Oral	[82]
CTA1-DD	NC8-pLP-S-CTA1-p14.5-CTA1-DD	Oral	[82]
SpyTag-SpyCatcher	SpyTag/SpyCatcher + ASFV epitopes + SpyTag-SpyCatcher	i.m.	[12]

i.m.: intramuscular immunization; i.n.: intranasal immunization.

4. Prospects of Antigenic Epitopes and Adjuvants in the Development of Novel ASFV Vaccines and Diagnostic Reagents

Currently, techniques for identifying B-cell epitopes are well established. However, obtaining high-quality monoclonal antibodies is challenging, especially those with neutralizing activity that can recognize conformational epitopes. For the recognition of conformational B-cell antigenic epitopes, monoclonal antibodies recognizing the whole virus are more effective than those recognizing recombinant proteins. Novel ASFV detection methods, such as ELISA and flow cytometry, rely on the identification of these B-cell epitopes. By recognizing and utilizing these epitopes, we can develop more sensitive and specific detection methods to facilitate early detection and diagnosis of ASFV infection, thereby effectively controlling the spread of ASF. The OIE recommended ELISA as the preferred serological method for ASF diagnosis [83]. Despite its lower sensitivity compared to highly sensitive early pathogen detection methods such as PCR, qPCR, multiplex PCR, LAMP, and NGS, the ELISA demonstrated significant advantages in large-scale sample testing. It was particularly suitable for detecting pig herds recovering from subacute and latent ASF infections [84]. Most commercialized ELISA kits were based on a single viral protein, which could lead to false-negative results. However, ASFV tandem proteins based on multiple B-cell epitopes showed significant improvements in sensitivity in the development of ELISA kits [85,86]. It enhanced the recognition of different types of antibodies in pigs, reduced the possibility of inaccurate negative or positive results, and allowed a comprehensive assessment of ASFV exposure by k3 derived from 27 multiple peptides of 11 ASFV proteins or the antigenic dominant domains from p30, p54, and p72 [85,86]. Compared with currently available commercial detection methods, innovative methods using multiple B-cell epitopes of multiple ASFV proteins can achieve higher sensitivity and specificity.

The identification of T-cell epitopes is crucial for research on cellular immune mechanisms and the development of subunit peptide vaccines. Apart from p30, p54, and p72, there are few reports on the research of T-cell antigenic epitopes of other proteins [23]. Since T-cells only recognize antigenic peptides presented by the Major Histocompatibility Complex (MHC) molecules on the surface of Antigen Presenting Cells (APCs), the recognition of T-cell epitopes is more challenging [87]. The development of ASFV vaccines that focused solely on humoral immune responses was insufficient, as inactivated vaccines had been shown to be ineffective in providing protection against ASFV challenge [68]. Attenuated or low-virulence ASFV strains induced protective immunity in pigs against virulent ASFV strains [88], but vaccinated pigs often experienced adverse side effects, such as chronic viremia [3]. Additionally, the use of live attenuated ASFV raised significant safety concerns [3]. Furthermore, the activation of CD8 T cells aided pigs in combating ASFV infection [89], underscoring the crucial role of antigen-specific T cell immune responses in ASFV vaccine development. This necessitates further investigation into ASFV's T-cell epitopes.

Although inactivated and subunit vaccines for ASFV have a high safety profile and have been used in conjunction with some adjuvants, they have not demonstrated robust protective effects in current research [3,90,91]. This indicates that there are many unknowns that need to be addressed with protective antigens or epitopes, adjuvants, and non-structural viral protein antigens in the development of new ASFV vaccines. Although the only attenuated live vaccine approved by Vietnam demonstrated that 93.34% of the 5958 randomly selected immunized pigs met the technical requirements, as reported in the June 2023 Global Disease Monitoring Report from the Swine Health Information Center, potential biosecurity risks such as reversion to virulence necessitated caution [91]. Therefore, the development of new nucleic acid vaccines, genetically engineered vaccines, or multi-peptide vaccines in combination with new adjuvants is an important direction for current ASFV vaccine development. When choosing a vaccine adjuvant, many factors need to be considered, with safety being the first. A good adjuvant must be safe, well-tolerated, and easy to produce; it should have good pharmaceutical properties (such as pH, osmotic pressure, endotoxin levels, etc.) and long shelf life; and finally, it should be economically viable.

Reverse vaccinology is an application that aids in the development of novel epitope vaccines based on pathogen genome sequencing [92]. Compared to traditional vaccines, epitope vaccines are safer, non-toxic, stable, and can more directly elicit immune responses against pathogenic microorganisms [93]. Epitope vaccines based on multiple epitope peptides can also overcome the problem of low conservation between epitopes of different genotype strains and elicit stronger immune responses. However, for ASFV, the development of diagnostics and vaccines based on multiple epitopes is still insufficient [94,95]. As multi-epitope vaccines are based on the selection of antigenic epitopes and the immune response of computer-screened epitopes, they are suitable for the development of universal vaccines against different ASFV genotypes, accelerating the vaccine design process and reducing its cost.

5. Conclusions

African swine fever infection causes high mortality in pigs, resulting in significant economic losses to the pig industry. Research on African swine fever vaccines has been ongoing in recent decades. The antigenic epitopes recognized by T cells and B cells of the immune system play a key role in the antiviral immune response, and therefore the resolution and precise characterization of the T-cell antigenic epitopes and B-cell antigenic epitopes of ASFV can provide an important basis for vaccine development. Here, we describe the major potential antigenic proteins of ASFV and methods for pinpointing T- and B-cell epitopes of ASFV antigens, providing detailed localization data for these epitopes. These will hold promise for the development of safe and effective ASF vaccines. Vaccines, combined with accurate, efficient, and early diagnostic techniques, could provide the basis for the prevention, control, and eradication of ASF.

Author Contributions: Conceptualization, Q.W., K.Y., W.L. and C.L.; Writing—original draft preparation, Q.W., C.L., B.Z., J.Z. and Z.L.; Resources and investigation, R.G., T.G. and F.Y.; Writing—review and editing, Q.W., D.Z. and Y.T. All authors have read and agreed to the published version of the manuscript.

Funding: This work was supported by the Wuhan Science and Technology Bureau (Grant No. 2023020302020573) and the Hubei Province Innovation Centre of Agricultural Sciences and Technology (No. 2024-620-000-001-013).

Institutional Review Board Statement: Not applicable.

Informed Consent Statement: Not applicable.

Data Availability Statement: No new data were created or analyzed in this study. Data sharing is not applicable to this article.

Acknowledgments: We express our gratitude to the authors of the cited works for their contribution of data to the final version of this manuscript.

Conflicts of Interest: The authors have no conflicts of interest to declare.

References

1. Dixon, L.K.; Sun, H.; Roberts, H. African swine fever. *Antiviral Res.* **2019**, *165*, 34–41. [CrossRef] [PubMed]
2. Penrith, M.L. African swine fever. *Onderstepoort J. Vet. Res.* **2009**, *76*, 91–95.
3. Vu, H.L.X.; McVey, D.S. Recent progress on gene-deleted live-attenuated African swine fever virus vaccines. *NPJ Vaccines* **2024**, *9*, 60. [CrossRef]
4. Wu, K.; Liu, J.; Wang, L.; Fan, S.; Li, Z.; Li, Y.; Yi, L.; Ding, H.; Zhao, M.; Chen, J. Current State of Global African Swine Fever Vaccine Development under the Prevalence and Transmission of ASF in China. *Vaccines* **2020**, *8*, 531. [CrossRef]
5. Sereda, A.D.; Balyshev, V.M.; Kazakova, A.S.; Imatdinov, A.R.; Kolbasov, D.V. Protective Properties of Attenuated Strains of African Swine Fever Virus Belonging to Seroimmunotypes I–VIII. *Pathogens* **2020**, *9*, 274. [CrossRef]
6. Schafer, A.; Franzoni, G.; Netherton, C.L.; Hartmann, L.; Blome, S.; Blohm, U. Adaptive Cellular Immunity against African Swine Fever Virus Infections. *Pathogens* **2022**, *11*, 274. [CrossRef] [PubMed]
7. Canter, J.A.; Aponte, T.; Ramirez-Medina, E.; Pruitt, S.; Gladue, D.P.; Borca, M.V.; Zhu, J.J. Serum Neutralizing and Enhancing Effects on African Swine Fever Virus Infectivity in Adherent Pig PBMC. *Viruses* **2022**, *14*, 1249. [CrossRef]
8. Escribano, J.M.; Galindo, I.; Alonso, C. Antibody-mediated neutralization of African swine fever virus: Myths and facts. *Virus Res.* **2013**, *173*, 101–109. [CrossRef] [PubMed]
9. Gomez-Puertas, P.; Escribano, J.M. Blocking antibodies inhibit complete African swine fever virus neutralization. *Virus Res.* **1997**, *49*, 115–122. [CrossRef]
10. Xu, Z.; Hu, Y.; Li, J.; Wang, A.; Meng, X.; Chen, L.; Wei, J.; Tong, W.; Kong, N.; Yu, L.; et al. Screening and identification of the dominant antigens of the African swine fever virus. *Front. Vet. Sci.* **2023**, *10*, 1175701. [CrossRef]
11. Zhang, G.; Liu, W.; Yang, S.; Song, S.; Ma, Y.; Zhou, G.; Liang, X.; Miao, C.; Li, J.; Liu, Y.; et al. Evaluation of humoral and cellular immune responses induced by a cocktail of recombinant African swine fever virus antigens fused with OprI in domestic pigs. *Virol. J.* **2023**, *20*, 104. [CrossRef] [PubMed]
12. Song, J.; Wang, M.; Zhou, L.; Tian, P.; Sun, Z.; Sun, J.; Wang, X.; Zhuang, G.; Jiang, D.; Wu, Y.; et al. A candidate nanoparticle vaccine comprised of multiple epitopes of the African swine fever virus elicits a robust immune response. *J. Nanobiotechnol.* **2023**, *21*, 424. [CrossRef]
13. Jenson, J.S.; Childerstone, A.; Takamatsu, H.; Dixon, L.K.; Parkhouse, R.M. The cellular immune recognition of proteins expressed by an African swine fever virus random genomic library. *J. Immunol. Methods* **2000**, *242*, 33–42. [CrossRef] [PubMed]
14. Lacasta, A.; Ballester, M.; Monteagudo, P.L.; Rodriguez, J.M.; Salas, M.L.; Accensi, F.; Pina-Pedrero, S.; Bensaid, A.; Argilaguet, J.; Lopez-Soria, S.; et al. Expression library immunization can confer protection against lethal challenge with African swine fever virus. *J. Virol.* **2014**, *88*, 13322–13332. [CrossRef]
15. Yang, G.; Wang, J.; Sun, P.; Qin, J.; Yang, X.; Chen, D.; Zhang, Y.; Zhong, N.; Wang, Z. SARS-CoV-2 epitope-specific T cells: Immunity response feature, TCR repertoire characteristics and cross-reactivity. *Front. Immunol.* **2023**, *14*, 1146196. [CrossRef]
16. Huang, C.; Cao, C.; Xu, Z.; Lin, Y.; Wu, J.; Weng, Q.; Liu, Z.; Jin, Y.; Chen, P.; Hua, Q. A blocking ELISA based on virus-like nanoparticles chimerized with an antigenic epitope of ASFV P54 for detecting ASFV antibodies. *Sci. Rep.* **2023**, *13*, 19928. [CrossRef]
17. Xie, C.; Yao, R.; Xia, X. The advances of adjuvants in mRNA vaccines. *NPJ Vaccines* **2023**, *8*, 162. [CrossRef]
18. Lavelle, E.C.; Ward, R.W. Mucosal vaccines—Fortifying the frontiers. *Nat. Rev. Immunol.* **2022**, *22*, 236–250. [CrossRef] [PubMed]
19. Pulendran, B.; S Arunachalam, P.; O'Hagan, D.T. Emerging concepts in the science of vaccine adjuvants. *Nat. Rev. Drug Discov.* **2021**, *20*, 454–475. [CrossRef]
20. Liu, S.; Ding, P.; Du, Y.; Ren, D.; Chen, Y.; Li, M.; Sun, X.; Wang, S.; Chang, Z.; Li, R.; et al. Development and characterization of monoclonal antibodies against the extracellular domain of African swine fever virus structural protein, CD2v. *Front. Microbiol.* **2022**, *13*, 1056117. [CrossRef]
21. Wang, A.; Chen, Z.; Zhou, J.; Chen, Y.; Liu, Y.; Liu, H.; Liang, C.; Zhu, X.; Zhang, Y.; Xin, C.; et al. Development and characterization of monoclonal antibodies against p37 protein of African swine fever virus. *Int. J. Biol. Macromol.* **2024**, *264*, 130689. [CrossRef]

22. Mima, K.A.; Katorkina, E.I.; Katorkin, S.A.; Tsybanov, S.Z.; Malogolovkin, A.S. In silico prediction of B- and T-cell epitopes in the CD2v protein of african swine fever virus (African swine fever virus, Asfivirus, Asfarviridae). *Vopr. Virusol.* **2020**, *65*, 103–112. [CrossRef]
23. Huang, Y.; Zhai, W.; Wang, Z.; He, Y.; Tao, C.; Chu, Y.; Pang, Z.; Zhu, H.; Jia, H. Analysis of the Immunogenicity of African Swine Fever F317L Protein and Screening of T Cell Epitopes. *Animals* **2024**, *14*, 1331. [CrossRef]
24. Heimerman, M.E.; Murgia, M.V.; Wu, P.; Lowe, A.D.; Jia, W.; Rowland, R.R. Linear epitopes in African swine fever virus p72 recognized by monoclonal antibodies prepared against baculovirus-expressed antigen. *J. Vet. Diagn. Invest.* **2018**, *30*, 406–412. [CrossRef] [PubMed]
25. Miao, C.; Yang, S.; Shao, J.; Zhou, G.; Ma, Y.; Wen, S.; Hou, Z.; Peng, D.; Guo, H.; Liu, W.; et al. Identification of p72 epitopes of African swine fever virus and preliminary application. *Front. Microbiol.* **2023**, *14*, 1126794. [CrossRef]
26. Tesfagaber, W.; Wang, W.; Wang, L.; Zhao, R.; Zhu, Y.; Li, F.; Sun, E.; Liu, R.; Bu, Z.; Meng, G.; et al. A highly efficient blocking ELISA based on p72 monoclonal antibody for the detection of African swine fever virus antibodies and identification of its linear B cell epitope. *Int. J. Biol. Macromol.* **2024**, *268*, 131695. [CrossRef]
27. Ren, D.; Ding, P.; Liu, S.; Zhang, N.; Chen, Y.; Li, Q.; Fan, L.; Chang, Z.; Zhang, G. Development and characterization of recombinant ASFV CD2v protein nanoparticle-induced monoclonal antibody. *Int. J. Biol. Macromol.* **2022**, *209*, 533–541. [CrossRef]
28. Jiang, W.; Jiang, D.; Li, L.; Wang, J.; Wang, P.; Shi, X.; Zhao, Q.; Liu, B.; Ji, P.; Zhang, G. Identification of Two Novel Linear B Cell Epitopes on the CD2v Protein of African Swine Fever Virus Using Monoclonal Antibodies. *Viruses* **2022**, *15*, 131. [CrossRef] [PubMed]
29. Liu, H.; Wang, A.; Yang, W.; Liang, C.; Zhou, J.; Chen, Y.; Liu, Y.; Zhou, Y.; Zhang, G. Expression of extracellular domain of ASFV CD2v protein in mammalian cells and identification of B cell epitopes. *Virus Res.* **2023**, *323*, 199000. [CrossRef]
30. Jia, R.; Zhang, G.; Bai, Y.; Liu, H.; Chen, Y.; Ding, P.; Zhou, J.; Feng, H.; Li, M.; Tian, Y.; et al. Identification of Linear B Cell Epitopes on CD2V Protein of African Swine Fever Virus by Monoclonal Antibodies. *Microbiol. Spectr.* **2022**, *10*, e0105221. [CrossRef]
31. Song, J.; Wang, M.; Du, Y.; Wan, B.; Zhang, A.; Zhang, Y.; Zhuang, G.; Ji, P.; Wu, Y.; Zhang, G. Identification of a linear B-cell epitope on the African swine fever virus CD2v protein. *Int. J. Biol. Macromol.* **2023**, *232*, 123264. [CrossRef]
32. Lu, W.; Bai, Y.; Zhang, S.; Zhao, X.; Jin, J.; Zhu, X.; Wang, R.; Wu, Y.; Zhang, A.; Zhang, G.; et al. An Intracellular Epitope of ASFV CD2v Protein Elicits Humoral and Cellular Immune Responses. *Animals* **2023**, *13*, 1967. [CrossRef]
33. Neilan, J.G.; Zsak, L.; Lu, Z.; Burrage, T.G.; Kutish, G.F.; Rock, D.L. Neutralizing antibodies to African swine fever virus proteins p30, p54, and p72 are not sufficient for antibody-mediated protection. *Virology* **2004**, *319*, 337–342. [CrossRef] [PubMed]
34. Murgia, M.V.; Mogler, M.; Certoma, A.; Green, D.; Monaghan, P.; Williams, D.T.; Rowland, R.R.R.; Gaudreault, N.N. Evaluation of an African swine fever (ASF) vaccine strategy incorporating priming with an alphavirus-expressed antigen followed by boosting with attenuated ASF virus. *Arch. Virol.* **2019**, *164*, 359–370. [CrossRef]
35. Petrovan, V.; Yuan, F.; Li, Y.; Shang, P.; Murgia, M.V.; Misra, S.; Rowland, R.R.R.; Fang, Y. Development and characterization of monoclonal antibodies against p30 protein of African swine fever virus. *Virus Res.* **2019**, *269*, 197632. [CrossRef]
36. Tian, P.; Sun, Z.; Wang, M.; Song, J.; Sun, J.; Zhou, L.; Jiang, D.; Zhang, A.; Wu, Y.; Zhang, G. Identification of a novel linear B-cell epitope on the p30 protein of African swine fever virus using monoclonal antibodies. *Virus Res.* **2024**, *341*, 199328. [CrossRef] [PubMed]
37. Rodriguez, F.; Alcaraz, C.; Eiras, A.; Yanez, R.J.; Rodriguez, J.M.; Alonso, C.; Rodriguez, J.F.; Escribano, J.M. Characterization and molecular basis of heterogeneity of the African swine fever virus envelope protein p54. *J. Virol.* **1994**, *68*, 7244–7252. [CrossRef]
38. Desmet, C.; Coelho-Cruz, B.; Mehn, D.; Colpo, P.; Ruiz-Moreno, A. ASFV epitope mapping by high density peptides microarrays. *Virus Res.* **2024**, *339*, 199287. [CrossRef] [PubMed]
39. Zheng, N.; Li, C.; Hou, H.; Chen, Y.; Zhang, A.; Han, S.; Wan, B.; Wu, Y.; He, H.; Wang, N.; et al. A Novel Linear B-Cell Epitope on the P54 Protein of African Swine Fever Virus Identified Using Monoclonal Antibodies. *Viruses* **2023**, *15*, 867. [CrossRef]
40. Zhao, H.; Wang, G.; Dong, H.; Wu, S.; Du, Y.; Wan, B.; Ji, P.; Wu, Y.; Jiang, D.; Zhuang, G.; et al. Identification of a Linear B Cell Epitope on p54 of African Swine Fever Virus Using Nanobodies as a Novel Tool. *Microbiol. Spectr.* **2023**, *11*, e0336222. [CrossRef]
41. Dodantenna, N.; Cha, J.W.; Chathuranga, K.; Chathuranga, W.A.G.; Weerawardhana, A.; Ranathunga, L.; Kim, Y.; Jheong, W.; Lee, J.S. The African Swine Fever Virus Virulence Determinant DP96R Suppresses Type I IFN Production Targeting IRF3. *Int. J. Mol. Sci.* **2024**, *25*, 2099. [CrossRef]
42. Qi, X.; Feng, T.; Ma, Z.; Zheng, L.; Liu, H.; Shi, Z.; Shen, C.; Li, P.; Wu, P.; Ru, Y.; et al. Deletion of DP148R, DP71L, and DP96R Attenuates African Swine Fever Virus, and the Mutant Strain Confers Complete Protection against Homologous Challenges in Pigs. *J. Virol.* **2023**, *97*, e0024723. [CrossRef]
43. O'Donnell, V.; Risatti, G.R.; Holinka, L.G.; Krug, P.W.; Carlson, J.; Velazquez-Salinas, L.; Azzinaro, P.A.; Gladue, D.P.; Borca, M.V. Simultaneous Deletion of the 9GL and UK Genes from the African Swine Fever Virus Georgia 2007 Isolate Offers Increased Safety and Protection against Homologous Challenge. *J. Virol.* **2017**, *91*. [CrossRef]
44. Li, C.; Si, X.Y.; Wang, X.G.; Yan, Z.W.; Hou, H.Y.; You, L.Q.; Chen, Y.L.; Zhang, A.K.; Wang, N.; Sun, A.J.; et al. Preparation and epitope analysis of monoclonal antibodies against African swine fever virus DP96R protein. *BMC Vet. Res.* **2024**, *20*, 191. [CrossRef]
45. Liu, H.; Zhu, Z.; Feng, T.; Ma, Z.; Xue, Q.; Wu, P.; Li, P.; Li, S.; Yang, F.; Cao, W.; et al. African Swine Fever Virus E120R Protein Inhibits Interferon Beta Production by Interacting with IRF3 To Block Its Activation. *J. Virol.* **2021**, *95*, e0082421. [CrossRef]

46. Liu, R.; Sun, Y.; Chai, Y.; Li, S.; Li, S.; Wang, L.; Su, J.; Yu, S.; Yan, J.; Gao, F.; et al. The structural basis of African swine fever virus pA104R binding to DNA and its inhibition by stilbene derivatives. *Proc. Natl. Acad. Sci. USA* **2020**, *117*, 11000–11009. [CrossRef]
47. Frouco, G.; Freitas, F.B.; Coelho, J.; Leitao, A.; Martins, C.; Ferreira, F. DNA-Binding Properties of African Swine Fever Virus pA104R, a Histone-Like Protein Involved in Viral Replication and Transcription. *J. Virol.* **2017**, *91*. [CrossRef]
48. Chen, Q.; Liu, L.; Guo, S.; Li, L.; Yu, Y.; Liu, Z.; Tan, C.; Chen, H.; Wang, X. Characterization of the monoclonal antibody and the immunodominant B-cell epitope of African swine fever virus pA104R by using mouse model. *Microbiol. Spectr.* **2024**, *12*, e0140123. [CrossRef]
49. Zhu, Z.; Li, S.; Ma, C.; Yang, F.; Cao, W.; Liu, H.; Chen, X.; Feng, T.; Shi, Z.; Tian, H.; et al. African Swine Fever Virus E184L Protein Interacts with Innate Immune Adaptor STING to Block IFN Production for Viral Replication and Pathogenesis. *J. Immunol.* **2023**, *210*, 442–458. [CrossRef]
50. Ramirez-Medina, E.; Vuono, E.; Rai, A.; Pruitt, S.; Espinoza, N.; Velazquez-Salinas, L.; Pina-Pedrero, S.; Zhu, J.; Rodriguez, F.; Borca, M.V.; et al. Deletion of E184L, a Putative DIVA Target from the Pandemic Strain of African Swine Fever Virus, Produces a Reduction in Virulence and Protection against Virulent Challenge. *J. Virol.* **2022**, *96*, e0141921. [CrossRef]
51. Tesfagaber, W.; Lan, D.; Wang, W.; Zhao, R.; Yin, L.; Yang, M.; Zhu, Y.; Sun, E.; Liu, R.; Lin, W.; et al. Identification of two novel B cell epitopes on E184L protein of African swine fever virus using monoclonal antibodies. *Virus Res.* **2024**, *346*, 199412. [CrossRef] [PubMed]
52. Urbano, A.C.; Ferreira, N.; Jordao, N.; Boinas, F.; Martins, C.; Ferreira, F. Targeted mutagenesis of the beta-strand DNA binding region of African swine fever virus histone-like protein (pA104R) impairs DNA-binding activity and antibody recognition. *Antiviral Res.* **2024**, *221*, 105784. [CrossRef] [PubMed]
53. Song, J.; Wang, M.; Zhou, L.; Tian, P.; Sun, J.; Sun, Z.; Guo, C.; Wu, Y.; Zhang, G. A novel conserved B-cell epitope in pB602L of African swine fever virus. *Appl. Microbiol. Biotechnol.* **2024**, *108*, 78. [CrossRef]
54. Shi, L.F.; Ren, H.; Zhang, B.; Shi, S.Y.; Shao, H.C.; Xing, H.; Zhao, Y.Y.; Lin, Z.Z.; Zhang, Y.; Han, S.; et al. Preparation and epitope mapping of monoclonal antibodies against African swine fever virus p22 protein. *Int. J. Biol. Macromol.* **2024**, *255*, 128111. [CrossRef]
55. Zhang, S.J.; Niu, B.; Liu, S.M.; Zhu, Y.M.; Zhao, D.M.; Bu, Z.G.; Hua, R.H. Identification of Two Linear Epitopes on MGF_110-13L Protein of African Swine Fever Virus with Monoclonal Antibodies. *Animals* **2024**, *14*, 1951. [CrossRef]
56. Burmakina, G.; Malogolovkin, A.; Tulman, E.R.; Xu, W.; Delhon, G.; Kolbasov, D.; Rock, D.L. Identification of T-cell epitopes in African swine fever virus CD2v and C-type lectin proteins. *J. Gen. Virol.* **2019**, *100*, 259–265. [CrossRef]
57. Liu, Q.; Ma, B.; Qian, N.; Zhang, F.; Tan, X.; Lei, J.; Xiang, Y. Structure of the African swine fever virus major capsid protein p72. *Cell Res.* **2019**, *29*, 953–955. [CrossRef]
58. Sun, W.; Zhang, H.; Fan, W.; He, L.; Chen, T.; Zhou, X.; Qi, Y.; Sun, L.; Hu, R.; Luo, T.; et al. Evaluation of Cellular Immunity with ASFV Infection by Swine Leukocyte Antigen (SLA)-Peptide Tetramers. *Viruses* **2021**, *13*, 2264. [CrossRef]
59. Yang, J.; Li, S.; Feng, T.; Zhang, X.; Yang, F.; Cao, W.; Chen, H.; Liu, H.; Zhang, K.; Zhu, Z.; et al. African Swine Fever Virus F317L Protein Inhibits NF-kappaB Activation To Evade Host Immune Response and Promote Viral Replication. *mSphere* **2021**, *6*, e0065821. [CrossRef]
60. Xu, L.; Hao, F.; Jeong, D.G.; Chen, R.; Gan, Y.; Zhang, L.; Yeom, M.; Lim, J.W.; Yu, Y.; Bai, Y.; et al. Mucosal and cellular immune responses elicited by nasal and intramuscular inoculation with ASFV candidate immunogens. *Front. Immunol.* **2023**, *14*, 1200297. [CrossRef]
61. Dodantenna, N.; Ranathunga, L.; Chathuranga, W.A.G.; Weerawardhana, A.; Cha, J.W.; Subasinghe, A.; Gamage, N.; Haluwana, D.K.; Kim, Y.; Jheong, W.; et al. African Swine Fever Virus EP364R and C129R Target Cyclic GMP-AMP to Inhibit the cGAS-STING Signaling Pathway. *J. Virol.* **2022**, *96*, e0102222. [CrossRef]
62. Zhai, W.; Huang, Y.; He, Y.; Chu, Y.; Tao, C.; Pang, Z.; Wang, Z.; Zhu, H.; Jia, H. Immunogenicity Analysis and Identification of Potential T-Cell Epitopes in C129R Protein of African Swine Fever Virus. *Microorganisms* **2024**, *12*, 1056. [CrossRef]
63. Lokhandwala, S.; Petrovan, V.; Popescu, L.; Sangewar, N.; Elijah, C.; Stoian, A.; Olcha, M.; Ennen, L.; Bray, J.; Bishop, R.P.; et al. Adenovirus-vectored African Swine Fever Virus antigen cocktails are immunogenic but not protective against intranasal challenge with Georgia 2007/1 isolate. *Vet. Microbiol.* **2019**, *235*, 10–20. [CrossRef]
64. Zajac, M.D.; Trujillo, J.D.; Yao, J.; Kumar, R.; Sangewar, N.; Lokhandwala, S.; Sang, H.; Mallen, K.; McCall, J.; Burton, L.; et al. Immunization of pigs with replication-incompetent adenovirus-vectored African swine fever virus multi-antigens induced humoral immune responses but no protection following contact challenge. *Front. Vet. Sci.* **2023**, *10*, 1208275. [CrossRef]
65. Levast, B.; Awate, S.; Babiuk, L.; Mutwiri, G.; Gerdts, V.; van Drunen Littel-van den Hurk, S. Vaccine Potentiation by Combination Adjuvants. *Vaccines* **2014**, *2*, 297–322. [CrossRef]
66. Pikalo, J.; Porfiri, L.; Akimkin, V.; Roszyk, H.; Pannhorst, K.; Kangethe, R.T.; Wijewardana, V.; Sehl-Ewert, J.; Beer, M.; Cattoli, G.; et al. Vaccination With a Gamma Irradiation-Inactivated African Swine Fever Virus Is Safe But Does Not Protect against a Challenge. *Front. Immunol.* **2022**, *13*, 832264. [CrossRef]
67. Andrianarivo, A.G.; Rowe, J.D.; Barr, B.C.; Anderson, M.L.; Packham, A.E.; Sverlow, K.W.; Choromanski, L.; Loui, C.; Grace, A.; Conrad, P.A. A POLYGEN-adjuvanted killed Neospora caninum tachyzoite preparation failed to prevent foetal infection in pregnant cattle following i.v./i.m. experimental tachyzoite challenge. *Int. J. Parasitol.* **2000**, *30*, 985–990. [CrossRef]
68. Blome, S.; Gabriel, C.; Beer, M. Modern adjuvants do not enhance the efficacy of an inactivated African swine fever virus vaccine preparation. *Vaccine* **2014**, *32*, 3879–3882. [CrossRef]

69. Hofacre, C.L.; Rosales, A.G.; Costa, M.D.; Cookson, K.; Schaeffer, J.; Jones, M.K. Immunity and Protection Provided by Live Modified Vaccines Against Paratyphoid Salmonella in Poultry-An Applied Perspective. *Avian Dis.* **2021**, *65*, 295–302. [CrossRef]
70. Mancera Gracia, J.C.; Pearce, D.S.; Masic, A.; Balasch, M. Influenza A Virus in Swine: Epidemiology, Challenges and Vaccination Strategies. *Front. Vet. Sci.* **2020**, *7*, 647. [CrossRef]
71. Renukaradhya, G.J.; Meng, X.J.; Calvert, J.G.; Roof, M.; Lager, K.M. Live porcine reproductive and respiratory syndrome virus vaccines: Current status and future direction. *Vaccine* **2015**, *33*, 4069–4080. [CrossRef]
72. Lokhandwala, S.; Waghela, S.D.; Bray, J.; Martin, C.L.; Sangewar, N.; Charendoff, C.; Shetti, R.; Ashley, C.; Chen, C.H.; Berghman, L.R.; et al. Induction of Robust Immune Responses in Swine by Using a Cocktail of Adenovirus-Vectored African Swine Fever Virus Antigens. *Clin. Vaccine Immunol.* **2016**, *23*, 888–900. [CrossRef]
73. Cadenas-Fernandez, E.; Sanchez-Vizcaino, J.M.; Kosowska, A.; Rivera, B.; Mayoral-Alegre, F.; Rodriguez-Bertos, A.; Yao, J.; Bray, J.; Lokhandwala, S.; Mwangi, W.; et al. Adenovirus-vectored African Swine Fever Virus Antigens Cocktail Is Not Protective against Virulent Arm07 Isolate in Eurasian Wild Boar. *Pathogens* **2020**, *9*, 171. [CrossRef]
74. O'Hagan, D.T.; Ott, G.S.; Nest, G.V.; Rappuoli, R.; Giudice, G.D. The history of MF59((R)) adjuvant: A phoenix that arose from the ashes. *Expert. Rev. Vaccines* **2013**, *12*, 13–30. [CrossRef]
75. Cadenas-Fernandez, E.; Sanchez-Vizcaino, J.M.; van den Born, E.; Kosowska, A.; van Kilsdonk, E.; Fernandez-Pacheco, P.; Gallardo, C.; Arias, M.; Barasona, J.A. High Doses of Inactivated African Swine Fever Virus Are Safe, but Do Not Confer Protection against a Virulent Challenge. *Vaccines* **2021**, *9*, 242. [CrossRef]
76. Leitao, A.; Malur, A.; Cornelis, P.; Martins, C.L. Identification of a 25-aminoacid sequence from the major African swine fever virus structural protein VP72 recognised by porcine cytotoxic T lymphocytes using a lipoprotein based expression system. *J. Virol. Methods* **1998**, *75*, 113–119. [CrossRef]
77. Leitao, A.; Malur, A.; Cartaxeiro, C.; Vasco, G.; Cruz, B.; Cornelis, P.; Martins, C.L. Bacterial lipoprotein based expression vectors as tools for the characterisation of African swine fever virus (ASFV) antigens. *Arch. Virol.* **2000**, *145*, 1639–1657. [CrossRef]
78. Liu, W.; Jiang, P.; Song, T.; Yang, K.; Yuan, F.; Gao, T.; Liu, Z.; Li, C.; Guo, R.; Xiao, S.; et al. A Recombinant Chimera Vaccine Composed of LTB and Mycoplasma hyopneumoniae Antigens P97R1, mhp390 and P46 Elicits Cellular Immunologic Response in Mice. *Vaccines* **2023**, *11*, 1291. [CrossRef]
79. Zhang, H.; Zhao, S.; Zhang, H.; Shen, Y.; Zhang, P.; Shan, H.; Cai, X. Orally administered recombinant Lactobacillus expressing African swine fever virus antigens that induced immunity responses. *Front. Microbiol.* **2022**, *13*, 1103327. [CrossRef]
80. Liao, Z.; Huang, J.; Lo, P.C.; Lovell, J.F.; Jin, H.; Yang, K. Self-adjuvanting cancer nanovaccines. *J. Nanobiotechnol.* **2022**, *20*, 345. [CrossRef]
81. Lu, H.; Zhou, X.; Wu, Z.; Zhang, X.; Zhu, L.; Guo, X.; Zhang, Q.; Zhu, S.; Zhu, H.; Sun, H. Comparison of the mucosal adjuvanticities of two Toll-like receptor ligands for recombinant adenovirus-delivered African swine fever virus fusion antigens. *Vet. Immunol. Immunopathol.* **2021**, *239*, 110307. [CrossRef]
82. Huang, Q.; Niu, T.; Zou, B.; Wang, J.; Xin, J.; Niu, H.; Li, N.; Jiang, Y.; Bao, J.; Zhang, D.; et al. Lactobacillus plantarum Surface-Displayed ASFV (p14.5) Can Stimulate Immune Responses in Mice. *Vaccines* **2022**, *10*, 355. [CrossRef]
83. Gallardo, C.; Fernandez-Pinero, J.; Arias, M. African swine fever (ASF) diagnosis, an essential tool in the epidemiological investigation. *Virus Res.* **2019**, *271*, 197676. [CrossRef]
84. Li, L.; Qiao, S.; Li, G.; Tong, W.; Dong, S.; Liu, J.; Guo, Z.; Zheng, H.; Zhao, R.; Tong, G.; et al. The Indirect ELISA and Monoclonal Antibody against African Swine Fever Virus p17 Revealed Efficient Detection and Application Prospects. *Viruses* **2022**, *15*, 50. [CrossRef]
85. Afayibo, D.J.A.; Zhang, Z.; Sun, H.; Fu, J.; Zhao, Y.; Amuda, T.O.; Wu, M.; Du, J.; Guan, G.; Niu, Q.; et al. Establishment of an ELISA Based on a Recombinant Antigenic Protein Containing Multiple Prominent Epitopes for Detection of African Swine Fever Virus Antibodies. *Microorganisms* **2024**, *12*, 943. [CrossRef]
86. Li, D.; Zhang, Q.; Liu, Y.; Wang, M.; Zhang, L.; Han, L.; Chu, X.; Ding, G.; Li, Y.; Hou, Y.; et al. Indirect ELISA Using Multi-Antigenic Dominants of p30, p54 and p72 Recombinant Proteins to Detect Antibodies against African Swine Fever Virus in Pigs. *Viruses* **2022**, *14*, 2660. [CrossRef]
87. Rossjohn, J.; Gras, S.; Miles, J.J.; Turner, S.J.; Godfrey, D.I.; McCluskey, J. T cell antigen receptor recognition of antigen-presenting molecules. *Annu. Rev. Immunol.* **2015**, *33*, 169–200. [CrossRef]
88. Wang, Z.; Zhang, J.; Li, F.; Zhang, Z.; Chen, W.; Zhang, X.; Sun, E.; Zhu, Y.; Liu, R.; He, X.; et al. The attenuated African swine fever vaccine HLJ/18-7GD provides protection against emerging prevalent genotype II variants in China. *Emerg. Microbes Infect.* **2024**, *13*, 2300464. [CrossRef]
89. Argilaguet, J.M.; Perez-Martin, E.; Nofrarias, M.; Gallardo, C.; Accensi, F.; Lacasta, A.; Mora, M.; Ballester, M.; Galindo-Cardiel, I.; Lopez-Soria, S.; et al. DNA vaccination partially protects against African swine fever virus lethal challenge in the absence of antibodies. *PLoS ONE* **2012**, *7*, e40942. [CrossRef]
90. Urbano, A.C.; Ferreira, F. African swine fever control and prevention: An update on vaccine development. *Emerg. Microbes Infect.* **2022**, *11*, 2021–2033. [CrossRef]
91. Borca, M.V.; Ramirez-Medina, E.; Silva, E.; Rai, A.; Espinoza, N.; Velazquez-Salinas, L.; Gladue, D.P. ASF Vaccine Candidate ASFV-G-ΔI177L Does Not Exhibit Residual Virulence in Long-Term Clinical Studies. *Pathogens* **2023**, *12*, 805. [CrossRef] [PubMed]

92. Shawan, M.; Sharma, A.R.; Halder, S.K.; Arian, T.A.; Shuvo, M.N.; Sarker, S.R.; Hasan, M.A. Advances in Computational and Bioinformatics Tools and Databases for Designing and Developing a Multi-Epitope-Based Peptide Vaccine. *Int. J. Pept. Res. Ther.* **2023**, *29*, 60. [CrossRef] [PubMed]
93. Tang, M.; Cai, J.H.; Diao, H.Y.; Guo, W.M.; Yang, X.; Xing, S. The progress of peptide vaccine clinical trials in gynecologic oncology. *Hum. Vaccin. Immunother.* **2022**, *18*, 2062982. [CrossRef] [PubMed]
94. Simbulan, A.M.; Banico, E.C.; Sira, E.; Odchimar, N.M.O.; Orosco, F.L. Immunoinformatics-guided approach for designing a pan-proteome multi-epitope subunit vaccine against African swine fever virus. *Sci. Rep.* **2024**, *14*, 1354. [CrossRef]
95. Nguyen, T.L.; Samuel Leon Magdaleno, J.; Rajjak Shaikh, A.; Choowongkomon, K.; Li, V.; Lee, Y.; Kim, H. Designing a multi-epitope candidate vaccine by employing immunoinformatics approaches to control African swine fever spread. *J. Biomol. Struct. Dyn.* **2023**, *41*, 10214–10229. [CrossRef]

pathogens MDPI

Brief Report

Glycerol Monolaurate Inhibits Wild-Type African Swine Fever Virus Infection in Porcine Macrophages

Joshua A. Jackman [1,*], Erik Arabyan [2], Hovakim Zakaryan [2] and Charles C. Elrod [3,4,*]

1 School of Chemical Engineering and Translational Nanobioscience Research Center, Sungkyunkwan University, Suwon 16419, Republic of Korea
2 Laboratory of Antiviral Drug Discovery, Institute of Molecular Biology of NAS, Yerevan 0014, Armenia; erik.arabyan.bio@gmail.com (E.A.); h_zakaryan@mb.sci.am (H.Z.)
3 Natural Biologics Inc., Newfield, NY 14867, USA
4 Department of Animal Science, Cornell University, Ithaca, NY 14853, USA
* Correspondence: jjackman@skku.edu (J.A.J.); cce1@cornell.edu (C.C.E.)

Abstract: Naturally abundant antimicrobial lipids, such as fatty acids and monoglycerides, that disrupt membrane-enveloped viruses are promising mitigants to inhibit African swine fever virus (ASFV). Among mitigant candidates in this class, glycerol monolaurate (GML) has demonstrated particularly high antiviral activity against laboratory-adapted ASFV strains. However, there is an outstanding need to further determine the effects of GML on wild-type ASFV strains, which can have different virulence levels and sensitivities to membrane-disrupting compounds as compared to laboratory-adapted strains. Herein, we investigated the antiviral effects of GML on a highly virulent strain of a wild-type ASFV isolate (Armenia/07) in an in vitro porcine macrophage model. GML treatment caused a concentration-dependent reduction in viral infectivity, and there was a sharp transition between inactive and active GML concentrations. Low GML concentrations had negligible effect on viral infectivity, whereas sufficiently high GML concentrations caused a >99% decrease in viral infectivity. The concentration onset of antiviral activity matched the critical micelle concentration (CMC) value of GML, reinforcing that GML micelles play a critical role in enabling anti-ASFV activity. These findings validate that GML can potently inhibit wild-type ASFV infection of porcine macrophages and support a biophysical explanation to guide antimicrobial lipid performance optimization for pathogen mitigation applications.

Keywords: African swine fever virus; antiviral; mitigation; glycerol monolaurate; monoglyceride

Citation: Jackman, J.A.; Arabyan, E.; Zakaryan, H.; Elrod, C.C. Glycerol Monolaurate Inhibits Wild-Type African Swine Fever Virus Infection in Porcine Macrophages. *Pathogens* **2023**, *12*, 1193. https://doi.org/10.3390/pathogens12101193

Academic Editors: Lauro Velazquez-Salinas, Gisselle N. Medina and Elizabeth Ramirez-Medina

Received: 4 August 2023
Revised: 8 September 2023
Accepted: 22 September 2023
Published: 25 September 2023

1. Introduction

The African swine fever virus (ASFV) is the causative agent of a highly lethal hemorrhagic disease with near 100% mortality in newly exposed pig populations and is a major agricultural biosecurity risk [1,2]. In recent years, ASFV outbreaks have significantly affected pig production worldwide, especially in parts of Asia where an epidemic occurred in 2018–2021, and have impacted global food and feed markets [3]. From a biosecurity perspective, ASFV is challenging to stop because there are no currently approved vaccines or therapeutics [4,5]. Hence, preventing ASFV transmission is critical and mainly relies on containment, sanitation, and surface disinfection [6–8]. Within this scope, there has been growing attention to the role that feed and drinking water can play as transmission vectors in contributing to ASFV disease spread [9] and in developing additive-based chemical mitigation strategies to inhibit ASFV in these matrices [10]. While formaldehyde is a widely used crosslinking additive to inhibit viral pathogens in such contexts, regulatory actions have banned its usage in certain jurisdictions, such as the European Union [Regulation (EU) 2018/183] (see also discussion in Ref. [11]), and there are ongoing efforts to develop new classes of regulatory acceptable mitigants with antiviral properties.

One promising target for antiviral mitigant development is the phospholipid membrane envelope that surrounds infectious ASFV particles [12]. Envelope targeting has two main advantages: (1) the lipid bilayer structure of the envelope is conserved across different virus strains so that one mitigant can broadly work against multiple virus strains in principle and potentially against different viruses; and (2) viruses cannot easily mutate to become resistant since the membrane components in the envelope are derived from host cell membranes and are not encoded in the viral genome [13,14]. Nevertheless, compared to other single-enveloped pig viruses, such as porcine epidemic diarrhea virus (PEDV) and porcine reproductive and respiratory syndrome virus (PRRSV), ASFV is a more rugged, double-enveloped virus belonging to the nucleocytoplasmic large DNA virus (NCLDV) family (Ref. [15]), and experimental testing on ASFV is thus needed to validate the antiviral efficacy of anti-ASFV mitigant candidates.

The need to test antiviral mitigants against ASFV while maintaining a high biosafety level has led to the development of the non-virulent ASFV BA71V strain, which is adapted to infect commonly used cell lines but neither infects porcine cells nor causes disease in pigs [16]. The ASFV BA71V strain has proven effective for testing membrane-disrupting antiviral mitigant candidates such as antimicrobial lipids (fatty acids and monoglycerides [17,18]) and rigid amphipathic fusion inhibitors (RAFIs) [19–21]. However, experimental data indicate that some membrane-disrupting mitigants can exhibit strain-specific antiviral activities. For example, certain RAFIs were shown to inhibit the ASFV BA71V strain by inhibiting virus-cell attachment, whereas the same compounds did not affect cellular attachment of the virulent, wild-type ASFV Armenia/07 strain isolate that infects porcine macrophages [19]. These findings underscore that the BA71V strain may be an effective screening tool (as are potentially other NCLDV surrogates under development as well [22,23]) to identify anti-ASFV mitigants, but further validation against virulent ASFV strains is needed to support translation.

Herein, we evaluated the antiviral properties of glycerol monolaurate (GML) against the virulent ASFV Armenia/07 strain in order to validate the potential antiviral efficacy of GML against circulating, wild-type ASFV strains. GML is a regulatory acceptable, food-grade monoglyceride that disrupts phospholipid membranes and has shown antiviral efficacy against a wide range of enveloped viruses, including in vivo treatment effects to ameliorate PEDV and Seneca Valley virus (SVV) infections in pigs [24,25]. Of note, GML has previously been shown to abrogate non-virulent ASFV BA71V strain infectivity in drinking water and had a higher level of antiviral activity than other tested medium-chain fatty acids [20]. Furthermore, GML blunted ASFV BA71V strain infectivity in feed while additionally causing conformational changes in viral surface proteins [20]. Building on these findings, our objective in the present study was to further explore the feasibility of GML to inhibit ASFV Armenia/07 infection in a porcine macrophage model, especially in terms of elucidating concentration-dependent effects that relate to the biophysical characteristics of GML as an antiviral mitigant.

2. Materials and Methods

2.1. Cell and Virus Preparations

The virulent ASFV Armenia/07 strain was used in all experiments, as previously described [26]. Viral titer quantification was conducted by the hemadsorption (HAD) assay, and titer levels are expressed in units of 50% hemadsorption doses ($HADU_{50}$) per mL. In addition, primary porcine alveolar macrophages (PAMSs) were obtained and prepared following established protocols [27]. Prior to antiviral testing, the PAMs were maintained at 37 °C in Dulbecco's modified Eagle's medium that was supplemented with 10% fetal bovine serum, 2 mM L-glutamine, 100 IU/mL penicillin, and 100 µg/mL streptomycin. All kits and reagents were obtained from Sigma-Aldrich (Darmstadt, Germany) unless otherwise specified.

2.2. Cytotoxicity Assay

The effect of GML on PAM cell viability was investigated by the 3-(4,5-dimethylthiazol-2-yl)-2,5-diphenyltetrazolium bromide (MTT) assay. Cells in a 96-well cell culture plate (1×10^4 cells per well) were treated with different GML concentrations (500, 250, 125, 63, or 31 μM in a two-fold dilution series). Treated cells were incubated for up to 72 h at 37 °C in a 5% CO_2 environment. After incubation, the medium was removed and MTT solution was added. The microplates were incubated at 37 °C for 3 h after adding MTT solution, followed by purple formazan extraction by MTT solvent. The colorimetric measurements were performed on a microplate reader at 570 nm. The percentage of viable cells was calculated at each GML concentration as [(ODT/ODC) × 100%], whereby ODT and ODC correspond to the absorbance (optical density) of treated and control cells, respectively.

2.3. Antiviral Assay

Suspensions of ASFV Armenia/07 were prepared at a multiplicity of infection (MOI) of 0.5 or 1 $HADU_{50}$ per well and were treated with different GML concentrations (250, 125, 63, or 31 μM in a two-fold dilution series) for one hour at room temperature along with negative control (virus-only without GML). GML is understood to quickly disrupt phospholipid membranes (<15 min), and the time period was selected to ensure sufficient incubation time in line with past reports [20].

The treated virus samples were then added to infect PAM cells seeded at 2×10^5 cells per well in a 24-well plate and were cultured for 48 or 72 h. Cell culture supernatants were collected at 24, 48, and/or 72 h post-infection as appropriate, and viral titer in the supernatant was quantified by HAD assay upon porcine erythrocyte addition [27]. ASFV presence was quantified by counting the formation of erythrocyte rosettes around infected macrophages and was expressed in $HADU_{50}$/mL units accordingly.

2.4. Statistical Analysis

All statistical tests were performed using the GraphPad Prism 8 software package (San Diego, CA, USA). One-way analysis of variance (ANOVA) with Dunnett's multiple comparisons test (versus virus-only control) was used. Statistical significance was computed in terms of multiplicity-adjusted p values, and $p < 0.05$ indicates the cutoff for statistical significance.

3. Results and Discussion

We selected the ASFV Armenia/07 strain as the model virulent strain for assessing the antiviral activity of GML, which is a saturated monoglyceride that is an esterified adduct of 12-carbon-long lauric acid and glycerol. GML was originally identified to possess potent antimicrobial activity for inhibiting bacteria compared to other medium-chain fatty acids and monoglycerides, and more recently it has been explored for inhibiting enveloped viruses as well [28,29]. From biophysical experiments, it is understood that GML mainly disrupts phospholipid membranes at concentrations above its critical micelle concentration (CMC) (Ref. [30]), and membrane disruption is understood to be the basis for its antiviral activity against enveloped viruses [31]. Our selection of the ASFV Armenia/07 strain in particular was further motivated by two main factors: (1) this strain isolate is highly virulent in pigs and similar to the wild-type ASFV Georgia/07 strain that initially caused an outbreak in the Caucasus region [32,33]; and (2) the strain has been used previously to test other membrane-disrupting antiviral molecules such as RAFIs [19]. Accordingly, ASFV virus suspensions (MOI: 0.5 $HADU_{50}$ per well) in aqueous solution were mixed with different GML concentrations for a treatment period, followed by adding the virus–GML mixtures to infect PAM cells. The infection process was allowed to continue, and cell culture media supernatants were collected at 48 and 72 h post-infection in order to measure the extent of infectious virus replication. The GML concentration range was chosen based on the range of antiviral potency observed in past testing with the ASF BA71V strain (31–250 μM) [20] and also bridges the CMC range of GML (~60 μM) [34].

At 48 h post-infection, the virus-only control titer was 5.1 ± 0.1 log($HADU_{50}$/mL), and the titer of the virus sample treated with 31 μM GML was similar at around 5.0 ± 0.3 log($HADU_{50}$/mL) (Figure 1A). In marked contrast, virus samples treated with 63, 125, or 250 μM GML had significantly reduced titers around 2.6 ± 0.1, 2.4 ± 0.1, and 2.3 ± 0.1 log($HADU_{50}$/mL), respectively. Thus, GML had a negligible antiviral activity until reaching a critical concentration of 63 μM, which is around the CMC value of GML. This concentration-dependent finding agrees well with past ASFV BA71V testing results [20] and supports that GML micelles are the principal unit involved in disrupting ASFV particles.

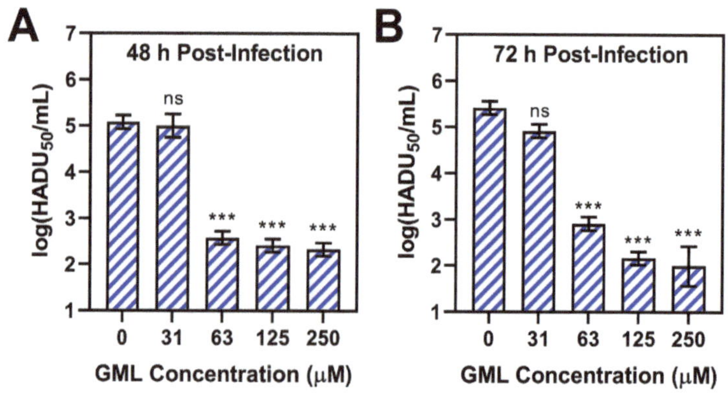

Figure 1. Antiviral activity of GML to inhibit Armenia/07 ASFV infection of porcine macrophages in vitro. The virus suspension was treated with different concentrations of GML (31–250 μM) prior to PAM cell infection. The 0 μM GML data point corresponds to the virus-only control. Viral titers of cell culture supernatants were measured (**A**) 48 h or (**B**) 72 h post-infection by hemadsorption assay. Data are reported in units of log 50% hemadsorption doses ($HADU_{50}$) per mL and presented as mean ± standard deviation from three independent experiments ($n = 3$ per group). The markers *** and ns indicate $p < 0.001$ and $p > 0.05$, respectively, as compared to the virus-only control.

Similar test results were also recorded at 72 h post-infection (Figure 1B). In this case, the virus-only control titer was 5.4 ± 0.1 log($HADU_{50}$/mL), whereas the titer of the virus sample treated with 31 μM GML was 4.9 ± 0.1 log($HADU_{50}$/mL). While the average titer difference was 0.5 log units, the difference was not statistically significant. By contrast, virus samples treated with 63, 125, or 250 μM GML had significantly reduced titers around 2.9 ± 0.1, 2.2 ± 0.1, and 2.0 ± 0.4 log($HADU_{50}$/mL), respectively.

We also performed similar experiments at a higher MOI (1 $HADU_{50}$ per well) and observed a similar trend in the concentration-dependent antiviral effects of GML (Figure S1). At 24 h post-infection, the virus-only control titer was ~5.5 log($HADU_{50}$/mL), and a similar titer was also recorded for the virus sample treated with 31 μM GML. Conversely, virus samples treated with 63, 125, or 250 μM GML tended to have reduced titers of around ~4.7, ~4.2, and ~3.7 log($HADU_{50}$/mL), respectively. At 48 h infection, a similar trend was again observed. For the virus-only control and virus sample treated with 31 μM GML, the titers were around ~6.3 log($HADU_{50}$/mL), whereas virus samples treated with 63, 125, or 250 μM GML tended to have decreased titers of around ~4.2, ~4.4, and ~3.9 log($HADU_{50}$/mL), respectively. Taken together, these data reinforce that GML was only active at and above 63 μM, whereas a two-fold reduction in the GML concentration led to an insignificant antiviral effect.

In addition to changes in viral titer, we also plotted the percentages of viral infectivity relative to the virus-only control (Figure 2). This quantification approach allowed us to obtain unitless indicators of viral infectivity reduction in order to plot data obtained with the virulent ASFV Armenia/07 strain here and data obtained with the non-virulent ASFV

BA71V strain in a past study [20]. While the absolute viral titers cannot be compared directly because different methods were used depending on the virus strain and cell type, we focused on calculating the percentage change in viral infectivity relative to the appropriate untreated control. We first analyzed the antiviral data collected with the ASFV Armenia/07 strain at 72 h post-infection and focused on ≥63 μM GML treatment conditions, within which range statistically significant antiviral effects were observed. Within this range, 63, 125, and 250 μM GML treatments caused significant drops in viral infectivity of around 99.7% ± 0.10%, 99.9% ± 0.02%, and 99.9% ± 0.06%, respectively.

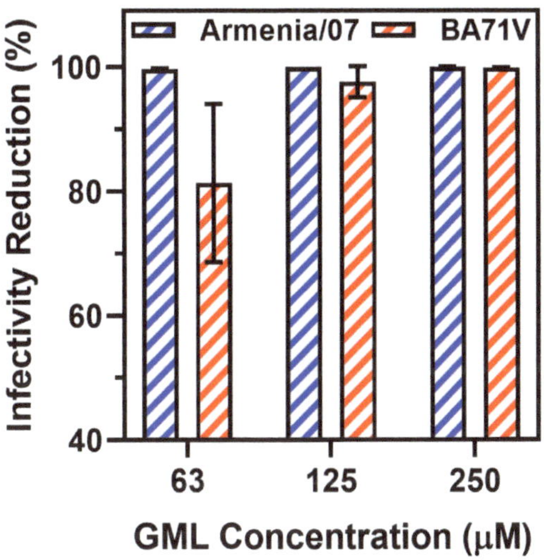

Figure 2. Quantitative analysis of GML inhibitory effect in vitro on virulent (Armenia/07) and non-virulent (BA71V) ASFV strains. The concentration-dependent inhibitory effects of GML are expressed in terms of the degree of infectivity reduction relative to virus-only controls. The Armenia/07 strain titers were measured in terms of $HADU_{50}$/mL units, and the BA71V strain titers were measured in terms of 50% tissue culture infective dose ($TCID_{50}$) per mL units (quantitative analysis was done on raw data from Figure S1 of Ref. [20]). Data are presented as mean ± standard deviation from three independent experiments (n = 3 per group).

These data were plotted alongside previously obtained antiviral data that evaluated the effects of GML treatment on ASFV BA71V strain infectivity in a permissive Vero cell model by using a cytopathic effect assay. In those experiments, 31 μM GML treatment had a negligible effect on viral infectivity relative to the virus-only control, whereas 63, 125, and 250 μM GML treatments caused significant drops in viral infectivity of around 81.3% ± 12.7%, 97.6% ± 2.5%, and 99.8% ± 0.06%, respectively.

Collectively, the data support that GML only exhibited antiviral activity against both ASFV strains at and above 63 μM GML concentration, which supports that GML micelles are the main membrane-disrupting species to inhibit ASFV. This finding fits with past biophysical studies that showed that GML only disrupts phospholipid bilayers at and above its CMC, whereas GML monomers had a negligible effect [30,35]. Furthermore, these findings support that GML exhibits similar levels of antiviral potency (i.e., the lowest GML concentration at which antiviral activity occurs to a significant extent) against the virulent ASFV Armenia/07 and the non-virulent ASFV BA71V strains, and this concentration dependency supports that GML micelles mainly contribute to antiviral activity against both virus strains. This latter conclusion is reinforced by two findings: (1) the concentration onset at which antiviral activity occurs; and (2) the sharp transition between inactive and active GML concentrations.

While acknowledging that the antiviral tests involving ASFV Armenia/07 and ASFV BA71V strains were conducted on different cell lines and with different assays and that a direct comparison between antiviral tests on the two strains should not be made for this reason, we may also briefly comment on the different degrees of viral inactivation that GML exhibited against each strain. If we treat a 3 log drop in viral infectivity (99.9%) reduction as a performance cutoff, then the data support that treatment of ASFV Armenia/07 with ≥125 μM GML meets this performance threshold (≥63 μM GML for a 99% threshold). In addition, treatment with 250 μM GML met the 99% performance threshold for inhibiting the ASFV BA71V strain. Across the two independent sets of experimental data, it should be emphasized that the effect of GML on reducing ASFV infectivity in both strains is quite high. For SVV, it has been reported that GML treatment yielded a maximum reduction of only ~80% in in vitro experiments, yet still resulted in good in vivo performance in a porcine model in terms of reducing clinical symptoms, viral loads, and organ damage as well as promoting positive inflammatory responses [25]. GML exhibited a similar inhibitory effect on PRRSV, with around 80% reduction in viral infectivity in vitro [36].

In addition to antiviral tests, we also investigated the concentration-dependent effects of GML on PAM cell viability (Figure 3). While the envisioned antiviral applications of GML are mainly aimed at ex vivo mitigation (e.g., in drinking water), these cell cytotoxicity experiments were directed at confirming sufficient PAM cell viability in the presence of GML within the tested concentration range and also at distinguishing between the effects of GML on abiotic membranes, such as those of enveloped viruses that lack reparative capacity as described above, vs. on biotic membranes, such as those of mammalian cells that have reparative capacity (as previously discussed in the context of other classes of membrane-disrupting antivirals, such as RAFIs [13,37]).

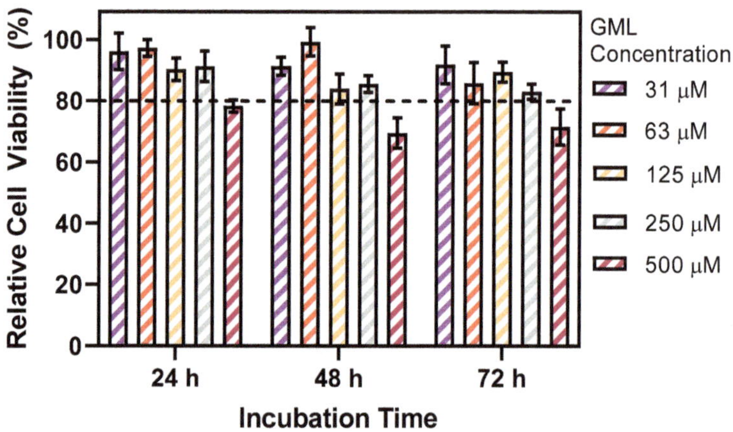

Figure 3. Effect of GML concentration on PAM cell viability. Different GML concentrations were incubated with PAM cells for 24, 48, or 72 h prior to MTT analysis. Data are expressed in terms of relative cell viability compared to negative control (no GML) and are presented as mean ± standard deviation from three independent experiments (n = 3 per group). The horizontal dashed line corresponds to a 20% cutoff drop in relative cell viability.

While ASFV infectivity decreased significantly when treated with GML at and above its CMC, PAM cell viability remained similarly high in this GML concentration range, and appreciable cytotoxicity (>20% drop in viability) only occurred at higher GML concentrations outside the antiviral test range. After 24 h incubation, cell viability upon treatment with 31–250 μM GML was >90% and only dropped to <80% viability at 500 μM GML. This trend is generally consistent with past reports describing how the 50% cell cytotoxicity values of GML against human lung fibroblasts and skin keratinocytes did not match the

CMC and were in the range of ~300 μM (Ref. [38]), supporting that the effects of GML on PAM cell viability are not CMC-dependent. Similar trends in cell viability (>80% viability up to 250 μM GML) were observed after 48 and 72 h post-incubation. These findings support that the CMC-dependent antiviral activity of GML to inhibit ASFV is related to GML-micelle-induced irreparable membrane disruption of the viral envelope, while further suggesting that GML micelles have less deleterious effects on PAM cells with reparative membrane capacity.

4. Conclusions

In summary, the findings in this study support that GML can inhibit wild-type ASFV strains and that the mechanism of antiviral activity depends on GML micelle formation. While there has been recent discussion about how micelle formation may not be an absolute prerequisite for the antiviral activity of surfactant-like molecules against enveloped viruses in all cases and modest antiviral activity can sometimes be observed at slightly lower concentrations below CMC (as described in the context of searching for Triton X-100 replacements [39]), the results obtained in the present study indicate that micelle formation is a key contributing factor to the antiviral activity of GML, which is also consistent with the membrane biophysics literature. From a translational perspective, these findings further emphasize the importance of organizing GML into supramolecular assemblies, whereby the effects of GML in self-assembled nanostructures are greater than the effects of GML monomers. Above CMC, GML micelles are spontaneously formed self-assembled nanostructures; however, they can collapse upon dilution, e.g., upon administration into an animal or injection into a drinking water line. Thus, incorporating GML into dilution-stable nanostructures, such as solid lipid nanoparticles [40], may represent a future opportunity to harness its antiviral activity in additional types of supramolecular assemblies beyond micelles and may support the use of GML in pathogen-mitigation applications across feed and drinking water matrices.

Supplementary Materials: The following supporting information can be downloaded at: https://www.mdpi.com/article/10.3390/pathogens12101193/s1, Figure S1: Antiviral activity of GML to inhibit Armenia/07 ASFV infection of porcine macrophages in vitro at higher multiplicity of infection.

Author Contributions: Conceptualization, J.A.J. and C.C.E.; Methodology, J.A.J., E.A., H.Z. and C.C.E.; Investigation, J.A.J., E.A. and H.Z.; Writing—original draft preparation, J.A.J. and C.C.E.; Writing—review and editing, J.A.J., H.Z. and C.C.E.; Supervision, J.A.J. and C.C.E. All authors have read and agreed to the published version of the manuscript.

Funding: Natural Biologics Inc. provided funding to conduct the experiments in this study. J.A.J. acknowledges additional funding support from the National Research Foundation of Korea (NRF) grant funded by the Korean government (MSIT) (No. 2020R1C1C1004385). This work was partially supported by the SKKU Global Research Platform Research Fund, Sungkyunkwan University, 2023.

Institutional Review Board Statement: Not applicable.

Informed Consent Statement: Not applicable.

Data Availability Statement: The data presented in this study are available upon reasonable request from the corresponding authors.

Conflicts of Interest: C.C.E. is employed by and J.A.J. serves as a board member of the company Natural Biologics Inc. The other authors declare no conflict of interest.

References

1. Galindo, I.; Alonso, C. African swine fever virus: A review. *Viruses* **2017**, *9*, 103. [CrossRef] [PubMed]
2. Gaudreault, N.N.; Madden, D.W.; Wilson, W.C.; Trujillo, J.D.; Richt, J.A. African swine fever virus: An emerging DNA arbovirus. *Front. Vet. Sci.* **2020**, *7*, 215. [CrossRef] [PubMed]
3. You, S.; Liu, T.; Zhang, M.; Zhao, X.; Dong, Y.; Wu, B.; Wang, Y.; Li, J.; Wei, X.; Shi, B. African swine fever outbreaks in China led to gross domestic product and economic losses. *Nat. Food* **2021**, *2*, 802–808. [CrossRef] [PubMed]

4. Arabyan, E.; Kotsynyan, A.; Hakobyan, A.; Zakaryan, H. Antiviral agents against African swine fever virus. *Virus Res.* **2019**, *270*, 197669. [CrossRef]
5. Penrith, M.L.; Kivaria, F.M. One hundred years of African swine fever in Africa: Where have we been, where are we now, where are we going? *Transbound. Emerg. Dis.* **2022**, *69*, e1179–e1200. [CrossRef]
6. Barongo, M.B.; Bishop, R.P.; Fèvre, E.M.; Knobel, D.L.; Ssematimba, A. A mathematical model that simulates control options for African swine fever virus (ASFV). *PLoS ONE* **2016**, *11*, e0158658. [CrossRef]
7. Mutua, F.; Dione, M. The context of application of biosecurity for control of African swine fever in smallholder pig systems: Current gaps and recommendations. *Front. Vet. Sci.* **2021**, *8*, 689811. [CrossRef]
8. Beato, M.S.; D'Errico, F.; Iscaro, C.; Petrini, S.; Giammarioli, M.; Feliziani, F. Disinfectants against African swine fever: An updated review. *Viruses* **2022**, *14*, 1384. [CrossRef]
9. Niederwerder, M.C.; Stoian, A.M.; Rowland, R.R.; Dritz, S.S.; Petrovan, V.; Constance, L.A.; Gebhardt, J.T.; Olcha, M.; Jones, C.K.; Woodworth, J.C. Infectious dose of African swine fever virus when consumed naturally in liquid or feed. *Emerg. Infect. Dis.* **2019**, *25*, 891. [CrossRef]
10. Phillips, F.C.; Rubach, J.K.; Poss, M.J.; Anam, S.; Goyal, S.M.; Dee, S.A. Monoglyceride reduces viability of porcine epidemic diarrhoea virus in feed and prevents disease transmission to post-weaned piglets. *Transbound. Emerg. Dis.* **2022**, *69*, 121–127. [CrossRef]
11. Śmiechowska, M.; Newerli-Guz, J.; Skotnicka, M. Spices and seasoning mixes in European Union—Innovations and ensuring safety. *Foods* **2021**, *10*, 2289. [CrossRef] [PubMed]
12. Wang, N.; Zhao, D.; Wang, J.; Zhang, Y.; Wang, M.; Gao, Y.; Li, F.; Wang, J.; Bu, Z.; Rao, Z.; et al. Architecture of African swine fever virus and implications for viral assembly. *Science* **2019**, *366*, 640–644. [CrossRef] [PubMed]
13. Vigant, F.; Santos, N.C.; Lee, B. Broad-spectrum antivirals against viral fusion. *Nat. Rev. Microbiol.* **2015**, *13*, 426–437. [CrossRef]
14. Yoon, B.K.; Jeon, W.-Y.; Sut, T.N.; Cho, N.-J.; Jackman, J.A. Stopping membrane-enveloped viruses with nanotechnology strategies: Toward antiviral drug development and pandemic preparedness. *ACS Nano* **2021**, *15*, 125–148. [CrossRef] [PubMed]
15. Hernáez, B.; Guerra, M.; Salas, M.L.; Andrés, G. African swine fever virus undergoes outer envelope disruption, capsid disassembly and inner envelope fusion before core release from multivesicular endosomes. *PLoS Pathog.* **2016**, *12*, e1005595. [CrossRef] [PubMed]
16. Rodríguez, J.M.; Moreno, L.T.; Alejo, A.; Lacasta, A.; Rodríguez, F.; Salas, M.L. Genome sequence of African swine fever virus BA71, the virulent parental strain of the nonpathogenic and tissue-culture adapted BA71V. *PLoS ONE* **2015**, *10*, e0142889. [CrossRef]
17. Jackman, J.A.; Boyd, R.D.; Elrod, C.C. Medium-chain fatty acids and monoglycerides as feed additives for pig production: Towards gut health improvement and feed pathogen mitigation. *J. Anim. Sci. Biotechnol.* **2020**, *11*, 44. [CrossRef]
18. Jackman, J.A.; Lavergne, T.A.; Elrod, C.C. Antimicrobial monoglycerides for swine and poultry applications. *Front. Anim. Sci.* **2022**, *3*, 1019320. [CrossRef]
19. Hakobyan, A.; Galindo, I.; Nañez, A.; Arabyan, E.; Karalyan, Z.; Chistov, A.A.; Streshnev, P.P.; Korshun, V.A.; Alonso, C.; Zakaryan, H. Rigid amphipathic fusion inhibitors demonstrate antiviral activity against African swine fever virus. *J. Gen. Virol.* **2018**, *99*, 148–156. [CrossRef]
20. Jackman, J.A.; Hakobyan, A.; Zakaryan, H.; Elrod, C.C. Inhibition of African swine fever virus in liquid and feed by medium-chain fatty acids and glycerol monolaurate. *J. Anim. Sci. Biotechnol.* **2020**, *11*, 114. [CrossRef]
21. Niederwerder, M.C.; Dee, S.; Diel, D.G.; Stoian, A.M.M.; Constance, L.A.; Olcha, M.; Petrovan, V.; Patterson, G.; Cino-Ozuna, A.G.; Rowland, R.R.R. Mitigating the risk of African swine fever virus in feed with anti-viral chemical additives. *Transbound. Emerg. Dis.* **2021**, *68*, 477–486. [CrossRef] [PubMed]
22. Palowski, A. The Use of a Risk-Free In Situ Non-Animal (RISNA) Surrogate Assay for Evaluating Inactivation Strategies of African Swine Fever Virus in Feed Ingredients in Real-World Demonstrations. Master's Thesis, University of Minnesota, Minneapolis, MN, USA, 2021.
23. Palowski, A.; Balestreri, C.; Urriola, P.E.; van de Ligt, J.L.G.; Sampedro, F.; Dee, S.; Shah, A.; Yancy, H.F.; Shurson, G.C.; Schroeder, D.C. Survival of a surrogate African swine fever virus-like algal virus in feed matrices using a 23-day commercial United States truck transport model. *Front. Microbiol.* **2022**, *13*, 1059118. [CrossRef] [PubMed]
24. Liu, Z.; Zhu, L.; Zhao, X.; Liu, J.; Cheng, H.; Zhang, L.; Tang, H.; Sun, X.; Hu, Y.; Xu, Z. Effects of oral of administration of monoglycide laurate on virus load and inflammation in PEDV infected porcine. *Front. Vet. Sci.* **2022**, *9*, 98038. [CrossRef] [PubMed]
25. Su, B.; Wang, Y.; Jian, S.; Tang, H.; Deng, H.; Zhu, L.; Zhao, X.; Liu, J.; Cheng, H.; Zhang, L.; et al. In vitro and in vivo antiviral activity of monolaurin against Seneca Valley virus. *Front. Vet. Sci.* **2023**, *10*, 980187. [CrossRef] [PubMed]
26. Arabyan, E.; Hakobyan, A.; Hakobyan, T.; Grigoryan, R.; Izmailyan, R.; Avetisyan, A.; Karalyan, Z.; Jackman, J.A.; Ferreira, F.; Elrod, C.C.; et al. Flavonoid library screening reveals kaempferol as a potential antiviral agent against African swine fever virus. *Front. Microbiol.* **2021**, *12*, 736780. [CrossRef]
27. Carrascosa, A.L.; Bustos, M.J.; de Leon, P. Methods for growing and titrating African swine fever virus: Field and laboratory samples. *Curr. Protoc. Cell Biol.* **2011**, *53*, 26.14.1–26.14.25. [CrossRef]
28. Kabara, J.J.; Swieczkowski, D.M.; Conley, A.J.; Truant, J.P. Fatty acids and derivatives as antimicrobial agents. *Antimicrob. Agents Chemother.* **1972**, *2*, 23–28. [CrossRef]

29. Welch, J.L.; Xiang, J.; Okeoma, C.M.; Schlievert, P.M.; Stapleton, J.T. Glycerol monolaurate, an analogue to a factor secreted by lactobacillus, is virucidal against enveloped viruses, including HIV-1. *mBio* **2020**, *11*, 10–1128. [CrossRef]
30. Yoon, B.K.; Jackman, J.A.; Kim, M.C.; Cho, N.-J. Spectrum of membrane morphological responses to antibacterial fatty acids and related surfactants. *Langmuir* **2015**, *31*, 10223–10232. [CrossRef]
31. Hierholzer, J.C.; Kabara, J.J. In vitro effects of monolaurin compounds on enveloped RNA and DNA viruses. *J. Food Saf.* **1982**, *4*, 1–12. [CrossRef]
32. Rowlands, R.J.; Michaud, V.; Heath, L.; Hutchings, G.; Oura, C.; Vosloo, W.; Dwarka, R.; Onashvili, T.; Albina, E.; Dixon, L.K. African swine fever virus isolate, Georgia, 2007. *Emerg. Infect. Dis.* **2008**, *14*, 1870. [CrossRef] [PubMed]
33. Sunwoo, S.-Y.; Pérez-Núñez, D.; Morozov, I.; Sánchez, E.G.; Gaudreault, N.N.; Trujillo, J.D.; Mur, L.; Nogal, M.; Madden, D.; Urbaniak, K.; et al. DNA-protein vaccination strategy does not protect from challenge with African swine fever virus Armenia 2007 strain. *Vaccines* **2019**, *7*, 12. [CrossRef] [PubMed]
34. Yoon, B.K.; Park, S.; Ma, G.J.; Kolahdouzan, K.; Zhdanov, V.P.; Jackman, J.A.; Cho, N.-J. Competing interactions of fatty acids and monoglycerides trigger synergistic phospholipid membrane remodeling. *J. Phys. Chem. Lett.* **2020**, *11*, 4951–4957. [CrossRef] [PubMed]
35. Tan, S.W.; Jeon, W.-Y.; Yoon, B.K.; Jackman, J.A. Mechanistic evaluation of antimicrobial lipid interactions with tethered lipid bilayers by electrochemical impedance spectroscopy. *Sensors* **2022**, *22*, 3712. [CrossRef]
36. Yang, L.; Wen, J.; Zhang, Y.; Liu, Z.; Luo, Z.; Xu, L.; Lai, S.; Tang, H.; Sun, X.; Hu, Y.; et al. The antiviral activity of caprylic monoglyceride against porcine reproductive and respiratory syndrome virus in vitro and in vivo. *Molecules* **2022**, *27*, 7263. [CrossRef]
37. Vigant, F.; Hollmann, A.; Lee, J.; Santos, N.C.; Jung, M.E.; Lee, B. The rigid amphipathic fusion inhibitor dUY11 acts through photosensitization of viruses. *J. Virol.* **2014**, *88*, 1849–1853. [CrossRef]
38. Yoon, B.K.; Jackman, J.A.; Park, S.; Mokrzecka, N.; Cho, N.-J. Characterizing the membrane-disruptive behavior of dodecylglycerol using supported lipid bilayers. *Langmuir* **2019**, *35*, 3568–3575. [CrossRef]
39. Farcet, J.-B.; Karbiener, M.; Zelger, L.; Kindermann, J.; Kreil, T.R. Detergent-mediated virus inactivation in biotechnological matrices: More than just CMC. *Int. J. Mol. Sci.* **2023**, *24*, 7920. [CrossRef]
40. Tan, J.Y.B.; Yoon, B.K.; Cho, N.-J.; Lovrić, J.; Jug, M.; Jackman, J.A. Lipid nanoparticle technology for delivering biologically active fatty acids and monoglycerides. *Int. J. Mol. Sci.* **2021**, *22*, 9664. [CrossRef]

pathogens MDPI

Brief Report

Evaluation of the Efficacy of Commercial Disinfectants against African Swine Fever Virus

Lorraine Frost [1,*], Matthew Tully [1], Linda Dixon [1], Hayley M. Hicks [1], Jessica Bennett [1], Isobel Stokes [1,2], Laura Marsella [1,3], Simon Gubbins [1] and Carrie Batten [1]

1 The Pirbright Institute, Pirbright, Woking GU24 0NF, UK; matt.tully@pirbright.ac.uk (M.T.); linda.dixon@pirbright.ac.uk (L.D.); hayley.hicks@pibright.ac.uk (H.M.H.); jessica.bennett@pirbright.ac.uk (J.B.); is00330@surrey.ac.uk (I.S.); laura.marsella@syngenta.com (L.M.); simon.gubbins@pirbright.ac.uk (S.G.); carrie.batten@pirbright.ac.uk (C.B.)
2 Department of Microbial Sciences, School of Biosciences, Faculty of Health and Medical Sciences, Stag Hill Campus, University of Surrey, Guildford GU2 7XH, UK
3 Jealott's Hill International Research Centre, Syngenta, Bracknell RG42 6EY, UK
* Correspondence: lorraine.frost@pirbright.ac.uk

Abstract: African swine fever (ASF) is an economically important disease due to high morbidity and mortality rates and the ability to affect all ages and breeds of pigs. Biosecurity measures to prevent the spread of the causative agent, African swine fever virus (ASFV), include prescriptive cleaning and disinfection procedures. The aim of this study was to establish the biocidal effects of twenty-four commercially available disinfectants including oxidizing agents, acids, aldehydes, formic acids, phenol, and mixed-class chemistries against ASFV. The products were prepared according to the manufacturer's instructions and a suspension assay was performed with ASFV strain, BA71V using Vero cells (African green monkey cells) to test efficacy in reducing ASFV infection of cells. Generally, disinfectants containing formic acid and phenolic compounds, as well as oxidizing agents reduced viral titers of ASFV by over 4 $\log_{10}$ at temperatures ranging from 4 °C to 20 °C. Hydrogen peroxide, aldehyde, and quaternary ammonium compounds containing disinfectants were cytotoxic, limiting the detection of viral infectivity reductions to less than 4 $\log_{10}$. These preliminary results can be used to target research on disinfectants which contain active ingredients with known efficacy against ASFV under conditions recommended for the country where their use will be applied.

Keywords: African swine fever virus; disinfectant; inactivation; control

Citation: Frost, L.; Tully, M.; Dixon, L.; Hicks, H.M.; Bennett, J.; Stokes, I.; Marsella, L.; Gubbins, S.; Batten, C. Evaluation of the Efficacy of Commercial Disinfectants against African Swine Fever Virus. *Pathogens* **2023**, *12*, 855. https://doi.org/10.3390/pathogens12070855

Academic Editors: Lauro Velazquez-Salinas, Gisselle N. Medina and Elizabeth Ramirez-Medina

Received: 17 May 2023
Revised: 15 June 2023
Accepted: 19 June 2023
Published: 21 June 2023

1. Introduction

African swine fever (ASF) is a hemorrhagic disease affecting domestic and wild pigs. The disease is caused by the African swine fever virus (ASFV); an enveloped, large DNA virus, the only known member of the family *Asfaviridae* [1]. Highly virulent strains can cause disease with morbidity and mortality rates near 100%. ASF is endemic in sub-Saharan Africa and Sardinia with the most recent incursion beginning in Georgia in 2007 which led to unprecedented spread [2]. Between 2007 and 2012, ASF spread through Armenia, Azerbaijan, and Russia before spreading into mainland Europe. In mid-2018, ASF entered China and spread throughout the region [3,4]. In July 2021, ASFV was reported in the Dominican Republic, signifying the first outbreak in the Americas in over 40 years [5]. The disease has now been reported in 45 countries including EU member states, parts of Asia, Oceania, and the Americas [6]. Since there is no suitable ASFV vaccine, control of ASF is supported through stamping out in affected holdings, cleaning, and disinfection and the application of biosecurity measures. Globally, millions of pigs destined for the pork market have been destroyed resulting in huge economic losses. China, the world's biggest producer of pork products, has estimated economic losses between 2018 and 2019 to be USD 111.2 billion [7]. Of these losses cleaning and disinfection represented a comparatively low-level expenditure of approximately USD 55 million and represents an area where a relatively small investment could lessen the impact of ASF [7]. Fomites contaminated with blood or excretions represent a major risk of secondary infections and, therefore, some countries have developed prescribed protocols for cleaning and disinfection [8,9]. Recently, two reviews on the control of ASFV using disinfectants were published, identifying the need for more data on the virucidal efficacy of certain chemical compounds [10,11]. In 2022, in response to the incursion of ASFV into the Americas, The National Institute of Food and Agriculture (NIFA) announced a USD 5 M Investment in agricultural biosecurity program to prevent, detect, and respond to potential spread of ASFV into the USA [12].

Countries have developed different rules for approving disinfectants for use during notifiable disease outbreaks. In the USA, this is set out by the Environmental Protection Agency (EPA) detailing testing methods described in ASTM E1053 (ASTM, 2020), published by the American Society for Testing and Materials (ASTM) [13]. In the UK, a disinfectant approved by the Department for Environment, Food and Rural Affairs (Defra) must be used in the event of a notifiable disease outbreak. There are four specific disease orders: The foot-and-mouth disease order, the diseases of poultry order, the tuberculosis in animals order and the swine disease order, which specifically covers swine vesicular disease virus (SVDV). All other notifiable diseases, which include other swine diseases such as ASF are covered under the "General order" (GO) category which uses *Salmonella enterica* serovar Enteritidis NCTC 13665 as the challenge organism [14,15]. Testing differs between the two countries: the USA requires a quantitative carrier test where the virus is dried on hard non-porous surfaces, whereas the UK requires a quantitative suspension test. Additionally, the USA requires a lower pass threshold of viral titer reduction ($\geq$3 log) compared to the UK ($\geq$4 log) [13,15].

The Pirbright Institute performs tests on behalf of Defra for disinfectants seeking approval under the diseases of swine order (test organism, SVDV) and the foot-and-mouth disease order (test organism foot-and-mouth disease virus (FMDV)) using a suspension test [15]. An ASFV disinfectant suspension test was developed and offered commercially in 2019, in response to requests from manufacturers for evidence of efficacy against ASFV. Twenty-four commercially available disinfectants were tested at The Pirbright Institute, this study discusses these test outcomes [15].

2. Materials and Methods

Disinfectants within this study were tested for ASF virucidal efficacy by suspension method [16]. Test conditions including dilution of disinfectant, test temperature, and contact time were stipulated by the manufacturer.

2.1. Disinfectant Test

Briefly, a cell culture-adapted, non-pathogenic ASFV strain, BA71V was grown in Vero cells (African green monkey cells, ECACC 84113001) [17]. Viral samples were prepared in cells cultured in Dulbecco's modified Eagle's medium (DMEM) containing 1% L-glutamine and 10% heat-inactivated fetal bovine serum (FBS). Disinfectants were stored according to the manufacturer's instructions and prepared directly before testing. Dilutions of each disinfectant were used as specified (Table 1).

Table 1. Virucidal efficacy of 24 disinfectants against African swine fever virus under various temperatures, contact times and dilutions.

Category	Disinfectant	Active Ingredient	Temperature (°C)	Contact Time (min)	Dilution	Log Reduction
Iodophor	A	IOD (1–3%), surfactant, acid	20	30	1:750 1:1500 1:2000 1:3000 1:4000	>4.0 >3.1 1.91 (CI 1.79–2.03) 0.45 (CI 0.33–0.57 0.17 (CI 0.03–0.30)
Peroxygen (except hydrogen peroxide)	B	MPS (50%), SDIC (5%), acid, surfactant	4	10	1:800	>3.3
	C	MPS (≤50%), SDIC (≤3%), acid, surfactant, inorganic buffer	4	30	1:799	>4.5
	D	MPS (<55%), acid, surfactant, inorganic buffer	10	ND	1:800	>4.4
	E	Sodium percarbonate (25–50%), acid, surfactant, sodium salt	10	10	1:1000 1:3000 1:5000 1:10,000 1:100,000	>3.9 >3.9 2.41 (CI 2.06–2.78) 1.97 (CI 1.64–2.31) 0.14 (CI −0.20–0.48)
	F	MPS (≤100%), surfactant, alcohol	20	5	1:800	>4.2
	G	MPS (30–60%), acid, surfactant	20	30	1:800	>4.2
	H	MPS (50–100%), citric acid (2.5–10%), surfactants	20	30	1:800	>3.8
Hydrogen peroxide	I	Hydrogen peroxide (49–49.9%)	10	ND	1:7.25	>2.2 *
Hydrogen peroxide and peracetic acid	J	Hydrogen peroxide (≤22%), peracetic acid (≤4%), acids, amine oxide	21	ND	1:50	>3.2 *
Aldehyde + QAC	K	GA (≤20%), DDAC (≤10%)	4	ND	1:399	>3.6 *
	L	GA (22%), DDAC (9%), ADBAC (14.5%)	10	10	1:800	>2.2 *
	M	GA (6.25%), ADBAC (5%), DDAC (7.55%), Pine oil (2%)	10	10	1:400	>3.3 *

Table 1. *Cont.*

Category	Disinfectant	Active Ingredient	Temperature (°C)	Contact Time (min)	Dilution	Log Reduction
	N	GA (10–15%), DDAC (3–5%), surfactant, acid, alcohol	10	30	1:800	>3.4 *
	O	GA (10.7%), DDAC (8%), ADBAC (17%), alcohol	20	30	1:1500	>3.1 *
	P	GA (10–25%), ADBAC (2.5–10%), formaldehyde (10–25%)	20	30	1:400	>3.3 *
QAC	Q	DDAB (10%)	4 20	1 1	1:800 1:800	>2.3 * >2.4 *
Aldehyde	R	GA (9.9%), formaldehyde (9.8%), surfactant	20	30	1:400	>3.3 *
	S	Formaldehyde (39%)	20	30	1:1785	0.95 (CI 0.84–1.05)
Formic Acid	T	Formic acid (>48%), carboxylic acid, surfactant	10	30	1:2000	>3.2
	U	Formic acid (60–70%), surfactant	20	30	1:800	>4.2
Phenol compound	V	Biphenyl-2-ol (1–3%), alcohol, acids, surfactants	10	10 30	1:400	2.94 (CI 2.73–3.15) 4.48 (CI 3.62—5.35)
	W	Mixed chlorocresols (20–40%), xylenol (1–10%), acid, alcohol, surfactant, solvent	10	10	1:400	>4.1
	X	Chlorocresol (≤25%), alcohol, acid, surfactant, solvent	20	30	1:2000	>4.2

* Due to cytotoxicity of the tested disinfectant the detection limit did not allow detection of higher virus reduction. Abbreviations: IOD = Iodine, CI = confidence interval, MPS = potassium bis (peroxymonosulphate_bis (sulphate), SDIC = sodium dichloroisocyanurate/troclosene sodium, ND = Not determined, QAC = quaternary ammonium compounds, GA = glutaraldehye, DDAC = didecyldimethylammonium chloride, ADBAC = Alkyldimethylbenzylammonium chloride, DDAB = didecyldimethylammonium bromide.

An initial 10× disinfectant starting concentration was diluted in 400 parts per million (ppm) calcium carbonate to simulate hard water conditions. A total of 100 µL of each 10× disinfectant was diluted in 800 µL hard water and 100 µL BA71V stock for the time and temperature specified (Table 1). A cytotoxicity control was evaluated for each concentration of disinfectant by diluting the 100 µL of 10× disinfectant in 800 µL hard water but using 100 µL of cell culture media instead of BA71V, a neutralization control was prepared in the same way. Cytotoxicity was assessed on the integrity and appearance of the cell sheet. A neutralizer, described below, was used to arrest the virucidal activity of the disinfectant at the end of the contact time. Disinfectant activity was neutralized by serial dilution (10-fold) with either phosphate-buffered saline (PBS) containing 1% penicillin/streptomycin, 1% L-glutamine, 1% FBS, and 0.01% phenol red (used for disinfectants A, B, H, J–S and U–X) or 0.05 M carbonate bicarbonate buffer (Merck, Rahway, NJ, USA) containing 1% FBS (used for disinfectants C–G, I and T). Suppression of disinfectant activity was evaluated by adding 100 μL neutralization control to 800 μL neutralizer and then adding 100 μL positive control. Negative controls without virus or disinfectant, positive controls with virus and without disinfectant, and reference controls using a standard 0.56% formaldehyde were included in each experiment. All controls were subject to neutralization following incubation times.

2.2. Plaque Assay

Following treatment, a plaque assay in Vero cells was used to detect ASFV. Using confluent cell monolayers in 6-well plates, 200 μL of each serially diluted product was added in triplicate to different wells. The cells were incubated in 5% CO_2 at 37 °C for 1 h and were overlaid with 2 mL of 1.375% Eagle's overlay supplemented with 4% FBS, 1% Avicel solution, 0.1% penicillin/streptomycin and 0.1% L-glutamine. After incubation for 6 days, the virus titer was determined as plaque-forming units per ml (pfu mL^{-1}) following staining with crystal violet solution.

2.3. Result Determination

Titer reductions (TR) were calculated:

$$TR = a - b,$$

where a = Titer of virus positive control and b = Titer of virus after exposure to disinfectant.

For disinfectants A, E, and V, the reduction in titer was estimated using a generalized linear model with Poisson errors and a log link function. The response variable was the number of plaques and the explanatory variable was disinfectant concentration. Model assumptions were checked by examining model residuals and calculating the variance to mean ratio, none of which suggested a substantial deviation from a Poisson distribution. Negative binomial and quasi-Poisson generalized linear models (GLMs), which relax the assumption about the relationship between mean and variance implicit in using a Poisson model, were also considered. The estimated reductions in titer were similar for all models, giving confidence that the results presented are robust.

Where no plaques were identified, titer reduction is provided as above the titer obtained in the positive control excluding the limit of detection (1.3 $\log_{10}$) and cytotoxicity.

Results were calculated using the positive control titer specific to each disinfectant test. The average titer for all positive controls used within the study was 5.24 $\log_{10}$ pfu mL^{-1} with a standard deviation of 0.50. A contact time is provided if the difference in results for the positive control and the neutralization control are $\leq$0.5 $\log_{10}$.

3. Results and Discussion

Disinfectants representing oxidizing agents, acids, aldehydes and mixed-class chemistries were tested. The most prevalent were oxidizing agents (peroxygens, iodo-phors), and mixed-class chemistries (aldehyde plus QACs, hydrogen peroxide plus peracetic acid).

One iodophor-based disinfectant was tested (A). This disinfectant reduced the titer of ASFV by over 4 $\log_{10}$ at a dilution of 1:750 within 30 min at 20 °C. A dose–response was observed, with efficacy falling above dilution 1:1500. Iodophors are generally non-toxic and their activity is not affected by hard water; however, they can stain some surfaces and can be expensive. A review of iodophor-based disinfectants recorded as approved for both SVDV and GO on the Defra-approved list indicates similar approved working dilutions for all listed iodophor disinfectants (1:100–1:150 SVDV and approx. 1:50 GO) [18]. Our results concur with previous results which showed that iodophor disinfectants have similar or greater efficacy on ASFV than SVDV [11]. There are currently no iodophor-based disinfectants listed on the "Disinfectants approved for use against African swine fever virus in farm settings" list held by the US EPA [19].

The majority of the peroxygen disinfectants submitted listed potassium peroxymonosulphate (MPS) as the main active component. Peroxygen disinfectants, except those based on hydrogen peroxide and peracetic acid, produced similar results at a dilution of 1:800

independent of temperature (4 °C, 10 °C or 20 °C). Disinfectant F inactivated over 4 log_{10} ASFV within 5 min. Disinfectant E, containing sodium percarbonate (25–50%), was tested at multiple dilutions and the results gave a dose–response. Dilutions of less than 1:3000 inactivated over 3.9 log_{10} of ASFV within 10 min at 10 °C.

There are many MPS-based disinfectants on the Defra-approved list. These disinfectants share similar approved dilution rates for FMDV (1:1200–1:1300) and also for GO (1:49–1:100). Wales and Davies determined that for one MPS containing disinfectant. ASFV is moderately less susceptible than FMDV [11]. They concluded that as the GO-approved dilutions were more concentrated than the FMDV-approved dilutions, using a MPS disinfectant at the GO dilution, would likely be effective [11]. Our data, which tested an MPS at 1:800 would at least concur with this assumption. There is currently one MPS-containing disinfectant on the US EPA-approved list and several others that list oxidizing agents as their main active ingredient.

The US EPA approved list includes one hydrogen peroxide product at a working dilution of 1:64; however, Gabbert et al., tested one hydrogen peroxide product with results indicating low efficacy, reducing ASF viral titer by less than 2 log_{10}. In this study, disinfectants containing hydrogen peroxide could not be neutralized at the dilutions requested by the manufacturers and additionally were found to be cytotoxic. Therefore, additional tests would be required to form confident conclusions for contact time or effectiveness against ASFV for hydrogen peroxide-based disinfectants.

Within the aldehyde plus QAC category, two disinfectants had a basic glutaraldehyde and QAC composition whilst four incorporated additional constituents (Table 1), all were cytotoxic. Disinfectant K could not be neutralized at a dilution of 1:399. Except for disinfectant L, all products reduced the ASFV titer by over 3 log_{10} within 30 min at 10 or 20 °C. No conclusions on effective dilution could be drawn as cytotoxicity affected the maximum reportable titer reductions. Virocid, listed on the US EPA approved list contains active ingredients of aldehyde, QAC, and alcohol, most similar to product O and is approved at a concentration of 1:200 also demonstrating a greater than 3 log_{10} reduction.

One disinfectant contained only QAC, and two disinfectants listed aldehydes as active ingredients. Disinfectants Q and R were cytotoxic at 1:800 and 1:400 dilutions, respectively. Product S, containing only formaldehyde and tested at 1:1785, reduced the ASFV titer by 0.9 log_{10}, lower dilutions were too cytotoxic to obtain a result.

Two disinfectants contained formic acid. Disinfectant T, tested at 10 °C and a dilution of 1:2000, reduced the ASFV titer by over 3.2 log_{10}. Disinfectant U, tested at 20 °C and a dilution of 1:800 reduced the ASFV titer by over 4.2 log_{10}. Lower dilution rates tested for both disinfectants were unable to be neutralized. No further data could be found for formic acid-based disinfectants and, therefore, no assumptions regarding maximum effective dilution could be made.

Finally, two phenol-based disinfectants reduced the ASFV titer by over 4 log_{10} at a dilution of 1:400 (V and W) and one reduced the titer by over 4.2 at a dilution of 1:2000 (X). Phenol-based disinfectants have a wide spectrum effect, are stable during storage, and maintain efficacy in the presence of organic matter but are toxic and corrosive.

In conclusion, decontamination including the cleaning and disinfection of premises is an important measure to halt the spread of ASFV and allow for the repopulation of previously infected premises. Disinfectants should have a wide range of use, be affordable, easy to store and prepare, and be non-toxic. Data on a range of chemistries is, therefore, required to enable an informed choice that is fit for the intended use. Knowledge gaps highlighted by Beato and co-authors listed limited data for aldehydes, with mention of glutaraldehyde, phenol and iodine compounds, and alcohols [10]. Our data suggest that iodophor and phenol-containing disinfectants are effective against ASFV. We showed that glutaraldehyde disinfectants are effective, but higher dilutions could be tested to confirm a higher than 4 log_{10} reduction in viral titer. Information in this study is preliminary and could be used by other markets to target research on disinfectants which contain active ingredients with known efficacy. This targeted research should utilize test procedures appropriate to the individual countries rules and broaden the disinfectant chemistries listed on country-specific approved lists. The study data was based on disinfectants where the manufacturers are actively seeking approval to market from their competent authority. Further studies on the disinfectant effectiveness against ASFV on different surfaces, under different conditions and potentially against different circulating ASFV strains, would benefit from being investigated.

Author Contributions: Conceptualization, C.B. and L.F.; methodology, L.F.; validation, L.F. and I.S.; formal analysis, S.G.; investigation, L.F., M.T., H.M.H., J.B., I.S. and L.M.; resources, C.B.; data curation, L.F.; writing—original draft preparation, L.F.; writing—review and editing, L.F., C.B. and S.G.; visualization, L.F.; supervision, C.B.; project administration, L.F.; funding acquisition, C.B. and L.D. All authors have read and agreed to the published version of the manuscript.

Funding: The authors would like to acknowledge support from BBSRC UKRI (BBS/E/I/00007036 and BBS/E/I/00007037) and Defra (SE1517).

Institutional Review Board Statement: Not applicable.

Informed Consent Statement: Not applicable.

Data Availability Statement: The data presented in this study are available from the corresponding author. The data are not publicly available due to confidentiality agreements between the manufacturers submitting disinfectants and the testing laboratory.

Acknowledgments: The authors also thank Ana Reis and Christopher Netherton for continued technical assistance and the support from the cell culture team at The Pirbright Institute.

Conflicts of Interest: The authors declare no conflict of interest.

References

1. Alonso, C.; Borca, M.; Dixon, L.; Revilla, Y.; Rodriguez, F.; Escribano, J.M. ICTV Virus Taxonomy Profile: Asfarviridae. *J. Gen. Virol.* **2018**, *99*, 613–614. [CrossRef] [PubMed]
2. Rowlands, R.J.; Michaud, V.; Heath, L.; Hutchings, G.; Oura, C.; Vosloo, W.; Dwarka, R.; Onashvili, T.; Albina, E.; Dixon, L.K. African Swine Fever Virus Isolate, Georgia, 2007. *Emerg. Infect. Dis.* **2008**, *14*, 1870–1874. [CrossRef] [PubMed]
3. Ma, J.; Chen, H.; Gao, X.; Xiao, J.; Wang, H. African Swine Fever Emerging in China: Distribution Characteristics and High-Risk Areas. *Prev. Vet. Med.* **2020**, *175*, 104861. [CrossRef] [PubMed]
4. Zhou, X.; Li, N.; Luo, Y.; Liu, Y.; Miao, F.; Chen, T.; Zhang, S.; Cao, P.; Li, X.; Tian, K.; et al. Emergence of African Swine Fever in China, 2018. *Transbound. Emerg. Dis.* **2018**, *65*, 1482–1484. [CrossRef] [PubMed]
5. Gonzales, W.; Moreno, C.; Duran, U.; Henao, N.; Bencosme, M.; Lora, P.; Reyes, R.; Núñez, R.; De Gracia, A.; Perez, A.M. African Swine Fever in the Dominican Republic. *Transbound. Emerg. Dis.* **2021**, *68*, 3018–3019. [CrossRef] [PubMed]
6. African Swine Fever (ASF)-Situation Report 20. Available online: https://www.woah.org/app/uploads/2022/09/asf-report20.pdf (accessed on 20 October 2022).
7. You, S.; Liu, T.; Zhang, M.; Zhao, X.; Dong, Y.; Wu, B.; Wang, Y.; Li, J.; Wei, X.; Shi, B. African Swine Fever Outbreaks in China Led to Gross Domestic Product and Economic Losses. *Nat. Food* **2021**, *2*, 802–808. [CrossRef] [PubMed]
8. FAD PReP/USDA Standard Operating Procedures: 15. Cleaning and Disinfection. Available online: https://www.aphis.usda.gov/animal_health/emergency_management/downloads/sop/sop_cd.pdf (accessed on 20 October 2022).
9. The European Commission: Commission Delegated Regulation (EU) 2020/688. Available online: https://eur-lex.europa.eu/legal-content/EN/TXT/?uri=CELEX:32020R0688 (accessed on 20 October 2022).
10. Beato, M.S.; D'Errico, F.; Iscaro, C.; Petrini, S.; Giammarioli, M.; Feliziani, F. Disinfectants against African Swine Fever: An Updated Review. *Viruses* **2022**, *14*, 1384. [CrossRef] [PubMed]
11. Wales, A.D.; Davies, R.H. Disinfection to Control African Swine Fever Virus: A UK Perspective. *J. Med. Microbiol.* **2021**, *70*, 001410. [CrossRef] [PubMed]
12. NIFA Invests Nearly $5M in Agricultural Biosecurity Program | National Institute of Food and Agriculture. Available online: https://www.nifa.usda.gov/about-nifa/announcements/nifa-invests-nearly-5m-agricultural-biosecurity-program (accessed on 20 April 2023).
13. Standard Practice to Assess Virucidal Activity of Chemicals Intended for Disinfection of Inanimate, Nonporous Environmental Surfaces. Available online: https://www.astm.org/e1053-20.html (accessed on 26 April 2023).
14. Get Your Disinfectant Approved by Defra–GOV.UK. Available online: https://www.gov.uk/guidance/get-your-disinfectant-approved-by-defra (accessed on 26 April 2023).

15. APHA. The Pirbright Institute Disinfectant Efficacy Tests. Available online: https://assets.publishing.service.gov.uk/government/uploads/system/uploads/attachment_data/file/1041676/disinfectants-efficacy-tests.pdf (accessed on 20 October 2022).
16. Frost, L.; Batten, C. African Swine Fever Virus Plaque Assay and Disinfectant Testing. *Methods Mol. Biol.* **2022**, *2503*, 187–194. [CrossRef] [PubMed]
17. Yáñez, R.J.; Rodríguez, J.M.; Nogal, M.L.; Yuste, L.; Enríquez, C.; Rodriguez, J.F.; Viñuela, E. Analysis of the Complete Nucleotide Sequence of African Swine Fever Virus. *Virology* **1995**, *208*, 249–278. [CrossRef] [PubMed]
18. Defra Disinfectants Approved for Use in England, Scotland and Wales. Department of Environment, Food and Rural Affairs. Available online: http://disinfectants.defra.gov.uk/DisinfectantsExternal/Default.aspx?Module=ApprovalsList_SI (accessed on 20 October 2022).
19. Disinfectants Approved for Use against African Swine Fever Virus in Farm Settings. Available online: http://ppis.ceris.purdue.edu/.ErrorsinthePPISdatacanbereportedatthiswebsiteusingtheDataQualitymenuoption (accessed on 20 April 2023).

pathogens

MDPI

Article

Epidemiological Assessment of African Swine Fever Spread in the Dominican Republic

Rachel A. Schambow [1,*], Syed Hussain [1], Maria C. Antognoli [2], Silvia Kreindel [2], Raysa Reyes [3] and Andres M. Perez [1]

1 Center for Animal Health and Food Safety, College of Veterinary Medicine, University of Minnesota, St. Paul, MN 55108, USA; hussa056@umn.edu (S.H.); aperez@umn.edu (A.M.P.)
2 International Services Action Programs, Animal Plant Health Inspection Service, United States Department of Agriculture, Fort Collins, CO 80526, USA; celia.antognoli@usda.gov (M.C.A.); silvia.kreindel@usda.gov (S.K.)
3 Instituto del Estudio de las Enfermedades Zoonóticas, Universidad Autónoma de Santo Domingo, Santo Domingo 10904, Dominican Republic; rreyes03@uasd.edu.do
* Correspondence: scham083@umn.edu

Abstract: Since African Swine Fever (ASF) was detected in the Dominican Republic in July 2021, it has negatively impacted the country's swine industry. Assessing the epidemiological situation is crucial to helping local authorities and industry stakeholders control the disease. Here, data on 155 reported outbreaks in the Dominican Republic from November 2022 to June 2023 were evaluated. Descriptive spatiotemporal analysis was performed to characterize disease distribution and spread, and between-herd R_0 was calculated for the study period. The Knox test and a space–time permutation model were used to evaluate clustering. Data on clinical presentation, biosecurity measures, and suspected reasons for introduction were categorized and summarized. The majority (78%) of outbreaks occurred on backyard farms which generally had low biosecurity. Across farm types, the majority of pigs were still alive at the time of depopulation. Spatiotemporal findings and R_0 estimates suggest an endemic pattern of disease geographically located centrally within the country. Clustering was detected even at small temporal and spatial distances due to outbreaks amongst neighboring backyard farms. These results provide critical information on the current state of the ASF epidemic in the Dominican Republic and will aid government officials and swine industry leaders in developing effective ASF control strategies.

Keywords: African Swine Fever; Dominican Republic; epidemiology; spatial; temporal; quantitative; biosecurity; pig; swine

Citation: Schambow, R.A.; Hussain, S.; Antognoli, M.C.; Kreindel, S.; Reyes, R.; Perez, A.M. Epidemiological Assessment of African Swine Fever Spread in the Dominican Republic. *Pathogens* **2023**, *12*, 1414. https://doi.org/10.3390/pathogens12121414

Academic Editor: Makoto Ozawa

Received: 2 November 2023
Revised: 16 November 2023
Accepted: 30 November 2023
Published: 1 December 2023

1. Introduction

Since its reintroduction into the Dominican Republic in 2021, African Swine Fever (ASF) has caused considerable disruption to the country's swine industry. ASF is caused by the ASF virus (ASFV), a double-stranded DNA arbovirus of the family Asfiviridae that produces hemorrhagic fever in infected pigs [1,2]. ASFV only infects members of the Suidae family, including domestic pigs and wild boar, and does not affect humans nor is it a food safety threat. ASF is a reportable disease to the World Organization for Animal Health [3]. Since 1995, it was only known to be present in Africa and the island of Sardinia off the coast of Italy [3–5]. However, since 2007, ASF genotype 2 has been spreading throughout Europe, Asia, Southeast Asia, and in 2021, it has spread to the Dominican Republic and Haiti, which make up the island of Hispaniola [1,6–8]. As of the latest WOAH ASF Situation Report (published 14 August 2023), 50 countries have reported ASF occurrences since 2021, and 1.5 million domestic pigs have been lost to the disease or culling [9]. In many European countries, only or mainly wild boar have been affected by ASF [10]. Because outbreaks are not reported in many parts of the world, including Africa, the true number of ASF-caused deaths in pigs is likely underrepresented [9,10]. ASF is mainly managed by preventing introduction to farms and countries through high biosecurity, and infected farms are typically depopulated to prevent further disease spread [6].

ASF was first detected in the Dominican Republic in 1978 in the midst of a wave of global ASF spread [5,11]. In response, the Dominican Republic and Haitian officials, with military support, depopulated the island's entire pig population to eradicate the disease. Hispaniola was free of ASF until it was detected again in July 2021 in the Dominican Republic and shortly after in Haiti in September 2021 [7,8]. Reports suggest that the initial introduction to the Dominican Republic may have occurred as early as April 2021 [12]. Currently, the Dominican Republic government employs a strategy of surveillance, depopulation, and indemnification of culled animals to support ASF control and eradication [13,14]. From 30 June 2021 to 30 September 2023, approximately $15.7 million USD was spent by Banco Agricola and $13 million by the United States Department of Agriculture Animal and Plant Health Inspection Service (USDA APHIS) on indemnification to affected producers in an effort to incentivize passive reporting of cases [15].

The swine industry is highly important to the Dominican Republic, being the second highest protein consumed following poultry and generating significant economic activity [13]. According to the 2022 census of the Ministry of Agriculture, the Dominican Republic has over 1300 moderate-to-high biosecurity commercial farms (Técnificadas and semi-téchnificadas), over 1600 small commercial farms (No técnificada), and approximately 13,000 backyard farms (transpatio) [16]. These backyard farms rely on pork production for economic support as living "piggy banks" and as a food source. Combined, the industry has an estimated inventory of over 500,000 head including all production stages as of December 2022 [16]. The Dominican Agribusiness Council estimated that the Dominican Republic swine industry has approximately $500 million USD in investments and generates around 58,000 jobs [13]. In 2022, local production was estimated to have decreased by 21% compared to 2021 because of the impact from ongoing ASF outbreaks [13].

Epidemiological analyses, including spatiotemporal techniques, can be highly useful in understanding ASF occurrences in the Dominican Republic and supporting the development of tailored prevention and control strategies. Spatial techniques, such as mapping and cluster detection, can help in identifying high-risk regions where ASF may be more prevalent, while temporal analysis allows for the identification of important trends in ASF prevalence [17,18]. Combined with descriptive analysis of the affected farms and their biosecurity measures, the potential reasons and risk factors for ASF introduction on farms can be better understood, ultimately aiding government officials and the swine industry in ASF control.

The objective of this work was to use descriptive epidemiology and spatiotemporal analysis to evaluate the ASF epidemic in the Dominican Republic, using data collected from November 2022 to June 2023. The study also aimed to identify potential biosecurity measures and reasons that may have contributed to ASF introduction on these farms. This work provides useful information on disease trends and producer behaviors for government officials and private sector stakeholders, ultimately to provide a better understanding of the current outbreak and to support ASF control efforts in the Dominican Republic. Additionally, these results will aid other ASF-affected and unaffected areas by providing important information for ASF preparedness and control.

2. Materials and Methods

2.1. Dataset

Data on farm outbreaks were made available from November 2022 to June 2023 as a part of ongoing government-operated outbreak verification, depopulation, and indemnity efforts. Outbreaks were identified through passive surveillance. According to WOAH, an outbreak is characterized as the identification of one or more cases within an epidemiological unit [19]. In the Dominican Republic, any premise containing pigs is defined as an epidemiological unit, i.e., any commercial or backyard farm, and an outbreak is defined as at least one ASF case identified on a farm.

Veterinary officers from the Dominican Republic Ministry of Agriculture collected the data analyzed here using a questionnaire (Supplementary File S1) as a part of the routine verification process of reported outbreaks. The data were not collected for the present study, and the questionnaire was designed by the veterinary authority for their use. At the time of their in-person visit to affected farms, veterinary officers collected information from affected producers and their site including date of the outbreak, location (latitude/longitude and administrative province, municipality, and section), farm type as defined below, clinical signs by production group (piglets, nursery, boar, sow, fattening), biosecurity measures related to farm and barn access, water source (recorded as aqueduct, river, stream, well, dam, or other), personnel and truck biosecurity, pig breeding and management, and an open-ended question asking the possible origin of the contagion according to producer. Farm types were consistent with those from the 2022 census [16], and were defined by the veterinary authority and are translated to English below (original versions are available in Supplementary File S1). Each site's farm type classification was performed by the veterinary authority.

Familiar Traspatio or backyard farms: Farms without biosecurity and whose production is for family consumption and/or is marketed and consumed limited within the location. Maximum of 25 pigs including up to two sows.

Comercial No Técnificadas (commercial, non-technical, CNT): Commercial farms with little to no degree of technology and with little to no biosecurity, with more than 25 pigs or more than two sows.

Semi-técnificadas (semi-technical, ST): Commercial farms with a moderate degree of technology and a moderate level of biosecurity.

Técnificadas (technical): Commercial farms with a high degree of technology and a high level of biosecurity.

In total, data containing information on all 155 outbreaks reported and verified by veterinary officials were made available by the Dominican Republic Incident Command for ASF for analysis. Received data were cleaned and prepared for analysis by organizing into a generic database, removing duplicate observations, verifying latitude/longitude coordinates using ArcGIS Pro (version 3.1.3, ESRI Inc., Redlands, CA, USA) and province/municipality boundaries, and by translating Spanish comments into English with support from bilingual native Spanish speakers.

2.2. Descriptive Analysis

Descriptive analysis was performed in Microsoft Excel, version 2016, and R version 4.3.1. The mean, median, standard deviation, and range were calculated for pig inventory by farm type. An overall assessment of biosecurity level was not performed, but the presence/absence of individual measures as recorded by veterinary officials was summed for each farm type. Water sourced from aqueducts, wells, or other domestic water sources were categorized as "drinking water", and water sourced from rivers and streams were categorized as "non-drinking water" (no farms were reported using dams). Open-ended responses (clinical signs and reason for introduction suspected by the producer) were analyzed using qualitative coding techniques [20]. Specifically, each text response was imported into Microsoft Excel and read, then an appropriate abbreviation of the comment, or a code, was written next to each response. A native Spanish speaker reviewed translations and codes. Codes were then reviewed and overlapping codes were combined into themes representing those ideas (e.g., "birds", "cows", "dogs", and "horses" were combined into "non-pig animals"; or "movement of buyers" and "movement of workers" were combined into "People"). Codes were summed for the number of times they appeared by production type (clinical signs) and by farm (suspected reason). Clinical signs were reported by 103 farms (66%; missing information from 52 farms) and categorized into 14 types of signs. Suspected reasons were reported by 122 farms (79%; missing information from 33 farms) and condensed from an initial list of 22 codes down to eleven themes.

2.3. Descriptive Spatiotemporal Analysis

Two outbreaks (1% of all data) had missing date information and were omitted from temporal analyses (n = 153 with complete spatiotemporal information). An epidemic curve was produced by plotting the number of cases by week for the duration of the study period. For each week starting with week three, a three-week moving average was estimated by summing the current and preceding two weeks' case count and dividing by three to produce a rolling average. Seasonal analysis was not performed due to having an insufficient time period of data (approximately eight months).

The locations of all outbreaks (n = 155) were plotted using ArcGIS Pro and summed by province to produce a choropleth map depicting the number of cases by province. The mean center of all outbreaks was also calculated. Map breaks were selected using the natural breaks (Jenks) setting with an additional category representing provinces where no outbreaks were reported. The exact locations of farms were not provided here to protect their confidentiality. Additional choropleth maps were produced using the same settings for approximately every three months of the period, as November–December 2022, January–March 2023, and April to June 2023. Two outbreaks (1 in Barahona and 1 in Santo Domingo provinces) did not have dates and thus could not be included in these approximately tri-monthly choropleth maps.

2.4. Cluster Detection

Global autocorrelation for all outbreaks was assessed in ArcGIS Pro using Moran's I, a measure of overall clustering of spatial data [21,22]. The null hypothesis of the Moran's I statistic states that the feature being analyzed is randomly distributed throughout the spatial study area, and values of Moran's I statistic may fall between −1.0 (similar features are dispersed) and +1.0 (similar features are clustered). Moran's I was estimated using the inverse distance setting (so that nearby neighboring features have a larger influence compared to those farther away) using a default threshold (the Euclidean distance, ensuring every feature has at least one neighbor) and Euclidean distance (distance estimated as a straight line between points).

ClusterSeer (v. 2.5, BioMedware, Ann Arbor, MI, USA) was used to evaluate the interactions between space and time and outbreak clustering using the Knox test (n = 153 due to 2 outbreaks with missing date information). The Knox test is a comparison of the proportion of cases in the study population that occurred in close proximity in both space and time to the proportion expected if cases were equally and homogenously distributed throughout the study space and time period [23]. Values of 1, 3, 5, 10, 15, 20, and 30 km and 1, 3, 5, 7, 9, 14, 21, and 28 days were used as critical values for spatial and temporal closeness, respectively. All combinations were analyzed using the Knox test to obtain their associated p-value (estimated via the generation of 999 Monte Carlo simulations) and observed-to-expected (O/E) ratio.

2.5. Space–Time Permutation Model

A space–time permutation model was performed in SaTScan (v10.0.2, Kulldorff, M. and Information Management Services, Boston and Calverton, MD, USA) to identify significant space–time clusters of cases. This model is appropriate to use where only case data are available and control or population data are unavailable, as was true for the present study [24]. Generally, this technique involves computing a cylindrical window where the circular base corresponds to the spatial area and the height corresponds to the time period. Within each cylinder, the O/E ratio is computed for the proportion of expected number of cases, assuming a homogenous distribution across space and time, compared to the observed number of cases. Cylinders are moved across the study period, centered on each outbreak location, and the O/E ratio is computed. Temporal and spatial window sizes for the analysis were selected based on visual inspection of O/E ratio plots produced from the Knox test. A temporal window of seven days (also chosen to account for "Monday effects") and spatial window of ten km were used for the space–time permutation model.

2.6. Estimation of Farm-Level Reproduction Ratio (R_0)

The reproduction ratio (R_0), or the average number of secondary infections caused by one infected unit throughout the duration of the infectiousness period, was estimated at the country and farm level as the average number of secondary infected farms caused by one infected farm throughout the Dominican Republic [25]. $R_0 > 1$ indicates that each infection causes more than one new infection and the outbreak will spread in the population. $R_0 < 1$ indicates that each infection causes less than one new infection, and the outbreak will decline. $R_0 = 1$ indicates an infectious individual causes only one new case, and the outbreak size does not increase or decrease which can lead to endemicity [26]. The entire Dominican Republic was considered the study area, and an infection was considered as an outbreak farm. R_0 was estimated using all outbreaks with complete dates (n = 153) following the method previously described for use in the absence of population data [27,28]:

$$R_0 = 1 + D \times \ln \frac{\frac{C_t}{C_0}}{t} \quad (1)$$

where D = the duration of the infectious period of the farm (days), C_t = the number of cases detected at time t, C_0 = the number of cases detected at the start of the study period. By considering t as t_d = the time to double the number of cases, then $C_t = 2$. Then, Equation (1) is simplified and computed as:

$$R_0 = 1 + \frac{D}{t_d} \ln 2 \quad (2)$$

R_0 was estimated for each doubling interval during the study period. The duration of ASF infection at individual pig level has been estimated to be highly variable depending on virus strain, isolate, and experimental procedures used. Estimates of the minimum and maximum duration of the infectious period (D) for an individual pig range from two to 40 days [29–31]. Considering that the infectious period on the farm level may differ greatly from the individual pig level and have high variability due to various factors (size of farm, time of detection and depopulation, etc.), R_0 was estimated across a range of values of D for an individual farm from one to 40 days. A locally estimated scatterplot smoothing (LOESS) line was fit in R using the package ggplot2 to each series to visualize the trends.

3. Results

3.1. Descriptive Factors

The dataset contained 155 outbreaks, of which 121 were from farms classified as backyard farms (78%), 30 were from CNT farms (19%), four were ST farms (3%) and none were technical farms. The total number of pigs reported (alive or dead) was 18,910. Despite being the largest proportion of farms, only 4% (857 pigs) of all pig inventory was in backyard farms; conversely, the four ST farms contained 49% (9234) of total pigs (Table 1). The remaining 47% of pigs (8819) were on CNT farms. The median total number of pigs for backyard, CNT, and ST farms was five, 72, and 1974, respectively. Considering all pigs, 97% (18,368) were reported as still alive at the farm visit. By farm type, 74% of backyard pigs, 96% of CNT pigs, and 100% of ST were reported alive.

Clinical signs and/or necropsy findings were reported for 103 unique farms (66%) for one or more production groups present on the farm (Table 2). This information was not recorded for 52 farms in the dataset. Of those farms where clinical presentation was recorded, dead pigs, anorexia/off-feed, and fever were the most reported signs. This order was consistent for piglets and sows, while for nursery, boars, and fattening groups, they were ordered, in highest to lowest frequency, as dead, fever, and anorexia. Diarrhea, lethargy, red skin, and vomiting were also frequently reported across farm types. On eight unique farms, no symptoms were reported in some groups.

Table 1. Inventory of all, live, and dead pigs on backyard (n = 121), commercial non-technical (CNT, n = 30), and ST (semi-technical, n = 4) farms. SD = standard deviation of the mean.

Inventory Type	Farm Type	Total Number (% of All Pigs)	Mean	SD	Median	Minimum	Maximum
All Pigs	Backyard	857 (4)	7	6.2	5	1	24
	CNT	8819 (47)	294	564	72	28	2855
	ST	9234 (49)	2309	1439	1974	1015	4271
Live Pigs	Backyard	631 (3)	5	5.1	3	0	24
	CNT	8503 (45)	283	552	72	12	2811
	ST	9234 (49)	2309	1439	1974	1015	4271
Dead Pigs	Backyard	226 (1)	2	3.7	0	0	18
	CNT	316 (1)	11	22	0	0	96
	ST	0 (0)	0	0	0	0	0

Table 2. Clinical signs and necropsy findings reported by farms (n = 103 unique farms) by production stage. "No symptoms" refers to farms where "asymptomatic" or "no signs" was reported and does not represent missing information.

Clinical Sign or Necropsy Finding	Production Type (Number of Farms) * Number Reporting Sign (% By Production Stage) **					
	Piglet (63)	Nursery (35)	Boars (39)	Sows (60)	Fattening (40)	Total
Dead	26 (41)	22 (63)	19 (49)	26 (43)	11 (28)	104 (44)
Anorexia	25 (40)	11 (31)	10 (26)	24 (40)	20 (50)	90 (38)
Fever	12 (19)	14 (40)	11 (28)	20 (33)	16 (40)	70 (30)
Diarrhea	12 (19)	6 (17)	7 (18)	11 (18)	7 (18)	43 (18)
Lethargy	9 (14)	7 (20)	4 (10)	13 (22)	10 (25)	39 (17)
Red skin	5 (8)	5 (14)	5 (13)	4 (7)	4 (10)	30 (13)
Epistaxis	4 (6)	3 (9)	5 (13)	8 (13)	3 (8)	23 (10)
No symptoms	3 (5)	0 (0)	1 (3)	4 (7)	0 (0)	8 (3)
Sudden death (specifically)	4 (6)	0 (0)	2 (5)	1 (2)	0 (0)	7 (3)
Vomiting	1 (2)	0 (0)	2 (5)	3 (5)	1 (3)	7 (3)
Hemorrhage (unspecified)	1 (2)	1 (3)	0 (0)	1 (2)	0 (0)	5 (2)
Nervous signs	1 (2)	0 (0)	0 (0)	1 (2)	2 (5)	4 (2)
Splenomegaly	0 (0)	0 (0)	0 (0)	2 (3)	1 (3)	3 (1)
Respiratory	0 (0)	0 (0)	0 (0)	0 (0)	1 (3)	1 (0.4)

* Number of farms sums greater than 103 due to some farms reporting multiple production types. ** Percentage by farm type of findings reported as percentage of production type reporting the presence of that finding. This does not sum to 1 due to most farms reporting multiple findings.

The biosecurity measures present varied by farm type (Table 3). No backyard farms reported the presence of measures related to farm access, barn access, staff and personnel protocols, pig breeding, and general farm management, and a minority had closed farm entrances, an unloading ramp for pigs, a perimeter fence in good condition, dedicated carts and work items for each barn, a cement floor with disinfection, good pen drainage, a dedicated feed truck, and used one needle per pig for vaccines and injections. No CNT farms reported having a clean/dirty line to the farm, a visitor log, a vehicle disinfection arch, disinfection of goods and equipment onto the farm, training for staff, or staff having their own pigs. Less than half reported having one or more of 22 different measures related to farm and barn access, staff and personnel protocols, and general farm management. For breeding, about half of backyard farms reported using a live boar, but only 16 reported producing having an exclusive boar for their own use. A higher proportion of CNT farms were reported as using live boars for breeding than backyard farms, but most used exclusive boars. All ST farms reported having an unloading ramp with a clean unloading/loading area, cement barn floors with disinfection, adequate pen drainage, feeders in good condition, using live boars but all with exclusive use or producing their own semen, burying mortality on-farm, and using one needle per pig for injections. The source of water used for pigs was not recorded on any ST farms and most CNT farms, but for backyard farms where it was recorded, the majority used drinking water sources (aqueduct, well, or house water) compared to non-drinking water sources (river or stream).

Table 3. Biosecurity measures present and absent by farm type, CNT = commercial non-technical, ST = semi-technical.

Category	Biosecurity Factor	Number (%)		
		Backyard	CNT	ST
Farm Access	Clean/dirty line with change of clothes and boot covers	0 (0)	0 (0)	2 (50)
Farm Access	Has a shower	0 (0)	5 (17)	3 (75)
Farm Access	Has closed entrance doors	2 (2)	10 (33)	3 (75)
Farm Access	Has visitor log	0 (0)	0 (0)	1 (25)
Farm Access	Has vehicle disinfection arch	0 (0)	0 (0)	1 (25)
Farm Access	Has foot baths and change of disinfectant	0 (0)	1 (3)	1 (25)
Farm Access	Disinfects goods and equipment brought on farm	0 (0)	0 (0)	3 (75)
Farm Access	Has unloading ramp	1 (0.8)	10 (33)	4 (100)
Farm Access	Perimeter fence is in good condition	1 (0.8)	4 (13)	3 (75)
Barn Access	Has cleaning and disinfection of the entrance	0 (0)	4 (13)	3 (75)
Barn Access	Each barn has dedicated staff	0 (0) *	6 (20)	3 (75)
Barn Access	Each barn has dedicated carts and work items	1 (0.83)	4 (13)	3 (75)
Barn Access	Has rodent control	0 (0)	2 (7)	3 (75)
Barn Access	Has bird curtain	0 (0)	5 (17)	1 (25)
Barn Access	Has clean staff clothing	0 (0)	1 (3)	2 (50)
Barn Access	Has footbath with frequent change of disinfectant	0 (0)	1 (3)	1 (25)
Barn Access	Has cement floor with periodic disinfection	6 (5.0)	12 (40)	4 (100)
Barn Access	Has feeders in good condition and periodic disinfection	0 (0)	6 (20)	4 (100)
Staff	Staff receive biosecurity training	0 (0) *	0 (0)	2 (50)
Staff	Staff have their own pigs	0 (0) *	0 (0)	1 (25)
Staff	Staff have dedicated and clean clothes	0 (0) *	2 (7)	3 (75)
Staff	Staff required to shower upon entering the barn or pig house	0 (0) **	3 (10)	3 (75)
Staff	Staff bring food into the barns	32 (27) *	26 (87)	3 (75)
Breeding	Use natural mount	67 (55)	26 (87)	4 (100)
Breeding	Use artificial insemination	0 (0)	1 (3)	3 (75)
Breeding	Produces their own semen	25 (21)	21 (70)	4 (100)
Breeding	Has boars for exclusive use	16 (13)	22 (73)	4 (100)

Table 3. *Cont.*

Category	Biosecurity Factor	Number (%)		
		Backyard	CNT	ST
Breeding	Produces their own replacement pigs	28 (23)	21 (70)	3 (75)
General	Buries mortality on farm	34 (28)	21 (70)	4 (100)
General	Feed enters with its own truck	9 (7)	16 (53)	3 (75)
General	Trucks are disinfected at the entrance and exit	0 (0)	4 (13)	3 (75)
General	Other species are present on premise	27 (23) *	18 (60)	1 (25)
General	Uses one needle per pig for injections of drugs or vaccines	7 (6)	5 (17)	4 (100)
General	Has weeds on the perimeter or close to the food	62 (51)	18 (60)	0 (0)
General	Pens drain adequately without causing flooding	1 (0.8)	5 (17)	4 (100)
General	Has clean animal loading and unloading area	0 (0)	6 (20)	4 (100)
Water source for pigs ***	Drinking water	67 (55)	10 (33)	***
	Non-drinking water	4 (3)	0 (0)	***

* 1 missing response. ** 2 missing responses. *** Missing responses were 44 (backyard), 26 (CNT), and 4 (ST).

The reason for the ASF introduction at their farm, proposed by the producers, was summarized into 11 categories (Table 4). A total of 122 farms (79%) provided a suspect reason. The most common reason given, entirely from backyard farms, was close proximity to other known outbreaks and nearby farms. Considering all farm types, movements of pigs between farms or allowing pigs to move free-range was the second most reported reason and first when considering only CNT farms. Swill-feeding, also known as garbage feeding, was the third most cited reason, second overall for backyard farms. Only two of the four ST farms provided a reason, both suggesting movements of herons specifically (grouped under "non-pig animals"). Movements or contact with people, feed (not specifically reported as swill-feeding), sharing breeding boars between multiple farms, and movements or presence of non-pig animals (including dogs, livestock species, birds) were also commonly reported.

Table 4. Summary of reasons proposed by producers for introduction of African Swine Fever (ASF) at their farm (n = 122), by farm type (CNT = commercial non-technical, ST = semi-technical).

Reason for ASF Introduction Suspected by Producer	Number of Farms (%) *			
	Backyard (n = 103)	CNT (n = 17)	ST (n = 2)	Total (n = 122)
Proximity to other outbreaks and farms	43 (42)	0 (0)	0 (0)	43 (35)
Movements of pigs between farms or free-range domestic pigs	26 (25)	8 (47)	0 (0)	34 (28)
Swill or garbage feeding	32 (31)	0 (0)	0 (0)	32 (26)
Movements of and contact with people	10 (10)	4 (24)	0 (0)	14 (12)
Feed (unspecified source)	11 (11)	2 (12)	0 (0)	13 (11)
Sharing breeding boar	8 (8)	4 (24)	0 (0)	12 (10)
Non-pig animals	5 (5)	3 (18)	2 (100)	10 (8)
Water	2 (2)	1 (6)	0 (0)	3 (3)
Vehicles/Trucks	0 (0)	2 (12)	0 (0)	2 (2)
Carcass disposal method	1 (1)	0 (0)	0 (0)	1 (0.8)
Near landfill	1 (1)	0 (0)	0 (0)	1 (0.8)

* Percentage calculated as the number of farms reporting that reason out of that farm type or all (n = 122) that reported a suspected reason. Percentages do not sum to one, due to some farms reporting multiple suspected reasons.

3.2. Descriptive Spatiotemporal Analysis

The number of reported cases by week (Figure 1) varied from zero to 14 with a mean of 4.8 (SD = 3.2) and median of five. The three-week rolling average varied from one to seven (mean = 4.5, sd = 1.7, median = 5). Two one-week periods (2–8 April and 30 April–6 May) had no cases reported. Overall, ASF cases appeared relatively evenly distributed throughout the study period, with no clear trends.

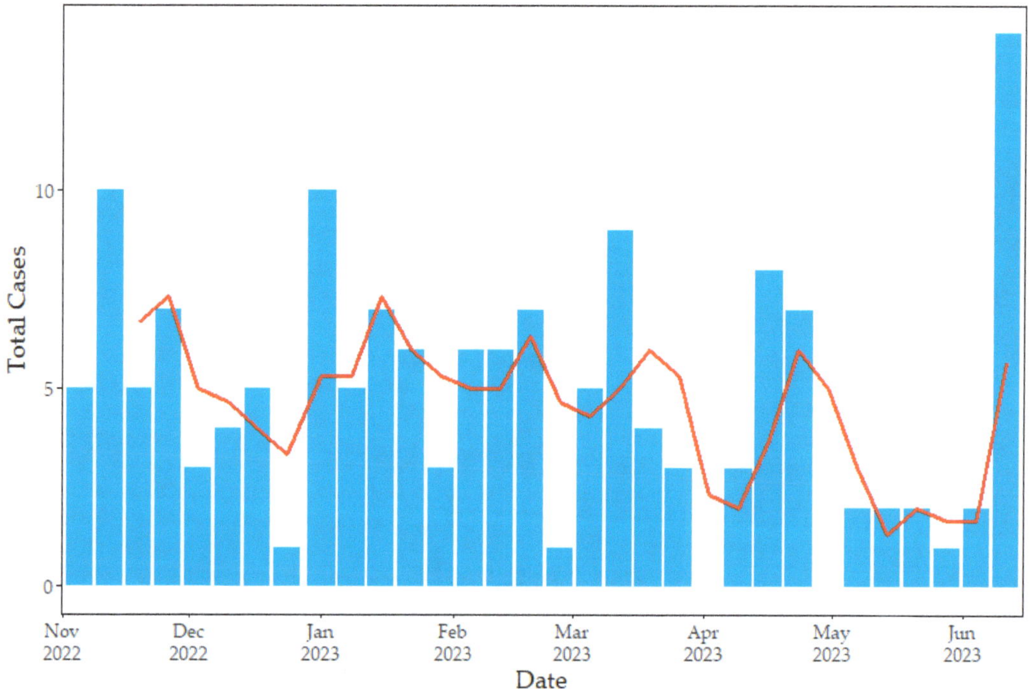

Figure 1. Weekly number of African Swine Fever outbreaks (blue bars) and three-week moving average (red line) reported in the Dominican Republic from 11 November 2022 to 17 June 2023.

Geographically, cases were mainly located in central regions. The provinces reported a range of 0 to 19 cases. The provinces of Monte Plata (*n* = 19), Santiago (*n* = 16), and María Trinidad Sánchez (*n* = 14) had the highest cumulative number of reported cases across the study period. Ten provinces had no reported ASF cases. The mean center (Figure 2a, blue circle) of all case locations was in the province of Monseñor Nouel.

3.3. Between-Farm Reproduction Ratio

Five time intervals were identified where the number of infections doubled (Figure 3; at days 4, 6, 28, 70, and 159 of the study period), resulting in five estimations of R_0. For values of D $\geq$ 6 days, R_0 was above two for the first two timepoints (corresponding to the first 6 days of the study period). For values of D < 32 days, R_0 was estimated at below two by day 28 of the study period. For all values of D, R_0 was below two for the remaining timepoints and approaching one. For example, in Figure 3, the value of R_0 at day 159 for D = 1, 6, 20, or 40 days was 1.01, 1.05, 1.16, and 1.31, respectively.

(a)

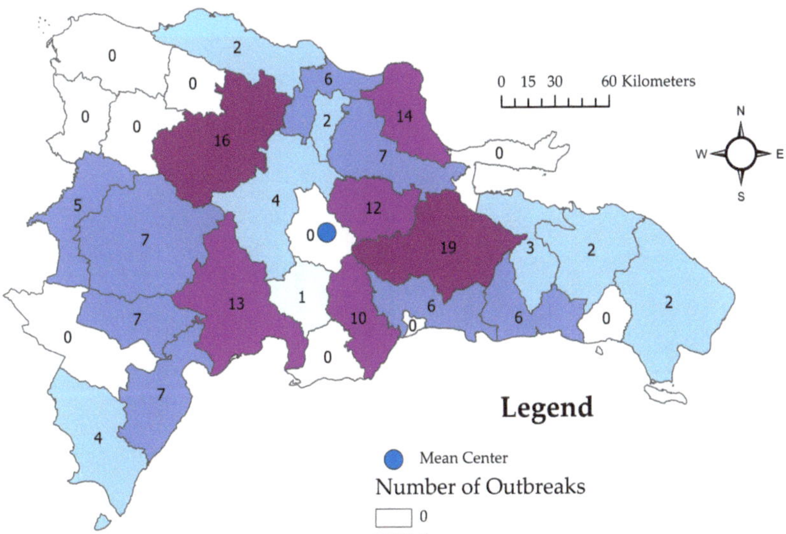

(b)

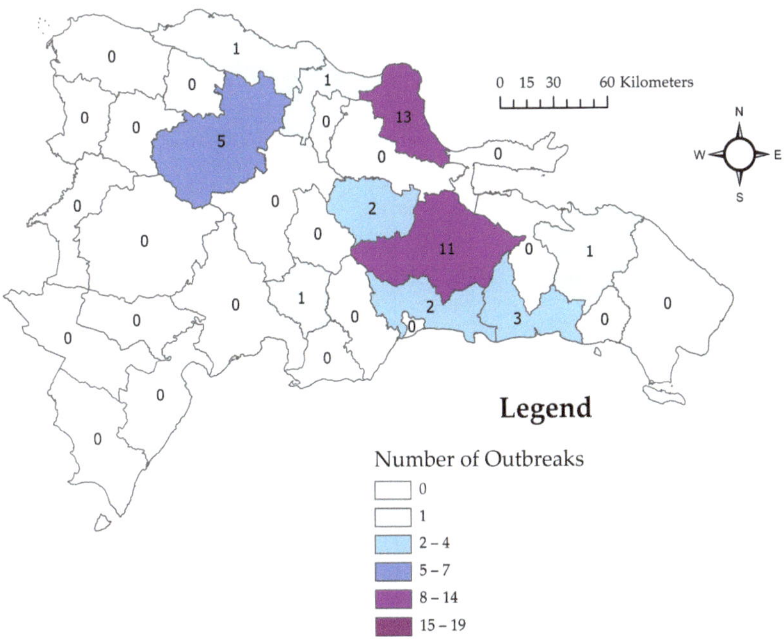

Figure 2. *Cont.*

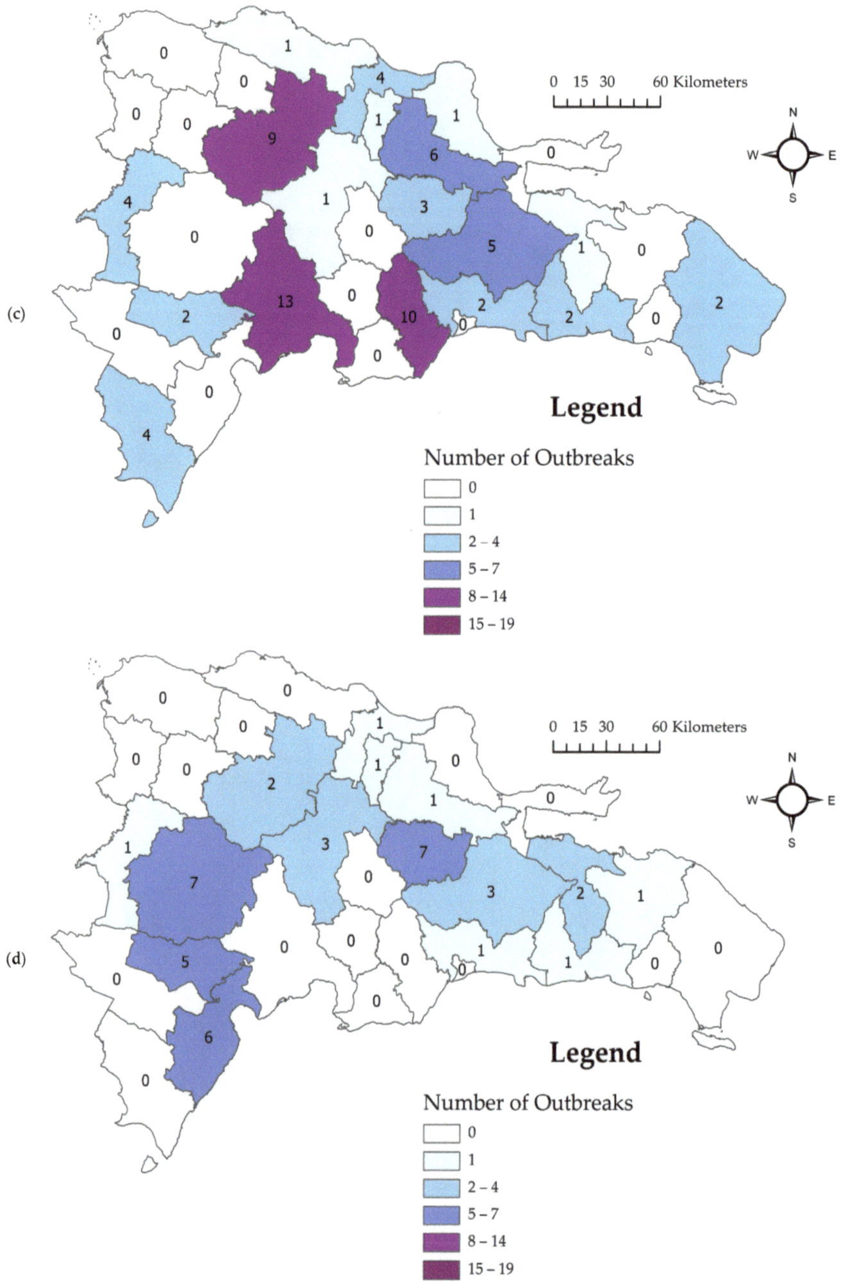

Figure 2. Choropleth maps of (**a**) total number of African Swine Fever outbreaks (n = 155 total) in the Dominican Republic reported by province (darker colors represent higher numbers of outbreaks) and overall mean center, and (**b**) number of African Swine Fever outbreaks by province from November to December 2022 (n = 40), (**c**) January to March 2023 (n = 71), and (**d**) April to June, 2023 (n = 42). Note that 1 outbreak from Barahona province and 1 outbreak from Santo Domingo province did not have dates and are only represented in (**a**).

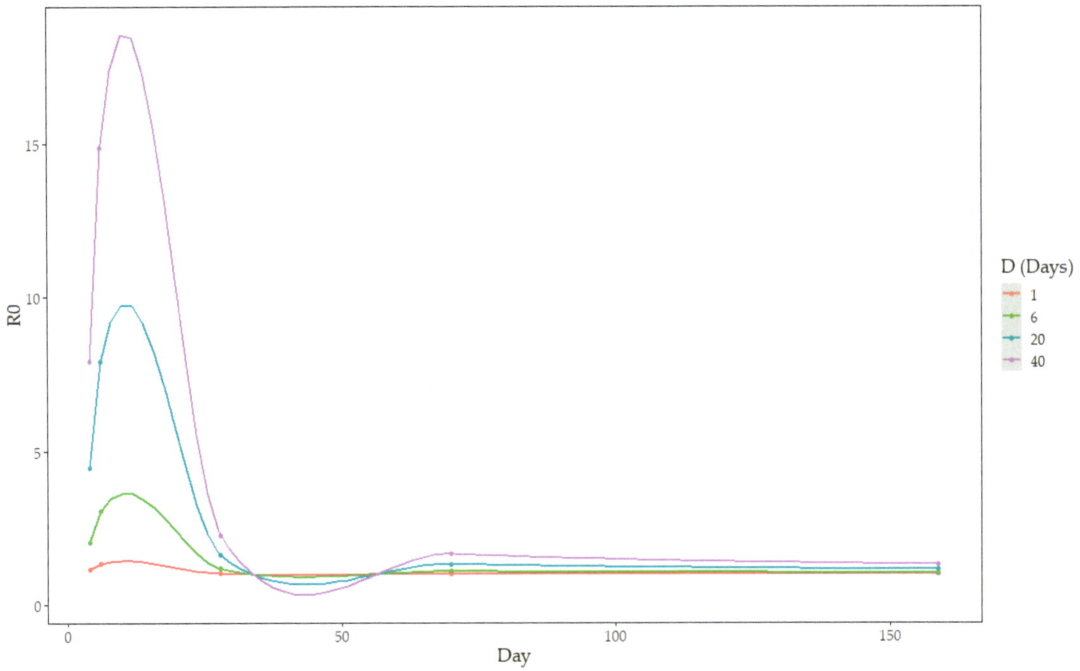

Figure 3. Values of between-herd reproduction ratio (R_0, points) and their locally estimated scatterplot smoothing estimations (lines) across the study period for different durations of the infectious period of the farm (D).

3.4. Cluster Analysis

The Global Moran I's autocorrelation statistic was significant (I = 0.71, $p < 0.001$), which is consistent with a clustered pattern of cases across the total study period. All combinations of spatial and temporal cutoff values assessed using the Knox test for clustering were significant ($p < 0.001$), supporting clustering of cases across a range of space and time values. O/E ratio values are depicted in Figure 4. O/E ratio values varied from 26.26 (spatial cutoff = 1 km, temporal cutoff = 1 day) to 1.42 (spatial cutoff = 30 km, temporal cutoff = 28 days). A flattening of O/E ratio values was observed with spatial cutoff values of ten km and temporal cutoff values of seven to nine days.

3.5. Space–Time Permutation Model

Nine significant space–time clusters ($p < 0.05$) were identified from the space–time permutation model (Table 5). These clusters occurred between January and June 2023 (within 7 day windows). Two clusters were identified in January, two in February, two in March, one in April, and two in June, all in 2023. The number of locations per cluster ranged from four to seven, and the radius of clusters ranged from zero to 2.23 km. Clusters with a radius of zero km involved locations with the same latitude and longitude coordinates as recorded by the veterinary officer at the time of their visit. All locations identified in the clusters were backyard farms.

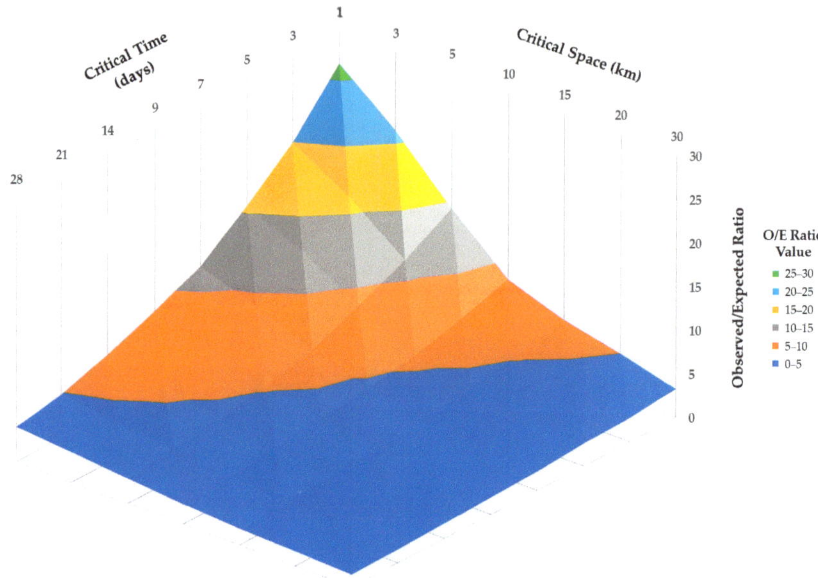

Figure 4. Surface map of observed to expected ratio values (*y*-axis, dark blue = 0—5, orange = 5—10, gray = 10—15, yellow = 15—20, light blue = 20—25, and green = 25—30) from Knox test across combinations of spatial cutoff values (*x*-axis, 1, 3, 5, 10, 15, 20, and 30 km) and temporal cutoff values (*z*-axis, 1, 3, 5, 7, 9, 14, 21, and 28 days).

Table 5. Significant clusters ($p < 0.05$) of African Swine Fever outbreaks reported in the Dominican Republic between 11 November 2022 and 17 June 2023.

Cluster	Radius (km)	Timeframe	Number of Cases	O/E Ratio	*p*-Value
1	0.76	1 January 2023 to 7 January 2023	6	15.3	<0.001
2	2.23	22 January 2023 to 28 January 2023	6	21.3	<0.001
3	0.00042	11 June 2023 to 17 June 2023	7	10.9	<0.001
4	0.21	5 March 2023 to 11 March 2023	4	30.6	<0.001
5	0	5 February 2023 to 11 February 2023	4	25.5	0.001
6	0	12 February 2023 to 18 February 2023	4	25.5	0.001
7	0.63	23 April 2023 to 29 April 2023	4	21.9	0.003
8	0	12 March 2023 to 18 March 2023	4	17	0.011
9	0	11 June 2023 to 17 June 2023	5	10.9	0.014

4. Discussion

ASF continues to be an important disease threat globally and to the Dominican Republican swine industry. The present study is one of few currently published sources to evaluate the epidemiological dynamics of ASF spread in the Dominican Republic. These results provide an important description of the ongoing situation that can be used by government officials and industry stakeholders to improve ASF control and management.

The location of cases in relatively central regions of the Dominican Republic (Figure 2) is consistent with pig production in the country and previous reports of ASF outbreaks [14,16,32]. The relatively stable incidence of ASF cases throughout the study period (Figure 1), lacking any clear upward or downward trend, and R_0 values approaching 1 (Figure 3) are consistent with endemic patterns of disease [26]. The amount of data was insufficient to identify clear temporal trends, such as seasonal analysis, or geographic spread. The R_0 values here are consistent with between-herd field estimates in domestic pigs from Uganda (ranging from 1.58 to 3.24 depending on the calculation method used) [33], Ukraine (1.65) [34], and the Russian Federation (2–3) [35]. However, it should be noted that the estimation of R_0 is based on assumptions of homogeneous mixing of the population (same number of contacts for all individuals), which may in reality differ based on varying practices by the farm type, and the estimated duration of infectiousness [36], which may be difficult to estimate at farm level. Additionally, underreporting of cases may lead to an underestimation of R_0, but this can be difficult to assess and could vary over time depending on economic factors and producer awareness.

Results from the Knox test indicate significant clustering of outbreak even when considering small spatial distances (1, 3, and 5 km) and temporal cutoff values (1, 3, 5, 7, and 9 days). In this dataset, this is because many outbreaks were recorded with the same or very similar date and location, likely reflecting a pattern of disease spread where multiple neighboring backyard farms may become infected with ASF nearly simultaneously or in close succession. This is consistent with findings from the space–time permutation model, in which the radius of significant space–time clusters was 2.23 km or less. In four of the clusters, the geographic location of the involved backyard farms was nearly indistinguishable, as represented by the same latitude and longitude coordinates being recorded for those farms (which was validated during data cleaning through discussions with data collectors). Results from the spatiotemporal and cluster analyses and observations of biosecurity measures and producer behaviors may indicate the need to redefine the epidemiological unit in settings where farms are located adjacent or overlapping one another, and management practices lead to the common sharing of land, animals, feed, and other potential disease introduction sources. Practically, the unit may be better defined as a geographic area, allowing for improved management by veterinary authority and epidemiologists.

The clinical signs and necropsy findings present on some farms are consistent with the previously reported presentation for ASF in the Dominican Republic. In previous works in 2021 and 2022, a heterogenous presentation of ASF was reported, from acute, hemorrhage disease in some outbreaks to a more subacute and nonspecific disease on other farms [32,37]. Of the 103 farms that had clinical sign information in this dataset, mortality, anorexia, and fever were the most reported signs across production types. Eight farms had asymptomatic infections reported. All reported hemorrhagic signs combined (red skin, epistaxis, unspecified hemorrhage) were the fourth most reported symptom. Almost no necropsy signs were reported, but this is likely because conducting a necropsy was not part of the official control program or case definition used to verify these outbreaks.

Despite reported mortalities, a relatively large number of pigs were reported as still alive on farms at the time of depopulation for indemnity purposes (Table 1). This could be due to early reporting of ASF outbreaks, which may be encouraged if incentivized through indemnity and other support. Another possibility is that the reported heterogenous presentation may result in outbreaks with lower mortality, but this is still unknown. Surviving but infected pigs that are not reported may be more likely to be moved or sent to slaughter, especially in cases where producers may perceive a slow or insufficient indemnity response by government officials or not wish their farm to be depopulated, which ultimately further drives the spread of ASF. Understanding the factors behind this high number of live pigs may help to reduce the risk of infected pigs being dispersed.

The biosecurity measures present varied considerably across farm types. As expected, backyard farms had fewer biosecurity measures present, such as lacking secure perimeters for farms and barns, protocols and training for farm workers (which in the case of backyard farms, are likely the owners themselves), consistent cleaning and disinfection, and protocols for breeding and animal management. CNT farms had more measures present compared to backyard farms, such as an increased percentage of farms reporting having showers, truck disinfection, animal loading and unloading areas with ramps, separation via clean-dirty lines and closed farm entrances, dedicated equipment, and pest control. ST farms had the most biosecurity measures present. This may contribute to the low number of reported outbreaks in ST farms, but this is difficult to assess without population data for each farm type. Overall, lack of biosecurity measures likely contributes significantly to the continued spread of ASF, especially through closely and densely located backyard farms.

The reason for their ASF introduction, as suspected by producers, is mostly consistent with known risks for ASF spread [1,6]. It should be noted that these responses do not reflect the results of outbreak investigations conducted by veterinary officials, but they do provide an indication of producers' behaviors and management actions. Being near a known outbreak or other farms was the most commonly listed reason (Table 4, 35% of 122 respondents) and solely by backyard farms. This is consistent with results from cluster analysis and space–time permutation model, where small spatial distances resulted in significant clusters with small radiuses. Other management practices, such as moving pigs, swill-feeding, and sharing breeding boars, were also frequently mentioned by producers, and providing key messaging and support targeting these behaviors will help to support improved biosecurity and management, ultimately to limit ASF spread.

Under-reporting may be an important limitation for this data set, but this is difficult to estimate or quantify. Case reporting amongst some producers in the Dominican Republic may be largely driven by economic factors to reduce financial losses from ASF outbreaks. Producers may compare the sale price amongst carnicerias, more informal meat markets and butcher shops as compared to slaughter plants, to indemnity rates provided after government-sponsored depopulation. Higher prices from carnicerias may incentivize producers to sell infected pigs before high mortality from ASF occurs, while when sale prices decrease, higher indemnity rates may be preferred. A similar effect of indemnity rates and market price on passive surveillance has been reported in Vietnam [38]. Similarly presenting swine diseases may also play a role in underreporting. Because other diseases such as Classical Swine Fever and Porcine Reproductive and Respiratory Syndrome (PRRS) are also present in the Dominican Republic, small backyard farms may simply dispose of dead pigs and repopulate without suspecting ASF. Other factors that have been previously described to affect passive reporting of veterinary diseases by farmers include producer knowledge and awareness of the disease and reporting systems, uncertainty about when to report, mistrust of veterinary authorities, and fear of social consequences of reporting [39]. The current analysis was also limited by the lack of control or population data, which is hard to access or not readily available in the Dominican Republic swine industry, so a risk factor analysis could not be performed. Future work should aim to collect uninfected population data for further analysis and comparison. Finally, not all farms had all information reported, such as clinical signs or suspected introduction reasons, resulting in a smaller sample size for those variables.

In summary, these results provide a recent update on spatiotemporal dynamics of ASF in the Dominican Republic and ongoing practices of pig holders that drive ASF spread. The information reported here will help swine industry stakeholders and government officials in developing ASF control strategies. Outbreaks were primarily centrally located within the country and backyard farms, though the vast majority of individual affected animals were located in the 30 CNT and 4 ST farms. Clustering was high due to many neighboring backyard farms being detected with ASF almost simultaneously. Future works with access to control or population data will allow for more investigation into the specific, important risk factors present in the Dominican Republic and provide further useful information for developing prevention and control strategies.

Supplementary Materials: The following supporting information can be downloaded at: https://www.mdpi.com/article/10.3390/pathogens12121414/s1, File S1: Questionnaire used by the Dominican Republic Ministry of Agriculture veterinary officers for data collection and verification of passively reported African Swine Fever farm outbreaks.

Author Contributions: Conceptualization, M.C.A., S.K., R.R. and A.M.P.; Formal analysis, R.A.S., S.H. and A.M.P.; Funding acquisition, A.M.P.; Supervision, A.M.P.; Validation, R.A.S. and S.H.; Writing—original draft, R.A.S.; Writing—review and editing, R.A.S., M.C.A., S.K. and A.M.P. All authors have read and agreed to the published version of the manuscript.

Funding: This research has been funded by a cooperative agreement with the USDA Animal and Plant Health Inspection Service (AP23IS000000C004).

Institutional Review Board Statement: Not applicable.

Informed Consent Statement: Not applicable.

Data Availability Statement: Data are the property of the Dominican Republic government and have been shared with us to conduct the research here to support ASF control activities in the country. Any requests should be directed to the Dominican Republic Incident Command for ASF.

Acknowledgments: The authors would like to acknowledge and thank the veterinary officials and staff who conducted the outbreak investigations and supported this data collection.

Conflicts of Interest: The authors declare no conflict of interest. Funding was provided by USDA, and USDA staff (C.A. and S.K.) had a role in the design of the study, in the collection and interpretation of data, and in the reviewing of the manuscript.

References

1. Blome, S.; Franzke, K.; Beer, M. African swine fever—A review of current knowledge. *Virus Res.* **2020**, *287*, 198099. [CrossRef] [PubMed]
2. Alonso, C.; Borca, M.; Dixon, L.; Revilla, Y.; Rodriguez, F.; Escribano, J.M.; ConsortiumYR, I.R. ICTV Virus Taxonomy Profile: Asfarviridae. *J. Gen. Virol.* **2018**, *99*, 613–614. [CrossRef] [PubMed]
3. World Organisation for Animal Health. *African Swine Fever*; OIE Technical Disease Card; World Organisation for Animal Health: Paris, France, 2019; p. 5.
4. Franzoni, G.; Dei Giudici, S.; Loi, F.; Sanna, D.; Floris, M.; Fiori, M.; Sanna, M.L.; Madrau, P.; Scarpa, F.; Zinellu, S.; et al. African Swine Fever Circulation among Free-Ranging Pigs in Sardinia: Data from the Eradication Program. *Vaccines* **2020**, *8*, 549. [CrossRef] [PubMed]
5. Danzetta, M.L.; Marenzoni, M.L.; Iannetti, S.; Tizzani, P.; Calistri, P.; Feliziani, F. African Swine Fever: Lessons to Learn From Past Eradication Experiences. A Systematic Review. *Front. Vet. Sci.* **2020**, *7*, 296. [CrossRef] [PubMed]
6. Dixon, L.K.; Stahl, K.; Jori, F.; Vial, L.; Pfeiffer, D.U. African Swine Fever Epidemiology and Control. *Annu. Rev. Anim. Biosci.* **2020**, *8*, 221–246. [CrossRef]
7. USDA-APHIS. *USDA Statement on Confirmation of African Swine Fever in the Dominican Republic*; USDA: Washington, DC, USA, 2021.
8. USDA-APHIS. *USDA Statement on Confirmation of African Swine Fever in Haiti*; USDA: Washington, DC, USA, 2021.
9. World Organisation for Animal Health. *African Swine Fever: Situation Report 39*; World Organisation for Animal Health: Paris, France, 2023.
10. EFSA Panel on Animal Health and Welfare (AHAW); More, S.; Miranda, M.A.; Bicout, D.; Bøtner, A.; Butterworth, A.; Calistri, P.; Edwards, S.; Garin-Bastuji, B.; Good, M.; et al. Scientific Opinion on the African swine fever in wild boar. *EFSA J.* **2018**, *16*, e05344. [CrossRef] [PubMed]
11. USDA. *History Highlight: APHIS Supports the Fight Against African Swine Fever*; USDA: Washington, DC, USA, 2022.
12. Jéssica Kayamori, L.; Trincado, C.; Souza, J.; Rojas, H.; Beltran-Alcrudo, D. *Informe Misión Apoyo Técnico Emergencia Peste Porcina Africana*; Food and Agriculture Organization of the United Nations: Rome, Italy, 2021.
13. Mayol, V. *Dominican Republic Swine Voluntary Report-ASF Outbreak Creates Increased Opportunities for US Swine Meat Exports to the DR*; United States Department of Agriculture Foreign Agricultural Service: Santo Domingo, Dominican Republic, 2022; p. 8.
14. Schambow, R.; Reyes, R.; Morales, J.; Diaz, A.; Perez, A.M. A qualitative assessment of alternative eradication strategies for African swine fever in the Dominican Republic. *Front. Vet. Sci.* **2022**, *9*, 1054271. [CrossRef]
15. Sandoval, R. Inter-American Institute for Cooperation on Agriculture. Personal Communication, 2023.
16. Arias, J. *Inventario Nacional Porcino*; Convenio Ministerio de Agricultura—OIRSA: Santo Domingo, Dominican Republic, 2022; p. 47.
17. Kanankege, K.S.T.; Alvarez, J.; Zhang, L.; Perez, A.M. An Introductory Framework for Choosing Spatiotemporal Analytical Tools in Population-Level Eco-Epidemiological Research. *Front. Vet. Sci.* **2020**, *7*, 339. [CrossRef]
18. Ward, M.P.; Carpenter, T.E. Analysis of time–space clustering in veterinary epidemiology. *Prev. Vet. Med.* **2000**, *43*, 225–237. [CrossRef]
19. World Organisation for Animal Health. Terrestrial Animal Health Code: Glossary. Available online: https://www.woah.org/en/what-we-do/standards/codes-and-manuals/terrestrial-code-online-access/?id=169&L=1&htmfile=glossaire.htm (accessed on 10 February 2023).

20. Cresswell, J. *Research Design: Qualitative, Quantitative, and Mixed Methods Approaches*, 3rd ed.; SAGE Publications: Thousand Oaks, CA, USA, 2009.
21. Ward, M.P.; Carpenter, T.E. Techniques for analysis of disease clustering in space and in time in veterinary epidemiology. *Prev. Vet. Med.* **2000**, *45*, 257–284. [CrossRef]
22. Moran, P.A.P. Notes on Continuous Stochastic Phenomena. *Biometrika* **1950**, *37*, 17–23. [CrossRef] [PubMed]
23. George, K. Detection of Low Intensity Epidemicity. *Br. J. Prev. Soc. Med.* **1963**, *17*, 121. [CrossRef]
24. Kulldorff, M.; Heffernan, R.; Hartman, J.; Assunção, R.; Mostashari, F. A Space–Time Permutation Scan Statistic for Disease Outbreak Detection. *PLoS Med.* **2005**, *2*, e59. [CrossRef] [PubMed]
25. Diekmann, O.; Heesterbeek, J.A.P.; Metz, J.A.J. On the definition and the computation of the basic reproduction ratio R0 in models for infectious diseases in heterogeneous populations. *J. Math. Biol.* **1990**, *28*, 365–382. [CrossRef] [PubMed]
26. Dohoo, I.; Martin, W.; Stryhn, H. *Veterinary Epidemiologic Research*, 2nd ed.; VER Inc.: Glendale, CA, USA, 2010.
27. Iglesias, I.; Perez, A.M.; SÁNchez-VizcaÍNo, J.M.; MuÑOz, M.J.; MartÍNez, M.; De La Torre, A. Reproductive ratio for the local spread of highly pathogenic avian influenza in wild bird populations of Europe, 2005–2008. *Epidemiol. Infect.* **2011**, *139*, 99–104. [CrossRef] [PubMed]
28. Iglesias, I.; Muñoz, M.J.; Montes, F.; Perez, A.; Gogin, A.; Kolbasov, D.; de la Torre, A. Reproductive Ratio for the Local Spread of African Swine Fever in Wild Boars in the Russian Federation. *Transbound. Emerg. Dis.* **2016**, *63*, e237–e245. [CrossRef]
29. de Carvalho Ferreira, H.C.; Backer, J.A.; Weesendorp, E.; Klinkenberg, D.; Stegeman, J.A.; Loeffen, W.L.A. Transmission rate of African swine fever virus under experimental conditions. *Vet. Microbiol.* **2013**, *165*, 296–304. [CrossRef]
30. Guinat, C.; Gogin, A.; Blome, S.; Keil, G.; Pollin, R.; Pfeiffer, D.U.; Dixon, L. Transmission routes of African swine fever virus to domestic pigs: Current knowledge and future research directions. *Vet. Rec.* **2016**, *178*, 262–267. [CrossRef]
31. Guinat, C.; Porphyre, T.; Gogin, A.; Dixon, L.; Pfeiffer, D.U.; Gubbins, S. Inferring within-herd transmission parameters for African swine fever virus using mortality data from outbreaks in the Russian Federation. *Transbound. Emerg. Dis.* **2018**, *65*, e264–e271. [CrossRef]
32. Gonzales, W.; Moreno, C.; Duran, U.; Henao, N.; Bencosme, M.; Lora, P.; Reyes, R.; Núñez, R.; De Gracia, A.; Perez, A.M. African swine fever in the Dominican Republic. *Transbound. Emerg. Dis.* **2021**, *68*, 3018–3019. [CrossRef]
33. Barongo, M.B.; Ståhl, K.; Bett, B.; Bishop, R.P.; Fèvre, E.M.; Aliro, T.; Okoth, E.; Masembe, C.; Knobel, D.; Ssematimba, A. Estimating the Basic Reproductive Number (R0) for African Swine Fever Virus (ASFV) Transmission between Pig Herds in Uganda. *PLoS ONE* **2015**, *10*, e0125842. [CrossRef] [PubMed]
34. Korennoy, F.I.; Gulenkin, V.M.; Gogin, A.E.; Vergne, T.; Karaulov, A.K. Estimating the Basic Reproductive Number for African Swine Fever Using the Ukrainian Historical Epidemic of 1977. *Transbound. Emerg. Dis.* **2017**, *64*, 1858–1866. [CrossRef] [PubMed]
35. Gulenkin, V.M.; Korennoy, F.I.; Karaulov, A.K.; Dudnikov, S.A. Cartographical analysis of African swine fever outbreaks in the territory of the Russian Federation and computer modeling of the basic reproduction ratio. *Prev. Vet. Med.* **2011**, *102*, 167–174. [CrossRef] [PubMed]
36. Thomas, J.C.; Weber, D.J. *Epidemiologic Methods for the Study of Infectious Diseases*; Oxford University Press: New York, NY, USA, 2001.
37. Ramirez-Medina, E.; O'Donnell, V.; Silva, E.; Espinoza, N.; Velazquez-Salinas, L.; Moran, K.; Daite, D.A.; Barrette, R.; Faburay, B.; Holland, R.; et al. Experimental Infection of Domestic Pigs with an African Swine Fever Virus Field Strain Isolated in 2021 from the Dominican Republic. *Viruses* **2022**, *14*, 1090. [CrossRef]
38. Nguyen-Thi, T.; Pham-Thi-Ngoc, L.; Nguyen-Ngoc, Q.; Dang-Xuan, S.; Lee, H.S.; Nguyen-Viet, H.; Padungtod, P.; Nguyen-Thu, T.; Nguyen-Thi, T.; Tran-Cong, T.; et al. An Assessment of the Economic Impacts of the 2019 African Swine Fever Outbreaks in Vietnam. *Front. Vet. Sci.* **2021**, *8*, 686038. [CrossRef]
39. Gates, M.C.; Earl, L.; Enticott, G. Factors influencing the performance of voluntary farmer disease reporting in passive surveillance systems: A scoping review. *Prev. Vet. Med.* **2021**, *196*, 105487. [CrossRef]

pathogens

MDPI

Article

Factors Affecting the Spread, Diagnosis, and Control of African Swine Fever in the Philippines

Chia-Hui Hsu [1,*], Rachel Schambow [1], Maximino Montenegro [2], Ruth Miclat-Sonaco [3] and Andres Perez [1]

1 Center for Animal Health and Food Safety, College of Veterinary Medicine, University of Minnesota, Saint Paul, MN 55108, USA; scham083@umn.edu (R.S.); aperez@umn.edu (A.P.)
2 Pig Improvement Company (PIC) Philippines, Pasig City 1605, Philippines; max.montenegro@genusplc.com
3 National Livestock Program, Office of the Undersecretary for Livestock, Department of Agriculture, Elliptical Road, Diliman, Quezon City 1101, Philippines; rmiclat@ati.da.gov.ph
* Correspondence: hsu00124@umn.edu

Abstract: African Swine Fever (ASF) is a highly contagious disease that threatens the swine industry globally. Since its introduction into the Philippines in 2019, ASF has spread extensively in both commercial and backyard farms. Here, using a mix of qualitative and quantitative methods, including conjoint and SWOT analyses, world café discussions, and multivariable regression models, the most important factors that influence the spread, diagnosis, and control of ASF in the Philippines were identified. Research findings suggest that swill or contaminated feed, inadequate biosecurity protocols, and movement of personnel were the top risk factors favoring ASF spread among farms in general. For commercial farms, contaminated vehicles and personnel were also important, whereas for backyard farms, the introduction of new pigs, environmental contamination, and poor feeding quality were relevant risk factors. Notable clinical signs of ASF in pigs include reduced feed intake, huddled behavior, and reluctance to stand. This study highlights the need for timely reporting, trust-building initiatives, and enhanced biosecurity measures to effectively manage ASF outbreaks in the country. Results here contribute to the knowledge of factors affecting ASF spread in the Philippines and can help design prevention and control measures in ASF-infected countries while enhancing preparedness in countries free from the disease.

Keywords: African Swine Fever; disease control; conjoint analysis; SWOT analysis; risk factors; the Philippines

Citation: Hsu, C.-H.; Schambow, R.; Montenegro, M.; Miclat-Sonaco, R.; Perez, A. Factors Affecting the Spread, Diagnosis, and Control of African Swine Fever in the Philippines. *Pathogens* **2023**, *12*, 1068. https://doi.org/10.3390/pathogens12081068

Academic Editor: Natasha N. Gaudreault

Received: 29 June 2023
Revised: 6 August 2023
Accepted: 10 August 2023
Published: 21 August 2023

1. Introduction

African Swine Fever (ASF) is a hemorrhagic disease of swine caused by infection with the ASF virus (ASFV), a double-stranded DNA arbovirus [1,2]. ASF is a disease reportable to the World Organisation for Animal Health (WOAH) due to the significant risk it poses to the global swine industry. ASF only infects swine and can cause up to 100% mortality. Pigs infected by ASF may exhibit clinical signs such as anorexia, high fever, hemorrhages in the skin, bloody diarrhea, pneumonia, abortion in pregnant sows, and sudden death [1,2]. Since its detection outside of Africa in 2007, ASF p72 genotype II has spread rapidly across Europe and Asia, notably being identified in China in 2018 [3]. Subsequently, from 2019 to 2023, it has spread throughout many Southeast Asian countries including Vietnam (2019), Cambodia (2019), Laos (2019), the Philippines (2019), Myanmar (2019), Timor-Leste (2019), Indonesia (2019), Malaysia (2021), Thailand (2022), and Singapore (2023).

Since its first appearance in the province of Rizal [4], the Philippines in July 2019, ASF has swiftly propagated across the country, causing extensive outbreaks in all 17 administrative regions. Spread followed a seasonal pattern, with ASF outbreaks being more frequent in the latter half of each year [5]. Pork production in the Philippines was heavily impacted. In 2021, there was a 41.7% reduction (approximately 2.01 million heads) of the total registered population on commercial pig farms, with subsequent inflation of pork prices [6]. Backyard farms suffered losses as well, dropping from 7.97 million head in 2020 to 6.91 million in the third quarter of 2021. In March 2023, the country's registered swine population reached approximately 10.18 million heads, roughly 20.5% lower compared to the same quarter in 2020 [6,7]. This ongoing situation, transitioning from epidemic to endemic status for the disease, has raised significant concerns among the general public. These concerns include increasing pork prices and, for some consumers, food security, despite ASFV not infecting humans [8]. To effectively address the ASF epidemic and ensure continued commercial activity in the swine market, there is a need to identify the distinctive risk factors associated with the Philippine context and allocate resources to selectively target the most impactful factors and risks.

As in many countries affected by ASF, much of the data and experience in controlling the disease has not been formally recorded. Qualitative tools and analyses can support the elicitation of the opinion of individuals that have been engaged in control activities to extract knowledge and evidence on the risk factors affecting ASF in the Philippines. Conjoint analysis is a research method that offers a useful approach to understanding the preferences and trade-offs made by stakeholders [9,10]. Originally, this tool was developed for market research to understand consumer preferences, but it has been adopted for use in veterinary sciences to quantify the preferences of respondents and conduct risk assessments [11,12]. Conjoint analysis is particularly useful in that it can help estimate the relative importance of various factors when a large number of variable combinations are present. Rather than evaluate every attribute combination (which would be virtually impossible with high numbers of attributes and levels), statistical methods are used to select a representative subset of attribute combinations. These combinations are used to produce a survey where participants commonly rank or rate the options presented. Survey responses are typically analyzed using regression techniques, which vary depending on the survey method used [10]. In the case of ASF in the Philippines, because data on ASF spread is often lacking or unavailable, conjoint analysis can aid the quantitative estimation of the relative importance of risk factors in ASF transmission using expert stakeholder opinion.

Strengths, weaknesses, opportunities, and threats (SWOT) analysis can also support the identification of risk factors. SWOT analysis is a qualitative technique commonly used to systematically evaluate the advantages and disadvantages of a program [13]. Strengths and weaknesses represent internal attributes of the program itself that may affect its success, while opportunities and threats are external attributes present in the outside environment. Strengths and opportunities help the program achieve its goal, and, conversely, weaknesses and threats hinder the program's success. Like conjoint analysis, SWOT analysis was originally developed for business management but has been adapted by the health sciences to evaluate health control programs [14]. This tool has previously been used by the veterinary sciences, such as the evaluation of alternative control strategies for ASF [15]. In the Philippine context, SWOT analysis can support the identification of important weaknesses and threats to controlling ASF.

The objective of this study was to gain insights into the risk factors associated with ASF in the Philippines by utilizing SWOT and conjoint analysis. Additionally, we sought to assess the ASF clinical presentation and industry control measures in both backyard farms and commercial farms, relying on the expertise of professionals in the swine industry. While acknowledging the inherent limitations of expert opinion-based research, these findings provide valuable insights that could potentially contribute to the enhancement of surveillance plans and policy development aimed at addressing ASF risks. Ultimately, this work endeavors to contribute to ongoing efforts in improving ASF control in the Philippines, with potential implications for safeguarding the swine industry and public health.

2. Materials and Methods

2.1. General Approach

To support the identification of ASF risk factors, a group of veterinary experts specializing in the Philippine swine industry and with field experience with ASF was selected through a committee evaluation process. A workshop using a modified world café approach was organized with the aim of facilitating knowledge sharing and education among these participants, as well as identifying research gaps in ASF control. Conjoint analysis and SWOT analysis were used to gather expert insights and elicit their opinion. These methodologies allowed the experts to engage in brainstorming sessions and collaboratively identify the current perspectives on the spread, diagnosis, and control of ASF in the Philippines, accounting for its endemic status for over four years.

2.2. Expert Selection

To select participants for the workshop, a committee was assembled by the Philippine College of Swine Practitioners (PCSP), the Department of Agriculture Bureau of Animal Industry (DA-BAI), and the Agricultural Training Institute-International Training Center on Pig Husbandry (ATI-ITCPH). The committee utilized specific criteria to evaluate the suitability of applicants. To be eligible for selection, applicants were required to have 1–2 years of experience in the Philippine swine industry, but preference was given to those having 5 to 10 years of experience. Additionally, applicants were required to either currently hold or have the potential to become decision makers, such as regulatory authorities in their respective areas. The committee also specifically aimed to have at least five participants from each of the Department of Agriculture Regional Field Offices, the DA-BAI Policy Division, local governmental units, and the private sector. Finally, the committee gave special consideration to ensuring balanced representation from each region. After an initial open call for applicants, approximately 100 applicants were considered in total, and 25 were selected for the final workshop. This process took place between February and March 2023, and the workshop was held in May 2023.

A group of 25 selected veterinarians from different spatial locations across the Philippines (Figure 1) attended the in-person workshop organized in Batangas in May 2023. The National Capital Region (NCR) had 7 participants, Region IV-A had 6 participants, and the remaining regions were evenly distributed. Ten participants were from the Philippine veterinary services, and 15 were from the private Philippine swine industry. The participants had a median of 20 years of experience in swine veterinary practice or service within the country. All participants indicated that they had some degree of field experience with ASF and that they have seen the disease.

2.3. Conjoint Analysis

Conjoint analysis was used to identify and prioritize factors associated with the introduction and transmission of ASF in the Philippines. A list of nine risk factors important to African Swine Fever introduction and spread were identified based on existing research [16–18]. Each factor was dichotomized into a high- or low-risk response (Table 1).

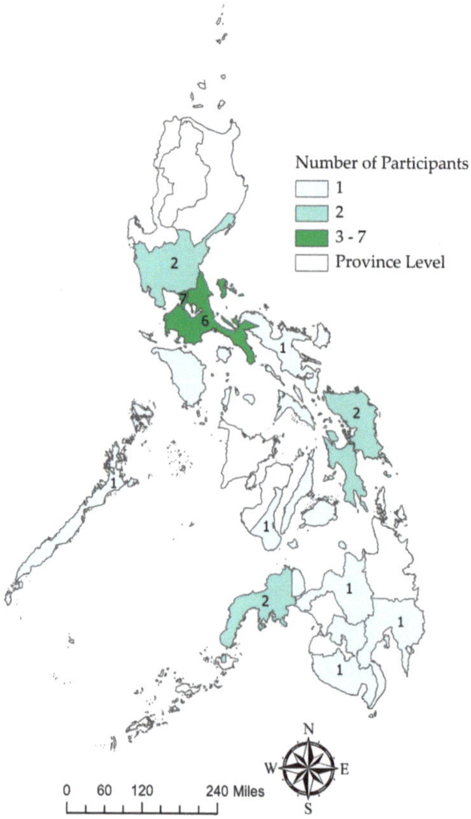

Figure 1. Geographic distribution of workshop participants in the Philippines.

Table 1. Risk factors and their levels associated with ASF introduction into farms for conjoint analysis.

No.	Risk Factors	High-/Low-Risk Response
1	Swill-fed or potential contamination of feed ingredients	Yes/No
2	Lack of double fencing	No double fencing/Have double fencing
3	Presence of flies and ticks	Yes/No
4	Presence of small and domestic mammals (e.g., rats, dogs, cats, or other farm animals)	Yes/No
5	Absence of protocols for changing clothes, separate entries and exits, disinfection of objects restrictions on food introduction, and external individuals accessing the farm	No appropriate protocols/Have appropriate protocols
6	Allowance for cars and trucks to enter premises	Cars and trucks can enter premises/Cars and trucks cannot enter premises
7	Non-closed herd with recent introduction of new animals (requiring importation of pigs) without a quarantine station within 1 km from premises or sharing of personnel	Non-closed herd/Closed herd
8	Movement of personnel (including vets, inseminators, and technicians) between this farm and other farms without trusted biosecurity measures	Personnel without trusted biosecurity measures/Personnel with trusted biosecurity measures
9	Area with presence of feral pigs	Yes/No

A ranking-based survey was developed to estimate the individual influence of each factor on the risk of an ASF outbreak. Because it would be virtually impossible for participants to rank all combinations of these factors, a factorial design was generated to select a subset of farm scenarios. The scenario selection process was performed using the orthogonal design and conjoint plan methodology available in IBM SPSS Statistics v28.0.0.0. The final conjoint experimental design incorporated 13 different scenarios, with best- and worst-case scenarios included as control groups (Figure 2).

Farm	Swill-Fed	Double fencing	Flies/ ticks	Other animals	Appropriate protocol for disinfection	Cars / Trucks enter	Non-closed herd (import new pigs)	Trusted biosecurity Personnel	Feral pigs	RANK (1-13)*	ASF Outbreak (Y/N)
A	Yes	Have	No	Yes	No protocol	Yes	Close herd	With trusted biosecurity	No		
B	Yes	Have	No	No	Have protocol	Yes	Non-closed herd	Without trusted biosecurity	No		
C	No	No	No	Yes	Have protocol	No	Close herd	Without trusted biosecurity	No		
D	No	No	No	No	No protocol	Yes	Close herd	With trusted biosecurity	Yes		
E	No	No	Yes	Yes	Have protocol	Yes	Non-closed herd	With trusted biosecurity	No		
F	Yes	Have	Yes	Yes	Have protocol	No	Close herd	With trusted biosecurity	Yes		
G	Yes	No	No	No	Have protocol	No	Non-closed herd	With trusted biosecurity	Yes		
H	Yes	No	Yes	No	No protocol	No	Close herd	Without trusted biosecurity	No		
I	No	Have	No	Yes	No protocol	No	Non-closed herd	Without trusted biosecurity	Yes		
J	Yes	No	Yes	Yes	No protocol	Yes	Non-closed herd	Without trusted biosecurity	Yes		
K	No	Have	Yes	No	Have protocol	Yes	Close herd	Without trusted biosecurity	Yes		
L	No	Have	Yes	No	No protocol	No	Non-closed herd	With trusted biosecurity	No		
M	No	Have	No	No	Have protocol	No	Close herd	With trusted biosecurity	No		

***Where 13 means the HIGHEST Risk and 1 means the LOWEST Risk**

Figure 2. Scenarios for ranking-based conjoint analysis of ASF introduction risk.

The participants were Individually surveyed to assess and prioritize the risk and likelihood of ASF outbreaks in 13 hypothetical farm scenarios (Figure 2) using the Qualtrics platform. The participants were tasked with ranking these farm scenarios based on two criteria: (1) the level of risk associated with each farm ranging from 1 (representing the lowest risk of an ASF outbreak) to 13 (the highest risk); and (2) the likelihood of an ASF outbreak occurring using a binary response format (Yes or No).

2.4. *Quantitative ASF Information Collection through World Café Discussion*

A modified Delphi approach was implemented in a world café format discussion to foster open and creative dialogue among the participants and validate received responses. The participants were divided into four groups to ensure comprehensive discussions. After a designated time, participants were asked to switch tables and engage in discussions on the additional sets of questions. Each round involved a randomly ordered sharing of ideas, and a designated reporter summarized the group's consensus while seeking agreement or dissent from other groups. A modified Delphi method was employed to gather further information on key parameters and to validate the answers received from previous groups.

The modified world café sessions included quantitative questions that focused on three main areas, namely, (a) the routes through which ASF enters farms; (b) clinical manifestations and necropsy findings associated with ASF in the field; and (c) strategies and policies for ASF control in backyard and commercial farming. For the details of quantitative questions, please refer to the Supplementary Materials provided.

2.5. *SWOT Analysis*

SWOT analysis was used to gather insights and perspectives on the current status, diagnosis, and control policies of ASF in the Philippines. Participants were asked to identify strengths, weaknesses, opportunities, and threats associated with these factors as part of the previously described modified world café discussion. The SWOT table was summarized by the authors and presented to the participants during the workshop for their agreement and further discussion. Throughout the process, all discussions and outcomes from the modified world café sessions and SWOT analysis were documented for reference and subsequent analysis.

2.6. Statistical Analyses

2.6.1. Logistic and Ordinal Regression Models

Participants' responses from the conjoint analysis survey (Figure 2) were analyzed using ordinal and binary logistic regressions. The response variables were whether participants believed a scenario farm ID (A to M) would lead to an ASF outbreak or not (Yes/No data for logistic regression) and the participants' individual ranking of perceived risk for each scenario (1–13 rankings data for ordinal regression). We created datasets for the analysis by matching each response with the corresponding scenario farm ID and its risk factor combination (Table 1). The dichotomous risk factors associated with each scenario were used as predictor variables in the regression models to predict the participant response.

Binary logistic regression and ordinal logistic regression models were performed in RStudio version 4.2.2 using the MASS package [19]. As an output of the regression analysis, regression coefficients were obtained for each risk factor, which provides an estimate of the relative weight each factor has on the responses. These coefficients were used to understand the significance and impact of the different risk factors under investigation on the probability of an ASF introduction on Philippine farms.

The Hosmer and Lemeshow goodness of fit (GOF) was performed using the ResourceSelection package [20]. The GOF test evaluates if a logistic regression model fits the observed data, determining if it could accurately represent the relationship between independent variables (risk factors) and the dependent variable (ASF occurrence, in this case).

2.6.2. Statistical Analyses World Café Discussion and SWOT Analysis

Discussions from the SWOT analysis were summarized using a SWOT (Strengths, Weaknesses, Opportunities, Threats) analysis table. We compiled the discussions from four groups and allowed sufficient time for participants to reconsider and reflect on their opinions. Our focus was on documenting the information that achieved consensus and was supported by multiple opinions, aiming to maximize the consensus in the SWOT analysis and seek detailed explanations for each characteristic. While this approach does not involve calculating the SWOT analysis numerically, it allowed us to capture the majority of opinions effectively.

For the quantitative questions in the world café sessions, each table was required to reach a consensus within their randomized group. Subsequently, median values and normalized median scores were estimated for questionnaire results in this session.

3. Results

3.1. Conjoint Analysis

The logistic regression analysis identified swill or contaminated feed as the most significant concern ($p < 0.001$) across all 13 scenarios (Table 2). Following that, the absence of a protocol for disinfection ($p = 0.0017$) and personnel lacking trusted biosecurity measures ($p = 0.0019$) were also identified as significant risks. In the ordinal logistic regression model, all risk factors showed significant associations with increasing the probability of a scenario being higher ranked (i.e., higher perceived risk of ASF introduction by the participants; Table 2). The common odds ratios provide information on the magnitude of these associations, indicating how much the odds of moving up to a higher category change with each unit increase in the predictor variable. The highest common odds ratio, with a value of 19.84 (95% CI 12.09–33.22), was observed for swill-fed or contaminated feed, followed by personnel lacking trusted biosecurity measures (11.11, 95% CI 6.88–18.26) and absence of a protocol for disinfection (7.54, 95% CI 4.73–12.21). These results from the ordinal logistic regression were consistent with the binary logistical regression, with the top three risk factors being swill-fed/contaminated feed, personnel, and the absence of a protocol for disinfection.

Table 2. Conjoint analysis results: logistic regression and ordinal logistic regression models.

	Logistic Regression Model						Ordinal Logistic Regression					
Description	Coefficient	CI (95%)	Odds Ratio	SE	*p*-Value	Sig.	Coefficient	SE	Common Odds Ratio	CI (95%)	*p*-Value	Sig.
Swill-fed/Contaminated feed	1.8740	(1.28, 2.54)	6.5143	0.3167	<0.001	***	2.988	0.258	19.8459	(12.09, 33.22)	3.92×10^{-31}	***
Absence of double fencing	0.4526	(−0.13, 1.09)	1.5723	0.3067	0.14001		1.611	0.235	5.0078	(3.18, 7.99)	6.57×10^{-12}	***
Presence of flies and ticks	0.7355	(0.15, 1.38)	2.0865	0.3080	0.01696	*	1.936	0.237	6.93	(4.38, 11.12)	3.56×10^{-16}	***
Presence of other animals	0.8089	(0.26, 1.38)	2.2454	0.2866	0.00476	**	1.558	0.231	4.749	(3.04, 7.53)	1.57×10^{-11}	***
Absence of a protocol for disinfection	0.9682	(0.38, 1.61)	2.6332	0.3092	0.00174	**	2.02	0.242	7.538	(4.73, 12.21)	6.61×10^{-17}	***
Cars/trucks enter	0.4669	(−0.07, 1.02)	1.595	0.2779	0.09293		1.401	0.232	4.059	(2.59, 6.44)	1.45×10^{-9}	***
Non-close herd	0.3551	(−0.20, 0.92)	1.4263	0.2829	0.20937		1.693	0.231	5.4357	(3.48, 8.61)	2.45×10^{-13}	***
Personnel without trusted biosecurity	0.9593	(0.37, 1.60)	2.6099	0.3094	0.00193	**	2.408	0.249	11.11	(6.88, 18.26)	3.51×10^{-22}	***
Area with feral pigs	0.4925	(−0.06, 1.05)	1.6364	0.2825	0.08126		1.463	0.233	4.3189	(2.75, 6.87)	3.40×10^{-10}	***

* $p < 0.05$; ** $p < 0.01$; *** $p < 0.001$.

The Hosmer and Lemeshow GOF test resulted in a p-value of 0.8487, suggesting a good model fit for the data.

3.2. World Café Discussion

The activity of estimating infection routes by participants revealed distinct patterns for commercial farms and backyard farms, as indicated by the summary statistics from the four discussion groups (Table 3). For commercial farms, participants believed contaminated vehicles and people were the primary risk factors (49.8% of the hypothetical cases). In contrast, participants indicated that backyard farms face a more diverse range of risks. These include importing new pigs, environmental contamination, vehicles, and concerns regarding feed quality. The participants considered wild boars to pose the lowest risk among all the factors considered in both types of farms.

Table 3. Consensus reached from the elicitation of expert opinion on the probability of ASF introduction route for commercial and backyard farms.

Infection Route	Commercial (Normalized Median, %)	Backyard (Normalized Median, %)
Introduction of sick pigs	9.9	15.8
Environmental contamination (water courses, rice fields next to the farm, contact with backyard farms, etc.)	17.7	13
Contaminated vehicles entering the farm	29.6	16.4
Contaminated people entering the farm	20.2	22.6
Feed	7.9	13.6
Wild boars	0	5.1
Rodents, flies, and other potential vectors	14.8	13.6
Other	0	0
Total	100	100

In terms of clinical presentation, the most prominent clinical signs typically observed in pigs with ASF include a drop in feed intake, huddled pigs, and reluctance to stand up (Table 4). Participants reported that these clinical signs should typically be detected by swine producers. However, they indicated that hematemesis (vomiting blood) and constipation followed by diarrhea have a lower probability of being detected by producers.

For necropsy findings, participants reported that splenomegaly (100%), kidney hemorrhages (80%), and lymphadenomegaly (65%) are among the most prominent observations. Conversely, they felt that the likelihood of producers noticing these signs is low (25%). Participants reported that they less frequently observe pneumonia, hydropericardium, hydrothorax, and shock lung/ARDS in the field, with a median percentage below 15%.

Participants also provided their estimates of the timeline of clinical signs and necropsy finding appearance during ASF infection (Figure 3). According to respondents, affected pigs initially show reduced activity and decreased feed consumption (Day 6). As ASF progresses, they develop fever (Day 7), huddle together (Day 7), and exhibit erythema, or reddening of the skin (Day 8.5). In pigs that survive to later stages, bleeding from the nose, mouth, or rectum (Day 14) and cyanosis (Day 14.5), a bluish discoloration of the skin and mucous membranes may occur.

Table 4. Consensus reached from the elicitation of expert opinion on the frequency and the likelihood seeing clinical signs and necropsy findings.

Clinical Sign	Median Frequency (%)	Median Likelihood of Producer Seeing (%)
Drop in feed consumption	99.5	100
Huddled pigs and other signs of fever	85	100
Reluctances to stand up and move (abnormal recumbence)	85	100
Reddish in the skin (Erythema)	65	100
Abortion	65	100
Increase in mortality or sudden death	64.5	100
Diarrhea with fresh blood (hematochezia)	35	100
Cyanosis (blue areas) of ears and limbs	30	100
Pigs bleeding from nose	28	100
Vomiting with blood (hematemesis)	12.5	75
Constipation followed by diarrhea	10	25
Vomiting	5	100
Necropsy Findings	Median Frequency (%)	Median Likelihood of Producer Seeing (%)
Splenomegaly	100	25
Kidney hemorrhages	80	25
Lymphadenomegaly	65	25
Lymph node hemorrhage or necrosis	50	25
Hemorrhagic intestinal contents	50	66.7
Hydropericardium	11	25
Hydrothorax	11	25
Shock lung/acute respiratory distress syndrome	10	25
Pneumonia	10	25

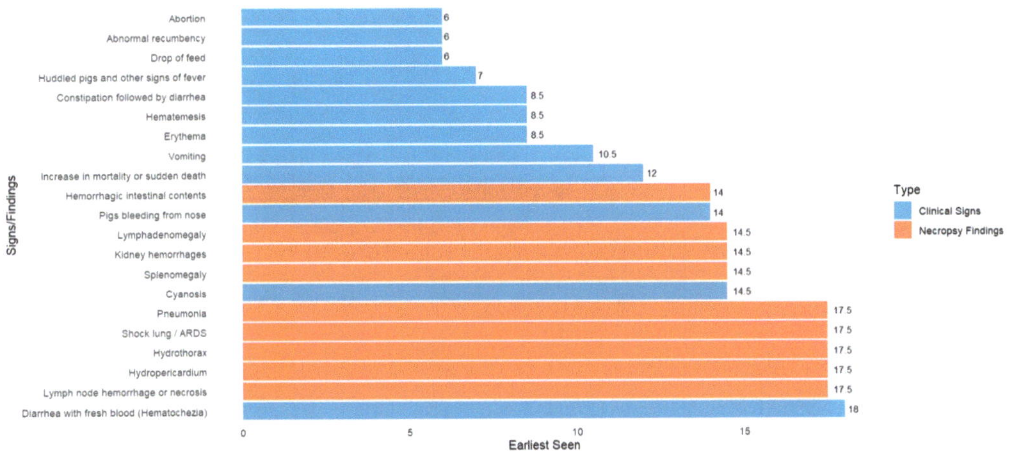

Figure 3. Consensus reached from the elicitation of expert opinion on the earliest timeline of clinical signs and necropsy findings during ASF infection.

Notably, participants reported that ASF clinical signs may be observed as early as Day 6. For necropsy findings, the earliest significant findings are usually observed approximately 14 days after ASF infection (Figure 3, coral orange bars).

Participants also reported on strategies and policies implemented for ASF control by producers. According to estimates from the participants, both commercial and backyard farmers frequently choose to sell their pigs (95%) once ASF is detected to prevent economic losses from total depopulation (Figure 4), which is mandated by Philippine government policies. Notably, the experts estimated that total depopulation is higher in backyard farms (80%) than in commercial farms (50%), but reporting rates of ASF cases are higher among backyard farmers (20%) compared to commercial farms (10%).

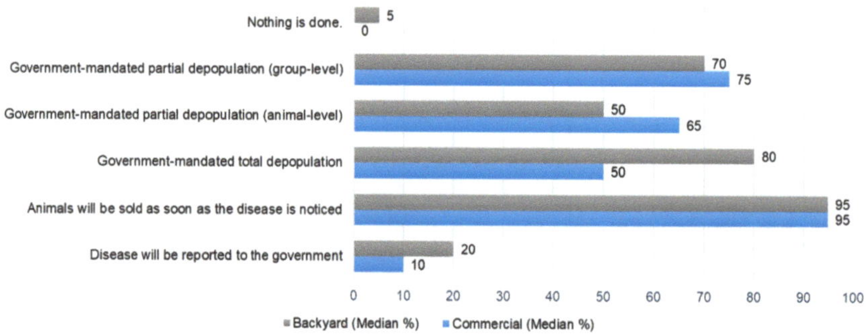

Figure 4. Consensus reached from the elicitation of expert opinion on the estimated control actions by commercial and backyard producers following an on-farm ASF detection.

3.3. SWOT Analysis

The SWOT analysis (Table 5) of ASF control in the Philippines identified strengths such as the National Control Policy and collaboration between government levels. Weaknesses included a lack of compensation and trust, the absence of a traceability system, and resource constraints. Opportunities included using social media and responsible technology, while threats encompassed movement risks and limited veterinarians.

Table 5. Summary of SWOT results from world café discussion.

Strengths	Weaknesses
1. National Control Policy and ASF task force. 2. Collaboration between different levels of government (central level and local governmental units). 3. Evidence-based approach. 4. Philippines Statistics Authority has a regular report of inventory of swine and farmers, including type of production.	1. Farmers face difficulties due to the absence of adequate compensation and a lack of trust in the government's support. This leads to resistance to reporting ASF cases due to the absence of incentives and negative public opinion. 2. The value chain lacks an established traceability system, making it difficult to track and map the routes of hog traders (viajeros), which hinders effective control measures. 3. Lack of resources. There are significant resource constraints in terms of manpower and testing capacity. Limited manpower affects the implementation of control measures, and although the testing capacity has improved, the efficiency of diagnostic PCR testing (RADDL) results in delayed reporting, which hampers timely control efforts.
Opportunities	**Threats**
1. Social media such as Facebook fan page or TikTok for dissemination of ASF prevention information. 2. Rapid adoption of responsible technology in diagnostics. 3. Environmental compliance and regulatory practices for related industries. 4. Recent vaccine trials. 5. Improvement in the execution of biosecurity measures and culture.	1. Risk factors such as human/trade movement and vectors in the Philippines. 2. Dwindling number of new swine veterinarians. 3. Issues with slaughterhouse compliance, tampering with documents, and border control corruption. 4. Time to detection of ASF.

4. Discussion

Despite ASF spreading within the country since 2019, field data describing the clinical presentation of the disease and the factors affecting disease spread and control in the

Philippines are limited, or where present, difficult to access. This may be explained, at least in part, by the challenges associated with collecting and reporting accurate data, such as knowledge of and access to affected farms, the potentially sensitive nature of collected data, and the availability of necessary resources and personnel for data collection. To better understand ASF in the Philippines, a group of practitioners and government officers with field experience with the disease was assembled, and a mixture of qualitative and quantitative methods were used to gather their collective opinion on important features of ASF outbreaks and spread. This paper presents, for the first time, information on the factors affecting the spread, diagnosis, and control of ASF in the Philippines.

Results of the conjoint analysis suggested that swill or contaminated feed would be the most significant factor influencing ASF spread, with a 6.51-fold increase in the odds of ASF occurrence compared to its absence (Table 2). This finding is consistent with previous research in Asia [21,22], indicating that despite government bans, the risk of swill feeding remains significant. However, when examining Table 3 and discussing the risk of ASF transmission in the Philippines, the perceived importance of swill feeding was not as high. The follow-up discussion revealed that the discrepancy was because of a reduction in the frequency of swill-feeding in the Philippines since the beginning of the epidemic. Thus, even though swill feeding is considered the most important factor affecting ASF spread in the country because most farms have reduced the use of swill for feeding, in the event of an outbreak, it is likely that other routes may be responsible for disease spread.

The second highest concern in the swine industry relates to human behavior, particularly the movement of personnel (such as veterinarians, technicians, and workers) between farms without proper biosecurity measures. This factor was substantiated by the data presented in Table 3, and the concern is the entry of contaminated individuals onto the farm, surpassing other risks. Additionally, during the world café discussion and SWOT analysis, the presence of hog traders, also known as middle men, intermediaries or "viajero" locally, was emphasized in the supply chain. Participants agreed that backyard farmers or small commercial producers, who lack direct access to customers, heavily rely on intermediaries or hog traders to facilitate pig sales. This reliance increases the risk of disease transmission, especially if proper biosecurity measures are not followed between farms.

Previous research has shown that the problem of wild boar and its relationship with ASF is underestimated in Asia, and the Philippines has been categorized as a medium-risk level [23]. Based on our conjoint results and world café findings (Table 3), the risk associated with wild boar or feral pigs is not a major concern, at least indicated by the consensus from the world café discussion. In the NCR region, the chance of domestic pigs encountering wild boar is low. However, in regions with higher wild pig populations like Ilocos (Region I), Cagayan Valley (Region II), and Cordillera Administrative Region, an increased likelihood of wild boar contact may exist [24]. Due to the lack of effective control measures for freely roaming cats and dogs [25,26], stray cats and dogs could act as potential mechanical vector animals for ASFV in the Philippines. The accessibility of backyard farms and the shift towards open fencing make it easy for stray animals to enter farms, access swine feeds and carcasses, and move between different farms within the cities. To prevent access from scavengers such as cats and dogs, burying swine carcasses with lime is suggested [27]. These roaming animals may potentially contribute to the spread of ASF, but as of yet, no studies have documented a major role of these animals in ASF spread. Improved biosecurity measures on farms will help keep out unwanted pets and wildlife from swine areas.

The participants also believed that swine producers could recognize most clinical signs associated with ASF in the Philippines. These findings are consistent with previous pathology research [28]. The necropsy findings provide additional confirmation, with splenomegaly, kidney hemorrhages, and lymph node enlargement being the most commonly observed indicators of the disease. However, the participants felt that pneumonia, hydropericardium, and acute respiratory distress syndrome are less frequently identified. This may be because they are associated with later stages of ASF that are not commonly seen. The timing of culling and necropsy procedures, which primarily occur during the early phases of ASF infection, may minimize the frequency of these signs being observed. This early intervention allows for the rapid detection and depopulation of affected swine, reducing the likelihood of severe necropsy findings during the examination by veterinarians. Drop in feed consumption seems to be an important early sign of infection and may be observed by producers as early as 6 days post infection (Figure 3, Table 4). While not specifically explored in the workshop activities, the likelihood of producers noticing specific clinical signs or necropsy findings may be due to differences in producers' individual competency and training.

In terms of control strategies, the participants felt that commercial farms typically lean towards partial depopulation, whereas backyard farms have a higher rate of total depopulation. This may reflect an advantage of commercial farms, as typically they possess better resources and capabilities to manage and mitigate losses during ASF outbreaks. They have established protocols and resources in place to handle the disposal of infected pigs and implement necessary biosecurity measures. In contrast, backyard farmers often rely on government support, particularly for compensation, during ASF outbreaks. Additionally, the smaller pig populations in backyard farms make it easier for them to conduct total culling compared to partial depopulation.

The experts' estimate of a low reporting rate for both commercial and backyard farms (Figure 4) highlights the need to improve ASF reporting for all farm types, such as by providing incentives for timely reporting. As observed during the workshop discussions, this untimely reporting is mainly due to a lack of trust among farmers regarding the effectiveness of government assistance. For example, farmers in both commercial and backyard settings bear the financial burden of burying dead pigs and processing affected animals, which can decrease their willingness to accurately report ASF cases. Concerns about the potential spread of ASF may lead farmers to resist the entry of government officers or veterinarians. These disparities in resources and perceptions highlight the challenges farmers face in managing ASF outbreaks effectively. Weaknesses identified in the SWOT analysis, such as the lack of compensation and trust, further contribute to farmers' perception of inadequate support. To address these issues, the government should develop an early reporting system and prioritize trust-building initiatives between swine farmers and government partnerships.

This research is limited by the potential introduction of sampling bias through the expert selection process, which may inadvertently favor certain perspectives and exclude insights from other stakeholders in the swine industry, including that of producers, for example. However, because much of the experience gained by veterinarians while fighting the disease in the field is not captured in official records (due to underreporting or limitations in the quality of data collected) or scientific publications, elicitation of their opinion is often the most valuable alternative to collect, organize, and share their collective experience with the disease in the field. Although efforts were made to ensure diversity following selection criteria, the final selection of 25 veterinarians, while reflecting a significant portion of industry opinion, may not fully represent the entire range of perspectives and experiences related to ASF in the Philippines. As a result, the findings and conclusions of this research may not encompass the complete spectrum of knowledge and viewpoints on ASF control. Despite these limitations and potential sources of bias, the publication here offers an opportunity to access and evaluate, in an organized way, the consensus that may have been reached in the community of veterinary practitioners in the Philippines working on the control of the disease in the field. For that reason, these results provide valuable input to supplement the assessment of "hard" outbreak data that may be collected in the field, which, by nature, may also be biased and limited.

In summary, the study findings and expert estimates suggest that human behavior might be the most important factor affecting the spread of ASF. The elicitation of the opinion of practitioners with field experience fighting ASF in the Philippines suggested that the most important factors for ASF introduction were swill feeding, movement of personnel without proper biosecurity, and the absence of disinfection protocols, while wild boar are only a concern in regions with higher feral pig populations. This consensus suggests that any effort made to avoid the entrance of vehicles or individuals into susceptible farms may be highly impactful in preventing disease spread in the Philippines. Overall, these findings contribute to a better understanding of ASF spread, diagnosis, and control. They may support government policy development in the Philippines and contribute to enhancing preparedness in ASF-free areas worldwide.

Supplementary Materials: The following supporting information can be downloaded at: https://www.mdpi.com/article/10.3390/pathogens12081068/s1.

Author Contributions: Conceptualization, C.-H.H., R.S. and A.P.; Data Curation, C.-H.H. and A.P.; Formal Analysis, C.-H.H. and A.P.; Funding Acquisition, A.P.; Investigation, C.-H.H., M.M., R.M.-S. and A.P.; Methodology, C.-H.H., R.S. and A.P.; Project Administration, C.-H.H., M.M. and R.M.-S.; Resources, M.M. and R.M.-S.; Software, C.-H.H.; Supervision, A.P.; Validation, M.M. and R.M.-S.; Visualization, C.-H.H.; Writing—Original Draft, C.-H.H.; Writing—Review and Editing, R.S. and A.P. All authors have read and agreed to the published version of the manuscript.

Funding: This research has been funded in part by grants from the USDA Foreign Agricultural Service (EMP-2022-09) and the USDA Agricultural Research Services (NACA 58-8064-2-005).

Institutional Review Board Statement: The work here summarizes a series of anonymous opinions collected as part of an educational program and for that reason IRB review is not required.

Informed Consent Statement: Not applicable.

Data Availability Statement: The data, which are the property of the Philippines government and have been shared with us to conduct research here in support of training activities in the country, can be obtained on request from the corresponding author. For inquiries, kindly direct requests to the Philippine Bureau of Animal Industry and the International Training Center on Pig Husbandry of the Philippines.

Acknowledgments: The authors would like to express gratitude to the Bureau of Animal Industry and International Training Center on Pig Husbandry of the Philippines, for their invaluable assistance in facilitating the data analysis and organizing workshops in this study. Special thanks are also extended to Meng-Hsuan Wu for his assistance in statistical modeling and data processing using RStudio and to Mary Katherine O'Brien for her efforts in data collection.

Conflicts of Interest: The authors declare no conflict of interest.

References

1. Blome, S.; Franzke, K.; Beer, M. African Swine Fever—A Review of Current Knowledge. *Virus Res.* **2020**, *287*, 198099. [CrossRef] [PubMed]
2. Dixon, L.K.; Sun, H.; Roberts, H. African Swine Fever. *Antivir. Res.* **2019**, *165*, 34–41. [CrossRef] [PubMed]
3. Ge, S.; Li, J.; Fan, X.; Liu, F.; Li, L.; Wang, Q.; Ren, W.; Bao, J.; Liu, C.; Wang, H.; et al. Molecular Characterization of African Swine Fever Virus, China, 2018. *Emerg. Infect. Dis.* **2018**, *24*, 2131–2133. [CrossRef] [PubMed]
4. OIE. *African Swine Fever, Philippines (Immediate Notification)*; OIE: Manila, Philippines, 2019.
5. Hsu, C.-H.; Montenegro, M.; Perez, A. Space–Time Dynamics of African Swine Fever Spread in the Philippines. *Microorganisms* **2023**, *11*, 1492. [CrossRef] [PubMed]
6. Tambago, E.R. Philippine Hog Industry Roadmap Development Team. Available online: https://psa.gov.ph/content/swine-situation-report-january-march-2023 (accessed on 20 June 2023).
7. Swine Situation Report, January to March 2023. Available online: www.pcaf.da.gov.ph/wp-content/uploads/2022/06/Philippine-Hog-Industry-Roadmap-2022-2026.pdf (accessed on 20 June 2023).
8. Gaudreault, N.N.; Madden, D.W.; Wilson, W.C.; Trujillo, J.D.; Richt, J.A. African Swine Fever Virus: An Emerging DNA Arbovirus. *Front. Vet. Sci.* **2020**, *7*, 215. [CrossRef] [PubMed]
9. Rao, V.R. *Applied Conjoint Analysis*; Springer: Berlin/Heidelberg, Germany, 2014; ISBN 978-3-540-87752-3.
10. Applied Conjoint Analysis | SpringerLink. Available online: https://link-springer-com.ezp2.lib.umn.edu/book/10.1007/978-3-540-87753-0 (accessed on 20 June 2023).
11. Van Den Borne, B.H.P.; Van Soest, F.J.S.; Reist, M.; Hogeveen, H. Quantifying Preferences of Farmers and Veterinarians for National Animal Health Programs: The Example of Bovine Mastitis and Antimicrobial Usage in Switzerland. *Front. Vet. Sci.* **2017**, *4*, 82. [CrossRef] [PubMed]

12. Schettino, D.N.; Abdrakhmanov, S.K.; Beisembayev, K.K.; Korennoy, F.I.; Sultanov, A.A.; Mukhanbetkaliyev, Y.Y.; Kadyrov, A.S.; Perez, A.M. Risk for African Swine Fever Introduction into Kazakhstan. *Front. Vet. Sci.* **2021**, *8*, 605910. [CrossRef] [PubMed]
13. Teoli, D.; Sanvictores, T.; An, J. SWOT Analysis. In *StatPearls*; StatPearls Publishing: Treasure Island, FL, USA, 2022.
14. Do a SWOT Analysis. Available online: https://www.cdc.gov/publichealthgateway/phcommunities/resourcekit/evaluate/do-a-swot-analysis.html (accessed on 28 February 2022).
15. Schambow, R.; Reyes, R.; Morales, J.; Diaz, A.; Perez, A.M. A Qualitative Assessment of Alternative Eradication Strategies for African Swine Fever in the Dominican Republic. *Front. Vet. Sci.* **2022**, *9*, 1054271. [CrossRef] [PubMed]
16. Schettino, D.M.; Perez, A.; Lantigua, E.; Beemer, O.; Remmenga, M.; Vanicek, C.; Lopes, G.; Arzt, J.; Reyes, R. Enhanced Passive Surveillance for Early Detection of African and Classical Swine Fevers: -EN- Enhanced Passive Surveillance for Early Detection of African and Classical Swine Fevers -FR- Surveillance Passive Renforcée Pour La Détection Précoce Des Pestes Porcines Africaine et Classique -ES- Vigilancia Pasiva Reforzada Para La Pronta Detección de Las Pestes Porcinas Africana y Clásica. *Rev. Sci. Tech. OIE* **2023**, *42*, 149–160. [CrossRef]
17. Organisation des Nations Unies pour L'alimentation et L'agriculture (Ed.) *Good Practices for Biosecurity in the Pig Sector: Issues and Options in Developing and Transition Countries*; FAO Animal Production and Health Paper; FAO: Rome, Italy, 2010; ISBN 978-92-5-106507-5.
18. Bremang, A.; Ho, H.P.J.; Conan, A.; Tang, H.; Oh, Y.; Pfeiffer, D.U. *Guidelines for African Swine Fever (ASF) Prevention and Control in Smallholder Pig Farming in Asia*; Food and Agriculture Organization of the United Nations: Québec City, QC, Canada, 2022.
19. Venables, W.N.; Ripley, B.D. *Modern Applied Statistics with S*, 4th ed.; Springer: New York, NY, USA, 2002; ISBN 0-387-95457-0.
20. Lele, S.R.; Keim, J.L.; Solymos, P. ResourceSelection (Probability) Functions for Use-Availability 2019. Available online: https://cran.r-project.org/web/packages/ResourceSelection/index.html (accessed on 20 June 2023).
21. Cheng, J.; Ward, M.P. Risk Factors for the Spread of African Swine Fever in China: A Systematic Review of Chinese-language Literature. *Transbound. Emerg. Dis.* **2022**, *69*, e1289. [CrossRef] [PubMed]
22. Kedkovid, R.; Sirisereewan, C.; Thanawongnuwech, R. Major Swine Viral Diseases: An Asian Perspective after the African Swine Fever Introduction. *Porc. Health Manag.* **2020**, *6*, 20. [CrossRef] [PubMed]
23. Cadenas-Fernández, E.; Ito, S.; Aguilar-Vega, C.; Sánchez-Vizcaíno, J.M.; Bosch, J. The Role of the Wild Boar Spreading African Swine Fever Virus in Asia: Another Underestimated Problem. *Front. Vet. Sci.* **2022**, *9*, 844209. [CrossRef] [PubMed]
24. Distribution of Sus Philippensis. Red—Extant (Resident). Data Sourced from IUCN Red List, World Bank Country Polygons (Licensed under Creative Commons Attribution 4.0); Amante, C.; Eakins, B.W. ETOPO1 1 Arc-Minute Global Relief Model: Procedures, Data Sources and Analysis. NOAA Technical Memorandum NESDIS NGDC-24, 19p. 2009. Available online: https://commons.wikimedia.org/wiki/File:Range_Sus_philippensis.png (accessed on 20 June 2023).
25. Fishbein, D.B.; Miranda, N.J.; Merrill, P.; Camba, R.A.; Meltzer, M.; Carlos, E.T.; Bautista, C.F.; Sopungco, P.V.; Mangahas, L.C.; Hernandez, L.M.; et al. Rabies Control in the Republic of the Philippines: Benefits and Costs of Elimination. *Vaccine* **1991**, *9*, 581–587. [CrossRef] [PubMed]
26. Dizon, T.J.R.; Saito, N.; Inobaya, M.; Tan, A.; Reñosa, M.D.C.; Bravo, T.A.; Endoma, V.; Silvestre, C.; Salunga, M.A.O.; Lacanilao, P.M.T.; et al. Household Survey on Owned Dog Population and Rabies Knowledge in Selected Municipalities in Bulacan, Philippines: A Cross-Sectional Study. *PLoS Negl. Trop. Dis.* **2022**, *16*, e0009948. [CrossRef] [PubMed]
27. Carlson, J.; Fischer, M.; Zani, L.; Eschbaumer, M.; Fuchs, W.; Mettenleiter, T.; Beer, M.; Blome, S. Stability of African Swine Fever Virus in Soil and Options to Mitigate the Potential Transmission Risk. *Pathogens* **2020**, *9*, 977. [CrossRef] [PubMed]
28. Sánchez-Vizcaíno, J.M.; Mur, L.; Gomez-Villamandos, J.C.; Carrasco, L. An Update on the Epidemiology and Pathology of African Swine Fever. *J. Comp. Pathol.* **2015**, *152*, 9–21. [CrossRef] [PubMed]

pathogens

MDPI

Article

Understanding Smallholder Pigkeepers' Awareness and Perceptions of African Swine Fever and Its Control Measures in Ukraine

Lidiia Moskalenko [1,*], Katja Schulz [2], Vitalii Nedosekov [3], Kerli Mõtus [1] and Arvo Viltrop [1]

1 Institute of Veterinary Medicine and Animal Sciences, Estonian University of Life Science, 51014 Tartu, Estonia; kerli.motus@emu.ee (K.M.); arvo.viltrop@emu.ee (A.V.)

2 Institute of Epidemiology, Friedrich-Loeffler-Institut, Federal Research Institute for Animal Health, 17493 Greifswald-Insel Riems, Germany; katja.schulz@fli.de

3 Department of Epizootology, National University of Life and Environmental Science of Ukraine, 03041 Kyiv, Ukraine; nedosekov06@gmail.com

* Correspondence: lidiia.moskalenko@emu.ee

Abstract: African swine fever (ASF) has posed a significant threat to Ukrainian pig farming since its identification in 2012. In this study, recognising the pivotal role of pigkeepers in disease control, we conducted ten focus groups involving 52 smallholders across eight regions in Ukraine. Using participatory methods, we revealed their awareness of ASF signs, transmission routes, preventive measures, and the perceptions of stakeholders involved in ASF control. Furthermore, we identified the smallholders' acceptance of eradication and restriction measures, the perceived impact of zoning consequences, and their main sources of ASF information. Smallholders identified fever and skin haemorrhage as the most indicative signs of ASF and highlighted rodents as a primary transmission concern. Disinfection was seen as the most effective measure for preventing the introduction of ASF. Pigkeepers who perceived their stakeholder role in ASF control showed more trust in themselves and veterinarians than in central veterinary authorities. Farm-level ASF eradication measures were generally accepted; however, culling within the protection zone was least accepted, with economic losses listed as the most impactful consequence for pigkeepers. For ASF information, pigkeepers favour web searches and veterinarians, as well as traditional media and word-of-mouth news. This study provides valuable insights into refining the ASF communication strategies in Ukraine.

Keywords: African swine fever; smallholder pig farmers; participatory epidemiology; control and preventive measures; biosecurity; acceptability

Citation: Moskalenko, L.; Schulz, K.; Nedosekov, V.; Mõtus, K.; Viltrop, A. Understanding Smallholder Pigkeepers' Awareness and Perceptions of African Swine Fever and Its Control Measures in Ukraine. *Pathogens* **2024**, *13*, 139. https://doi.org/10.3390/pathogens13020139

Academic Editors: Lauro Velazquez-Salinas, Gisselle N. Medina and Elizabeth Ramirez-Medina

Received: 31 December 2023
Revised: 25 January 2024
Accepted: 29 January 2024
Published: 3 February 2024

1. Introduction

African swine fever (ASF) is a viral haemorrhagic disease that affects domestic pigs and wild boars and has serious socioeconomic consequences in affected countries. When ASF outbreaks occur, the impact can be significant, affecting not only the pig industry, but also other sectors of the economy [1]. The first incursion of the African swine fever virus (ASFV) into Ukraine dates back to 1977, when it was introduced to the southwestern part of the country, presumably through food waste from ships arriving at the Odesa city seaport. The outbreak was successfully contained within six months by implementing strict control measures and eradicating the entire pig population in the affected region [2]. In July 2012, an ASF outbreak was reported in a small holding in the Zaporizhzhia region, in the southeast of Ukraine. The most likely route of virus introduction was swill feeding [3]. No further outbreaks were reported until January 2014, when the virus was first identified in wild boars in the Luhansk region. This region borders the Russian Federation, where the last cases of ASF in wild boars were reported in December 2013. After the first detection of ASF in 2014, the disease spread throughout the domestic pig and wild boar populations, eventually affecting all administrative regions of Ukraine by the end of 2017. Ukrainian ASF surveillance identified an ongoing circulation of ASFV in both the wild boar and domestic pig populations [4].

In less than a decade, the pig sector of Ukraine has undergone significant transformations, characterised by a decrease in the number of domestic pigs within smallholdings. As of 2020, the pig population is estimated to be 3.3 million in the commercial sector and 2.43 million in smallholdings. While domestic pig herds in the smallholder sector are distributed throughout the entire country, their number is greatest in the western regions of Ukraine [5]. Ukrainian law defines smallholders as individuals within families whose products are used for subsistence within the household and sold at marketplaces or directly to other persons if the production exceeds family consumption needs [6]. Therefore, these pigs also provide an additional source of income for households. Pigkeepers must identify and register their households and pigs in the Unified State Register of Animals in Ukraine no later than 60 days after the animal is born [7]. The Ukrainian government implemented an order stipulating measures for the prevention and eradication of ASF. In the case of a suspicion of ASF in pigs, the pigkeeper must inform the veterinary office of their respective district or city. The chief veterinary officer is then responsible for initiating disease eradication measures within the affected farm and the surrounding region [8]. In response to the quarantine measures enforced to control and prevent ASF, the state provides compensation to smallholder pigkeepers [9], with payments applicable for pigs registered in the state animal register [10].

Throughout the ASFV epidemic during the last century in Ukraine, tackling the challenge of ASF control was particularly complex for smallholdings. This complexity stems from the frequent outbreaks that occur predominantly within these farms [4]. When addressing such epidemiological scenarios, the primary emphasis should be placed on control measures targeting the factors that contribute to the spread of ASFV. In this context, the virus seems to have been transmitted to new areas primarily because of the uncontrolled movement of live pigs and pig meat between villages [11,12]. Factors associated with human behaviour have certainly played an important role in many countries during ASF outbreaks, particularly in regions where smallholder farming is prevalent [13–16]. Pigkeepers play an important role in controlling and preventing ASF in domestic pigs through the implementation of biosecurity measures. Consequently, farmers' willingness to accept ASF control and eradication measures holds significant importance in ensuring the effectiveness of the control system. A fundamental aspect of this is that farmers have a strong understanding of the basic characteristics of the disease. Following the introduction of ASF into Ukraine in 2012, international campaigns have been implemented in various regions to promote the awareness of ASF, its transmission pathways, and adequate biosecurity measures [17–19].

This study targeted smallholder pigkeepers who primarily keep pigs for their own consumption. In recent ASF epidemics in Europe, controlling the disease in the domestic pig population has been particularly challenging in countries with substantial smallholder pig sectors [20,21]. This may be due to a lack of awareness of ASF among smallholders. Another potential factor contributing to this situation might be inadequate communication between the authorities responsible for disease control and the pigkeepers. This communication gap may lead to misunderstandings between authorities and pigkeepers, ultimately reducing the acceptance of control measures and compliance with them.

In this study, we aimed to investigate the awareness, perceptions, and attitudes of smallholder pigkeepers towards ASF and its control measures in Ukraine. More specifically, we focused on the awareness of ASF's clinical signs, transmission routes, and preventive measures. We aimed to reveal the pigkeepers' awareness of the stakeholders involved in ASF control and their level of trust in these stakeholders. We also sought to explore how smallholder pigkeepers accept eradication measures and restrictions during ASF outbreaks. In addition, our objective was to determine which sources of information on ASF were used and deemed important by pigkeepers.

2. Materials and Methods

2.1. Organization of Focus Groups (FGs)

The target group of this study was smallholder farmers rearing up to ten pigs on their farms in Ukraine. The study participants were recruited using convenience sampling. Participants were contacted via phone calls, personal contacts, or publicly available contacts from Facebook groups dedicated to swine welfare. During the invitation process to join the focus group (FG), all participants were informed about the study's organisation, study aims, and voluntary participation, including their ability to drop out without any reason at any time. When organising the focus group discussions (FGDs), the goal was to include smallholder pigkeepers from different parts of Ukraine, with three to seven participants per group. Prior to their first implementation in the field, FGDs were conducted and study questions administered at the Estonian University of Life Sciences by a volunteer FG consisting of students with a veterinary background. The participants' feedback was incorporated into the FGs' further implementation.

Meetings were conducted from May to October 2021 on the property of one participant. The FGs were moderated in Ukrainian and Russian by the same female facilitator. Meetings were conducted by the facilitator, who was trained in the methodology applied. Every participant provided positive oral consent to audio recording and anonymous use of the gathered data for scientific publications. To keep the FGs anonymous, participants' names and demographic information, such as age and sex, were not recorded. The facilitator then transcribed the meeting recordings.

2.2. Tasks of the FGD and Employed Participatory Tools

The meetings were organised into eight tasks. Each aimed at investigating various aspects related to ASF control in smallholder farms where domestic pigs are kept. To quantify the awareness, acceptability, and affinity of pigkeepers, we employed proportional piling and visualisation tools, drawing inspiration from previous studies [22–25]. Furthermore, we introduced a 10-bead scoring tool during Tasks 4 and 6, allowing the group to collectively score each listed item based on the questions presented. For this, 10 glass beads per listed item were provided to the participants, which they then allocated based on their evaluations. Qualitative findings from the recorded FGDs were descriptively incorporated into the analysis.

2.3. Task 1: Awareness of ASF Signs in Domestic Pigs

In this task, participants were asked to list signs that might lead them to suspect ASF within the herd. Using proportional piling, they were asked to express their opinions on which of these signs would lead them to suspect ASF in the herd. The group was instructed to reach a consensus on the allocation of 100 glass beads, with the distribution reflecting the perceived indicative value assigned to each sign.

Through the use of participatory tools and discussions, the facilitator regularly enquired into participants' statements and decisions to confirm their accurate understanding of the task and foster further discussions.

2.4. Task 2: Awareness of ASFV Transmission Routes

The group was tasked with listing all possible transmission routes for the introduction of ASFV into smallholdings. Then, they were asked to express their opinion on which of these transmission routes posed the highest risk of ASF introduction into smallholdings (proportional piling).

2.5. Task 3: Awareness and Attitude towards ASF Preventive Measures

The group was instructed to list preventive practices that could be applied to avoid the introduction of ASFV into a smallholding. Then, participants were asked to express their opinions on the effectiveness of the listed measures at preventing the introduction of ASF (proportional piling).

Subsequently, the participants were asked to express their personal opinions on how much they liked implementing these measures, setting aside their effectiveness. Each individual's opinion was expressed using face emojis (Figure 1), accompanied by verbal explanations for the given evaluation. The emojis were colour-coded, green signifying a positive response, yellow for neutral, and red for a negative response, to visually represent each participant's stance. For semi-quantitative analysis, each emoji was assigned a rank.

Figure 1. Visualization tool that was used to illustrate pigkeepers' personal opinions.

2.6. Task 4: Perception of and Trust towards Stakeholders Involved in ASF Control

For this task, participants were asked to name all the institutions and stakeholders contributing to the compliance, execution, and surveillance of control and preventive measures for ASF. Participants explained how they perceived the role of each listed stakeholder in ASF control. If 'pigkeepers' were not initially listed as stakeholders, the facilitator suggested their inclusion, recognising that participants might not have self-identified in relation to the question. This prompted further discussion in which participants could decide whether to include themselves. Then, they were asked to express their opinion on which of these stakeholders played a larger role in preventing and fighting ASF in the country (proportional piling).

Subsequently, participants were asked to express their trust in the capabilities of the listed stakeholders to fulfil their respective roles in ASF control. Using a 10-bead scoring tool, ten glass beads were designated to each stakeholder on the list. The group assigned scores to reflect their collective opinion on a scale ranging from zero (indicating complete mistrust) to 10 (complete trust).

2.7. Task 5: Acceptability of Farm-Level ASF Eradication Measures

The facilitator presented a list of four ASF eradication measures applied to outbreak herds and ensured uniform understanding among participants.

- Farm quarantine;
- Culling of all pigs on the farm;
- Destroying the feed and bedding materials on the farm;
- Cleaning and disinfection of the farm.

The participants were then invited to express their personal acceptance of each measure (face emojis) and to verbally articulate the rationale behind their assessments.

2.8. Task 6: Acceptability of Measures Applied in ASF Restricted Zones

The facilitator presented a list of three restriction measures applied during zoning in the case of an ASF outbreak and ensured uniform understanding among participants.

- Culling of all pigs in the protection zone;
- Restrictions on moving pigs in the protection zone;
- Restrictions on trading live pigs and pork products in the protection and surveillance zones.

The participants were asked to express their collective acceptance of each listed restriction measure (10-bead scoring tool).

2.9. Task 7: Consequences of ASF Zoning

The group was provided with information indicating that restricted zones were established around ASF-affected farms, which led to restrictions on the movement of pigs and trade of pigs and pork products. Participants were then asked to identify the consequences for pigkeepers when their farms fell within these restricted zones. Then, they were asked to allocate glass beads to express their opinion regarding which of these consequences had the greatest impact on pigkeepers (proportional piling).

2.10. Task 8: Sources of Information about ASF

In the last task, participants were asked to name all the sources from which they had received or expected to receive information about ASF. Then, were instructed to indicate the relative importance of the information sources (proportional piling).

2.11. Data Management and Analysis

After each FG meeting, the collected data were entered into Microsoft Excel (2019) spreadsheets and translated into English. After analysing the audio recordings, the entire dataset in the Excel file was rechecked to avoid possible insertion mistakes. The listed items with similar contents were then merged into a single entity by creating a common notation (see the example in Table 1).

Table 1. Merging of listed items related to the veterinarian into a common notation.

N of Groups Naming the Listed Item	Listed Item	Notation of the Merged Item
2	Veterinarian	Veterinarian
1	Vet workers	

Numerical data from the division of 100 glass beads were used to analyse the proportional piling tool. The sum of the weighted proportional piling score S_i for each listed item i was computed as follows:

$$S_i = \sum_{j=1}^{10} GB_{ij} \cdot \frac{N_j}{N_{total}}, \quad (1)$$

where GB_{ij} indicates the number of glass beads assigned to the listed item i by one group j; N_j is the number of items listed by one individual group j during the task; and N_{total} is the sum of all listed items in one task by all 10 groups.

For the analysis of face emojis, a numerical value was allocated to each response: +1 for positive, 0 for neutral, and −1 for negative (see Figure 1). The numerical rank values assigned by all voting participants to each listed item across the ten meetings were then summarised. Finally, the sum was divided by the total number of voting participants to obtain the average emoji score.

For the analysis of the 10-bead scoring tool, the mean score value for each listed item was calculated by summing the scores given by participants from all focus groups in which it was listed.

3. Results

3.1. Recruitment of Participants

In total, ten meetings were conducted with three to seven participants per FG. The pigkeepers (n = 52), 13 male and 39 female, were recruited from eight administrative regions (Table 2, Figure 2). The estimated age of the participants ranged from 20 to 80 years, with approximately 6 aged under 40 years old, 7 between 40 and 49 years old, 19 between 50 and 59 years, and 20 over 60 years. None of the participating pigkeepers had directly experienced the confirmed outbreak, although one participant was affected by the restrictions imposed in an ASF surveillance zone.

Table 2. Locations of ten focus group meetings and the number of participating pigkeepers.

Administrative Region [a]	No. of Participating Pigkeepers	Gender Composition within the Group	
		Male	Female
Kyiv	5	1	4
Odesa	3	2	1
Cherkasy	4	2	2
Cherkasy	3	1	2
Sumy	7	0	7
Kyiv	5	0	5
Chernihiv	7	0	7
Vinnytsa	7	1	6
Zhytomyr	5	4	1
Rivne	6	2	4
Total	52	13	39

[a] Highest administrative division of the country.

Figure 2. Map of Ukraine with regions in blue depicting where FGs were conducted.

3.2. *Awareness of ASF Signs in Domestic Pigs*

Among the 16 listed signs, fever was considered the most indicative sign that would incline pigkeepers to suspect ASF in the herd, ($n = 10$). Skin haemorrhage was ranked as the second most indicative sign ($n = 8$), with three groups mentioning that this could also be a characteristic sign of other diseases, such as erysipelas. Additionally, two groups that did not mention skin haemorrhages revealed that they lacked knowledge of the characteristic signs of ASF, yet would assume a viral infection in the case of high fever, and accordingly consult a local veterinarian. The other parameters are summarised in Table 3.

Table 3. ASF signs listed by pigkeepers and ranked according to their level of indicativeness.

Rank	Sign	*n* Groups Listing	Perceived Indicativeness (Score [a])
1	Fever	10	82.2
2	Skin haemorrhage	8	79.3
3	Loss of appetite	8	38.8
4	Lethargy	6	35.9
5	Difficulty standing (weakness)	3	16.0
6	Increased mortality	2	10.6
7	Enlarged lymph nodes	1	8.3
8	Lethargy, loss of appetite	1	6.4
9	Foam from the mouth	1	4.8
10	Diarrhea	1	3.8
11	Constipation	1	3.8
12	Red ears	1	2.3
13	Rickets	1	1.8
14	Swollen joints	1	0.0
14	Dyspnea	1	0.0
14	Head swelling	1	0.0

[a] Sum of the weighted proportional piling scores.

3.3. *Awareness of ASFV Transmission Routes*

Among the 18 listed transmission routes, rodents were identified as posing the highest risk for ASFV introduction to the pig herd ($n = 5$). Pigkeepers cited the easy access of rodents to pig holdings, transport vehicles, fields, forests, and barns where feed was stored. Shoes were ranked as the second riskiest route ($n = 8$). Pigkeepers explained that the virus could be transmitted by the dirty footwear of visitors, such as veterinarians, animal dealers, and pig owners. Transport vehicles were designated the third riskiest route ($n = 7$). During the discussions, all vehicles entering the village or territory of the pigkeepers' holdings were considered potential risk factors. The fourth riskiest route was human-related transmission and activity ($n = 6$). For both transport vehicles and humans, participants mentioned dealers who buy pigs and travel long distances as well as veterinarians who use the same pair of shoes and clothes for all their client visits throughout the day. Feed was identified as the fifth riskiest route of ASFV introduction (n = 6). Half of the groups cited using commercially sourced grain, which could potentially be exposed to multiple contacts, while the other half mentioned grains grown in their own fields; both could be at risk of rodent encounters during storage, as noted by pigkeepers. Beyond grains, the participants discussed other supplements to the pig diet, such as commercial feed, boiled vegetables, and herbs. One group mentioned providing kitchen leftovers. Wild animals, including wild boars, ranked eighth in terms of risk factors (n = 5). It was mentioned that wild animals may still find ways to enter villages or access the territories of pigkeepers, including their fields. Table 4 summarises the other transmission routes.

Table 4. Transmission routes, as listed by pigkeepers, ranked according to the perceived risk of ASFV introduction to the herd.

Rank	Transmission Route	*n* Groups Listing	Perceived Risk of ASFV Introduction (Score [a])
1	Rodents	5	52.7
2	Shoes	6	43.6
3	Transport vehicles	7	34.4
4	Human	6	30.9
5	Feed	6	27.5
6	Air	6	24.6
7	Insects	3	23.1
8	Wild animals incl. wild boar	5	17.6

Table 4. *Cont.*

Rank	Transmission Route	*n* Groups Listing	Perceived Risk of ASFV Introduction (Score [a])
9	Clothes	2	12.7
10	Other domestic animals	3	11.1
11	Dust	1	10.1
12	Bought new young animals	1	8.9
13	Shoes, clothes	2	8.1
14	All animals	1	6.6
15	Birds	1	5.3
16	Semen	2	4.4
17	Meat washing water	1	3.6
18	Through the meat of infected pigs	1	2.7

[a] Sum of the weighted proportional piling scores.

3.4. Awareness and Attitude towards ASF Preventive Measures

Among the 18 listed preventive measures (Table 5), disinfection of pig premises ranked first in terms of effectiveness in preventing the introduction of ASF into herds (n = 10). This was moderately favoured due to the chemical nature of the disinfectants, which could have an unpleasant smell and adverse effects on the lungs and eyes (e.g., chlorine). Some participants disclosed a preference for lime treatment on their walls because it helps maintain a clean environment for animals. Access bans for people were ranked as the second most effective measure (n = 6). This was highly favoured for its ability to isolate animals from potential sources of contamination and its ease of implementation. Changing clothes and shoes was ranked as the third most effective preventive measure (n = 6). Pigkeepers indicated that they typically have separate shoes for working with animals and in the field, but that having separate clothes is less common. The use of a 'vaccine' as a preventive measure was mentioned once by a group that claimed to have documents from a veterinarian that their pigs were vaccinated against ASF and classical swine fever (CSF) viruses.

Table 5. Preventive measures listed by pigkeepers, ranked based on their perceived effectiveness in preventing ASF introduction to the farm, and described in terms of pigkeepers' affinity towards their implementation.

Pigkeepers' Affinity (Score [a])	Preventive Measure	*n* Groups Listing	Perceived Effectiveness (Score [b])
0.26	Disinfection of pig premises	9	56.5
0.96	Access ban for people	6	44.7
0.46	Changing clothes and shoes	6	27.2
0.73	Control of rodents and insects	3	25.4
1.00	Vaccine	1	16.7
−0.95	Manure removal	3	12.3
1.00	Changing bedding	2	11.8
1.00	Heat treatment of feed	1	9.7
0.00	Disinfection mats	4	8.4
1.00	Minimization of contact between pigkeepers	1	6.9
−0.46	Cleaning and disinfecting of troughs	2	6.8
1.00	Keeping pigs in outdoors enclosures for fresh air and sunny weather	1	5.8
0.50	Washing of pigs with water	2	5.8
1.00	Hygienic procedures for humans (showering, sauna)	1	4.7
0.00	Disinfection of pig premises, disinfection mats	1	4.0
0.40	Treatment of pigs against mites (kerosene)	1	3.6
1.00	Keeping feed with limited access to rodents	1	2.9
0.67	All-in-all-out management system	1	2.2

[a] Average emoji score; [b] sum of the weighted proportional piling scores.

3.5. Perception of and Trust towards Stakeholders Involved in ASF Control

The participants listed 14 stakeholders who contributed to the development, enforcement, and implementation of ASF control measures. Pigkeepers were assigned the highest rank based on their perceived role in controlling ASF (n = 10). From the discussions, it appeared that pigkeepers were recognised as responsible for implementing measures to ensure pig welfare and adhering to instructions provided by veterinarians and authorities.

Pigkeepers expressed a high level of trust in their fellow pigkeepers, considering themselves reliable in carrying out their roles in ASF control. They also displayed considerable trust in the district veterinary hospital and local veterinarians (both private and official), recognising their competence and expertise in the field. At the same time, the level of trust in the central veterinary authority or village administration was relatively lower. During the group discussions, the pigkeepers' level of trust appeared to be linked to the perception that the highest level of veterinary authority and governmental bodies merely issued directives without sufficient engagement in or consideration of farmers' concerns. Furthermore, many participants expressed concerns that authorities at various levels might make efforts to avoid providing compensation for depopulated pigs. A summary of the other stakeholders is provided in Table 6.

Table 6. Stakeholders in ASF control, listed by pigkeepers and ranked by the perceived role of each stakeholder and trust regarding their capability to carry out their role in ASF control.

Trust (Score [a])	Stakeholder	n Groups Listing	Perceived Role (Score [b])
8.8	Pigkeepers	10 (6/4) [c]	118.0
4.6	Central veterinary authority	5	54.3
7.7	District veterinary hospital (district veterinary office)	3	38.2
7.0	Local private veterinarian	3	32.5
4.8	Village administration	6	17.5
10	Local official veterinarian	2	16.8
2.0	Ministry of Emergency Situations	3	15.7
4.4	Police	4	9.5
5.0	Professional hunters	1	6.4
4.0	Sanitary station	2	6.4
1.7	District Administration	3	6.1
3.0	Veterinary laboratory	1	3.6
0.0	District waste management service	1	2.0
10	Military	1	0.0

[a] 10-bead scoring tool; [b] sum of the weighted proportional piling scores; [c] independently/after the suggestion of the facilitator.

3.6. Acceptability of Farm-Level ASF Eradication Measures

The most widely accepted eradication measures were the cleaning and disinfection of the farm, followed by farm quarantine (Table 7). Both were seen as justified measures with the highest impact on virus eradication in the infected area, without requiring additional effort or investments. Culling all animals ranked third in terms of acceptance by the participants. While it was regarded as a justified measure, it was less favoured because of the established emotional bond with the pigs. Destroying feed and bedding materials on farms was the least accepted measure. Most groups claimed that the destruction of commercial feed and additives for pigs leads to economic losses. Expressed arguments highlighted the fact that commercial feed is typically stored in its original bags in separate barns, which reduces the probability of contamination. Some participants expressed a preference for having a document confirming feed contamination to justify its destruction. In addition, one group suggested that uncontaminated grains could be used for human consumption. The destruction of bedding materials was considered a justified measure for eradicating the disease.

Table 7. The acceptance of farm-level ASF eradication measures by pigkeepers.

On-Farm Eradication Measure	Acceptance (Score [a])
Cleaning and disinfection of the farm	1
Farm quarantine	0.9
Culling of all pigs on the farm	0.4
Destroying the feed and bedding materials on the farm	0.3

[a] Average emoji score.

3.7. *Acceptability of Measures Applied in ASF Restricted Zones*

Culling all the pigs in the protection zone was the least accepted control measure. Six groups evaluated this measure positively, mentioning its importance for preventing further spread of the virus. Others questioned the necessity of culling if pigs were healthy and showed no clinical signs of infection. The explanations for the lowest scores were linked to financial losses, potential stress, feelings of useless effort, and wasted time. Restrictions on moving animals and trading live pigs and pork products were broadly accepted by pigkeepers (Table 8).

Table 8. The acceptance of the ASF restriction measures applied in restricted zones among pigkeepers.

ASF Restriction Measure Applied in Zones	Acceptance (Score [a])
Restrictions on trading live pigs and pork products in the protection and surveillance zones	9.0
Restrictions on moving pigs in the protection zone	8.8
Culling of all pigs in the protection zone	4.9

[a] 10-bead scoring tool.

3.8. *Consequences of ASF Zoning*

During this task, 12 consequences arose from being in a zone with ASF restrictive measures applied in the event of an outbreak. Concerning the perceived impact of these consequences on pigkeepers, economic loss ranked first ($n = 7$), with participants discussing its importance in impacting the economy of households. The precautionary slaughter of pigs for consumption ranked third ($n = 7$). During the discussions, the participants disclosed that they would most likely process slaughtered pigs into heat-treated canned meat as soon as information on the establishment of restriction zones reached them (word-of-mouth news) to precede the culling of animals by veterinary authorities. The participants explained these actions as measures to prevent economic loss, as they did not believe that they could receive fair compensation for the culled pigs, emphasising that they would never slaughter a sick animal. At the same time, participants from one group argued that their neighbours could slaughter sick animals for heat-treated canned meat because the disease was not zoonotic. The other results are presented in Table 9.

Table 9. Consequences of being in a zone with ASF restriction measures applied, listed by pigkeepers and ranked by their perceived impact on pigkeepers.

Rank	Consequence	*n* Groups Listing	Perceived Impact (Score [a])
1	Economic loss	7	104.0
2	Psychological stress	6	68.1
3	Precautionary slaughter of pigs for consumption	7	52.3
4	Illegal sales of pigs	1	8.3
5	Discomfort of being in the quarantine zone	1	6.7
6	Prohibition on keeping pigs	1	5.0

Table 9. *Cont.*

Rank	Consequence	*n* Groups Listing	Perceived Impact (Score [a])
7	Prohibition on trading pork products	1	4.0
8	Forced disinfection of pig premises	1	1.7
9	Alcoholism	1	0.0
9	More free time	1	0.0
9	Death of pigs	1	0.0
9	You can't buy a pig right away	1	0.0

[a] Sum of the weighted proportional piling scores.

3.9. Sources of Information about ASF

Of the 11 listed sources providing information about ASF, sources accessed via Google web search were attributed the highest importance (n = 8), with participants discussing their quick and easy access to relevant information about ASF. Traditional media, including television news and radio, ranked third (n = 9) for staying informed of the broader picture of ASF at the national level. Spoken communication (word-of-mouth news) was the fourth most important source (n = 8) for staying updated on local news, as discussed. The accuracy of this information was questioned unless it was related to nearby outbreaks or restrictions imposed. Then, veterinarians ranked fifth (n = 3), with participants highly valuing the qualifications of local veterinarians. Furthermore, pigkeepers indicated that more sources in Ukraine with detailed explanations of ASF's signs, transmission routes, preventive measures, biosecurity procedures, surveillance, and legal frameworks are needed. A summary of the other sources of information about ASF that were discussed is presented in Table 10.

Table 10. Sources of information about ASF, listed by pigkeepers and ranked by their importance to pigkeepers.

Rank	Source of Information	*n* Groups Listing	Perceived Importance (Score [a])
1	Web search (Google)	8	126.8
2	Veterinary authority	3	64.7
3	Traditional media (television, radio)	9	51.5
4	Spoken communication (word-of-mouth news)	8	46.5
5	Veterinarian	3	44.8
6	Digital social networks (Viber, Facebook)	2	19.4
7	Posters, leaflets, brochures	3	13.6
8	Internet videos (Youtube)	2	8.9
9	Local newspaper	3	6.5
10	Specialized literature	1	5.0
11	Governmental websites	1	3.3

[a] Sum of the weighted proportional piling scores.

4. Discussion

It has been shown that a participatory approach can improve communication among different disciplines, foster mutual understanding, and consequently facilitate effective disease control [26–28]. Notably, participatory methods have been successfully employed in the evaluation of ASF surveillance [22,29] and to appraise the attitudes of farmers towards ASF control in Northern Uganda [30] and Tanzania [31], as well as in Lao PDR [32] and the Philippines [33]. Several participatory studies addressing ASF control have recently been conducted in Europe. The perceptions of farm managers regarding ASF and its control have been studied in Estonia [34]. Further insights into ASF control and surveillance in wild boar have been gained by studying the perceptions of hunters in Estonia, Latvia, and Lithuania [24,25,35].

Given the often limited resources and access to veterinary services, active involvement in implementing control measures and adherence to biosecurity practices by pigkeepers can greatly reduce the chances of ASF transmission. By doing so, pigkeepers not only protect their own livestock, but also contribute significantly to the broader containment of the disease. In this study, we aimed to gain insights into the awareness, perceptions, and attitudes of smallholder pigkeepers related to ASF and its control measures in Ukraine.

Regarding disease recognition in pigs, pigkeepers demonstrated adequate awareness of the early signs of ASF, which is expected to empower them to promptly report their suspicions to their veterinarians. Non-specific ASF signs such as fever, loss of appetite, and lethargy were mentioned as important indicators of ASF in pigs and were given a high ranking in the assessment. Similar results were obtained in a recent questionnaire study conducted across five regions of Ukraine, in which most respondents selected fever and lethargy as clinical signs of ASF in pigs [36]. Nevertheless, there is room for improvement, as several participants highlighted a gap in their understanding of the characteristic signs of ASF such as skin haemorrhage. This underscores the importance of veterinary authorities continuing their efforts to reach every pigkeeper to ensure that essential knowledge regarding ASF is effectively delivered and comprehended. In our study, participants did not name ASF postmortem findings. We assumed that this could be explained by how the question was presented to the participants. A broad term 'sign' was used without specifying ante- or postmortem signs. It may be that the pigkeepers might not connect the term 'sign' with lesions in internal organs or that they do not know this sign, or they understood the question to be related only to living animals. Several relevant ASF transmission routes that present a risk of virus introduction to domestic pigs were listed by the participants. Nevertheless, the relatively less important routes (mechanical vectors, e.g., rodents, airborne transmission) were given a higher ranking than the routes of higher epidemiological relevance (direct contact with a wild boar, indirect modes of transmission, e.g., swill feeding). It can be assumed that pigkeepers' opinions concerning the high risk of ASF introduction by rodents were influenced by national legislation regulating ASF control, which highlights deratisation as a crucial measure required in outbreak farms and within the protection zone [8]. There is no evidence that ASFV is spread over long distances by droplets or air [37]. Simultaneously, within a short range (such as within a pigsty), the transmission of ASFV through the air has been demonstrated [38]. The participants frequently cited air as a possible transmission route for ASFV. During discussions, pigkeepers often drew parallels between ASF and COVID-19, leading to the assumption that ASFV may also be capable of airborne transmission. This suggests that they might have been confused about COVID-19, thinking that ASFV, as a virus, might have the same routes of transmission. In turn, swill feeding, which is considered one of the main possible routes for ASFV introduction into pig herds, particularly in smallholdings [11,39–41], was rarely brought up by the participants during discussions. In a study by Muñoz-Gómez et al. [33], swill feeding was found to be a common practice on smallholder pig farms in Ukraine. The lack of awareness of the risks related to swill feeding among the study participants is highly worrying and emphasises the need to distribute relevant information to smallholders and ensure that the information truly reaches the target groups. None of the FGs named a theoretically important route of ASFV transmission: ticks. This may be due to the fact that the soft tick Ornithodoros verrucosus currently inhabiting the southern regions of Ukraine [42] is unlikely to be capable of transmitting ASFV [43]. Therefore, measures against ticks are not included in the national legislation regulating ASF control in Ukraine. Consequently, pigkeepers are not provided with information on this aspect by authorities, and in turn, they cannot name it as a transmission route. In general, the results demonstrate an important knowledge gap among smallholder pigkeepers concerning the routes of ASFV transmission, which could potentially affect the implementation of preventive measures against the disease.

Several studies indicated that farmers are more inclined to adopt biosecurity and disease control measures when they perceive them as important [44,45]. In another study, it was found that perceived strategy efficacy played a predominant role in the adoption of animal disease management strategies, especially in the context of biosecurity measures [46]. We assumed that if pigkeepers perceived certain measures as ineffective or unfavourable, their compliance with those measures would likely be lower; conversely, measures they favoured and found effective would likely have higher compliance rates. The participants in our study were mostly aware of the basic principles of disease control, such as disinfection measures (named in various formulations by all focus groups), and the limited access of people to the farm or changing clothes and shoes before entering the pigs' premises. Regarding disinfection measures, however, the participants could not distinguish between preventive measures and measures used to eradicate infection or for internal biosecurity. Disinfection of their pigs' premises was highly ranked as the most effective preventive measure, although it does not prevent the introduction of the virus to the herd. Several FGs named measures not related to ASF prevention, such as manure removal, bedding change, and washing the pigs with water, which indicates that possibly the participants might not fully understand the question and the concept of 'introduction of the disease.' Participants in one group mentioned the use of vaccines as a preventive measure against ASF, asserting that they possessed an official document regarding the vaccination of their pigs against both CSF and ASF (although this document was not presented to the facilitator). This finding is in line with a study by Muñoz-Gómez et al. [36], in which 21.6% of respondents in a questionnaire study among smallholder pigkeepers marked vaccination as an available preventive tool against ASF. To our knowledge, no ASF vaccine was officially available in the market or used in Ukraine before or during the study period. The participants' mention of an ASF vaccination was most likely attributable to their confusion with the regular vaccination efforts against CSF conducted by Ukrainian veterinary authorities on smallholder farms. None of the FGs listed fencing of farm perimeters as a preventive measure against ASF. A possible explanation for this could be that, in Ukraine, smallholdings in rural areas are typically surrounded by fences, and pig houses are located within these fenced areas. This is why the participants might not have considered this an extra biosecurity measure.

The implementation of preventive measures is likely influenced by farmers' attitudes, which, in turn, are shaped by their belief in the effectiveness of the measures, the effort required, and the discomfort caused by their implementation. In our study, we employed the straightforward question 'How do you like the measure?' to assess their attitudes towards preventive measures. Our results show that measures needing less extra effort and resources were most favoured, such as an access ban for people, the minimisation of contacts between pigkeepers, and the heat treatment of feed. In contrast, cleaning and disinfection were less favoured. This indicates that veterinary authorities should pay special attention to these measures when they are explained to smallholders, and, if possible, incentives should be provided to ensure compliance with the requirements.

Successfully controlling a disease within an animal population is a collaborative effort that requires the involvement of multiple stakeholders. It is crucial that all the parties involved have a clear understanding of their respective roles and responsibilities within this framework. Trust between counterparts is also important to ensure the swift exchange of unbiased information. This, in turn, forms the foundation for informed decisions and actions. The results of our study revealed a low awareness among smallholder pigkeepers regarding the stakeholders involved in ASF control in Ukraine. All focus groups mentioned veterinary professionals of various capacities, including those working for the government and private practices. Hunters were mentioned once, which could be explained by the insufficient interaction between hunters and pigkeepers regarding ASF control. Promoting a culture of cooperation by establishing clear communication channels and mechanisms for information exchange and mutual support among pigkeepers, including smallholders and other stakeholders, can improve disease surveillance and control in the country. These results highlight the need for extensive awareness-raising efforts among smallholder farmers regarding disease control systems in the country.

The evaluation of pigkeepers' trust in various parties regarding their role in disease control revealed that smallholders generally have more trust in private entities (such as the pigkeepers themselves) and veterinary professionals (including local official veterinarians, local private veterinarians, and district veterinary hospitals) than in other governmental institutions (such as the central veterinary authority, village administration, and police). The concept of social identity, which encompasses a shared sense of group membership and values, has been found to be positively correlated with farmers' trust in the government and their intention to report disease outbreaks [47]. The perception of not receiving compensation likely stems from the previous compensation system funded by local budget reserve funds, which was in place before the implementation of a resolution from the government [9] that stipulated the allocation of compensation funds from the state budget. Pigkeepers' distrust towards the government's ability to provide compensation, coupled with their propensity to neglect the registration of pigs—a prerequisite for securing compensation—may result in failed compliance with control measures. When an outbreak occurs, pigkeepers may be inclined to ignore control measures, recognising that they are unlikely to receive legal compensation. In the study by Cooper et al. [33], there were accounts of community members hiding pigs to prevent the culling of their herd due to the fear of insufficient compensation. Raising awareness among smallholder pigkeepers about the conditions for receiving compensation from the state could potentially improve their trust in the authorities. Enhanced communication and trust-building can lead to the increased effectiveness of control programs.

Acceptability of disease control measures by all parties involved is a crucial prerequisite for the successful implementation of measures and the effective control of diseases. Discussions centred around the feed destruction and culling of pigs on the farm and in the protection zone brought up emotional and strong opinions among pigkeepers. The emotional hardships faced by farmers and animal health workers resulting from control measures were discovered in the study by Cooper et al. [33], in which depopulation campaigns emerged as a dominant topic in discussions. Partial culling as a chosen ASF control measure in the country, with limited resources for compensation, demonstrated benefits to farmers and veterinary services in the study by Nga et al. [48]. As was suggested by Cooper et al. [33], the topic of human trauma arising from animal disease control measures is often overlooked, highlighting the need for greater global attention to the profound and far-reaching effects of stamping out strategies. Our findings emphasise the importance of continuous communication between stakeholders, providing detailed explanations concerning the reasons for existing disease control strategies and compensation rules.

The consequences that arise from pigkeepers being in ASF restricted zones were identified in this study as unfavourable, with the strongest adverse impact on the pigkeepers' economy and psychological wellbeing. Psychosocial impacts of ASF on farmers, along with the corresponding effects of control measures, were also found in the study by Cooper et al. [33]. This study's findings on the precautionary slaughter for their own consumption and illegal sales of pigs align with previous studies, which reported that smallholder farmers, especially those in economically challenged areas, are inclined toward the sale or slaughter of pigs to minimize their economic losses [11,40,49]. Our results highlight the significance of providing attractive reporting incentives and fostering trust between authorities and smallholder pigkeepers for the early detection of and response to disease outbreaks.

The listed sources of information about ASF by participants show that pigkeepers consider digital and traditional sources of information essential depending on the context of its usage. Either veterinary authorities or veterinarians were mentioned by groups, indicating their importance in the dissemination of reliable information. Furthermore, groups emphasized the necessity of increased sources that provide comprehensive information on different aspects related to ASF.

Our study describes the perceptions and opinions of pigkeepers who voluntarily participated, using a convenience sampling method, which may not fully represent those who declined to participate. The total number of 52 participants may raise a question as to the representativeness of our study. However, in participatory studies a statistically representative sample size has not been considered necessary. The sample size is determined by the heterogeneity of the answers given by the participants and justified upon reaching a saturation of answers, meaning that no new information can be obtained from the participants through the addition of new focus groups [50]. As we reached saturation in participants responses, we could assume that this study is representative in regards of the smallholders in the study area. To ensure balanced group interactions and encourage open discussions involving everyone, it is important for the facilitator to have good communication skills and an understanding of power dynamics. This helps create an environment in which diverse viewpoints are welcomed and valued [51]. Hence, the potential for biased results stemming from imbalanced group dynamics cannot be completely dismissed. To address this issue, the facilitator included all perspectives in the discussion, especially during activities designed to reach consensus. This helped prevent data gaps among participants who might have been less dominant. Furthermore, our findings represent the views of pigkeepers with no experience with ASF outbreaks. Not having participants with previous experience may be considered a weakness of the study. However, this can also be seen as a strength, as the focus groups were more homogeneous and the potential dominance of more experienced participants was avoided. Moreover, incorporating the perspectives of pigkeepers without direct experience with ASF increases the likelihood of identifying starting points for early interventions.

5. Conclusions

- Smallholder pigkeepers in Ukraine generally have high awareness about ASF's clinical features.
- There is a lack of knowledge about ASF transmission routes and, therefore, a limited understanding of preventive measures among pigkeepers.
- Smallholder pigkeepers in Ukraine acknowledge their role in disease control. Nevertheless, their trust in the government was relatively lower when compared to their trust in veterinarians. This trust disparity could create a barrier to effective collaboration during disease control efforts.
- In general, smallholder pigkeepers in Ukraine tended to accept the disease eradication measures implemented at the farm level during ASF outbreaks.
- Despite the existing compensation scheme for smallholder pigkeepers in Ukraine, the consequences of being in a zone with applied ASF restriction measures are seen as having a detrimental impact on pigkeepers' economic prospects and personal well-being.
- Some measures applied in the ASF restricted zones were not accepted by smallholders in Ukraine. Therefore, it is important to provide detailed explanations to improve their understanding and increase their acceptance of these measures.
- While the use of web sources is increasing, veterinarians, traditional media, and spoken communication continue to play important role in the dissemination of information regarding ASF to smallholder pigkeepers in Ukraine. Additionally, pigkeepers emphasized the need for more comprehensive information from different sources on various aspects related to ASF. We suggest implementing various activities, including educational campaigns for pigkeepers, to address the challenge of ASF control in Ukrainian smallholdings.

Author Contributions: Conceptualization, A.V., K.M., and L.M.; methodology, K.S., A.V., and K.M.; formal analysis, L.M., A.V., K.M., and V.N.; investigation, L.M.; data curation, A.V., K.M., and L.M.; writing—original draft preparation, L.M.; writing—review and editing, A.V., K.M., L.M., K.S., and V.N.; supervision, A.V. and K.M. All authors have read and agreed to the published version of the manuscript.

Funding: This study was funded by the Estonian University of Life Sciences under the research base funding grant number P170193 and by the Education and Youth Board of Estonia contract 5.1-2/21/7-4.

Institutional Review Board Statement: According to national legislation, ethical approval was not required for this study as it did not include clinical trials in humans or animal experiments. Personalised and sensitive personal information was not collected from the participants.

Informed Consent Statement: All participants were informed of the study objectives and participated voluntarily. During the introduction of each meeting, participants provided verbal consent of their participation and to the recording of the interview. All participants approved the publication of the summarised study results.

Data Availability Statement: The original data used for the analyses can be obtained from the corresponding author after approval by the responsible institutions in Ukraine and Estonia.

Acknowledgments: The authors thank the pigkeepers from Ukraine who participated in this study.

Conflicts of Interest: The authors declare no conflicts of interest.

References

1. WOAH (World Organisation for Animal Health (OIE)). Terrestrial Animal Health Manual Chapter 3.8.1 African Swine Fever (Infection with African Swine Fever Virus). 2019. Available online: https://www.woah.org/fileadmin/Home/eng/Health_standards/tahm/3.08.01_ASF.pdf (accessed on 4 May 2022).
2. Anonymous. The Report on the Ways of Introduction and Spread of ASF in the Odesa Region in 1977. SSI VNIIVViM, Pokrov, Russia. 64 p. "Otchet o Rezultatah Izucheniya Putej Zanosa I Rasprostraneniya Afrikanskoj Chumy Svinej (AChS) v Odesskoj Oblasti v 1977 Godu" (in Russian). 2014. Available online: https://old.fsvps.gov.ru/fsvps-forum/posts/downloadAttach/975.page (accessed on 15 September 2022).
3. EFSA. Scientific Opinion on African swine fever. *EFSA J.* **2014**, *12*, 3628.
4. WAHIS. World Animal Health Information System. 2022. Available online: https://wahis.woah.org/#/event-management (accessed on 15 September 2022).
5. The State Statistics Service of Ukraine. 2020. Available online: https://www.ukrstat.gov.ua/ (accessed on 28 March 2023).
6. Verkhovna Rada of Ukraine Law of Ukraine on personal household farming No, 7.4.2.-I.V. 2003. Available online: https://zakon.rada.gov.ua/laws/show/742-15#Text (accessed on 4 March 2023).
7. Ministry of Agriculture Policy and Food of Ukraine. Procedure No. 639 for Identification and Registration of Pigs. 2017. Available online: https://zakon.rada.gov.ua/laws/show/z0154-18#n15 (accessed on 4 March 2023).
8. Ministry of Agriculture Policy and Food of Ukraine. Order No. 111 on the Approval of the Instruction for the Prevention and Control of African Swine Fever. 2017. Available online: https://zakon.rada.gov.ua/laws/show/z0432-17#Text (accessed on 4 March 2023).
9. Cabinet of Ministers of Ukraine. Resolution of the Cabinet of Ministers of Ukraine No. 1050 on Amendments to the Procedures Approved by Resolutions of the Cabinet of Ministers of Ukraine No. 413, 2008 and No. 1003, 2012. 2018. Available online: https://zakon.rada.gov.ua/laws/show/1050-2018-%D0%BF#Text (accessed on 4 March 2023).
10. Verkhovna Rada of Ukraine. The law of Ukraine No. 1445-VI about Identification and Registration of Animals. 2009. Available online: https://zakon.rada.gov.ua/laws/show/1445-17#Text (accessed on 4 March 2023).
11. Costard, S.; Zagmutt, F.J.; Porphyre, T.; Pfeiffer, D.U. Small-scale pig farmers' behavior, silent release of African swine fever virus and consequences for disease spread. *Sci. Rep.* **2015**, *5*, 17074. [CrossRef] [PubMed]
12. Kukielka, E.A.; Martínez-López, B.; Beltrán-Alcrudo, D. Modeling the live-pig trade network in Georgia: Implications for disease prevention and control. *PLoS ONE* **2017**, *12*, e0178904. [CrossRef]
13. Zani, L.; Dietze, K.; Dimova, Z.; Forth, J.H.; Denev, D.; Depner, K.; Alexandrov, T. African swine fever in a Bulgarian backyard farm-a case report. *Vet. Sci.* **2019**, *6*, 94. [CrossRef]
14. Boklund, A.; Dhollander, S.; Chesnoiu Vasile, T.; Abrahantes, J.C.; Bøtner, A.; Gogin, A.; Villeta, L.C.G.; Gortázar, C.; More, S.J.; Papanikolaou, A.; et al. Risk factors for African swine fever incursion in Romanian domestic farms during 2019. *Sci. Rep.* **2020**, *10*, 10215. [CrossRef] [PubMed]
15. Ardelean, F.; Globig, A.; Gârdan Năvălici, A.I.; Blome, S.; Dietze, K.; Depner, K.; Zani, L. The course of African swine fever in Romanian backyard holdings—A case report. *Vet. Med. Sci.* **2021**, *7*, 2273–2279. [CrossRef] [PubMed]
16. Prodanov-Radulović, J.; Đurđević, B.; Petrović, J.; Mirčeta, J.; Polaček, V. African swine fever: A biosecurity challenge for domestic pig production in Serbia. *Arch. Vet. Med.* **2022**, *15*, 21–38. [CrossRef]
17. De Nardi, M.; Léger, A.; Stepanyan, T.; Khachatryan, B.; Karibayev, T.; Sytnik, I.; Tyulegenov, S.; Akhmetova, A.; Nychyk, S.; Sytiuk, M.; et al. Implementation of a regional training program on African swine fever as part of the cooperative biological engagement program across the Caucasus region. *Front. Vet. Sci.* **2017**, *4*, 164. [CrossRef]

18. FAO. Technical Assistance Project "Building the Potential for Early Detection and Response to African Swine Fever in UKRAINE". 2013. Available online: https://www.asf.vet.ua (accessed on 26 April 2023).
19. FAO. EBRD African Swine Fever: Risk Awareness Raising and Risk Mitigation in Ukraine Project. 2016. Available online: https://www.fao.org/support-to-investment/news/detail/en/c/416192/ (accessed on 26 April 2023).
20. EFSA (European Food Safety Authority); Anette, B.; Anette, B.; Theodora, C.V.; Klaus, D.; Daniel, D.; Vittorio, G.; Georgina, H.; Daniela, K.; Annick, L.; et al. Scientific report on the epidemiological analyses of African swine fever in the European Union (November 2018 to October 2019). *EFSA J.* **2020**, *18*, e05996. [CrossRef]
21. European Food Safety Authority (EFSA); Desmecht, D.; Gerbier, G.; Gortázar Schmidt, C.; Grigaliuniene, V.; Helyes, G.; Kantere, M.; Korytarova, D.; Linden, A.; Miteva, A.; et al. Scientific Opinion on the epidemiological analysis of African swine fever in the European Union (September 2019 to August 2020). *EFSA J.* **2021**, *19*, 6572. [CrossRef]
22. Calba, C.; Antoine-Moussiaux, N.; Charrier, F.; Hendrikx, P.; Saegerman, C.; Peyre, M.; Goutard, F.L. Applying participatory approaches in the evaluation of surveillance systems: A pilot study on African swine fever surveillance in Corsica. *Prev. Vet. Med.* **2015**, *122*, 389–398. [CrossRef]
23. Dunkle, S.; Mariner, J.C.; Ameri, A.A.; Hendrickx, S.; Jones, B.; Mehta, P.; Pissang, C. *Participatory Epidemiology: A Toolkit for Trainers*; Dunkle, S., Mariner, J.C., Eds.; PENAPH Participatory Epidemiology Network for Animal and Public Helath, ILRI the International Livestock Research Institute: Nairobi, Kenya, 2013. Available online: https://hdl.handle.net/10568/35216 (accessed on 15 November 2019).
24. Urner, N.; Mõtus, K.; Nurmoja, I.; Schulz, J.; Sauter-Louis, C.; Staubach, C.; Conraths, F.J.; Schulz, K. Hunters' Acceptance of Measures against African Swine Fever in Wild Boar in Estonia. *Prev. Vet. Med.* **2020**, *182*, 105121. [CrossRef]
25. Urner, N.; Seržants, M.; Užule, M.; Sauter-Louis, C.; Staubach, C.; Lamberga, K.; Oļševskis, E.; Conraths, F.J.; Schulz, K. Hunters' view on the control of African swine fever in wild boar. A participatory study in Latvia. *Prev. Vet. Med.* **2021**, *186*, 105229. [CrossRef]
26. Alders, R.G.; Ali, S.N.; Ameri, A.A.; Bagnol, B.; Cooper, T.L.; Gozali, A.; Hidayat, M.M.; Rukambile, E.; Wong, J.T.; Catley, A. Participatory Epidemiology: Principles, Practice, Utility, and Lessons Learnt. *Front. Vet. Sci.* **2020**, *7*, 876. Available online: https://www.frontiersin.org/article/10.3389/fvets.2020.532763 (accessed on 3 September 2023).
27. Catley, A. Participatory epidemiology: Reviewing experiences with contexts and actions. *Prev. Vet. Med.* **2020**, *180*, 105026. [CrossRef]
28. Chenais, E.; Fischer, K. Power, participation and interdisciplinary tensions: Introducing a special issue on methodological developments in participatory epidemiology. *Prev. Vet. Med.* **2021**, *196*, 105490. [CrossRef] [PubMed]
29. Chenais, E.; Sternberg-Lewerin, S.; Boqvist, S.; Emanuelson, U.; Aliro, T.; Tejler, E.; Cocca, G.; Masembe, C.; Ståhl, K. African swine fever in Uganda: Qualitative evaluation of three surveillance methods with implications for other resource-poor settings. *Front. Vet. Sci.* **2015**, *2*, 51. [CrossRef] [PubMed]
30. Chenais, E.; Boqvist, S.; Lewerin, S.S.; Emanuelson, U.; Ouma, E.; Dione, M.; Aliro, T.; Crafoord, F.; Masembe, C.; Ståhl, K. Knowledge, Attitudes and Practices Related to African Swine Fever Within Smallholder Pig Production in Northern Uganda. *Transbound. Emerg. Dis.* **2017**, *64*, 101–115. [CrossRef] [PubMed]
31. Fasina, F.O.; Kissinga, H.; Mlowe, F.; Mshang'a, S.; Matogo, B.; Mrema, A.; Mhagama, A.; Makungu, S.; Mtui-Malamsha, N.; Sallu, R.S.; et al. Drivers, risk factors and dynamics of african swine fever outbreaks, southern highlands, Tanzania. *Pathogens* **2020**, *9*, 155. [CrossRef] [PubMed]
32. Matsumoto, N.; Siengsanan-Lamont, J.; Halasa, T.; Young, J.R.; Ward, M.P.; Douangngeun, B.; Theppangna, W.; Khounsy, S.; Toribio, J.L.; Bush, R.D.; et al. The impact of African swine fever virus on smallholder village pig production: An outbreak investigation in Lao PDR. *Transbound. Emerg. Dis.* **2021**, *68*, 2897–2908. [CrossRef] [PubMed]
33. Cooper, T.L.; Smith, D.; Gonzales, M.J.C.; Maghanay, M.T.; Sanderson, S.; Cornejo, M.R.J.C.; Pineda, L.L.; Sagun, R.A.A.; Salvacion, O.P. Beyond Numbers: Determining the Socioeconomic and Livelihood Impacts of African Swine Fever and Its Control in the Philippines. *Front. Vet. Sci.* **2022**, *8*, 734236. [CrossRef] [PubMed]
34. Moskalenko, L.; Schulz, K.; Mõtus, K.; Viltrop, A. Pigkeepers' knowledge and perceptions regarding African swine fever and the control measures in Estonia. *Prev. Vet. Med.* **2022**, *208*, 105717. [CrossRef]
35. Stončiūtė, E.; Malakauskas, A.; Conraths, F.J.; Masiulis, M.; Sauter-Louis, C.; Schulz, K. The perceptions of Lithuanian hunters towards African swine fever using a participatory approach. *BMC Vet. Res.* **2022**, *18*, 401. [CrossRef]
36. Muñoz-Gómez, V.; Solodiankin, O.; Rudova, N.; Gerilovych, A.; Nychyk, S.; Hudz, N.; Ukhovska, T.; Sytiuk, M.; Polischuk, V.; Mustra, D.; et al. Supporting control programs on African swine fever in Ukraine through a knowledge, attitudes, and practices survey targeting backyard farmers. *Vet. Med. Sci.* **2021**, *7*, 1786–1799. [CrossRef]
37. La, A.; Zhang, Q.; Cicek, N.; Coombs, K.M. Current understanding of the airborne transmission of important viral animal pathogens in spreading disease. *Biosyst. Eng.* **2022**, *224*, 92–117. [CrossRef]
38. Olesen, A.S.; Lohse, L.; Boklund, A.; Halasa, T.; Gallardo, C.; Pejsak, Z.; Belsham, G.J.; Rasmussen, T.B.; Bøtner, A. Transmission of African swine fever virus from infected pigs by direct contact and aerosol routes. *Vet. Microbiol.* **2017**, *211*, 92–102. [CrossRef] [PubMed]
39. FAO. *African Swine Fever (ASF) Recent Developments and Timely Updates—Worrisome Dynamics: Steady Spread towards Unaffected Areas could Have Disastrous Impact*; Focus on No. 6. [Electronic Bulletin]; FAO: Rome, Italy, 2012. Available online: http://www.fao.org/docrep/016/ap372e/ap372e.pdf (accessed on 28 April 2023).

40. Nantima, N.; Ocaido, M.; Ouma, E.; Davies, J.; Dione, M.M.; Okoth, E.; Mugisha, A.; Bishop, R. Risk factors associated with occurrence of African swine fever outbreaks in smallholder pig farms in four districts along the Uganda-Kenya border. *Trop. Anim. Health Prod.* **2015**, *47*, 589–595. [CrossRef] [PubMed]
41. Heilmann, M.; Lkhagvasuren, A.; Adyasuren, T.; Khishgee, B.; Bold, B.; Ankhanbaatar, U.; Fusheng, G.; Raizman, E.; Dietze, K. African swine fever in Mongolia: Course of the epidemic and applied control measures. *Vet. Sci.* **2020**, *7*, 24. [CrossRef] [PubMed]
42. ECDC. European Centre for Disease Prevention and Control and European Food Safety Authority. Tick Maps. Stockholm. 2020. Available online: https://ecdc.europa.eu/en/disease-vectors/surveillance-and-disease-data/tick-maps (accessed on 3 May 2023).
43. de Oliveira, R.P.; Hutet, E.; Paboeuf, F.; Duhayon, M.; Boinas, F.; de Leon, A.P.; Filatov, S.; Vial, L.; Le Potier, M.-F. Comparative vector competence of the Afrotropical soft tick Ornithodoros moubata and Palearctic species, O. Erraticus and O. Verrucosus, for African swine fever virus strains circulating in Eurasia. *PLoS ONE* **2019**, *14*, e0225657. [CrossRef] [PubMed]
44. Casal, J.; De Manuel, A.; Mateu, E.; Martín, M. Biosecurity measures on swine farms in Spain: Perceptions by farmers and their relationship to current on-farm measures. *Prev. Vet. Med.* **2007**, *82*, 138–150. [CrossRef] [PubMed]
45. Toma, L.; Stott, A.W.; Heffernan, C.; Ringrose, S.; Gunn, G.J. Determinants of biosecurity behaviour of British cattle and sheep farmers-A behavioural economics analysis. *Prev. Vet. Med.* **2013**, *108*, 321–333. [CrossRef] [PubMed]
46. Valeeva, N.I.; van Asseldonk, M.A.P.M.; Backus, G.B.C. Perceived risk and strategy efficacy as motivators of risk management strategy adoption to prevent animal diseases in pig farming. *Prev. Vet. Med.* **2011**, *102*, 284–295. [CrossRef]
47. Scutt, K.; Walker, I.; Ernst, I. The influence of trust and social identity in farmers' intentions to report suspected emergency disease outbreaks. *Prev. Vet. Med.* **2023**, *212*, 105855. [CrossRef]
48. Nga, B.T.T.; Padungtod, P.; Depner, K.; Chuong, V.D.; Duy, D.T.; Anh, N.D.; Dietze, K. Implications of partial culling on African swine fever control effectiveness in Vietnam. *Front. Vet. Sci.* **2022**, *9*, 957918. [CrossRef]
49. Chenais, E.; Boqvist, S.; Emanuelson, U.; von Brömssen, C.; Ouma, E.; Aliro, T.; Masembe, C.; Ståhl, K.; Sternberg-Lewerin, S. Quantitative assessment of social and economic impact of African swine fever outbreaks in northern Uganda. *Prev. Vet. Med.* **2017**, *144*, 134–148. [CrossRef] [PubMed]
50. Guest, G.; Bunce, A.; Johnson, L. How Many Interviews Are Enough? *Field Methods* **2006**, *18*, 59–82. [CrossRef]
51. Fischer, K.; Schulz, K.; Chenais, E. "Can we agree on that"? Plurality, power and language in participatory research. *Prev. Vet. Med.* **2020**, *180*, 104991. [CrossRef] [PubMed]

pathogens

MDPI

Article

Whole Genome Sequencing Shows that African Swine Fever Virus Genotype IX Is Still Circulating in Domestic Pigs in All Regions of Uganda

Rodney Okwasiimire [1], Joseph F. Flint [2], Edrine B. Kayaga [1], Steven Lakin [3], Jim Pierce [3], Roger W. Barrette [3], Bonto Faburay [3], Dickson Ndoboli [1], John E. Ekakoro [2], Eddie M. Wampande [1] and Karyn A. Havas [2,*]

[1] Central Diagnostic Laboratory, College of Veterinary Medicine, Animal Resources and Biosecurity, Makerere University, Kampala P.O. Box 7062, Uganda; rodn.ok17@gmail.com (R.O.); edrinekayaga2@gmail.com (E.B.K.); dickson.ndoboli@gmail.com (D.N.); eddiewampande@gmail.com (E.M.W.)

[2] Department of Public and Ecosystem Health, College of Veterinary Medicine, Cornell University, Ithaca, NY 14853, USA; jff22@cornell.edu (J.F.F.); jekakoro@vet.k-state.edu (J.E.E.)

[3] Foreign Animal Disease Diagnostic Laboratory, National Veterinary Services Laboratories, Veterinary Services, Animal and Plant Health Inspection Services, United States Department of Agriculture, Greenport, NY 11957, USA; steven.lakin@usda.gov (S.L.); jim.l.pierce@usda.gov (J.P.); roger.w.barrette@usda.gov (R.W.B.); bonto.faburay@usda.gov (B.F.)

* Correspondence: kah47@cornell.edu

Abstract: Blood samples were collected from pigs at six abattoirs in the Kampala, Uganda metropolitan area from May 2021 through June 2022, and tested for African swine fever virus. Thirty-one samples with cycle threshold values < 26 from pigs with different geographic origins, clinical and pathologic signs, and *Ornithodoros moubata* exposure underwent whole genome sequencing. The p72 gene was used to genotype the isolates, and all were found to be genotype IX; whole genome sequences to previous genotype IX isolates confirmed their similarity. Six of the isolates had enough coverage to evaluate single nucleotide polymorphisms (SNPs). Five of the isolates differed from historic regional isolates, but had similar SNPs to one another, and the sixth isolate also differed from historic regional isolates, but also differed from the other five isolates, even though they are all genotype IX. Whole genome sequencing data provide additional detail on viral evolution that can be useful for molecular epidemiology, and understanding the impact of changes in genes to disease phenotypes, and may be needed for vaccine targeting should a commercial vaccine become available. More sequencing of African swine fever virus isolates is needed in Uganda to understand how and when the virus is changing.

Keywords: African swine fever virus; genotype; whole genome sequencing; Uganda

Citation: Okwasiimire, R.; Flint, J.F.; Kayaga, E.B.; Lakin, S.; Pierce, J.; Barrette, R.W.; Faburay, B.; Ndoboli, D.; Ekakoro, J.E.; Wampande, E.M.; et al. Whole Genome Sequencing Shows that African Swine Fever Virus Genotype IX Is Still Circulating in Domestic Pigs in All Regions of Uganda. *Pathogens* **2023**, *12*, 912. https://doi.org/10.3390/pathogens12070912

Academic Editor: Lawrence S. Young

Received: 16 June 2023
Revised: 1 July 2023
Accepted: 3 July 2023
Published: 6 July 2023

1. Introduction

The African swine fever virus (ASFV) is a double-stranded DNA arbovirus with a genome size of 170 to 190 kb, which encodes over 150 proteins, depending on the viral strain [1], and belongs to genus *Asfivirus*, family *Asfarviridae* [2]. ASFV causes an infectious disease called African swine fever (ASF), which is a highly contagious hemorrhagic disease that has been reported in both domestic and wild suids including warthogs, bush pigs, giant forest hogs, and wild boars [3]. Domestic pigs are highly susceptible, and case fatality rates can approach 100% [4]. Though the virus does not infect humans, ASF has led to food insecurity and enormous economic losses to farmers due to the high mortality rates of pigs during outbreaks [5,6] and disruptions of the market structure in the pig value chain [7].

The ASFV was first reported in Kenya in the early 20th century [8]. There have been 24 genotypes of the virus reported based on genome sequencing of the p72 segment of the virus since this initial discovery [9]. It is evident that ASFV variants can quickly spread across regions. Currently, there is a genotype II variant causing a global panzootic that was first diagnosed outside of Africa in the country of Georgia [10] and has spread widely since that introduction, impacting Europe, Asia, and the island of Hispaniola (www.wahis.woah.org, accessed on 27 February 2023). Given this background, the risk of introduction of new ASFV genotypes into Uganda is likely to occur, and ongoing monitoring is needed.

In Uganda, sequencing of the p72 gene of the virus from ASF outbreaks in Central Uganda in 2007 [11], between 2010 and 2013 from all districts in Uganda [12], and 2015 from the Central [13] and Eastern region [5] detected only genotype IX. An ASFV genotype X virus was found from a Ugandan pig in isolates that were part of the ASFV collection at the Institute for Animal Health in the Pirbright Laboratory [14,15]. The neighboring country of Kenya reported genotype IX in warthogs and X in domestic pigs [16]. Another neighbor, Tanzania, has multiple genotypes in domestic pigs. Isolates from pigs in Tanzania were sequenced and genotypes X, XV, and XVI were found [17], an outbreak in 2008 was caused by an ASFV genotype XV variant [18], and samples collected between 2015 and 2017 in Tanzania were found to be infected with ASFV genotypes II, IX, and X [19]. Sequenced samples from domestic pigs in the Democratic Republic of the Congo (DRC), a country on Uganda's western border, reported detection of genotypes IX [20], and genotype X [21]. Further, sequencing of samples collected from domestic pigs between 2005 and 2012 detected genotypes I, IX, and X in the DRC [22] Numerous genotypes are commonly reported among Uganda's neighbors. Given the diversity of genotypes in the region, it is important to track their presence in the country to understand when new genotypes emerge and why.

Previous work in Uganda has been constrained to outbreak locations and to collections of samples at global reference laboratories. Genotyping of isolates that were representatively collected and that were associated with a variety of characteristics (clinical signs, region, tick exposure, etc.) had not been done. Further, maintenance of knowledge about circulating strains is important for the development of appropriate vaccine candidates for a given region/country and for purposes of molecular epidemiology. The purpose of this study was to determine what genotypes were circulating based on samples of varying characteristics, and to establish whether any new ASFV genotypes were circulating in Uganda.

2. Materials and Methods

2.1. Sample Collection

A total of 1318 pigs had blood samples collected at six abattoirs around the Kampala metropolitan area in Central Uganda (Figure 1). In addition, pig sex, type (local breed, European breed, or cross-bred), clinical signs, and pathologic lesions were recorded at the time of sample collection. Traders were asked about the origin district of the pig as well. Abattoirs in metropolitan Kampala have been reported to receive pigs from all the regions of the country [23]. Uganda has four geographic regions, including the Central, Northern, Western and the Eastern (Figure 1). A stratified systematic sampling method was followed from May 2021 through June 2022. Total sample sizes were calculated to capture approximately 200 positive pigs. To detect the expected ASFV prevalence of pigs in Uganda, 11.5% [24], with 95% confidence and 5% error, 157 pigs would be needed (openepi.com; accessed July 2018). This would yield 18 positive pigs. To detect 200 positive pigs, at least 1200 pigs were sampled. The total sample size was stratified across abattoirs based on the annual number of pigs slaughtered at each site. This was then calculated to monthly sample sizes. Sites had two to four days per month randomly selected for sampling so as to not exceed 10 pigs sampled per day to allow enough time to collect samples from all of the pigs. On the day of sampling, pigs were sampled systematically until the sample size was met. The sampling team consisted of trained veterinarians to ensure proper capture of clinical signs and pathologic lesions, as well as appropriate sample collection. Blood samples were transported from the slaughterhouse to the Central Diagnostic Laboratory,

College of Veterinary Medicine, Animal Resources and Biosecurity, Makerere University using a cold chain where they were stored at −20 °C.

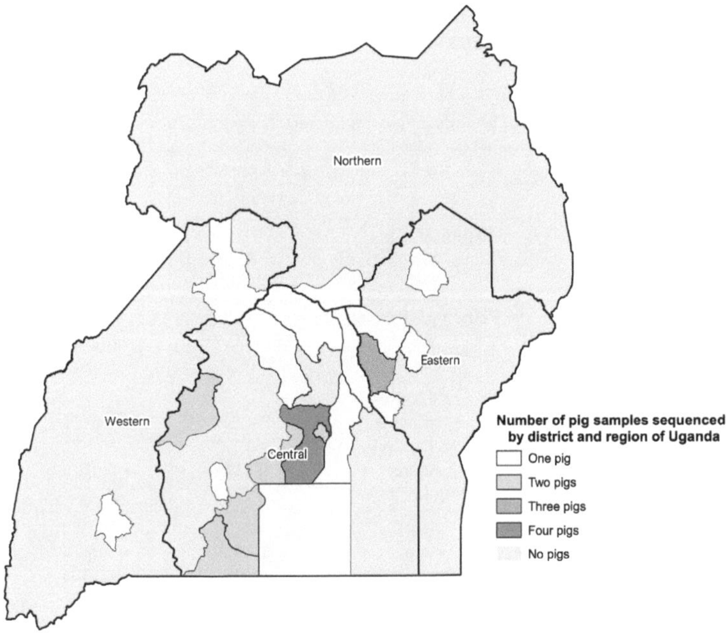

Figure 1. Summary of the number of pig blood samples collected from May 2021 through June 2022 sequenced by administrative district and region in Uganda.

2.2. Nucleic Acid Extraction, Amplification, and Detection of ASFV by qPCR

Whole blood was diluted 1:1 with 1X PBS and total DNA extraction was performed using the Qiagen DNeasy tissue and blood kits (Qiagen, Hilden, Germany). The US Department of Agriculture's (USDA) Foreign Animal Disease Diagnostic Laboratory's (FADDL) standard operating procedures (SOPs) [25] which follow the manufacturer's instructions were used during these extractions. The real-time PCR (qPCR) assay used has been previously described [26] and the FADDL ASF qPCR SOP [27] was again followed. The TaqMan® Fast Virus 1-Step Master Mix (Thermo Fisher Scientific, Waltham, MA, USA) along with the forward primer of 5′-CCTCGGCGAGCGCTTTATCAC-3′, reverse primer of 5′-GGAAACTCATTCACCAAATCCTT-3′, and probe of FAM-CGATGCAAGCTTTAT-MGB/NFQ ordered from Thermo Fisher Scientific (Waltham, MA, USA) were used in the qPCR procedure. The VetMax Xeno DNA internal positive control (IPC) (Thermo Fisher Scientific, Waltham, MA, USA) was used during the DNA extraction procedures and the VetMax Xeno IPC LIZ Assay (Thermo Fisher Scientific, Waltham, MA, USA) was used during the real time qPCR. This was done for each individual sample following FADDL SOPs. The qPCR assay was run on a QuantStudio 5 thermocycler (Thermo Fisher Scientific, Waltham, MA, USA) in 25 µL reaction volumes containing 20 µL of master mix and 5 µL of the extracted total DNA.

2.3. Blood Sample Selection

Blood samples used for sequencing were positive based on qPCR results and had a cycle threshold (Ct) value < 26. In total, 31 qPCR positive samples were sequenced. Samples from different regions of the country (See Figure 1), and different districts (for the central region), as well as those from pigs that had presented with and without clinical signs and pathologic lesions at the time of sampling at the abattoir were considered for selection. The samples sequenced also covered a range of *O. moubata* exposure status of the pigs (See Table 1). Table 1 also summarizes the distribution of pig sex and type by region, although they were not used for sample selection. It is important to note that the intent was to sequence a diverse set of positive samples, but not necessarily a representative set of samples as there is no national level data on disease prevalence across Uganda.

Table 1. Characteristics of the 31 blood samples collected from May 2021 through June 2022 that were selected for sequencing to determine the African swine fever virus genotype.

n = 31	Central	Eastern	Northern	Western
	# (%)	# (%)	# (%)	# (%)
Clinical signs				
Yes	8 (25.8)	2 (6.5)	1 (3.2)	0 (0.0)
No	12 (38.7)	6 (19.4)	0 (0.0)	2 (6.5)
Pathologic lesions				
Yes	16 (51.6)	8 (25.8)	1 (3.2)	2 (6.5)
No	4 (12.9)	0 (0.0)	0	0
Ornithodoros				
Negative	7 (22.6)	3 (9.7)	0 (0.0)	0 (0.0)
Weak Positive	9 (29.0)	3 (9.7)	1 (3.2)	1 (3.2)
Positive	1 (3.2)	0 (0.0)	0 (0.0)	0 (0.0)
Strong Positive	3 (9.7)	2 (6.5)	0 (0.0)	0 (0.0)
Sex				
Male	6 (19.35	3 (9.7)	1 (3.2)	2 (6.5)
Female	14 (45.2)	5 (16.1)	0 (0.0)	0 (0.0)
Pig type				
Local	1 (3.2)	3 (9.7)	0 (0.0)	0 (0.0)
European	9 (29.0)	3 (9.7)	1 (3.2)	0 (0.0)
Cross-bred	9 (29.0)	2 (6.5)	0 (0.0)	2 (6.5)

Clinical signs, pathologic lesions, *Ornithodoros moubata* exposure and region were used for sample selection.

2.4. African Swine Fever Genome Sequencing

Following sample selection, the extracted DNA previously used for qPCR testing was prepared for sequencing with minor modifications from the protocol previously described [28], adapted from the Nextera XT DNA Sample Preparation Guide (Illumina, San Diego, CA, USA, 2019) [29]. All the reagents used were supplied with the Nextera XT DNA Library Prep sequencing kit (Illumina, San Diego, CA, USA) unless otherwise stated. Briefly, the DNA was quantified on a Qubit 4 fluorometer and 1.0 ng from each DNA sample was fragmented and adapter sequences added to the ends to allow amplification by limited-cycle PCR in downstream steps. The incubation time for the fragmentation and tagmentation was increased to 15 min to allow for generation of DNA fragments of appropriate sequencing size. The sizes of the fragments produced were examined by capillary electrophoresis on a 5200 Fragment Analyzer System (Agilent, Santa Clara, CA, USA). For the PCR amplification, Nextera PCR Master Mix (NPM) Index 1 (i7) and 2 (i5) primers in a TruSeq index plate fixture were utilized. The PCR was carried out in a 96-well plate on a SimpliAmp™ thermocycler (Thermo Fisher Scientific, Waltham, MA, USA) following the limited PCR program outlined in the Nextera XT DNA Sample Preparation Guide. The sizes of the fragments produced were examined by capillary electrophoresis on a 5200 Fragment Analyzer System (Agilent, Santa Clara, CA, USA). The PCR products were cleaned using AMPure XP beads (Beckman Coulter, Indianapolis, IN, USA), washed with freshly prepared 80% ethanol on a magnetic stand and suspended into 50 µL resuspension buffer (RSB) supplied in the sequencing kit.

The resultant libraries were pooled in equal concentrations to create a pooled amplicon library (PAL) of 4 nM. The PAL was denatured according to manufacturer's instructions to create a diluted amplicon library (DAL) of 14 pM. The DAL was loaded into a thawed MiSeq

V3 600 cycle reagent cartridge for sequencing on the Illumina MiSeq platform (Illumina, Inc., San Diego, CA, USA). Sequencing occurred at the Makerere University College of Health Science Biomedical Research Centre (MakBRC, Kampala, Uganda) Sequencing Laboratory.

2.5. Sequence Data Analysis

Samples were analyzed using methods and software as previously described by Lakin et. al. [30]. Briefly, Illumina data were aligned to the ASFV Kenya Bus/2006 reference genome (GenBank accession KM111295.1) using the Burrows-Wheeler Aligner (v0.7.17) with options "-a-h 2-Y-M" (Li & Durbin, 2012). Freebayes parallel (v1.3.4) was used to call insertions and deletions for the Illumina data with the option "standard-filters" (Garrison and Marth, 2012). The publicly available vSNP pipeline (USDA, https://github.com/lakinsm/simple-snp, accessed on 27 December 2022) was used to visualize SNPs for the epidemiological analysis calculated using an open-source SNP caller (https://github.com/lakinsm/simple-snp, accessed on 27 December 2022). Variants were filtered to meet the following thresholds: a minimum depth of 10 observed alleles at a genomic location across the population of samples (DP > 10), a minimum observed alternate allele count of 7 at a given genomic location across the population of samples (AO > 7), and an alternative allele frequency greater than or equal to 70% at a given site within a given sample.

To construct the phylogenetic tree, a total of 46 ASFV genomes from public databases were aligned alongside the Illumina sequencing data using the Burrows-Wheeler Aligner. Reference genomes for Genotype IX (ASFV Kenya Bus/2006 KM111295.1) and X (ASFV Kenya Tk1/2005 NC_044945.1 and ASFV Kenya/1950 NC_044944.1) were used during alignment, and SNPs were called as described above. A SNP-based phylogenetic tree was generated using RAxML (v8.2.12).

2.6. Mapping

QGIS version 3.28.1 Firenze (qgis.org) was used for mapping. Uganda district shp files were downloaded from the United Nations (UN) Human High Commissioner for Refugees (https://data.unhcr.org/en/documents/details/83043, accessed on 30 March 2023) and regional data was downloaded from the Office for the Coordination of Humanitarian Affairs (https://data.humdata.org/dataset/cod-ab-uga, accessed on 29 June 2023). Districts were linked to data on the number of samples sequenced per district and mapped.

3. Results

3.1. Genotype Characterization

This study had sequencing results for 31 blood samples that were positive for ASFV with Ct values of less than 26. The pigs from which the blood samples were taken vary by region and district of origin, clinical presentation, and *O. moubata* tick exposure. The samples were taken from a larger set of 1318 blood samples representatively collected between May 2021 and June 2022. All 31 ASFV isolates were classified as genotype IX based on their p72 sequences and whole genome similarity to known genotype IX sequences.

3.2. Sequence Analysis

Of the 31 ASFV samples sequenced, six had sufficient depth of coverage across the genome (>7x) to characterize variants. Although these six isolates aligned closely to a Genotype IX virus found in Uganda in 2015, they established their own clade, suggesting further evolution of the virus (Figure 2). Further, one of these six new isolates (S24/2021) differed from the other five isolates.

All six ASFV genotype IX sequences characterized in this study diverged from Kenya genotype X sequences (Kenya/1950, Tk1/2005) by over 2500 SNPs (>1%) of the genome and grouped with historic genotype IX sequences from Kenya (Bus/2006) and Uganda (2015 isolates). The newly characterized genotype IX sequences shared approximately 100 single nucleotide polymorphisms (SNPs) with the historic genotype IX sequences but diverged from the 2015 Ugandan sequences by approximately 20–60 SNPs not previously described in the 2015 sequences. Five of the six isolates characterized in this study were closely related and clustered into the same clade, while one sequence (S24/2021) appeared to be more ancestrally related to the previously sequenced Ugandan isolates. The SNPs characterized occurred throughout the genome and were not isolated to any region, gene, or multigene family. Further sequencing is needed to thoroughly describe these isolates and to elucidate the level of difference, but the results did show that there were SNPs that were shared among five of this study's isolates, and the sixth isolate also had a unique SNP fingerprint.

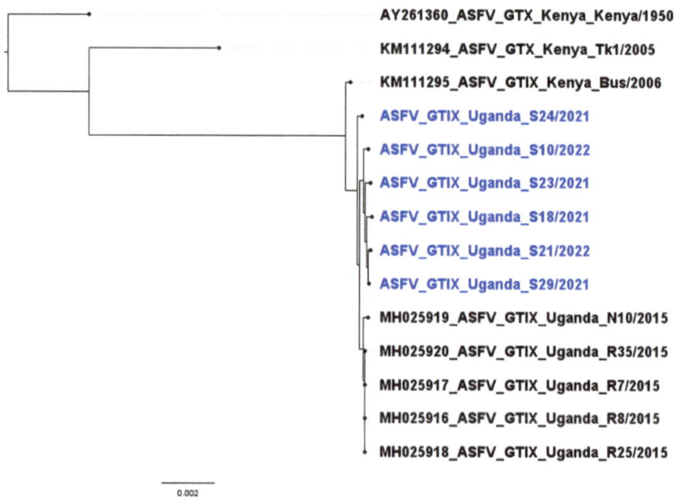

Figure 2. Phylogenetic tree of six isolates sampled from Uganda in 2021–2022 (blue) compared to historic genotype X isolates from Kenya and genotype IX isolates from Kenya and Uganda.

4. Discussion

This study sequenced 31 ASFV detected in blood samples collected between May 2021 through June 2022 from pig abattoirs in the Kampala metropolitan area of Uganda. All samples were identified as genotype IX based on p72 sequence analysis. This aligns with previous work done on in 2007 [11], 2013 [12], and 2015 [13,31]. There have also been two isolates classified as genotype X from Uganda, one from 1965 and another from 1995 [14]. Although we cannot definitively say that other genotypes are not present in Uganda, it appears as though genotype IX is a stable and common cause of ASFV despite the fact that neighboring countries that have a shared border with Uganda have various other genotypes circulating [16–22,32]. Given that the genotype II that was introduced into the Republic of Georgia [10] has spread globally in the same period of time, this suggests that there is not a rapid regional movement of variants, although studies that would target sequencing of isolates in high-risk areas of entry would better determine if there were any incursions of new genotypes.

There were six ASFV sequences that had enough depth of coverage to further evaluate. It was found that they created their own clade and one of the six differed from the other five variants and had its own SNP pattern (Supplementary Files S1 and S2). The differences detected among these isolates suggests that viruses continue to evolve within genotypes and specific geographical locations, as has been previously described for the genotype II epizootic ongoing since 2008 [33]. The impact of this evolution on considerations such as clinical presentation and pathologic presentation, as well as transmission efficacy from *O. moubata* or between pigs will require further study.

In this study the ASFV genotype IX was confirmed to be circulating widely in Uganda, but the work also revealed that the viruses in this genotype continue to evolve, creating diversity within the genotype and the country. There is a need for researchers to leverage whole genome sequencing and to develop a more robust database of African swine fever sequences for comparison to track this evolution and its impact. The p72 gene segment has allowed for genotyping [14] of ASFV, but with whole genome sequencing technology, more detailed comparisons of viruses and their evolution are possible. Such work will allow better understanding of relationships between the genome and disease presentation and is critical to fully leverage molecular epidemiology in outbreak responses, which can determine transmission dynamics and spatiotemporal trends.

Supplementary Materials: The following supporting information can be downloaded at: https://www.mdpi.com/article/10.3390/pathogens12070912/s1, Supplementary File S1: SNP table describing the variants used to build the phylogenetic tree, displaying a subset of mutations, ordered by relevance to the phylogenetic classification. Supplementary File S2: SNP table with variants ordered by genomic location, including all variants called within the Ugandan genomes relative to Genotype IX ASFV strain Kenya Bus/2006.

Author Contributions: Conceptualization, K.A.H., E.M.W., D.N. and B.F.; methodology, R.O., J.F.F. and E.B.K.; formal analysis, S.L., J.P. and R.W.B.; writing—original draft preparation, R.O.; writing—review and editing, all authors; visualization, J.E.E. and S.L.; supervision, K.A.H., E.M.W. and B.F.; project administration, D.N.; funding acquisition, K.A.H. and E.M.W. All authors have read and agreed to the published version of the manuscript.

Funding: This research was sponsored by the U.S. Department of the Defense, Defense Threat Reduction Agency, under grant number HDTRA1-20-1-0007. The content of the information does not necessarily reflect the position or the policy of the federal government, and no official endorsement should be inferred.

Institutional Review Board Statement: This study was endorsed by the College of Veterinary Medicine, Animal Resources, and Biosecurity, Makerere University Higher Degrees Research Committee (Reference number: SBLS.EWM.2020), approved by the Uganda National Council for Science and Technology (Registration number: NS266ES). Animal work was approved by Cornell University's Institutional Animal Care and Use Committee (Protocol number: 2019-0108), and the US Army Medical Research and Development Command's Animal Care and Use Review Office (Protocol number: CT-2020-38). Collection of pig biodata was determined to be exempt by the Cornell University's Institutional Review Board (Protocol number: 2012010020) with a concurrence from the Human Research Protection Office of the US Army Medical Research and Development Command (Log number: A-21116).

Informed Consent Statement: Not applicable.

Data Availability Statement: Sequences may be available upon request with the permission of the Ugandan Ministry of Agriculture, Animal Industries and Fisheries.

Acknowledgments: We would like to thank our colleagues: Edgar Kigozi, Fred Ashaba, Moses Kimuli and Simon Sendawula at the Makerere University College of Health Science Biomedical Research Centre (MakBRC) Sequencing Laboratory for the support. We would also like to thank Laura Goodman at the College of Veterinary Medicine, Cornell University for providing the Makerere University team sequencing training.

Conflicts of Interest: The authors declare no conflict of interest.

References

1. Dixon, L.K.; Chapman, D.A.G.; Netherton, C.L.; Upton, C. African Swine Fever Virus Replication and Genomics. *Virus Res.* **2013**, *173*, 3–14. [CrossRef]
2. Alonso, C.; Borca, M.; Dixon, L.; Revilla, Y.; Rodriguez, F.; Escribano, J.M.; ICTV Report Consortium. ICTV Virus Taxonomy Profile: Asfarviridae. *J. Gen. Virol.* **2018**, *99*, 613–614. [CrossRef] [PubMed]
3. Gaudreault, N.N.; Madden, D.W.; Wilson, W.C.; Trujillo, J.D.; Richt, J.A. African Swine Fever Virus: An Emerging DNA Arbovirus. *Front. Vet. Sci.* **2020**, *7*, 215. [CrossRef] [PubMed]
4. Chenais, E.; Sternberg-Lewerin, S.; Boqvist, S.; Liu, L.; LeBlanc, N.; Aliro, T.; Masembe, C.; Ståhl, K. African Swine Fever Outbreak on a Medium-Sized Farm in Uganda: Biosecurity Breaches and within-Farm Virus Contamination. *Trop. Anim. Health Prod.* **2017**, *49*, 337–346. [CrossRef] [PubMed]

5. Muhangi, D.; Masembe, C.; Emanuelson, U.; Boqvist, S.; Mayega, L.; Ademun, R.O.; Bishop, R.P.; Ocaido, M.; Berg, M.; Ståhl, K. A Longitudinal Survey of African Swine Fever in Uganda Reveals High Apparent Disease Incidence Rates in Domestic Pigs, but Absence of Detectable Persistent Virus Infections in Blood and Serum. *BMC Vet. Res.* **2015**, *11*, 106. [CrossRef]
6. Ogweng, P.; Masembe, C.; Mayega, J.F.; Keeya, I.; Tumuhe, C.; Okwasiimire, R.; Muwanika, V.B. Involvement of Key Stakeholders in Controlling Animal Diseases in Rural Settings: Experiences with African Swine Fever in Uganda. *Afr. J. Agric. Res.* **2020**, *16*, 1599–1610.
7. Ouma, E.; Dione, M.; Birungi, R.; Lule, P.; Mayega, L.; Dizyee, K. African Swine Fever Control and Market Integration in Ugandan Peri-Urban Smallholder Pig Value Chains: An Ex-Ante Impact Assessment of Interventions and Their Interaction. *Prev. Vet. Med.* **2018**, *151*, 29–39. [CrossRef]
8. Montgomery, R.E. On A Form of Swine Fever Occurring in British East Africa (Kenya Colony). *J. Comp. Pathol. Ther.* **1921**, *34*, 159–191. [CrossRef]
9. Njau, E.P.; Machuka, E.M.; Cleaveland, S.; Shirima, G.M.; Kusiluka, L.J.; Okoth, E.A.; Pelle, R. African Swine Fever Virus (ASFV): Biology, Genomics and Genotypes Circulating in Sub-Saharan Africa. *Viruses* **2021**, *13*, 2285. [CrossRef]
10. Rowlands, R.J.; Michaud, V.; Heath, L.; Hutchings, G.; Oura, C.; Vosloo, W.; Dwarka, R.; Onashvili, T.; Albina, E.; Dixon, L.K. African Swine Fever Virus Isolate, Georgia, 2007. *Emerg. Infect. Dis.* **2008**, *14*, 1870–1874. [CrossRef]
11. Gallardo, C.; Ademun, R.A.; Nieto, R.; Nantima, N.; Arias, M.; Martin, E.; Pelayo, V.; Bishop, R.P. Genotyping of African Swine Fever Virus (ASFV) Isolates Associated with Disease Outbreaks in Uganda in 2007. *Afr. J. Biotechnol.* **2011**, *10*, 3488–3497. [CrossRef]
12. Atuhaire, D.K.; Afayoa, M.; Ochwo, S.; Mwesigwa, S.; Okuni, J.B.; Olaho-Mukani, W.; Ojok, L. Molecular Characterization and Phylogenetic Study of African Swine Fever Virus Isolates from Recent Outbreaks in Uganda (2010–2013). *Virol. J.* **2013**, *10*, 247. [CrossRef] [PubMed]
13. Mwiine, F.N.; Nkamwesiga, J.; Ndekezi, C.; Ochwo, S. Molecular Characterization of African Swine Fever Viruses from Outbreaks in Peri-Urban Kampala, Uganda. *Adv. Virol.* **2019**, *2019*, 1463245. [CrossRef]
14. Bastos, A.D.S.; Penrith, M.-L.; Crucière, C.; Edrich, J.L.; Hutchings, G.; Roger, F.; Couacy-Hymann, E.; Thomson, G.R. Genotyping Field Strains of African Swine Fever Virus by Partial P72 Gene Characterisation. *Arch. Virol.* **2003**, *148*, 693–706. [CrossRef] [PubMed]
15. Nix, R.J.; Gallardo, C.; Hutchings, G.; Blanco, E.; Dixon, L.K. Molecular Epidemiology of African Swine Fever Virus Studied by Analysis of Four Variable Genome Regions. *Arch. Virol.* **2006**, *151*, 2475–2494. [CrossRef]
16. Gallardo, C.; Okoth, E.; Pelayo, V.; Anchuelo, R.; Martin, E.; Simon, A.; Llorente, A.; Nieto, R.; Soler, A.; Martin, R.; et al. African Swine Fever Viruses with Two Different Genotypes, Both of Which Occur in Domestic Pigs, Are Associated with Ticks and Adult Warthogs, Respectively, at a Single Geographical Site. *J. Gen. Virol.* **2011**, *92*, 432–444. [CrossRef] [PubMed]
17. Lubisi, B.A.; Bastos, A.D.S.; Dwarka, R.M.; Vosloo, W. Molecular Epidemiology of African Swine Fever in East Africa. *Arch. Virol.* **2005**, *150*, 2439–2452. [CrossRef]
18. Misinzo, G.; Magambo, J.; Masambu, J.; Yongolo, M.G.; Van Doorsselaere, J.; Nauwynck, H.J. Genetic Characterization of African Swine Fever Viruses from a 2008 Outbreak in Tanzania: Genotyping of 2008 Tanzanian ASFV. *Transbound. Emerg. Dis.* **2011**, *58*, 86–92. [CrossRef]
19. Yona, C.M.; Vanhee, M.; Simulundu, E.; Makange, M.; Nauwynck, H.J.; Misinzo, G. Persistent Domestic Circulation of African Swine Fever Virus in Tanzania, 2015–2017. *BMC Vet. Res.* **2020**, *16*, 369. [CrossRef]
20. Patrick, B.N.; Machuka, E.M.; Githae, D.; Banswe, G.; Amimo, J.O.; Ongus, J.R.; Masembe, C.; Bishop, R.P.; Steinaa, L.; Djikeng, A.; et al. Evidence for the Presence of African Swine Fever Virus in Apparently Healthy Pigs in South-Kivu Province of the Democratic Republic of Congo. *Vet. Microbiol.* **2020**, *240*, 108521. [CrossRef]
21. Bisimwa, P.N.; Ongus, J.R.; Tiambo, C.K.; Machuka, E.M.; Bisimwa, E.B.; Steinaa, L.; Pelle, R. First Detection of African Swine Fever (ASF) Virus Genotype X and Serogroup 7 in Symptomatic Pigs in the Democratic Republic of Congo. *Virol. J.* **2020**, *17*, 135. [CrossRef] [PubMed]
22. Mulumba-Mfumu, L.K.; Achenbach, J.E.; Mauldin, M.R.; Dixon, L.K.; Tshilenge, C.G.; Thiry, E.; Moreno, N.; Blanco, E.; Saegerman, C.; Lamien, C.E.; et al. Genetic Assessment of African Swine Fever Isolates Involved in Outbreaks in the Democratic Republic of Congo between 2005 and 2012 Reveals Co-Circulation of P72 Genotypes I, IX and XIV, Including 19 Variants. *Viruses* **2017**, *9*, 31. [CrossRef] [PubMed]
23. Atuhaire, D.K.; Ochwo, S.; Afayoa, M.; Norbert Mwiine, F.; Kokas, I.; Arinaitwe, E.; Ademun-Okurut, R.A.; Boniface Okuni, J.; Nanteza, A.; Ayebazibwe, C.; et al. Epidemiological Overview of African Swine Fever in Uganda (2001–2012). *J. Vet. Med.* **2013**, *2013*, 949638. [CrossRef]
24. Atuhaire, D.K.; Afayoa, M.; Ochwo, S.; Mwesigwa, S.; Mwiine, F.N.; Okuni, J.B.; Olaho-Mukani, W.; Ojok, L. Prevalence of African Swine Fever Virus in Apparently Healthy Domestic Pigs in Uganda. *BMC Vet. Res.* **2013**, *9*, 263. [CrossRef] [PubMed]
25. Foreign Animal Disease Diagnostic Laboratory. *Extraction of Total DNA from Tissues, Tissue Culture Supernatant and Whole Blood Using a Qiagen DNeasy Blood and Tissue Mini Kit*; National Veterinary Services Laboratory, Plum Island Animal Disease Center, US Department of Agriculture: Greenport, NY, USA, 2017.
26. Zsak, L.; Borca, M.V.; Risatti, G.R.; Zsak, A.; French, R.A.; Lu, Z.; Kutish, G.F.; Neilan, J.G.; Callahan, J.D.; Nelson, W.M.; et al. Preclinical Diagnosis of African Swine Fever in Contact-Exposed Swine by a Real-Time PCR Assay. *J. Clin. Microbiol.* **2005**, *43*, 112–119. [CrossRef]

27. Foreign Animal Disease Diagnostic Laboratory. *Preparation, Performance and Interpretation of the African Swine Fever RPCR Assay on the Applied Biosystem QuantStudio 5 REal-Time PCR System*; National Veterinary Services Laboratory, US Department of Agriculture: Greenport, NY, USA; Plum Island Animal Disease Center: New York, NY, USA, 2021.
28. Haendiges, J.; Gonzalez-Escalona, N.; Timme, R.; Balkey, M. Illumina DNA Prep (M) Tagmentation Library Preparation for Use on an Illumina MiSeq Sequencer V2. 2021. Available online: https://www.protocols.io/view/illumina-dna-prep-m-tagmentation-library-preparati-x54v9m7ezg3e/v2 (accessed on 29 June 2023).
29. Illumina. *Nextera XT DNA Library Prep Reference Guide*; Illumina: San Diego, CA, USA, 2019.
30. Lakin, S.M.; O'Donnell, V.; Xu, L.; Barrette, R.W.; Barnabei, J.; Nunez, R.; Holland, R.; Faburay, B.; Robbe-Austerman, S. Whole Genome Sequencing and Molecular Epidemiology of the 2021 African Swine Fever Virus Outbreak in the Dominican Republic. *Transbound. Emerg. Dis.* **2022**, *69*, 4059.
31. Masembe, C.; Sreenu, V.B.; Da Silva Filipe, A.; Wilkie, G.S.; Ogweng, P.; Mayega, F.J.; Muwanika, V.B.; Biek, R.; Palmarini, M.; Davison, A.J. Genome Sequences of Five African Swine Fever Virus Genotype IX Isolates from Domestic Pigs in Uganda. *Microbiol. Resour. Announc.* **2018**, *7*, e01018. [CrossRef]
32. Misinzo, G.; Jumapili, F.; Ludosha, M.; Mafie, E.; Silialis, J.; Mushi, R.; Viaene, W.; van Doorsselaere, J. Genotyping of African Swine Fever Virus from a 2009 Outbreak in Tanzania. *Res. Opin. Anim. Vet. Sci.* **2012**, *2*, 334–338.
33. Shen, Z.-J.; Jia, H.; Xie, C.-D.; Shagainar, J.; Feng, Z.; Zhang, X.; Li, K.; Zhou, R. Bayesian Phylodynamic Analysis Reveals the Dispersal Patterns of African Swine Fever Virus. *Viruses* **2022**, *14*, 889. [CrossRef]

pathogens

Brief Report

Genetic Characterization of the Central Variable Region in African Swine Fever Virus Isolates in the Russian Federation from 2013 to 2017

Ali Mazloum [1,*], Antoinette Van Schalkwyk [2,3], Roman Chernyshev [1], Andrey Shotin [1], Fedor I. Korennoy [1], Alexey Igolkin [1] and Alexander Sprygin [1]

1 Reference Laboratory for African Swine Fever Virus, FGBI "Federal Centre for Animal Health" (FGBI "ARRIAH"), 600901 Vladimir, Russia
2 Agricultural Research Council-Onderstepoort Veterinary Institute, 100 Old Soutpan Road, Onderstepoort 0110, South Africa
3 Department of Biotechnology, University of the Western Cape, Robert Sobukwe Road, Bellville 7535, South Africa
* Correspondence: ali.mazloum6@gmail.com

Abstract: African swine fever virus (ASFV), classified as genotype II, was introduced into Georgia in 2007, and from there, it spread quickly and extensively across the Caucasus to Russia, Europe and Asia. The molecular epidemiology and evolution of these isolates are predominantly investigated by means of phylogenetic analysis based on complete genome sequences. Since this is a costly and time-consuming endeavor, short genomic regions containing informative polymorphisms are pursued and utilized instead. In this study, sequences of the central variable region (CVR) located within the B602L gene were determined for 55 ASFV isolates submitted from 526 active African swine fever (ASF) outbreaks occurring in 23 different regions across the Russian Federation (RF) between 2013 and 2017. The new sequences were compared to previously published data available from Genbank, representing isolates from Europe and Asia. The sequences clustered into six distinct groups. Isolates from Estonia clustered into groups 3 and 4, whilst sequences from the RF were divided into the remaining four groups. Two of these groups (5 and 6) exclusively contained isolates from the RF, while group 2 included isolates from Russia as well as Chechnya, Georgia, Armenia, Azerbaijan and Ukraine. In contrast, group 1 was the largest, containing sequences from the RF, Europe and Asia, and was represented by the sequence from the first isolate in Georgia in 2007. Based on these results, it is recommended that the CVR sequences contain significant informative polymorphisms to be used as a marker for investigating the epidemiology and spread of genotype II ASFVs circulating in the RF, Europe and Asia.

Keywords: African swine fever virus; central variable region; single-nucleotide polymorphism; phylogenetic analysis

Citation: Mazloum, A.; Van Schalkwyk, A.; Chernyshev, R.; Shotin, A.; Korennoy, F.I.; Igolkin, A.; Sprygin, A. Genetic Characterization of the Central Variable Region in African Swine Fever Virus Isolates in the Russian Federation from 2013 to 2017. *Pathogens* **2022**, *11*, 919. https://doi.org/10.3390/pathogens11080919

Academic Editors: Lauro Velazquez-Salinas, Gisselle N. Medina and Elizabeth Ramirez-Medina

Received: 27 July 2022
Accepted: 11 August 2022
Published: 15 August 2022

1. Introduction

African swine fever (ASF) is a fatal viral disease of domestic pigs and wild boars of all ages. The causative agent is the only known double-stranded DNA arbovirus, ASF virus (ASFV), which belongs to the *Asfarviridae* family in the genus Asfivirus. The genome ranges from 170 to 190 kilobase pairs (kbp) and encodes more than 150 open reading frames (ORFs), depending on the viral strain [1,2]. Traditionally, the disease has been confined to sub-Saharan Africa and the Italian island of Sardinia, but sporadic epidemics have affected a number of countries throughout the 20th century. However, in 2007, ASFV was first reported in Georgia, and from there, it spread to Armenia and Azerbaijan and subsequently into the Russian Federation (RF), Ukraine and Belarus [3]. In 2014, four European Union countries, Lithuania, Poland, Latvia, and Estonia, reported ASF prior to its subsequent spread to Belgium, Bulgaria, Czech Republic, Hungary, Romania and Slovakia [4]. In 2018, the disease spread to China, the world's largest pig-producing country, resulting in devastating economic losses to the country [5]. In recent years, ASF has rapidly spread beyond China to neighboring countries, including Mongolia, Cambodia, North Korea, South Korea, the Philippines and India [6]. In July 2021, ASFV genotype II was reported in the Dominican Republic and Haiti, and additionally in Germany, Greece and Italy [7].

Currently, ASFV is classified into 24 genotypes based on sequence data from the C-terminal region of the open reading frame (ORF) B646L, encoding the major capsid protein p72 [8,9]. This gene region is frequently used to investigate the molecular epidemiology of ASF by determining and comparing the genotypes circulating in a region [9]. The additional differentiation of closely related viruses into sub-groups has been subsequently performed using the p54 locus (E183L), the central variable region (CVR) in the B602L gene [10], tandem repeat sequence (TRS) insertion in the intergenic regions (IGR) and multigene family (MGF) 505 9R/10R [11]. Additional genome markers, K145R, MGF 505-5R and O174L, have been used to differentiate isolates from Poland [12]. A novel 14 base pair (bp) TRS insertion of CAGTAGTGATTTTT was identified in the O174L gene of certain isolates from Poland [12]. Complete genome sequences from isolates in the RF recommended that MGF-360-10L, MGF-505-9R and I267L be included as additional genome markers, since they have the resolving capabilities of separating isolates from the RF, EU and China geographically into an eastern and western cluster [13]. The CVR region is frequently targeted for sequence analyses due to unique mutations capable of resolving phylogenies at a regional level, including clustering isolates into groups and sub-types [9,14]. Isolates from Africa were recently divided into various sub-types, based on informative polymorphisms observed within this gene region [15]. Despite the large number of ASFV genotype II outbreaks in Europe and Asia, variants in the CVR have only been described for isolates from Estonia, classifying these isolates into three uniquely distinctive CVR variant groups [16]. Based on the resolving power of this gene region in isolates from Africa and the polymorphisms observed in isolates from Estonia, it was hypothesized that this gene region could be used as a fast and cost-effective method to investigate the epidemiology, evolution and molecular relatedness of large numbers of isolates from the RF. The aim of this study was to characterize the CVR sequences of 55 ASFV isolates, each representing closely linked outbreaks from 23 different regions of the RF during 2013–2017, and subsequently determine the phylogenetic relationship between these isolates with ASFVs from Europe and Asia.

2. Materials and Methods

2.1. Ethics Statement

No animals were used during this study, but samples from clinically infected domestic pigs and wild boars were submitted for the laboratory confirmation of ASFV to the national reference laboratory at the Federal Center for Animal Health (FGBI "ARRIAH") in Vladimir, Russia.

2.2. Isolates and Virus Identification

In this study, 55 ASFV PCR-positive samples were selected as representatives of the 526 outbreaks reported in 23 different regions of the RF at different times during the period 2013–2017. A brief summary of these isolates is provided in Table 1.

Blood or organ tissue samples were collected from domestic pigs (DPs) and either hunted or dead wild boars (WBs). These samples were refrigerated and shipped to FGBI ARRIAH within 24 h of collection. Viral DNA was extracted using the DNeasy Blood & Tissue Kit (Qiagen, Germany) following the manufacturer's recommendations, and the presence of ASFV nucleic acids was determined via real-time PCR according to recommendations of the OIE [17].

Table 1. Brief characteristics of samples used in this study, including the collection year and location as well as the GenBank accession number and the group each sequence is assigned to.

No.	Collection Date	Host	Isolate Name	Region	Accession Number of the CVR Locus	Group
Seq1	2016	DP	Sudogda-Vladimir_16-DP	Vladimir	ON098019	5
Seq2	2016	DP	Arkhangelsk_16-DP	Arkhangelsk	ON098020	5
Seq3	2017	WB	Gorokhovets-Vladimir_17-WB/325	Vladimir	ON098021	5
Seq4	2017	WB	Orel_17-WB/337	Orel	ON098022	5
Seq5	2015	DP	Krasnodar_07/15	Krasnodar	ON098023	2
Seq6	2016	DP	Crimea_Martins_01/16-DP	Crimea	ON098024	2
Seq7	2016	WB	Sobinka-Vladimir_16-WB	Vladimir	ON098025	6
Seq8	2013	WB	Anino-Moscow_13-WB	Moscow	ON098026	6
Seq9	2015	WB	Sobinka-Vladimir_15-WB	Vladimir	ON098027	6
Seq10	2013	WB	Kashino-Tver_13-WB	Tver	ON098028	6
Seq11	2013	WB	Karamzino-Tver_13-WB	Tver	ON098029	6
Seq12	2013	WB	Shihobalovo-Vladimir_13-WB	Vladimir	ON098030	6
Seq13	2013	DP	Boguchary-Voronezh_13-DP/2051	Voronezh	ON098031	1
Seq14	2013	DP	Volgograd_13-DP/2078	Volgograd	ON098032	1
Seq15	2013	DP	Volgograd_13-DP/2059	Volgograd	ON098033	1
Seq16	2014	DP	Voronezh_Agro_14-DP	Voronezh	ON098034	1
Seq17	2014	WB	Grafskoe-Belgorod_14-WB	Belgorod	ON098035	1
Seq18	2014	WB	Odintsovo_02/14-Moscow_14-WB	Moscow	ON098036	1
Seq19	2014	WB	Gribovo-Kaluga_14-WB	Kaluga	ON098037	1
Seq20	2014	WB	Vasilenki-Kaluga_08/14-WB	Kaluga	ON098038	1
Seq21	2014	DP	Antonovo-Pskov_14-DP	Pskov	ON098039	1
Seq22	2015	WB	Ryazan_15-WB	Ryazan	ON098040	1
Seq23	2015	DP	Saratov_01/15	Saratov	ON098041	1
Seq24	2015	DP	Guskhrustalny-Vladimir_15-DP	Vladimir	ON098042	1
Seq25	2015	WB	Ryazan(Autumn)_15-WB	Ryazan	ON098043	1
Seq26	2015	DP	Bolhovsky-Orel_15-DP	Orel	ON098044	1
Seq27	2015	DP	Kurtnikovo-Moscow_15-DP	Moscow	ON098045	1
Seq28	2015	WB	Sashino-Vladimir_15-WB	Vladimir	ON098046	1
Seq29	2015	DP	Smolensk_15-DP	Smolensk	ON098047	1
Seq30	2015	DP	Kursk_15-DP	Kursk	ON098048	1
Seq31	2015	DP	Sokolckie_Vol-Krasnodar_15-DP	Krasnodar	ON098049	1
Seq32	2015	DP	Orel_15-DP	Orel	ON098050	1
Seq33	2017	DP	Krasnodar_07/17	Krasnodar	ON098051	1
Seq34	2016	DP	South_Osetia_16/DP/2325	South Ossetia	ON098052	1
Seq35	2016	WB	Ryazan_03/16	Ryazan	ON098053	1
Seq36	2016	WB	Ryazan_07/16-WB	Ryazan	ON098054	1
Seq37	2016	DP	Shatsky_-_Ryazan_16-DP	Ryazan	ON098055	1
Seq38	2016	DP	Orel-Mtsensk_16-DP	Orel	ON098056	1
Seq39	2016	DP	Kropotkin-Krasnodar_16-DP	Krasnodar	ON098057	1
Seq40	2016	WB	Crimea_16-WB	Crimea	ON098058	1
Seq41	2016	DP	Vrachovo-Moscow_16-DP	Moscow	ON098059	1
Seq42	2016	DP	Penza_16-DP	Penza	ON098060	1
Seq43	2017	DP	Kolchugino-Vladimir_17-DP/5662	Vladimir	ON098061	1
Seq44	2017	DP	Sobinka-Vladimir_17-DP/5660	Vladimir	ON098062	1
Seq45	2017	DP	Sobinka-_Vladimir_17-DP/328	Vladimir	ON098063	1
Seq46	2017	DP	Omsk_17-DP/5665	Omsk	ON098064	1
Seq47	2017	WB	Vyaznikovski-Vladimir_17-WB/330	Vladimir	ON098065	1
Seq48	2017	DP	Irkutsk_17-DP/447	Irkutsk	ON098066	1

Table 1. *Cont.*

No.	Collection Date	Host	Isolate Name	Region	Accession Number of the CVR Locus	Group
Seq49	2017	DP	South_Osetia_17-DP/2196	South Osetia	ON098067	1
Seq50	2017	WB	Crimea_17-WB/470	Crimea	ON098068	1
Seq51	2017	DP	Krasnoyarsk_10/2017	Krasnoyarsk	ON098069	1
Seq52	2017	DP	Omsk_10/2017	Omsk	ON098070	1
Seq53	2017	DP	Belgorod_10/17	Belgorod	ON098071	1
Seq54	2017	WB	Belgorod_12/17	Belgorod	ON098072	1
Seq55	2017	DP	Kaliningrad_12/17	Kaliningrad	ON098073	1

DP = domestic pig; WB = wild boar.

2.3. Sequence Alignment and Phylogenetic Analysis

A 233bp region of the CVR (B602L) gene was PCR-amplified as previously described using the primer pairs ORF9L-F (5′-AATGCGCTCAGGATCTGTTAAATCGG-3′) and ORF9L-R (5′-TCTTCATGCTCAAAGTGCGTATACCT-3′) [10,18]. These amplicons were submitted for Sanger sequencing at the FGBI "ARRIAH" institute using both primers incorporated during the generation of the amplicons. Both sequences were assembled to generate a consensus sequence representing the CVR gene of each isolate. Nucleotide sequences were aligned and compared to corresponding sequences from Genbank (Supplementary Table S1) using Bioedit v7.2.5 software (by Hall, T.A., CA, USA). The phylogenetic relatedness of these sequences was analyzed using Maximum Likelihood under General Time Reversal (GTR + GI = 4), with the consideration of all the sites to account for the gaps due to deletions in the analysis. The sequence of ASFV genotype I Liv13/33 was included as an outlier.

3. Results

The partial gene region of ORF B602L from 55 ASFV isolates, representing 526 outbreaks from 23 different regions within the RF during 2013–2017, were amplified and Sanger sequenced. The sequences were compared to data available from GenBank, representing additional genotype II ASFVs obtained from Europe and Asia. These sequences were aligned, and single-nucleotide polymorphisms (SNPs) were described using isolate Georgia-2007/1 (FR682468.2) as a reference. By comparing the new sequences with all the available data from Europe and Asia, the sequences were sub-divided into six distinct groups. These sub-divisions were based on five SNPs and one deletion and are subsequently described in detail (Figures 1, 2 and S1).

The first group (1) consisted of isolates that shared 100% sequence identity to Georgia 2007/1 (FR682468.2). This group included the majority of the new and old isolates from the RF as well as isolates from Europe and Asia with the exception of Estonia (Table 1, Supplementary Figure S1).

Sequences belonging to group 2 were characterized by two unique SNPs in contrast to the reference sequence Georgia 2007/1 (FR682468.2). A synonymous (C/T) SNP at position 480 and a non-synonymous (A/G) SNP at position 616 were described within these sequences (Supplementary Figure S1). The latter SNP resulted in threonine (T) exchange of an alanine (A) at position 206 of the complete B602L predicted protein (Supplementary Figure S2). This group contained 14 isolates from Eastern Europe, submitted between 2007 and 2014, with 12 sequences obtained from GenBank [19]. The sequences represent samples from Chechnya in 2007 (Che07; JX857524); Georgia in 2007 (Abk07; JX857523); Armenia in 2007 (Arm 07; JX857522); Azerbaijan in 2008 (Az08B; JX857530 and Az08D; JX857529); Ukraine in 2012 (Ukr12/Zapo; JX857535); and the RF between 2008 to 2016 (Oren08; JX857526, Ing08; JX857525, StPet09; JX857534; Kalmykia09; JX857533, Rostov 09; JX857532, Dagestan09; JX857531 as well as from this study: Krasnodar 07/15; ON098023 and Crimea 01\6 Martins; ON098024). Isolates from the RF belonging to this group were mainly submitted from the south-western regions of the RF (Figure 1).

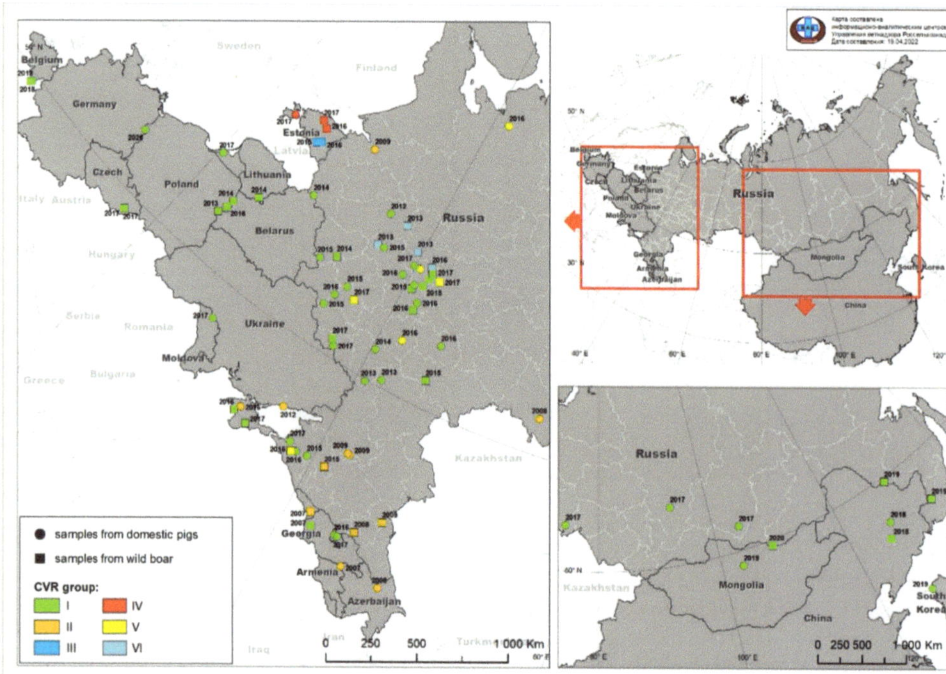

Figure 1. Distribution of the ASFVs from the territory of the RF, EU and Asia based on the sub-division of sequences using the CVR locus. Sequences represent ASFV samples obtained between 2007 and 2020, from either DPs (circles) or WBs (squares). The sub-division of CVR sequences into six groups is graphically presented as indicated by the key provided in the figure.

Groups 3 and 4 were previously identified in a study performed by Vilem et al., 2020. Group 3 was characterized by a 35 bp deletion at position 481, resulting in an amino acid deletion of CASMCADTNVDT (Supplementary Figures S1 and S2). Group 4 contained a non-synonymous (A/G) SNP at position 506, which resulted in a cysteine (C) to tyrosine (Y) exchange at the predicted amino acid position 193 of the complete B602L protein. All sequences analyzed in both of these groups were obtained from Genbank, representing isolates unique to Estonia submitted between 2014 and 2017 [16] (Figure 2).

Group 5 had a non-synonymous (A/G) mutation at position 601, resulting in a lysine (K) exchange of glutamic acid (E) at the predicted amino acid position 201 of the B602L protein (Supplementary Figures S1 and S2). This SNP was observed in six isolates selected and sequenced in this study, representing outbreaks from the RF submitted to the laboratory in 2016 and 2017 (Sudogda-Vladimir 16-DP, Arkhangelsk 16-DP, Tambov 2016, Krasnodar 2016, Gorokhovets-Vladimir 17-WB/325 and Orel 17-WB/337) (Figure 2).

Lastly, group 6 was characterized by a single (A/T) SNP at position 459, identified in six isolates selected and analyzed in this study, representing samples submitted between 2013 and 2016 in Russia (Anino-Moscow 13-WB, Kashino-Tver 13-WB, Karamzino-Tver 13-WB, Shihobalovo-Vladimir 13-WB, Sobinka-Vladimir 15-WB and Sobinka-Vladimir 16-WB) (Figure 2). This is a synonymous SNP involving leucine (L) at amino acid position 153 (Supplementary Figures S1 and S2).

As represented by the phylogenetic tree, all isolates belonging to genotype II were clustered into six different groups, based on the mutations identified in the CVR of the virus genome. Group 1 constituted the largest number of isolates and shared 100% identity to the Georgia 2007/1 sequence. Isolates from the RF could be divided into four distinct groups: 1, 2, 5 and 6. Isolates from Azerbaijan, Georgia, Armenia and Ukraine clustered in group 2 along with samples from the RF (Figure 2). The latter had samples that were unique to groups 5 and 6 (Figure 2). Isolates from Estonia were sub-divided into groups 3 and 4, resulting in both groups being unique to this country (Figure 2).

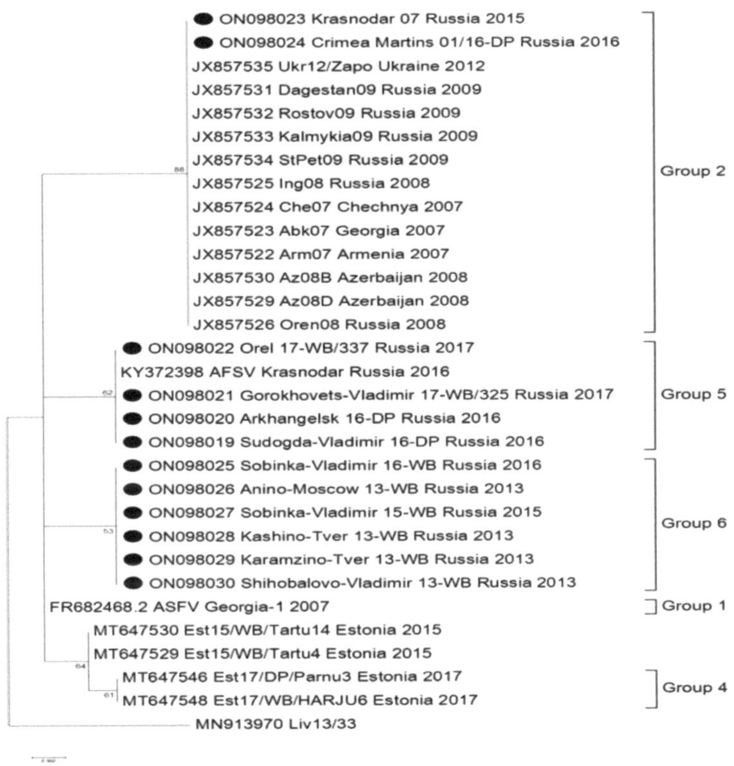

Figure 2. Maximum-likelihood phylogenetic tree based on the 281 bp partial sequence of the B602L gene (CVR) of the ASFV genome. Included in this analysis are the 55 sequences generated in this study from isolates within the RF, as well as sequences obtained from Genbank. Solid black circles are used to identify isolates from this study that belonged to groups other than group 1 (Georgia 2007/1); isolates that showed 100% identity to Georgia 2007/1 are not shown.

Three of the five SNPs utilized during the demarcation of the six groups were non-synonymous, resulting in T206A exchange in 16 isolates and E201K exchange in 7 isolates submitted from the RF (Supplementary Figure S2). In addition, C169Y exchange was uniquely described for 24 isolates from Estonia in 2017 (Supplementary Figure S2) [16]. The isolates from Estonia in 2015 and 2016 were the only sequences that had a deletion of one of the tetramer-tandem repeats (CADT, NVDT and CASM) containing only seven of the eight tetramers (Supplementary Figure S2). The synonymous SNP in group six reduced the number of groups based on amino acid differentiation to five, compared to the six groups described based on nucleotide analysis (Supplementary Figures S1 and S2).

4. Discussion

Since the introduction of ASF into Georgia in 2007, the disease has been spreading in an unprecedented manner across Eurasia. Fear of ASF emergence in the territory, either in domestic pigs or in wild boar populations, exists in many countries currently still free from the disease [20]. From 2007 to 04.04.2022, about 62,351 outbreaks/cases of ASF have been reported in the territory of the RF, Europe, Asia and the Caribbean [7]. Of these,

2139 outbreaks were reported in the RF. Studies on the epidemiology of ASFVs in Europe indicated that wild boars and the products of affected pigs were the largest contributing factors pertaining to the spread of the disease in this region [21]. The subsequent monitoring of outbreaks and tracking of virus movements using genetic tools are therefore imperative in an efficient ASF control strategy. The gold standard in unraveling the relationship between ASFVs is based on the elucidation of a complete genome sequence of individual isolates and subsequent comparative analysis involving multiple ASFVs [22]. However, this procedure is time-consuming, labor-intensive and expensive [23]. Single genomic loci could provide a fast and cost-effective medium to resolve ASFV isolates based on differences in informative SNPs and size variations. Potential loci that could be used to resolve the molecular epidemiology of closely related genotype II ASFVs in Russia, Europe and Asia include I267L, MGF 505-5R and K145R and the CVR locus (B602L) [10,12,13,19]. Based on the analysis of O174L, three variants of ASFV were identified in Poland, whilst a single SNP in the K145R gene identified two additional variants of the virus [12]. Additionally, the IGR (I73R/I329L) verified the circulation of three variants [12]. Isolates in Vietnam clustered in a single group identical to Georgia 2007 based on the sequence analysis of the CVR, while the same isolates were divided into three groups based on their IGR sequences [6]. The CVR gene region is frequently applied to the intra-genotype differentiation of ASFVs circulating in African countries [14,15].

In this study, the partial B602L gene containing the CVR locus of 55 ASFVs, selected to represent outbreaks from different regions of the RF, was determined and compared to previously published sequence data from Europe and Asia. Based on these sequence analyses, samples from the RF were sub-divided into four unique clusters. In addition, ASFVs from Europe and Asia were divided into six distinct groups based on the same region (Supplementary Figure S1) [16]. This is due to the unique sequences previously described in Estonia between 2015 and 2017 [16].

The data generated in this study clustered the isolates into four groups that mirror the spatial or temporal origins of the isolates from the RF (Figure 1). This suggests that the outbreaks were highly clonal and that this marker could be used to track the origin and spread of viruses in future epidemiological studies. The identification of genetically highly related strains observed over multiple years within the same geographical location is indicative of the localized circulation of ASFV in possibly wild boar populations, which has been suggested by previous studies [24].

Interestingly, there were four exceptions to the observed spatiotemporal clustering of the defined groups. Two isolates from group 2, St. Petersburg in 2009 and Orenburg in 2008, as well as two isolates in group 5, Arkhangelsk in 2016 and Krasnodar in 2016, were submitted between 1300 and 1500 km from the nearest isolate belonging to the same cluster (Figure 1). This vast distances between outbreaks suggest possible transboundary incursions from outside the area of study, possibly related to movement of domestic pigs or pig-based products, rather than the localized wild boar transmission pathway. Additionally, group 2 included isolates from Armenia, Georgia, Azerbaijan and Ukraine, which further supports the hypothesis of the possible involvement of human actions in the spread of the disease across international borders [19].

5. Conclusions

This is the first study to differentiate isolates from the territory of the RF, based on CVR sequences. The sequences were clustered into four groups that mirrored the spatial and/or temporal distribution of the outbreaks represented by these isolates from the RF. Based on these results, representatives of outbreaks submitted from the same regions between 2018 and 2022 will be analyzed based on the CVR. The aim of these studies will be to identify any novel mutations and to evaluate if the variants identified within this study are still circulating in the same regions or have spread to new regions.

Supplementary Materials: The following supporting information can be downloaded at: https://www.mdpi.com/article/10.3390/pathogens11080919/s1, Figure S1: Alignment of CVR (B602L) nucleotide sequences of isolates representing the six groups; Figure S2: Alignment of the predicted partial amino acidic sequence of pB602L from representative isolates of each of six groups; Table S1: List of isolates obtained from Genbank used in this study to compare with new isolates from the RF.

Author Contributions: A.M., A.V.S., A.I. and A.S. (Alexander Sprygin) designed the study and revised the draft. A.M., A.V.S., F.I.K. and R.C. performed the experiments. A.S. (Andrey Shotin) and A.I. collected the samples. A.M., A.V.S. and A.S. drafted the manuscript. All authors have read and agreed to the published version of the manuscript.

Funding: This work was supported by grant no. 075-15-2021-1054 from the Ministry of Education and Science of Russia to implement objectives of the Federal Scientific and Technical Program for the Development of genetic technologies during 2019–2027.

Institutional Review Board Statement: No animals were sacrificed for this study.

Informed Consent Statement: Not applicable.

Data Availability Statement: The datasets presented in this study were submitted to the Genbank database (ON098019-ON098073). The names of the isolates and accession number(s) can be found in the article/Table 1.

Acknowledgments: The authors thank Zinyakov Nikolay for their excellent technical assistance.

Conflicts of Interest: The authors declare that the research was conducted in the absence of any commercial or financial relationships that could be construed as a potential conflict of interest.

Abbreviations

ASFV: African swine fever virus; PCR, polymerase chain reaction; CVR, central variable region; kbp, kilobase pair; IGR, intergenic region; TRS, tandem repeat sequence; UPGMA, unweighted pair group method with arithmetic mean.

References

1. Chapman, D.A.; Tcherepanov, V.; Upton, C.; Dixon, L.K. Comparison of the genome sequences of non-pathogenic and pathogenic African swine fever virus isolates. *J. Gen. Virol.* **2008**, *89*, 397–408. [CrossRef] [PubMed]
2. Dixon, L.K.; Chapman, D.A.; Netherton, C.L.; Upton, C. African swine fever virus replication and genomics. *Virus Res.* **2013**, *173*, 3–14. [CrossRef] [PubMed]
3. Jurado, C.; Mur, L.; Pérez, A.M.S.; Cadenas-Fernández, E.; Martínez-López, B.; Sánchez-Vizcaíno, J.M.; Perez, A. Risk of African swine fever virus introduction into the United States through smuggling of pork in air passenger luggage. *Sci. Rep.* **2019**, *9*, 14423. [CrossRef] [PubMed]
4. Cwynar, P.; Stojkov, J.; Wlazlak, K. African swine fever status in Europe. *Viruses* **2019**, *11*, 310. [CrossRef]
5. Zhou, L.; Evan, Y.W.; Wang, S.; Sun, C. African swine fever epidemic in China. *Vet. Rec.* **2019**, *184*, 713. [CrossRef]
6. Nguyen, V.T.; Cho, K.; Mai, N.T.A.; Park, J.; Trinh, T.B.N.; Jang, M.; Nguyen, T.T.H.; Vu, X.D.; Nguyen, T.L.; Nguyen, V.D.; et al. Multiple variants of African swine fever virus circulating in Vietnam. *Arch. Virol.* **2022**, *167*, 1137–1140. [CrossRef]
7. Federal Service for Veterinary and Phytosanitary Surveillance. The Epidemic Situation of ASF in the Territory of the Russian Federation and the Countries of Europe and Asia. OIE Data from 2007 to 2022. Available online: https://fsvps.gov.ru/sites/default/files/files/iac/asf/04-04/04.png (accessed on 4 April 2022).
8. Bastos, A.D.; Penrith, M.L.; Crucière, C.; Edrich, J.L.; Hutchings, G.; Roger, F.; Couacy-Hymann, E.R.; Thomson, G. Genotyping field strains of African swine fever virus by partial p72 gene characterisation. *Arch. Virol.* **2003**, *148*, 693–706. [CrossRef]
9. Quembo, C.J.; Jori, F.; Vosloo, W.; Heath, L. Genetic characterization of African swine fever virus isolates from soft ticks at the wildlife/domestic interface in Mozambique and identification of a novel genotype. *Transbound. Emerg. Dis.* **2018**, *65*, 420–431. [CrossRef]
10. Gallardo, C.; Mwaengo, D.M.; Macharia, J.M.; Arias, M.; Taracha, E.A.; Soler, A.; Okoth, E.; Martín, E.; Kasiti, J.; Bishop, R.P. Enhanced discrimination of African swine fever virus isolates through nucleotide sequencing of the p54, p72, and pB602L (CVR) genes. *Virus Genes.* **2009**, *38*, 85–95. [CrossRef]
11. Mazloum, A.; Igolkin, A.S.; Vlasova, N.N. African Swine Fever Virus: Use genetic markers in analysis of its routes of spread. *Vet. Sci. Today* **2019**, *30*, 9–14. [CrossRef]

12. Mazur-Panasiuk, N.; Walczak, M.; Juszkiewicz, M.; Woźniakowski, G. The Spillover of African Swine Fever in Western Poland Revealed Its Estimated Origin on the Basis of O174L, K145R, MGF 505-5R and IGR I73R/I329L Genomic Sequences. *Viruses* **2020**, *12*, 1094. [CrossRef] [PubMed]
13. Mazloum, A.; Van, S.A.; Shotin, A.; Igolkin, A.; Shevchenko, I.; Gruzdev, K.N.; Vlasova, N. Comparative Analysis of Full Genome Sequences of African Swine Fever Virus Isolates Taken from Wild Boars in Russia in 2019. *Pathogens* **2021**, *10*, 521. [CrossRef] [PubMed]
14. Lubisi, B.A.; Bastos, A.D.; Dwarka, R.M.; Vosloo, W. Intra-genotypic resolution of African swine fever viruses from an East African domestic pig cycle: A combined p72-CVR approach. *Virus Genes* **2007**, *35*, 729–735. [CrossRef] [PubMed]
15. Sidi, M.; Zerbo, H.L.; Ouoba, B.L.; Settypalli, T.B.K.; Bazimo, G.; Ouandaogo, H.S.; Sie, B.; Guy, I.S.; Adama, D.D.T.; Savadogo, J.; et al. Molecular characterization of African swine fever viruses from Burkina Faso, 2018. *BMC Vet. Res.* **2022**, *18*, 1–11. [CrossRef]
16. Vilem, A.; Nurmoja, I.; Niine, T.; Riit, T.; Nieto, R.; Viltrop, A.; Gallardo, C. Molecular Characterization of African Swine Fever Virus Isolates in Estonia in 2014–2019. *Pathogens* **2020**, *9*, 582. [CrossRef]
17. World Organisation for Animal Health. Manual of Diagnostic Tests and Vaccines for Terrestrial Animals, OIE, Paris, France, 2019; Chapter 3.9.1. African Swine Fever. (Version Adopted in May 2021). Available online: https://www.woah.org/fileadmin/Home/eng/Health_standards/tahm/3.09.01_ASF.pdf (accessed on 4 April 2022).
18. Nix, R.J.; Gallardo, C.; Hutchings, G.; Blanco, E.; Dixon, L.K. Molecular epidemiology of African swine fever virus studied by analysis of four variable genome regions. *Arch. Virol.* **2006**, *151*, 2475–2494. [CrossRef]
19. Gallardo, C.; Fernández-Pinero, J.; Pelayo, V.; Gazaev, I.; Markowska-Daniel, I.; Pridotkas, G.; Nieto, R.; Fernández-Pacheco, P.; Bokhan, S.; Nevolko, O.; et al. Genetic variation among African swine fever genotype II viruses, eastern and central Europe. *Emerg. Infect. Dis.* **2014**, *20*, 1544–1547. [CrossRef]
20. Schulz, K.; Conraths, F.J.; Blome, S.; Staubach, C.; Sauter-Louis, C. African Swine Fever: Fast and Furious or Slow and Steady? *Viruses* **2019**, *11*, 866. [CrossRef]
21. Beltrán-Alcrudo, D.; Arias, M.; Gallardo, C.; Kramer, S.; Penrith, M.L. *African Swine Fever: Detection and Diagnosis–A Manual for Veterinarians*; Food and Agriculture Organization of the United Nations (FAO): Rome, Italy, 2017; 88p, Available online: https://www.fao.org/3/i7228e/i7228e.pdf (accessed on 4 April 2022).
22. Forth, J.H.; Forth, L.F.; Blome, S.; Höper, D.; Beer, M. African swine fever whole-genome sequencing—Quantity wanted but quality needed. *PLoS Pathog.* **2020**, *16*, e1008779. [CrossRef]
23. Ji, C.; Jiang, J.; Wei, Y.; Wang, Z.; Chen, Y.; Mai, Z.; Cai, M.; Qin, C.; Cai, Y.; Yi, H.; et al. A Method for the Analysis of African Swine Fever by Viral Metagenomic Sequencing. *Front. Vet. Sci.* **2021**, *8*, 766533. [CrossRef]
24. Malogolovkin, A.; Yelsukova, A.; Gallardo, C.; Tsybanov, S.; Kolbasov, D. Molecular characterization of African swine fever virus isolates originating from outbreaks in the Russian Federation between 2007 and 2011. *Vet. Microbiol.* **2012**, *158*, 415–419. [CrossRef]

pathogens

MDPI

Article

The First Eighteen Months of African Swine Fever in Wild Boar in Saxony, Germany and Latvia—A Comparison

Michael Richter [1], Katja Schulz [2,*], Tobias Elflein [1], Jens Achterberg [3], Edvīns Oļševskis [4,5], Mārtiņš Seržants [4], Kristīne Lamberga [4,5], Franz Josef Conraths [2] and Carola Sauter-Louis [2]

1 Department 25, Veterinary Affairs and Food Control, Landesdirektion Sachsen, 01099 Dresden, Germany
2 Institute of Epidemiology, Friedrich-Loeffler-Institut, Federal Research Institute for Animal Health, 17493 Greifswald-Insel Riems, Germany
3 Saxon State Ministry for Social Affairs and Cohesion, 01097 Dresden, Germany
4 Food and Veterinary Service, LV-1050 Riga, Latvia
5 Institute of Food Safety, Animal Health and Environment "BIOR", LV-1076 Riga, Latvia
* Correspondence: katja.schulz@fli.de; Tel.: +49-38351-7-1803

Abstract: African swine fever (ASF) emerged in Latvia in 2014. In 2020, the virus has been detected in the German federal state, Saxony. In both regions, the virus was probably introduced by infected wild boar coming from affected neighboring countries. As the current ASF control strategy at EU level had not yet been developed at the time of ASF introduction into Latvia, disease control measures in both study areas differed over time. Assessing the course of ASF in Saxony and the implemented control strategies, the first 18 months of the ASF epidemic were epidemiologically compared between Saxony and Latvia. ASF wild boar surveillance data were analyzed and the prevalence of ASF virus-positive wild boar was estimated. For estimating the wild boar density, the numbers of wild boar per km^2 were calculated for the respective geographical areas. The number of samples collected from hunted wild boar and wild boar found dead was higher in Saxony. The ASF virus prevalence in Latvia was significantly higher than in Saxony, indicating that Saxony has had more time for getting prepared for dealing with an ASF incursion. Experience from other countries and the rapid implementation of new control strategies may have helped Saxony deal with ASF.

Keywords: epidemiology; prevalence; wild boar population density; spread; control measures

Citation: Richter, M.; Schulz, K.; Elflein, T.; Achterberg, J.; Oļševskis, E.; Seržants, M.; Lamberga, K.; Conraths, F.J.; Sauter-Louis, C. The First Eighteen Months of African Swine Fever in Wild Boar in Saxony, Germany and Latvia—A Comparison. *Pathogens* **2023**, *12*, 87. https://doi.org/10.3390/pathogens12010087

Academic Editor: Makoto Ozawa

Received: 18 November 2022
Revised: 29 December 2022
Accepted: 2 January 2023
Published: 5 January 2023

1. Introduction

African swine fever (ASF) has, in recent years, become a threat to the global pig industry. The disease is caused by a large DNA virus that has no zoonotic potential and affects only members of the Suidae family [1]. In Eastern and Central Europe, infected wild boar populations play a major role in the epidemiology of the disease and constitute an important risk factor for disease introduction into domestic pig holdings [2–5]. Within the European Union (EU), the occurrence of ASF was restricted to Sardinia for many years. This changed after the disease entered the Baltic states and Poland in 2014 [6]. In Latvia, ASF of genotype II emerged in the wild boar population in June 2014, probably introduced by immigration of infected wild boar from an affected neighboring non-EU country, presumably Belarus [4]. Although ASF of this genotype had already been introduced into Georgia in 2007 and constituted thus a more tangible threat to countries in the EU, the disease emergence in 2014 hit Latvia relatively unexpectedly. In Belarus, only in 2013 but not in 2014, ASF-positive wild boar or outbreaks in domestic pig holdings were reported. Still, the first infected wild boar in Latvia were detected on the border with this country. Several control measures were swiftly put in place in Latvia. Among others, incentives were paid for hunting wild boar with the aim of reducing the ASF-susceptible population. The collection and safe disposal of detected wild boar carcasses was initiated to remove infection sources from the environment [7].

Germany started to increase ASF surveillance activities and to prepare control measures since 2014. At the latest, the introduction of ASF into west Poland in November 2019 [8] put the German veterinary authorities on alert. Thus, the detection of an ASF-positive wild boar carcass in the German federal state of Brandenburg in September 2020 was not completely unexpected [9]. The size of the affected area in western Poland, and the constant infection pressure across the long common border between Poland and Germany caused several independent ASF clusters in wild boar in eastern Germany within a couple of weeks or months [10]. Although an electric fence built close to the border between Poland and Saxony on German territory may have delayed the ASF entry at least to some extent, it was not surprising that the first ASF cases appeared in Saxony by the end of October 2020. After the first detection of an ASF-positive wild boar in the district of Görlitz, eastern Saxony, intensive carcass searches were organized. In addition, further fences were built to reduce or delay further disease spread. One year after the first detection of ASF in Saxony, the disease emerged in the district of Meißen, 65 km westwards to the initially affected region in the East. Due to the large distance and the usually expected speed of ASF spread (between the median of 8.2 and 16.3 km/year) [11], it was assumed that the cases in the district of Meißen were due to a separate introduction of ASF into this region, possibly by human activity [12].

Due to the apparently similar first introduction of ASF into both study areas through wild boar migration but also due to constant infection pressure from neighboring countries (in particular Belarus in the case of Latvia and Poland in the case of Saxony), we aimed to compare the first 18 months of the ASF epidemic in Saxony and Latvia. Although the emergence of ASF in Latvia almost seven years before the one in Germany was due to a similar introduction pathway, the conditions in 2014 were different. Thus, we also aimed to evaluate and discuss the effect of the implementation of control measures, which had been developed over time, also in the Czech Republic and in Belgium, the only countries that managed to eliminate ASF from wild boar in affected regions so far [13,14]. Thus, the study may help to assess the epidemiological situation, surveillance and control measures in a newly affected region such as Saxony and to revise the currently implemented measures if necessary.

2. Materials and Methods

2.1. Surveillance Data

For the analyses, ASF wild boar surveillance data from the German federal state of Saxony and from Latvia were used. Data from the first 18 months of the epidemic were analyzed. Consequently, Saxon data were used from 1 November 2020 through to 27 April 2022 and Latvian data were used from 25 June 2014 through to 30 November 2015.

Surveillance data were retrieved from the CSF/ASF wild boar surveillance database of the European Union (https://surv-wildboar.eu) (accessed on 11 July 2022). The data were used with the approval of the respective veterinary authorities in both countries. Each record referred to a single wild boar and included information about the age of the sampled animal, the test result (serology and/or virology), the date of sampling and the origin of the sample (from apparently healthy, hunted wild boar (active surveillance) or from wild boar shot sick, found dead or died through a road traffic accident (RTA) (passive surveillance)).

All data were analyzed descriptively. Data records originating from wild boar found dead, shot sick and killed in an RTA were merged and analyzed together. Thus, in the present study, all these samples were summarized under wild boar found dead/passive surveillance. In Saxony, 617 data records had no information on the origin of the sample, these samples were assigned to the samples originating from passive surveillance.

The number of investigated samples and prevalence estimates for wild boar that had tested positive for ASF virus (ASFV) were determined for both countries on the municipality level and on a monthly basis. Prevalence estimates were calculated separately for hunted wild boar and wild boar found dead. The estimates were generated by dividing the number of wild boar (either hunted or found dead) that had tested positive for ASFV by PCR by the total number of sampled wild boar (either hunted or found dead). The differences in the ASFV prevalence estimates in hunted wild boar and wild boar found dead between Saxony and Latvia was statistically compared using the Mann–Whitney U-Test test. A p-value < 0.05 was considered statistically significant.

The size of the ASF-affected municipalities was calculated by using ArcGIS ArcMap 10.8.1 (ESRI, Redlands, CA, USA). To determine the size of the affected area in the first 18 months of the epidemic, the areas of the municipalities with an ASFV prevalence above 0 were merged.

The 95% confidence intervals for the ASFV prevalence estimates per geographical area and per month were calculated according to Clopper and Pearson [15]. All analyses and figures for results on monthly basis were performed using R (https://www.r-project.org/) (accessed on 7 November 2022). Results on a geographical level were plotted on maps using ArcGIS ArcMap 10.8.1 (ESRI, Redlands, CA, USA).

2.2. Wild Boar Population Data

Hunting bag data from Saxony was provided on a district level and for the hunting season 2020/21. The data originated from the Saxon Game Monitoring, State Enterprise Saxony Forest, Upper Hunting Authority.

Latvian wild boar population data were estimated based on hunting bag data, which were available for each hunting management unit. For analyses, only data from the hunting season 2014/15 were used. A detailed description of this data set has been published [7]. For the comparison of the estimated wild boar population density between Latvia and Saxony, the number of wild boar/km^2 was used and mapped using ArcGIS ArcMap 10.8.1 (ESRI, Redlands, CA, USA).

3. Results

From Saxony, 38,078 data entries and from Latvia, 16,271 data records were available for analysis (Table 1). The composition of the data is presented in detail in Table 1.

Table 1. Composition of the analyzed ASF wild boar surveillance data from Saxony and Latvia for the first 18 months of ASF occurrence, as entered into the CSF/ASF wild boar surveillance database of the European Union (https://surv-wildboar.eu) (accessed on 11 July 2022).

		Saxony	**Latvia**
Study period		1 November 2020–27 April 2022	25 June 2014–30 November 2015
Number of analyzed data records		38,078	16,271
Size of the whole study area in km^2		18,416	64,589
Number of samples from active surveillance		34,304	15,134
Number of samples from passive surveillance	Not specified	617	0
	Found dead	1611	1130
	Shot sick	147	1
	RTA	1399	6
Age of the sampled wild boar	Not specified	35,469	1265
	<1 year	1183	5615
	1–2 years	1062	9389
	>2 years	364	2

In Saxony, most samples from active surveillance (4367) were examined in November 2021, 13 months after the introduction of ASF. In Latvia, tested samples from hunted wild boar peaked with 1843 specimens in August 2015, 15 months after ASFV introduction into the country (Figure 1). The lowest number of samples originating from active surveillance was taken in Latvia two months later than in Saxony. In Saxony, 986 samples were obtained in June 2021, and in Latvia, 150 samples were collected in March and April 2015. Unsurprisingly, in Latvia, the number of samples was lowest in June 2014, since the disease was first confirmed only on the 26th day of this month (Figure 1).

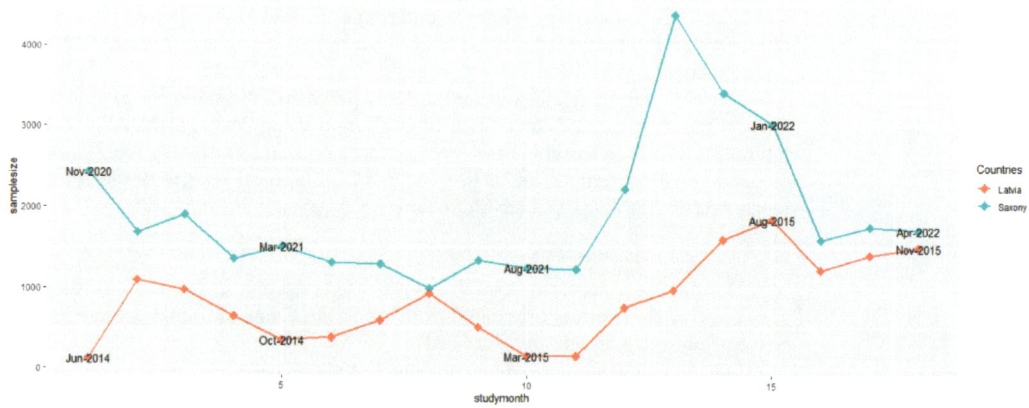

Figure 1. Numbers of samples from hunted wild boar (active surveillance) for African swine fever from Saxony and Latvia in the 18 study months.

In Saxony, 617 data records had no information on the origin of the sample, these samples were assigned to the samples originating from passive surveillance. Similar to the number of samples from active surveillance, the number of samples for passive surveillance in Saxony was almost the highest in November 2021 (301). Only in the first study month (November 2020), the number of samples was slightly higher (335) (Figure 2). Similarly, the trend in the sample size obtained from passive surveillance was comparable to the sample size from active surveillance in Latvia: With 178 and 149 samples, the largest numbers of samples from passive surveillance were investigated in July and August 2015 in this country (Figure 2). In Saxony, the lowest number of samples from passive surveillance was investigated in the last month of the study period (April 2022, 76 samples). Apart from June 2014, only 35 samples were taken in Latvia in January and October 2015. However, in October and November 2014 and in February 2015, the numbers did not exceed 40 samples (Figure 2).

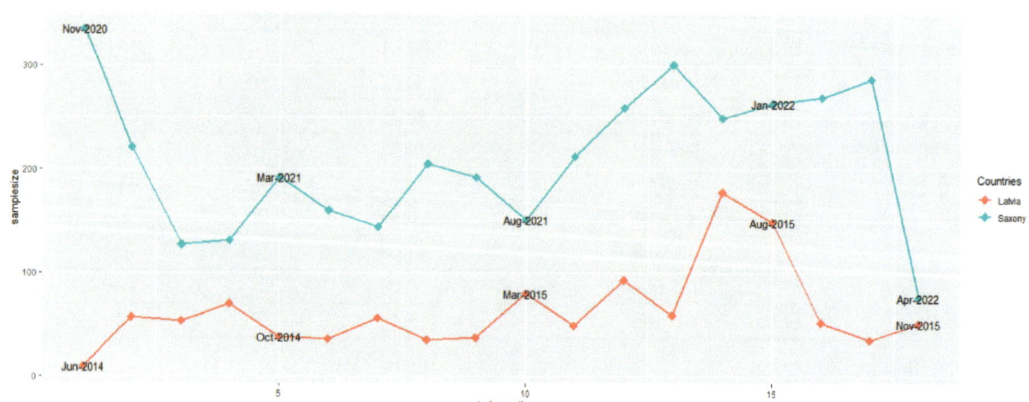

Figure 2. Numbers of samples from wild boar found dead (passive surveillance) for African swine fever from Saxony and Latvia in the 18 study months.

In Saxony, the majority of samples from hunted wild boar and from wild boar found dead were taken in the Northeast. The largest number of samples from hunted wild boar (748 samples) and from wild boar found dead (156) were taken in the municipality of Boxberg/Oberlausitz located in the district of Görlitz. In Lohsa, located in the district of Bautzen, a large number of samples from hunted wild boar (622 samples) was investigated (Figure 3). In Borna, a town in the district of Leipzig, many (138) samples were investigated from wild boar found dead. In 195 of the 419 Saxon municipalities, less than 50 samples were taken from hunted wild boar and in 407 municipalities less than 50 samples were obtained from wild boar found dead. In 77 municipalities, no samples from wild boar found dead were investigated in the first 18 months of the ASF epidemic in Saxony (Figure 3).

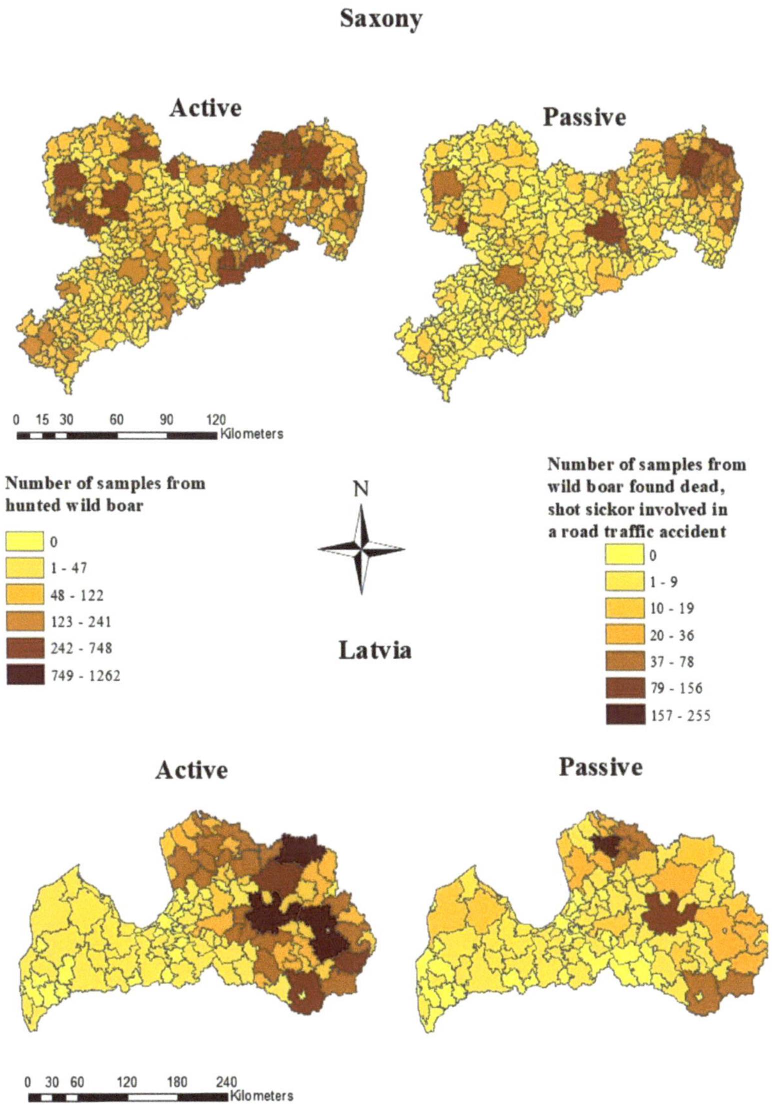

Figure 3. Numbers of samples from hunted wild boar (active surveillance) and wild boar found dead (passive surveillance) from Saxony and Latvia on municipality level.

In Latvia, the majority of samples from hunted wild boar were taken in the east of the country within the study period. In Rēzeknes novads and in Madonas novads, 1262 samples were collected. In 72 of 119 municipalities, less than 50 samples were investigated from hunted wild boar. Samples of wild boar found dead originated mainly from Madonas novads (121 samples) and Burtnieku novads (255 samples). In 116 of the municipalities, less than 50 samples were obtained and in 33 municipalities no samples were obtained (Figure 3).

The size of the area, where ASFV-positive wild boar were detected encompassed 12% of the whole study area (2163 km^2) in Saxony and 43% (27,538 km^2) in Latvia. In Saxony as well as in Latvia, the ASFV prevalence estimates were clearly higher in wild boar found dead than in hunted wild boar (Figures 4 and 5).

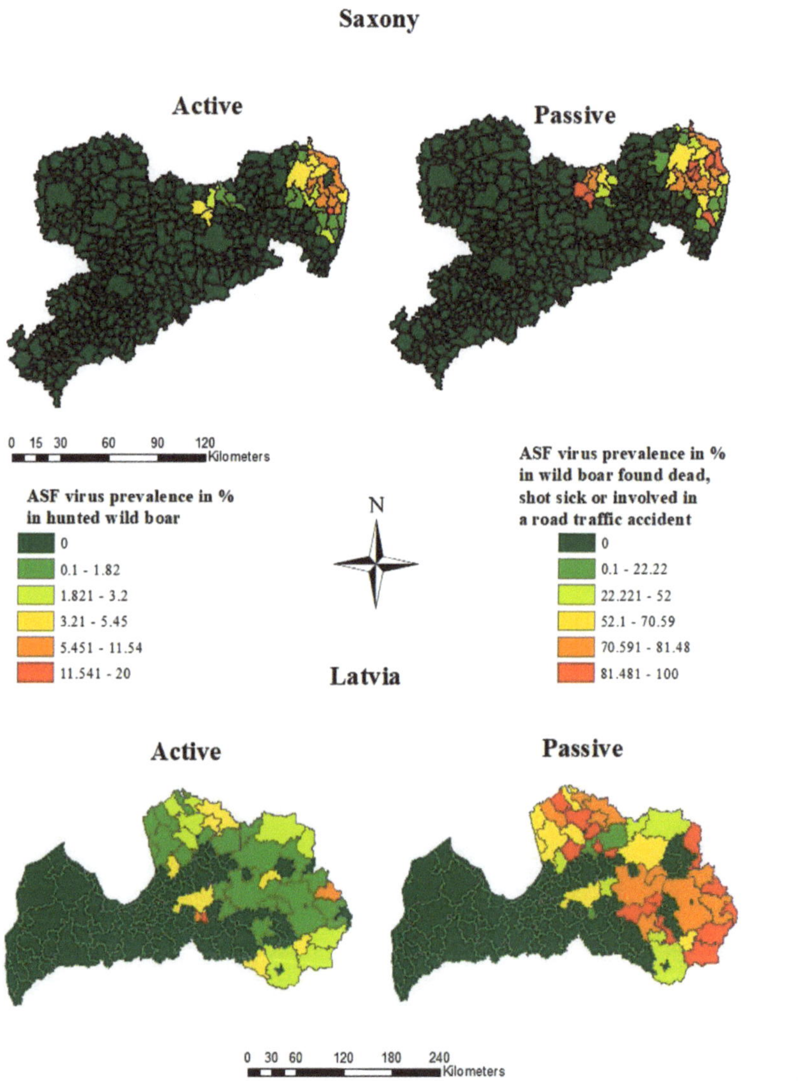

Figure 4. ASFV prevalence estimates in the municipalities of Saxony and Latvia in the first 18 study months of the ASF epidemic, calculated for hunted wild boar and wild boar found dead.

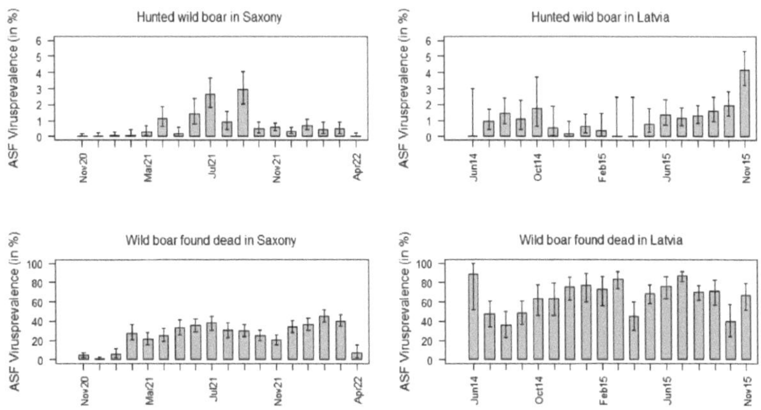

Figure 5. ASFV prevalence estimates in hunted wild boar and wild boar found dead in Saxony and in Latvia for each of the 18 months of the study period.

The highest prevalence estimates in hunted wild boar were found in the district of Görlitz (Figure 4), whereby the highest prevalence (19.2%; 95% CI: 10.9–30.1%) was detected in the municipality of Königshain, while the prevalence was only 4.0% (95% CI: 2.7–5.7%) in Boxberg/Oberlausitz. The highest ASFV prevalence in wild boar found dead (100.0%; 95% CI: 71.5–100.0%) was found in Ebersbach, which is located in the more centrally located district Meißen. However, 28 of the 41 affected municipalities were located in Görlitz, the district bordering Poland (Figure 4).

In Latvia, the highest ASFV prevalence in hunted wild boar (20.0%; 95% CI: 6.8–40.7% and 6.7%; 95 CI: 0.8–22.1%) was found in two small municipalities in the center of the country (Figure 4). In Ciblas novads, in the east of Latvia, the ASFV prevalence in hunted wild boar was 5.5% but with a narrower confidence interval (95% CI: 2.5–10.1%). In seven municipalities, located mainly in the east of the country, the ASFV prevalence in wild boar found dead was 100.0%. However, all 95% confidence intervals were very wide and covered at least 45.0%. In Beverīnas novads, the ASFV prevalence in wild boar found dead was 95.2% with a narrow confidence interval (83.8–99.4%). Overall, the prevalence exceeded 50.0% in 31 municipalities, with the numbers of taken samples varying greatly between municipalities (Figure 4).

In Saxony, the ASFV prevalence in hunted wild boar increased in the first few months of the study period, but it decreased significantly from September 2021 to October 2021. In the subsequent months, it did not exceed 1.0% anymore (Figure 5).

Except for a clear increase after the first three study months in February 2020, the ASFV prevalence in wild boar found dead did not change significantly over time. Yet, in April 2022, the prevalence dropped again to 7.0% (95% CI: 2.3–15.5%) (Figure 5).

Although the ASFV prevalence in Latvian hunted wild boar seemed to increase over time, only in the last study month (November 2015), the prevalence was clearly higher than in the other months (4.2%; 95% CI: 3.2–5.3%). Moreover, the ASFV prevalence in Latvian wild boar found dead did not vary considerably between the different study months (Figure 5).

The ASFV prevalence estimates in hunted wild boar did not differ significantly ($p = 0.145$) between Saxony (median 0.5%) and Latvia (median 1.0%) (Supplemental material Figure S1).

However, the prevalence estimates in Latvia (median 69.3%) were significantly higher ($p < 0.001$) than in the Saxon (median 28.8%) wild boar found dead (Supplemental material Figure S2).

In Saxony, hunting bag data was analyzed at a district level. The number of wild boar/km^2 was highest in the city of Dresden (4.4 wild boar/km^2). In the Eastern districts of Görlitz and Bautzen, the density was only 2.8 and 2.5 wild boar/km^2, respectively (Figure 6). In Latvia, the wild boar density based on hunting bag data was lower than in Saxony. The hunting management unit with the highest wild boar density (1.5 wild boar/km^2) was located in the west of the country (Figure 6). In the majority (393) of the 447 hunting management units, the number of wild boar /km^2 did not exceed 1.0 wild boar/km^2 (Figure 6).

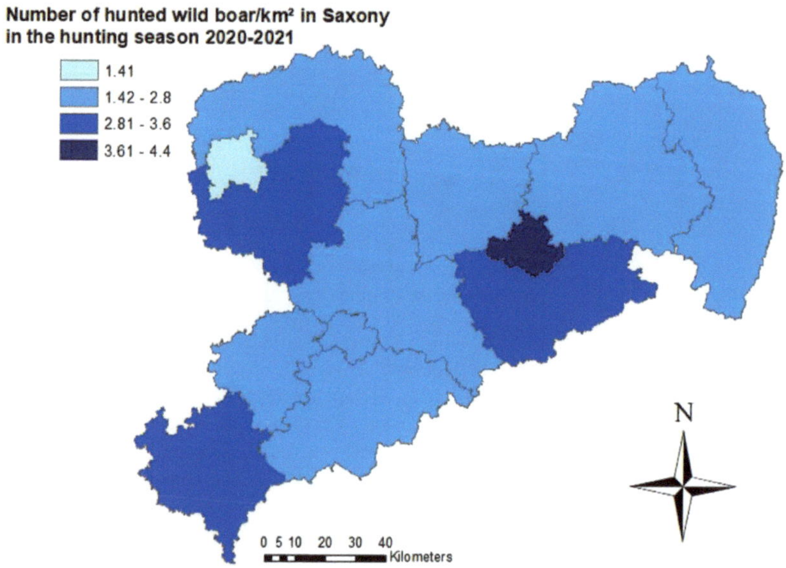

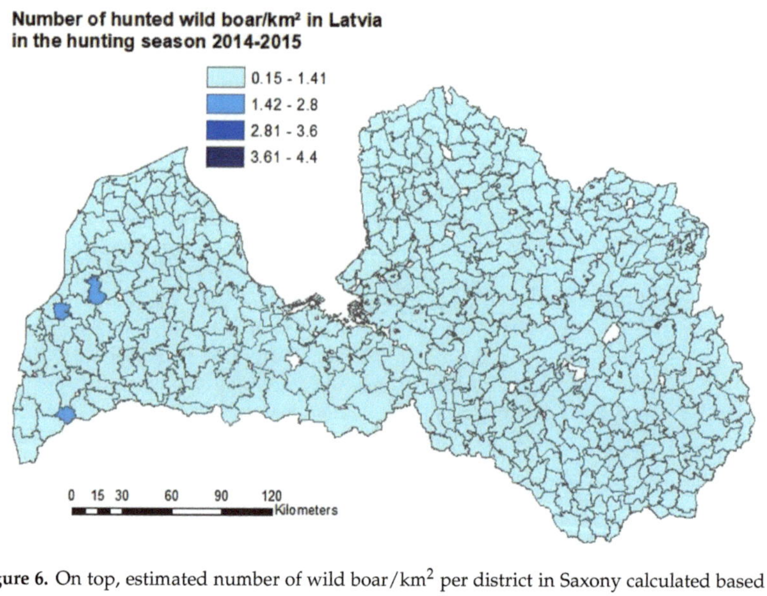

Figure 6. On top, estimated number of wild boar/km^2 per district in Saxony calculated based on hunting bag data from the hunting season 2020–2021. Below, estimated numbers of wild boar/km^2 per Latvian hunting management unit calculated on the basis of hunting bag data from the hunting season 2014–2015.

4. Discussion

In the present study, the first 18 months of the respective ASF epidemics in Saxony and Latvia were analyzed and compared. The aim of the study was to describe the epidemiological situation of ASF in Saxony and to analyze the currently recommended control strategies. To amplify any possible effect, the data was compared with data analyses from the first 18 months of the ASF epidemic in a country that already had been affected before the currently available set of surveillance and control measures had been developed in detail. Latvia was chosen for comparison because ASF had presumably been introduced through migrating wild boar from an affected neighboring country, which resembled the situation in Saxony. Thus, both areas suffered from constant infection pressure due to migrating wild boar, some of which may have been infected with ASFV [4,10]. In addition, the availability of comprehensive surveillance data and good knowledge of the epidemiological situation made the comparison of these two study areas possible.

Despite the considerably smaller size of the study area in Saxony compared to Latvia, the number of samples investigated in the first 18 months after ASFV introduction was in Saxony more than twice as large as in Latvia. This could be due to the larger wild boar population density in Saxony, but it has been influenced also by different control strategies implemented in the two areas. Latvia was one of the first countries in the EU that was hit by the new ASF epidemic in 2014 [2,4,16]. Before ASF, genotype II had entered the first EU Member States and even during the first period of the epidemic, a common EU strategy for disease control in wild boar had not yet been elaborated. Thus, not only in Latvia but also in Lithuania and Estonia, which were also affected in 2014, most probably by the introduction by infected wild boar migrating across the border of affected neighboring countries, the disease spread through the whole country within two to three years [16]. In Poland, ASF was introduced in 2014 and is now present in wild boar populations in several regions, including a large area that has a common border with Saxony in Germany, without an indication that the epidemic might fade out [17]. In contrast, Saxony could benefit from the experience that had already been gained in other countries and had helped to eliminate ASF from affected wild boar populations in the Czech Republic and in Belgium [13,14,18].

In Saxony, a huge number of samples retrieved through passive surveillance originated from wild boar killed through an RTA; whereas, in Latvia, only six samples were obtained from wild boar that had been killed in an RTA. Yet, Schulz et al. [19] attributed the small number of samples from wild boar killed in an RTA rather to a reporting bias than to the actual occurrence of 'road kills'. Due to the potential reporting bias, all wild boar sampled through passive surveillance (found dead, shot sick or involved in an RTA) were merged together and analyzed as originating from passive surveillance without including the exact cause of death of the wild boar. This approach may have biased the results. In a wild boar found dead, the probability of detecting ASFV is usually higher than in animals killed through an RTA [19–21]. Thus far, we are not aware of any conclusive evidence that ASF-infected animals have a higher chance to be killed in an RTA than healthy animals. Thus, merging data from wild boar found dead and killed in RTA may have led to underestimating the true ASFV prevalence in wild boar found dead. This could have been one of the reasons for the significantly lower ASFV prevalence in wild boar sampled in the course of passive surveillance in Saxony as compared to the ASFV prevalence in wild boar found dead in Latvia. In Saxony, almost the same numbers of samples originated from wild boar killed through an RTA than from wild boar found dead. Another reason for the lower ASFV prevalence in dead wild boar in Saxony could have been the active carcasses search that resulted in the detection of a larger number of dead wild boar compared to Latvia. In Latvia, active search for wild boar carcasses was originally not carried out systematically, since this measure had not yet been included in the control strategy at the time of disease introduction into Latvia. Furthermore, the period of time, during which the virus remained infectious in the carcasses and the contaminated environment, might have influenced the ASFV prevalence in the two areas. The temperature, soil composition and other factors affect the tenacity of the virus, thus the transmission rate and the prevalence [6,22]. The postmortem interval, i.e., the time between the death and the detection of these carcasses can influence the probability of transmission and thus the course of the epidemic [23].

The largest number of samples from both, active and passive surveillance was investigated in both countries 13–15 months after the introduction of ASF. Possibly, as the epidemic progressed, it became evident that the disease will not just disappear and more resources supporting the hunting effort and carcass searches were mobilized. Due to an increasing ASFV prevalence over time and the associated increase in mortality [24], it can be assumed that the number of dead wild boar generally increased, consequently resulting in a larger sample size. The increasing size of the affected area and the resulting larger number of ASFV-positive samples could also be seen in Poland [25]. The large sample size in November could also be observed in other countries [16] and is likely to be caused by the hunting season (driven hunts), which spans the winter months. A large number of samples from wild boar found dead in the summer months and also in November was comparable to the trend in the other two Baltic countries Estonia and Lithuania [16]. It is not surprising that the majority of samples were taken in the affected areas. Although the wild boar population density was found to be higher in other areas more centrally and westerly located, the hunting effort and the active carcasses search focused on the areas, where ASF had been newly introduced. EU legislation (Commission Implementing Regulation (EC) 2021/605) demands that all hunted wild boar and all detected carcasses need to be sampled in ASF-restricted areas, which obviously led to an increase in the sample size in affected areas.

The size of the affected area after the first 18 months of the ASF epidemic was clearly larger in Latvia than in Saxony. Due to the higher ASFV prevalence in Latvia, disease control was more difficult and thus, the virus had the chance to spread faster. One cornerstone of ASF control in Saxony was the erection of fences that were built up quickly and over long distances. In contrast, fences were not used in Latvia to control the spread of ASF in the wild boar population, since this measure had not yet been developed at that time. Saxony, therefore, had the advantage that the authorities could rely on the experience of other countries in controlling ASF, e.g., by fencing off affected areas and by implementing systematic wild boar carcass searches and disposing of the carcasses safely, thus reducing the amount of infectious ASFV present in the environment. At the time, when the first ASFV-positive wild boar was detected in Saxony, Belgium and the Czech Republic had already shown that intensive carcass searches supported by high financial rewards and fencing could reduce the spread of ASF and support disease elimination [13,14]. Thus, in contrast to Latvia, Saxony had the chance to prepare and to act immediately using measures that had been effective elsewhere, when ASF hit the region. It needs to be pointed out, however, that the number of affected wild boar is still growing and the affected areas increased in recent months. It seems, therefore, necessary to warrant that the required resources (personnel, finances, materials) remain continuously available until ASF is eliminated. This requires a high level of commitment from the State Government of Saxony, veterinary authorities, hunters and many others, as well as the critical evaluation of the measures and flexibility.

The rather low ASFV prevalence estimates in hunted wild boar and the high prevalence estimates in wild boar found dead in both study areas were not surprising and comparable to prevalence estimates in other countries [17,26,27]. The significantly higher ASFV prevalence estimates and the faster virus spread in Latvia probably supported the large number of outbreaks in domestic pigs in 2014 (32 outbreaks) and 2015 (10 outbreaks) (Animal Disease Information System, ADIS, of the European Union; accessed on 25 October 2022). Most of the affected farms in Latvia were small and poor biosafety/hygiene was identified as the most probable reason for virus introduction [4]. That could also be observed in Lithuania [28]. In Saxony, these kinds of pig holdings may be less frequent, which may have helped to prevent ASF outbreaks in domestic pig holdings in this region so far.

In Saxony, ASF seems to spread more slowly and despite the larger wild boar population density, a smaller area than in Latvia was affected by the disease after 18 months. Yet, ASF-positive wild boar carcasses are still regularly detected, indicating the constant need to further reduce the wild boar population in the areas at risk (areas geographically adjacent to an area with ASFV circulating in wild boar) and thus the number of susceptible hosts [29]. Further research is necessary to understand transmission dynamics and the maintenance of the virus within and between wild boar sub-populations, particularly if the wild boar population density is low [30].

Moreover, after one year, in which the virus only circulated in the eastern districts adjacent to the region affected in Poland, new cases emerged in the district of Meißen, approximately 65 km west of the initial cluster. This "jump" of ASF illustrates the continuous threat of a human-mediated disease spread. Even if the vast majority of people follow the rules meant to prevent ASF spread, ignorance or risky behavior of a single person can spark a new epidemic far away from previously ASF-affected regions and the best control measures cannot prevent such events. Therefore, to minimize these risks, it is inevitable to stay in contact with different groups of people, to actively distribute information and to raise awareness.

The course of the disease in Latvia, but also in Estonia, where it was thought for 1.5 years that the virus had been eliminated from the wild boar population [30,31], shows that ASF control in wild boar is very difficult. Thus far, no country in a comparable epidemiological situation has managed to successfully eliminate ASF from their country [10]. Although the epidemiological situation in Saxony cannot be compared to that in the Czech Republic or Belgium due to the constant infection pressure from western Poland, the comparison with Latvia has shown that the recently developed control strategies work, at least in principle, and have the potential to slow down the further spread of the disease. Accordingly, hope should not be given up that ASF can be successfully controlled through joint efforts and good national and international cooperation.

Supplementary Materials: The following supporting information can be downloaded from: https://www.mdpi.com/article/10.3390/pathogens12010087/s1, Figure S1: ASFV prevalence estimates in hunted wild boar from Saxony and Latvia. The horizontal lines that form the top of the boxes illustrate the 75th percentile. The horizontal lines that form the bottom of the boxes represent the 25th percentile. The horizontal lines that intersect the box are the estimated median ASFV prevalences in hunted wild boar. Whiskers represent maximum and minimum values that are no more than 1.5 times the span of the interquartile range. Open circles represent outliers, which are single values greater or less than the extremes indicated by the whiskers.; Figure S2: ASFV prevalence estimates in wild boar found dead from Saxony and Latvia. The horizontal lines that form the top of the boxes illustrate the 75th percentile. The horizontal lines that form the bottom of the boxes represent the 25th percentile. The horizontal lines that intersect the box are the estimated median ASFV prevalences in wild boar found dead. Whiskers represent maximum and minimum values that are no more than 1.5 times the span of the interquartile range.

Author Contributions: Conceptualization, M.R., K.S. and C.S.-L.; methodology, K.S. and C.S.-L.; formal analysis, K.S.; data curation, K.S., T.E., E.O. and M.S.; writing—original draft preparation, M.R., K.S., E.O. and K.L.; writing—review and editing, M.S., J.A., T.E., C.S.-L. and F.J.C.; supervision, F.J.C. and C.S.-L. All authors have read and agreed to the published version of the manuscript.

Funding: This research received no external funding.

Institutional Review Board Statement: No ethical approval was required for the present study as the research does not describe research on animals or humans. In the study, already existing official surveillance data, which have been collated on the basis of EU and national legislation, was analyzed without including any experiments involving animals or humans. No ethical approval from a research ethics committee was required for this type of data analysis.

Informed Consent Statement: Not applicable.

Data Availability Statement: The original data used for the analyses can be obtained from the corresponding author after approval by the responsible institutions in, Latvia and Saxony.

Conflicts of Interest: The authors declare no conflict of interest.

References

1. Alonso, C.; Borca, M.; Dixon, L.; Revilla, Y.; Rodriguez, F.; Escribano, J.M.; Consortium, I.R. ICTV Virus Taxonomy Profile: Asfarviridae. *J. Gen. Virol.* **2018**, *99*, 613–614. [CrossRef] [PubMed]
2. European Food Safety Authority; Cortinas Abrahantes, J.; Gogin, A.; Richardson, J.; Gervelmeyer, A. Scientific report on epidemiological analyses on African swinefever in the Baltic countries and Poland. *EFSA J.* **2017**, *15*, 4732.
3. Nurmoja, I.; Mõtus, K.; Kristian, M.; Niine, T.; Schulz, K.; Depner, K.; Viltrop, A. Epidemiological analysis of the 2015–2017 African swine fever outbreaks in Estonia. *Prev. Vet. Med.* **2018**, *181*, 104556. [CrossRef] [PubMed]
4. Olševskis, E.; Guberti, V.; Serzants, M.; Westergaard, J.; Gallardo, C.; Rodze, I.; Depner, K. African swine fever virus introduction into the EU in 2014: Experience of Latvia. *Res. Vet. Sci.* **2016**, *105*, 28–30. [CrossRef] [PubMed]
5. Boklund, A.; Dhollander, S.; Chesnoiu Vasile, T.; Abrahantes, J.C.; Bøtner, A.; Gogin, A.; Gonzalez Villeta, L.C.; Gortázar, C.; More, S.J.; Papanikolaou, A.; et al. Risk factors for African swine fever incursion in Romanian domestic farms during 2019. *Sci. Rep.* **2020**, *10*, 10215. [CrossRef]
6. Sauter-Louis, C.; Conraths, F.J.; Probst, C.; Blohm, U.; Schulz, K.; Sehl, J.; Fischer, M.; Forth, J.H.; Zani, L.; Depner, K.; et al. African Swine Fever in Wild Boar in Europe—A Review. *Viruses* **2021**, *13*, 1717. [CrossRef]
7. Schulz, K.; Olševskis, E.; Staubach, C.; Lamberga, K.; Seržants, M.; Cvetkova, S.; Conraths, F.J.; Sauter-Louis, C. Epidemiological evaluation of Latvian control measures for African swine fever in wild boar on the basis of surveillance data. *Sci. Rep.* **2019**, *9*, 4189. [CrossRef]
8. Mazur-Panasiuk, N.; Walczak, M.; Juszkiewicz, M.; Woźniakowski, G. The Spillover of African Swine Fever in Western Poland Revealed Its Estimated Origin on the Basis of O174L, K145R, MGF 505-5R and IGR I73R/I329L Genomic Sequences. *Viruses* **2020**, *12*, 1094. [CrossRef]
9. Sauter-Louis, C.; Forth, J.H.; Probst, C.; Staubach, C.; Hlinak, A.; Rudovsky, A.; Holland, D.; Schlieben, P.; Göldner, M.; Schatz, J.; et al. Joining the club: First detection of African swine fever in wild boar in Germany. *Transbound. Emerg. Dis.* **2020**, *68*, 1744–1752. [CrossRef]
10. Sauter-Louis, C.; Schulz, K.; Richter, M.; Staubach, C.; Mettenleiter, T.C.; Conraths, F.J. African swine fever: Why the situation in Germany is not comparable to that in the Czech Republic or Belgium. *Transbound. Emerg. Dis.* **2021**, *69*, 2201–2208. [CrossRef]
11. European Food Safety Authority; Boklund, A.; Cay, B.; Depner, K.; Földi, Z.; Guberti, V.; Masiulis, M.; Miteva, A.; More, S.; Olsevskis, E.; et al. Epidemiological analyses of African swine fever in the European Union (November 2017 until November 2018). *EFSA J.* **2018**, *16*, e05494. [CrossRef] [PubMed]
12. Sächsisches Staatsministerium für Soziales und Gesellschaftlichen Zusammenhalt. ASP-Ausbruch im Landkreis Meißen ist Keine Verschleppung Sondern Eigenständiges Seuchengeschehen. Available online: https://www.medienservice.sachsen.de/medien/news/1039046 (accessed on 22 June 2022).
13. Šatrán, P. From ASF infection in wild boar to eradication and free status recovery in the Czech Republic. In Proceedings of the 12th GF-TADs Standing Group of Experts on African Swine Fever for the Baltic and Eastern Europe Region (SGE ASF12), Prague, Czech Republic, 11 March 2019.
14. Federal Agency for the Safety of the Food Chain Belgium. African Swine Fever in Wild Boar in Belgium. Available online: https://ec.europa.eu/food/sites/food/files/animals/docs/reg-com_ahw_20180919_pres_asf_bel.pdf (accessed on 29 April 2021).
15. Clopper, C.J.; Pearson, E.S. The use of confidence or fiducial limits illustrated in the case of the binomial. *Biometrika* **1935**, *26*, 404–413. [CrossRef]
16. Schulz, K.; Olševskis, E.; Viltrop, A.; Masiulis, M.; Staubach, C.; Nurmoja, I.; Lamberga, K.; Seržants, M.; Malakauskas, A.; Conraths, F.J.; et al. Eight Years of African Swine Fever in the Baltic States: Epidemiological Reflections. *Pathogens* **2022**, *11*, 711. [CrossRef] [PubMed]
17. Frant, M.P.; Gal-Cisoń, A.; Bocian, Ł.; Ziętek-Barszcz, A.; Niemczuk, K.; Szczotka-Bochniarz, A. African Swine Fever (ASF) Trend Analysis in Wild Boar in Poland (2014–2020). *Animals* **2022**, *12*, 1170. [CrossRef] [PubMed]
18. European Commission. Strategic approach to the management of African Swine Fever for the EU. In *SANTE G3*; European Commission: Brussels, Belgium, 2020; Volume SANTE/7113/2015—Rev 12.
19. Schulz, K.; Conraths, F.J.; Staubach, C.; Viltrop, A.; Olševskis, E.; Nurmoja, I.; Lamberga, K.; Sauter-Louis, C. To sample or not to sample? Detection of African swine fever in wild boar killed in road traffic accidents. *Transbound. Emerg. Dis.* **2020**, *67*, 1816–1819. [CrossRef]
20. European Food Safety Authority. Risk assessment of African swine fever in the south-eastern countries of Europe. *EFSA J.* **2019**, *17*, 5861. [CrossRef]
21. European Food Safety Authority. Epidemiological analyses of African swine fever in the European Union (November 2018 to October 2019). *EFSA J.* **2018**, *18*, 5996.
22. Fischer, M.; Huehr, J.; Blome, S.; Conraths, F.J.; Probst, C. Stability of African Swine Fever Virus in Carcasses of Domestic Pigs and Wild Boar Experimentally Infected with the ASFV "Estonia 2014" Isolate. *Viruses-Basel* **2020**, *12*, 1118. [CrossRef]
23. Probst, C.; Gethmann, J.; Hohmann, U.; Knoll, B.; Amendt, J.; Teifke, J.P.; Conraths, F.J. Zersetzungsstadien bei Wildschweinkadavern—und wie die Liegezeit geschätzt werden kann. *Amtstierärztlicher Dienst Und Lebensm.* **2020**, *27*, 85–94.
24. Morelle, K.; Bubnicki, J.; Churski, M.; Gryz, J.; Podgorski, T.; Kuijper, D.P.J. Disease-Induced Mortality Outweighs Hunting in Causing Wild Boar Population Crash After African Swine Fever Outbreak. *Front. Vet. Sci.* **2020**, *7*, 378. [CrossRef]
25. Szymańska, E.J.; Dziwulaki, M. Development of African Swine Fever in Poland. *Agriculture* **2022**, *12*, 119. [CrossRef]

26. Olševskis, E.; Schulz, K.; Staubach, C.; Serzants, M.; Lamberga, K.; Pule, D.; Ozolins, J.; Conraths, F.J.; Sauter-Louis, C. African swine fever in Latvian wild boar-A step closer to elimination. *Transbound. Emerg. Dis.* **2020**, *67*, 2615–2629. [CrossRef] [PubMed]
27. Schulz, K.; Masiulis, M.; Staubach, C.; Malakauskas, A.; Pridotkas, G.; Conraths, F.J.; Sauter-Louis, C. African Swine Fever and Its Epidemiological Course in Lithuanian Wild Boar. *Viruses* **2021**, *13*, 1276. [CrossRef] [PubMed]
28. Malakauskas, A.; Schulz, K.; Kukanauskaitė, I.; Masiulis, M.; Conraths, F.J.; Sauter-Louis, C. African Swine Fever Outbreaks in Lithuanian Domestic Pigs in 2019. *Animals* **2022**, *12*, 115. [CrossRef] [PubMed]
29. Lange, M.; Reichold, A.; Thulke, H.-H. Modelling wild boar management for controlling the spread of ASF in the areas called white zones (zones blanche). *EFSA Support. Publ.* **2021**, *18*, 6573E. [CrossRef]
30. Schulz, K.; Schulz, J.; Staubach, C.; Blome, S.; Nurmoja, I.; Conraths, F.J.; Sauter-Louis, C.; Viltrop, A. African Swine Fever Re-Emerging in Estonia: The Role of Seropositive Wild Boar from an Epidemiological Perspective. *Viruses* **2021**, *13*, 2121. [CrossRef]
31. Schulz, K.; Staubach, C.; Blome, S.; Nurmoja, I.; Viltrop, A.; Conraths, F.J.; Kristian, M.; Sauter-Louis, C. How to Demonstrate Freedom from African Swine Fever in Wild Boar-Estonia as an Example. *Vaccines* **2020**, *8*, 336. [CrossRef]

pathogens

MDPI

Article

Evaluation of the Dose of African Swine Fever Virus Required to Establish Infection in Pigs Following Oral Uptake

Ann Sofie Olesen [1,*], Christina Marie Lazov [1,2], Francesc Accensi [3,4,5], Camille Melissa Johnston [1], Thomas Bruun Rasmussen [1], Anette Bøtner [2], Louise Lohse [1] and Graham J. Belsham [2]

1 Section for Veterinary Virology, Department of Virus & Microbiological Special Diagnostics, Statens Serum Institut, DK-2300 Copenhagen, Denmark; chlv@ssi.dk (C.M.L.); camj@ssi.dk (C.M.J.); tbru@ssi.dk (T.B.R.); lolo@ssi.dk (L.L.)

2 Section for Veterinary Clinical Microbiology, Department of Veterinary and Animal Sciences, University of Copenhagen, DK-1870 Frederiksberg C, Denmark; anettebotner@outlook.com (A.B.); grbe@sund.ku.dk (G.J.B.)

3 Research Combined Unit IRTA-UAB in Animal Health, Research Center in Animal Health (Centre de Recerca en Sanitat Animal CReSA), Campus of Autonomous University of Barcelona (UAB), 08193 Bellaterra, Spain; francesc.accensi@uab.cat

4 Departament de Sanitat i Anatomia Animals, Facultat de Veterinària, Campus of Autonomous University of Barcelona UAB, 08193 Bellaterra, Spain

5 WOAH Collaborating Centre for the Research and Control of Emerging and Re-Emerging Swine Diseases in Europe (IRTA-CReSA), 08193 Bellaterra, Spain

* Correspondence: asjo@ssi.dk

Abstract: African swine fever virus (ASFV) is known to be very stable within a protein-rich environment and indirect virus transmission can be mediated via oral uptake of different materials. However, experimental studies in pigs have shown that infection by ASFV via the oral route can be difficult to establish. Currently, there is a lack of studies using strict oral inoculations of pigs with different doses of ASFV. Therefore, we aimed to determine the dose of a European genotype II ASFV that is required to establish infection of pigs by the oral route. In this study, 24 pigs were divided into four groups of six. Three of the groups were fed with a low, medium or high dose of the ASFV POL/2015/Podlaskie virus. The pigs in the fourth group served as positive controls and were inoculated intranasally, just once, using the low dose of the virus. All the pigs inoculated intranasally with ASFV succumbed to the infection, while only three of the six pigs that were fed the high dose of the virus became infected. None of the 12 pigs that were fed with either the medium or low dose of the virus became infected, despite receiving up to thirteen doses each. In two of the pigs infected by intranasal inoculation, the presence of a variant form of the ASFV genome was detected. The results obtained in this study underline that ASFV infection is more difficult to establish via the oral route when compared to the intranasal route. The high dose needed in order to establish oral infection could have implications for future strategies using baited vaccines containing infectious live-attenuated ASFV.

Keywords: ASFV; dosing study; feeding; transmission; oral uptake

Academic Editors: Natasha N. Gaudreault and Wei Liu

Received: 10 December 2024
Revised: 21 January 2025
Accepted: 23 January 2025
Published: 27 January 2025

Citation: Olesen, A.S.; Lazov, C.M.; Accensi, F.; Johnston, C.M.; Rasmussen, T.B.; Bøtner, A.; Lohse, L.; Belsham, G.J. Evaluation of the Dose of African Swine Fever Virus Required to Establish Infection in Pigs Following Oral Uptake. *Pathogens* **2025**, *14*, 119. https://doi.org/10.3390/pathogens14020119

1. Introduction

African swine fever (ASF) is a hemorrhagic disease of both domestic pigs and wild boar. The disease is caused by infection with African swine fever virus (ASFV), a large dsDNA virus, the sole member of the *Asfarviridae* family [1,2]. Since 2007, a genotype II strain of ASFV has been spreading outside Africa, and thus, ASF is now reaching pandemic proportions in Europe, Asia, Oceania and the Americas [3], with a high case fatality rate in both domestic pigs and wild boar.

Although ASFV is an arbovirus, naturally transmitted by soft ticks of the genus *Ornithodoros*, it is also efficiently transmitted via direct contact between animals [1,4]. In Europe, however, most virus introductions into domestic pig herds seem to be mediated via indirect virus transmission, e.g., via pig meat or pork products, materials contaminated with carcass material or blood from infected suids or potentially via blood-feeding insects [5]. One likely route of transmission from these different materials to pigs is via oral uptake. For meat products, it has been demonstrated that ASFV can be transmitted to pigs indirectly via oral uptake of different products containing the virus [6–8], and infection of pigs has been demonstrated following ingestion of blood-fed stable flies [9]. Proof of efficient transmission via an environment contaminated with ASFV has not been demonstrated in experimental settings. Indeed, on the contrary, transmission from an environment contaminated with the virus has been limited or absent in such studies [10–12]. Studies using strict oral inoculation (i.e., oropharyngeal or via intragastric tube) with feed or insect larvae (used for food and feed) containing ASFV have also shown that it can be difficult to demonstrate infection. In one study, using doses from 1 to 10^8 $TCID_{50}$ in plant-based feed or liquid, the minimum infectious dose of ASFV for pigs in compound feed was reported to be 10^4 $TCID_{50}$, while in liquid it was found to be 1 $TCID_{50}$ [13]. However, in another study, using commercial feed spiked with $10^{4.3}$ to 10^5 $TCID_{50}$ of the virus, pigs did not become infected even after feeding with the virus-contaminated feed on 14 consecutive days [14]. In a study using insect larvae (*Tenebrio molitor* or *Hermetia illucens*), which had fed on infectious ASFV, infection of pigs could not be demonstrated after feeding 50 virus-fed larvae to each pig. Each *T. molitor* larva had fed on 5 μL serum from an infected pig (titer $10^{3.3}$ $TCID_{50}/5$ μL), while *H. illucens* had fed on feed containing a virus load of $10^{5.0}$ $TCID_{50}/g$ [15].

Given the currently available information, it seems as if it can be rather difficult to establish ASFV infection via the oral route. To our knowledge, except for one published study [13], ASFV dose studies have not used strict oral inoculations. For example, one dose study used intraoropharyngeal inoculation [16], while another used oronasal inoculation [17]. Therefore, we aimed to investigate the dose of a European genotype II ASFV needed to establish oral infection of pigs in our experimental settings. In such animal studies, thorough characterization of the pathogens used for inoculation is important. Therefore, we also aimed to use molecular methods to analyze the inoculation material and the viruses recovered from the infected pigs.

2. Materials and Methods

2.1. Pigs and Housing

Twenty-four male pigs, at six weeks of age, were included in this study. The pigs were Landrace x Large White and were obtained from a conventional swine herd in Catalonia, Spain. During the experiment, the pigs were housed, under high containment, at the Centre de Recerca en Sanitat Animal (IRTA-CReSA, Barcelona, Spain). On arrival at the IRTA-CReSA, the pigs were found to be healthy upon veterinary inspection. Water and a commercial diet for weaned pigs were provided ad libitum.

This study was approved by the Ethical and Animal Welfare Committee of the Generalitat de Catalunya (Autonomous Government of Catalonia; permit number: 12187). All the experimental procedures, animal care and maintenance were conducted in accordance with EU legislation on animal experimentation (EU Directive 2010/63/EU).

2.2. Virus and Inoculation Material

For the inoculation of the pigs, the highly virulent ASFV POL/2015/Podlaskie was used. This virus was isolated from spleen material from an ASFV-infected wild boar in Poland in 2015, essentially as described previously but with an additional passage (i.e., three passages in total) in porcine pulmonary alveolar macrophages (PPAMs) [18,19].

For the intranasal inoculation of the pigs, the third passage was diluted in phosphate-buffered saline (PBS, Thermo Fisher Scientific, Waltham, MA, USA) to a final concentration of 10^3 $TCID_{50}/2$ mL.

For the oral inoculation (feeding) of the pigs, the third passage virus was diluted to a final concentration of 10^3 $TCID_{50}/0.5$ mL, 10^4 $TCID_{50}/0.5$ mL and 10^5 $TCID_{50}/0.5$ mL, respectively, in PBS (Thermo Fisher Scientific) with 5% fetal bovine serum (FBS, Gibco, Thermo Fisher Scientific). For each dilution, ice cubes (0.5 mL, ~$1 \times 1 \times 1$ cm) were then prepared using silicone ice trays. After freezing at -80 °C, one ice cube was put into one soft cake (~50 g, diameter of cake ~2 cm) per pig. This allowed the ice cubes to thaw within the cakes prior to feeding of the pigs; however, this method retained the thawed virus suspension within the cakes.

Prior to the animal experiment, a small pilot study was performed to investigate the stability of infectious ASFV within the ice cubes using different concentrations of FBS. The pilot study was performed using the 10^3 $TCID_{50}$, 10^4 $TCID_{50}$ and 10^5 $TCID_{50}$ (per 0.5 mL) doses of the third passage virus in either 100% FBS, PBS with 5% FBS or PBS without FBS (Gibco, Thermo Fisher Scientific). The titers of both the freshly made virus dilutions and the thawed ice cubes of the three virus dilutions with the different concentrations of FBS were determined in primary cells as described just below. This pilot study was performed in order to ensure that the final virus dose provided to the pigs was as expected after the preparation procedures.

Back-titration of the samples from the pilot study and on the inoculums from the animal experiment (syringe with virus for intranasal inoculations, ice cubes with the three virus dilutions for oral inoculations) was carried out in PPAMs. Following 72 h of incubation at 37 °C (5% CO_2), virus-infected cells were stained using an immunoperoxidase monolayer assay (IPMA) [18,20]. Infected (red-colored) cells were identified using a light microscope and virus titers were calculated using the method described by Reed and Muench [21].

2.3. Study Design

The twenty-four pigs (numbered 51–74) were randomly allocated into four groups (labelled as groups 1–4), with six pigs in each. The pens were within two high containment stable units, termed boxes 5 and 6, respectively. Pigs 51–56 (group 1) and pigs 57–62 (group 2) were housed within two pens in box 5. These pens were separated by a ~2 m high solid metal wall to prevent direct contact between the groups. Pigs 63–68 (group 3) and pigs 69–74 (group 4) were housed in two similar pens in box 6. Box 5 and box 6 were completely separated from each other, with separate air supplies, equipment, clothing, etc. Personnel showered upon exiting each of the two boxes.

After an acclimatization period of one week, the pigs (in groups 1–3) were fed different doses of ASFV POL/2015/Podlaskie orally, while pigs in group 4 were inoculated intranasally with ASFV POL/2015/Podlaskie. Specifically, three groups of six pigs were fed cakes with 10^3 $TCID_{50}$ (pigs 51–56, group 1), 10^4 $TCID_{50}$ (pigs 57–62, group 2) or 10^5 $TCID_{50}$ (pigs 63–68, group 3) of the virus. For the oral inoculations of the pigs in groups 1–3, within each group, two pigs were separated from the remaining four pigs in the group and these two pigs were then fed with their own individual cake containing either the low, medium or high dose of the virus. This process was then repeated for the remaining four pigs in the group, using two animals at a time.

In total, 13 oral inoculations were planned per group, on days 0, 2, 4, 7, 9, 11, 14, 16, 18, 21, 23, 25 and 28 (Figure 1). Pigs 69–74 (group 4) were inoculated intranasally on day 0 with 10^3 $TCID_{50}$ virus administered using 1 mL per nostril.

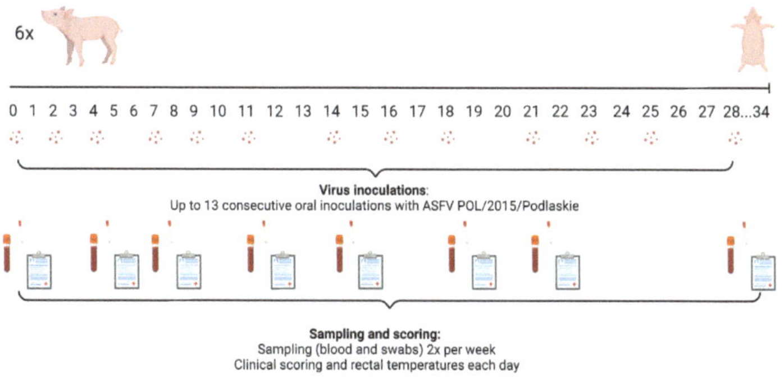

Figure 1. Study design for the pigs (in groups 1–3) that were administered ASFV orally in cakes. The serum tubes indicate that sampling (blood and swabs) was performed from the pigs on those days. Clinical scores and rectal temperatures were recorded daily. Created in BioRender (Toronto, ON, Canada) under the license Olesen, A. (2025) https://BioRender.com/c84u960 (accessed 10 December 2024).

2.4. Clinical Examinations and Euthanasia

The clinical signs and rectal temperatures were recorded for each pig on a daily basis. Euthanasia was performed by intravascular injection of pentobarbital following deep anesthesia, either when the pigs reached the pre-determined humane endpoints or at the end of the study period.

2.5. Sampling

EDTA-stabilized blood (EDTA-blood), unstabilized blood samples (for serum), and nasal and oral swabs were obtained from the pigs up to twice a week, on days 0, 4, 7, 11, 14, 18, 21, 28 and at euthanasia, as depicted in Figure 1. The nasal and oral swabs were added to 1 mL of PBS (Thermo Fisher Scientific). For sampling, the pigs were restrained using a wire snare.

The collected serum, EDTA-blood and swab samples were frozen at −80 °C until further use. Prior to analysis, the swab samples were vortexed and centrifuged briefly.

2.6. qPCR Analysis of EDTA-Blood and Swab Samples

Nucleic acids were isolated from the EDTA-blood samples and the supernatants from the nasal and oral swabs using the MagNA Pure 96 system (Roche, Basel, Switzerland) and analyzed for the presence of ASFV DNA, essentially as described previously [18,22], using the CFX Opus Real-Time PCR System (Bio-Rad, Hercules, CA, USA). The qPCR results are presented as viral genome copy numbers per mL. The genome copy numbers were calculated based on a standard curve made from assaying a 10-fold dilution series of the pVP72 plasmid [19]. A positive result was defined as giving a threshold cycle value (Cq) at which the FAM (6-carboxy fluorescein) dye emission was above background within 42 cycles.

2.7. Detection of Infectious Virus in Nasal Swabs

Nasal swab samples in which ASFV DNA was readily detected (Cq value below 30) were added to PPAMs seeded into NUNC 96-well plates (Thermo Fisher Scientific) and the virus was passaged once following the addition of antibiotics and filtration of the inoculum, essentially as described previously [12]. Following inoculation, the cells were incubated at 37 °C (in 5% CO_2) for 72 h, and the virus-infected cells were identified using IPMA, as described in Section 2.2.

2.8. Anti-ASFV Antibody Detection in Serum

Serum samples obtained from the pigs prior to inoculation and at euthanasia were tested for the presence of anti-ASFV antibodies using the Ingezim PPA Compac ELISA (®INGENASA INGEZIMPPA COMPAC K3 INGENASA, Madrid, Spain). The analysis was performed in accordance with the manufacturer's instructions.

In addition, an in-house indirect immunoperoxidase test (IPT) was used to test for the presence of anti-ASFV antibodies. Briefly, Vero cells were inoculated with a Vero-cell-adapted ASFV POL/2015/Podlaskie and fixed following 72 h of incubation at 37 °C. Following the addition of serum samples, protein-A-conjugated horseradish peroxidase (Sigma-Aldrich, St. Louis, MO, USA) together with hydrogen peroxide and 3-amino-9-ethyl carbazole (Sigma Aldrich) were used as a chromogenic substrate.

2.9. Preparation of Long PCR Products from Viral DNA

DNA was extracted from the virus sample used as the inoculum (3rd passage), as in Section 2.6, and used to generate overlapping long PCR products, as previously described [23]. Briefly, amplicons derived from the inoculum sample were amplified by long PCR using AccuPrime high-fidelity DNA polymerase (Thermo Scientific, Thermo Fisher Scientific, Waltham, MA, USA). The PCR products were analyzed using the Genomic DNA ScreenTape on a 4200 TapeStation (Agilent Technologies, Santa Clara, CA, USA) and their concentrations were estimated with the Quant-iT™ 1X dsDNA broad-range kit (Invitrogen) on a FLUOstar® Omega (BMG LABTECH, Mornington, VIC, Australia) instrument.

2.10. Variant Calling

The overlapping PCR products were pooled for the inoculum and sequenced using MiSeq (Illumina, San Diego, CA, USA) with a modified Nextera XT DNA library protocol with the MiSeq reagent kit v2 (300 cycles), resulting in 2 × 150 bp paired-end reads. The reads were trimmed using AdapterRemoval [24] by at least 30 bp at both the 5′ and 3′ ends to ensure primer removal, as well as for quality purposes (q30). Variant calling and annotation were performed using a combination of BWA-MEM, Samtools, Lo-Freq and SnpEff [25–28], as previously described [23], together with the ASFV POL/2015/Podlaskie reference sequence (MH681419.2). Variants were filtered for a minimum coverage of 50, frequency above 2%, and strand-bias Phred Score below 60.

2.11. Deletion Screening by PCR

The extracted DNA preparations from the EDTA-blood samples in which ASFV DNA was readily detected (Cq value below 30, corresponding to above 6.4 $\log_{10}$ genome copies/mL) were screened for the internal deletion event at pos. 6362–16,849 (10,487 bp) present within the population of viruses in the ASFV POL/Podlaskie/2015 inoculum, as described previously [23]. Briefly, the extracted DNAs were amplified by long PCR with primers spanning the deletion region at pos. 6188–17,145 (del-PCR) or primers located within this region at pos. 6708–7668 (noDel-PCR). As positive controls, we used extracted DNA from the spleen or EDTA-blood from ASFV-infected pigs, derived in previous studies [9,23], which had been shown to lack or to have the deletion, respectively. UltraPure™ DNase/RNase-Free Distilled Water was used as a negative control. The PCR products were analyzed using the Genomic DNA ScreenTape or D5000 ScreenTape on a 4200 TapeStation (Agilent Technologies) [23].

2.12. Nanopore Sequencing

The PCR products were cut out of agarose gels and purified using the GeneJET Gel Extraction Kit (Thermo Scientific) according to the manufacturer's instructions, and they were then sequenced on the Nanopore (Oxford Nanopore Technologies, Oxford, UK) using a standard ligation sequencing of amplicons with native barcoding protocol for the SQK-LSK109 and native barcoding expansion kits on a R9.4.1 flow-cell. The rads were filtered for size (3200–4200 bp) using SeqKit [29] and trimmed using chopper [30] by 30 bp at both the 5′ and 3′ ends to ensure primer removal, as well as for quality purposes (q9). The trimmed reads were subsequently mapped to the ASFV POL/2015/Podlaskie reference sequence using a combination of minimap2 [31] and Samtools [26].

3. Results

3.1. Preparation and Back-Titration of the Inoculation Material

A pilot study was performed with the three different dilutions of the third passage ASFV POL/2015/Podlaskie, i.e., a low (10^3 $TCID_{50}$), medium (10^4 $TCID_{50}$) and high (10^5 $TCID_{50}$) dose. The dilutions were made using 100% FBS, or PBS with 5% FBS or PBS without FBS, and they were titrated in PPAM with, or without, a freeze/thaw cycle (one freeze/thaw cycle mimicked how the ice cubes used for inoculation of the pigs were handled). The pilot study indicated a stabilizing effect of both 5% FBS and 100% FBS during the freeze/thaw cycle, and this stabilizing effect was especially apparent at the low (10^3) and medium (10^4) doses. Hence, the final inoculation material consisted of ice cubes made from the third passage ASFV POL/2015/Podlaskie diluted in PBS with 5% FBS.

Back-titration of the inoculation material in PPAMs yielded titers of between 4.50 and 4.75 $\log_{10}$ $TCID_{50}$ per 0.5 mL, with a mean of 4.68 $\log_{10}$ $TCID_{50}$ for the high doses. The medium doses yielded titers from 2.51 to 3.50 $\log_{10}$ $TCID_{50}$ per 0.5 mL, with a mean of 3.17 $\log_{10}$ $TCID_{50}$, and the low dose titers from 1.48 to 3.25 $\log_{10}$ $TCID_{50}$ per 0.5 mL, with a mean of 2.45 $\log_{10}$ $TCID_{50}$. Note that the back-titrations were performed on the inoculums following two additional freeze/thaw cycles when compared to the feeding of the pigs due to shipment of the samples from Barcelona to Denmark for laboratory analysis after the animal experiment.

3.2. Course of Infection in Intranasally Inoculated Pigs

Following intranasal inoculation with 10^3 $TCID_{50}$ of ASFV, all six inoculated pigs in group 4 (pigs 69–74) presented with a high fever (rectal temperature above 41 °C) from 4–5 dpi. The clinical signs at 6 and 7 dpi included moderate to severe depression, dyspnea and ataxia. Pigs 72 and 74 were found dead upon entering the pens at 6 dpi and 7 dpi, respectively. Prior to this, a high fever had been observed in both of the pigs for two consecutive days, along with mild (pig 72) to moderate depression (pig 74) and slight dyspnea (pig 74).

EDTA-blood was drawn from the heart of deceased pig 72 at 6 dpi; the same, however, was not possible from pig 74 at 7 dpi. The remaining four pigs were euthanized at 7 dpi for animal welfare reasons. The rectal temperatures, for each pig, are shown in Figure 2 (panel A). Measurements of viremia, by qPCR, confirmed that high levels of viral DNA were present in EDTA-blood from the infected animals (Figure 2, panel B). ASFV DNA was also detectable in nasal and oral swabs from all six pigs, with higher levels of viral DNA in the nasal swabs compared to the oral swabs (Figure 2, panels C and D). Using the inoculation of PPAMs, infectious virus was detected in nasal swabs obtained from all six pigs at euthanasia. No anti-ASFV-specific antibodies were detected in serum from the pigs at euthanasia (see Supplementary Table S1 for a summary of the data obtained for the pigs).

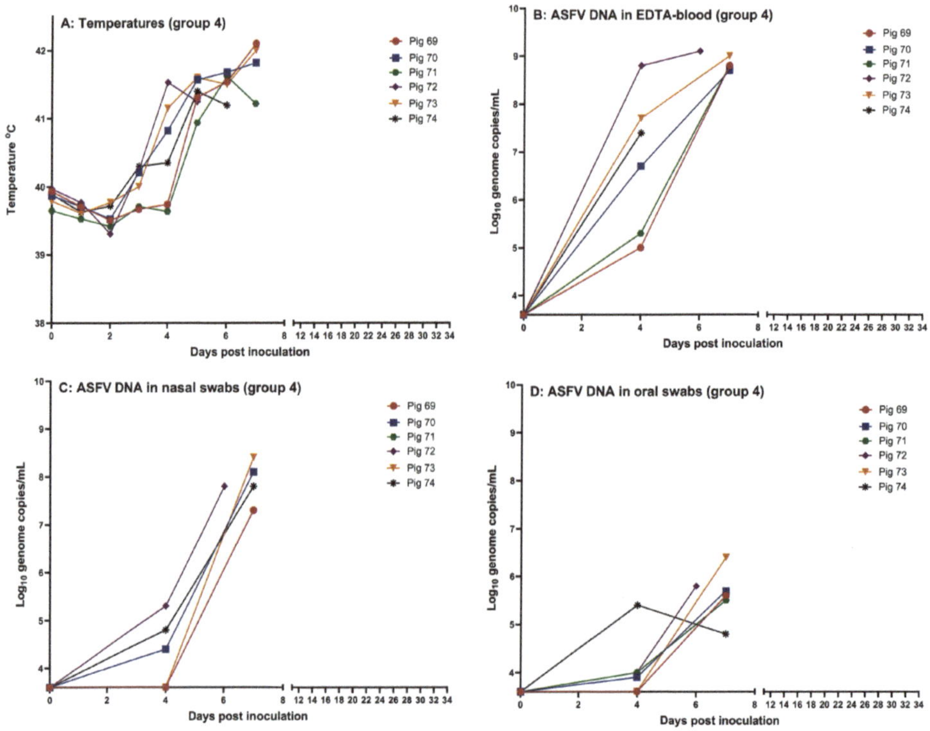

Figure 2. Data obtained from the intranasally inoculated pigs (pigs 69–74, group 4). (**A**) Rectal temperatures, (**B**) detection of ASFV DNA in EDTA-blood, (**C**) detection of ASFV DNA in nasal swabs, and (**D**) detection of ASFV DNA in oral swabs. The threshold for detection of ASFV DNA in the qPCR is Cq 42, which corresponds to 3.6 $\log_{10}$ genome copies/mL. Created using GraphPad Prism 9 (GraphPad Software, Boston, MA, USA).

3.3. Course of ASFV Infection in Orally Inoculated Pigs

Following feeding of the pigs in group 3 with the highest dose of the virus, ~10^5 $TCID_{50}$ (on days 0, 2 and 4), pigs 63 and 64 presented with a high fever from day 4 or 5. The clinical signs included depression, dyspnea and reddening of the skin. Two pigs reached the humane endpoints at day 7 and were euthanized. The rectal temperatures are shown in Figure 3 (panel A). As infection after feeding with the high dose was evident at this time point, no further feeding of pigs 65–68 with ASFV was performed on day 7 onwards. In addition, pig 66 had a high fever from day 10 and was euthanized on the following day when it presented with depression and dyspnea. Except for transient mild depression and diarrhea on some days, the remaining three pigs, 65, 67 and 68, despite three repeated oral inoculations, did not develop clinical signs that could indicate an infection with ASFV during the study period (Figure 3, panel B). These three pigs were euthanized on day 34, when the study was terminated. Assays for ASFV DNA in EDTA-blood provided results that were fully consistent with the clinical findings. High levels of ASFV DNA were readily detected in the blood samples obtained from pigs 63, 64 and 66, while no ASFV DNA was detected in EDTA-blood from the three remaining pigs in this group (Figure 3, panel B).

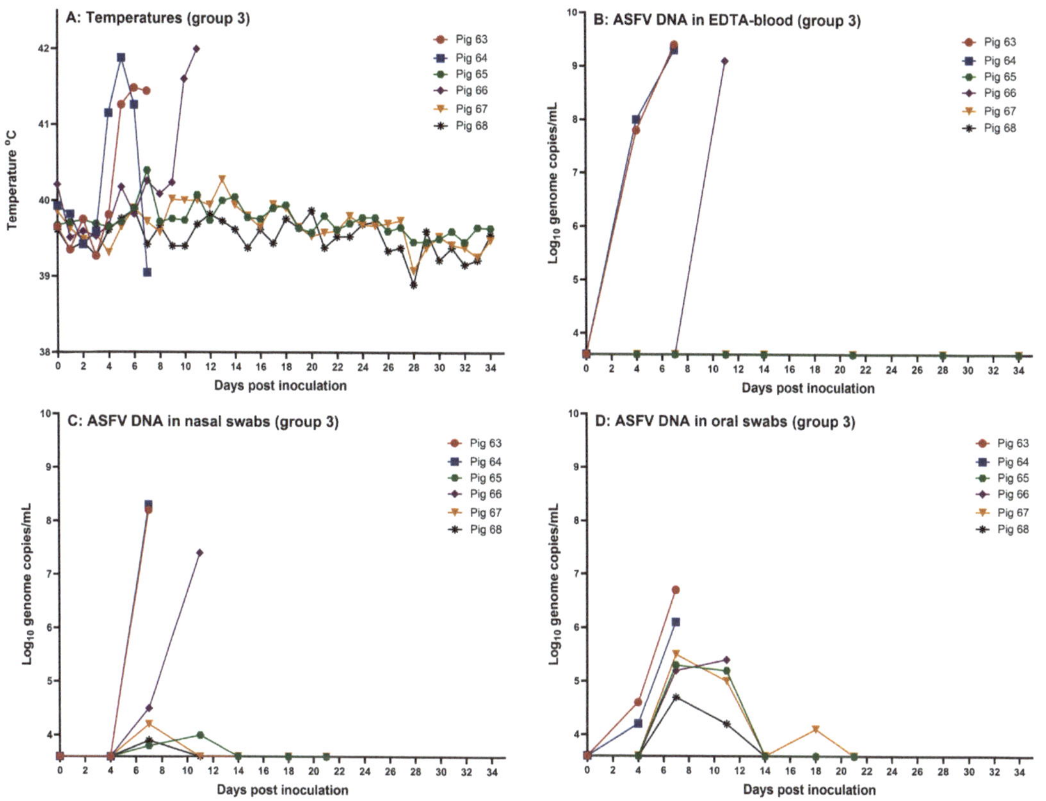

Figure 3. Data obtained from the pigs fed the high dose of the virus orally (pigs 63–68, group 3). (**A**) Rectal temperatures, (**B**) detection of ASFV DNA in EDTA-blood, (**C**) detection of ASFV DNA in nasal swabs, and (**D**) detection of ASFV DNA in oral swabs. The threshold for detection of ASFV DNA in the qPCR is Cq 42, which corresponds to 3.6 $\log_{10}$ genome copies/mL. Created using GraphPad Prism 9 (GraphPad Software).

No anti-ASFV-specific antibodies were detected in the serum obtained at euthanasia from any of the six pigs within group 3 (see Supplementary Table S1).

Among the nasal and oral swabs, the highest levels of ASFV DNA were found in the samples from the three infected pigs, while lower levels of the viral DNA were found in the swabs obtained from the remaining three pigs in the pen. As also observed in the intranasally infected pigs, higher levels of viral DNA were present in the nasal swabs than in the oral swabs from the three infected pigs. However, for the three uninfected pigs (pigs 65, 67 and 68), the opposite was observed, i.e., higher levels of ASFV DNA were detected in oral swabs obtained from these pigs than in nasal swabs (Figure 3, panels C and D and Supplementary Table S2). Infectious ASFV was also detected in the nasal swabs obtained from pigs 63, 64 and 66 at euthanasia (see Supplementary Table S1).

Following feeding with either the low dose of ASFV (group 1) or the medium dose (group 2), respectively, one pig (pig 56, from group 1) that was fed the low dose presented with a high fever (and nasal mucoid discharge) on days 9 and 10. This pig was euthanized on day 11 when it appeared slightly depressed. However, no ASFV DNA was detected in the blood samples obtained from this pig (Supplementary Table S2). The remaining 11 pigs did not present with a high fever (defined as a rectal temperature above 41 °C) or with any clinical signs indicative of ASFV infection, and no ASFV DNA was detected in their blood.

These pigs were euthanized on day 34, when the study was terminated. At this time point, they had each received 13 consecutive oral doses of ASFV. During the time course of the experiment, very low levels of viral DNA were detected in a few swab samples obtained from the pigs in the medium and low dose oral groups. Specifically, from the medium dose group (group 2), 3.7 and 4.5 $\log_{10}$ genome copies/mL were detected in two mouth swabs from pigs 59 and 62 on day 4, and 4.2 $\log_{10}$ genome copies/mL were present in a nasal swab from pig 61 at day 7. In the low dose oral group (group 1), 3.8 and 4.0 $\log_{10}$ genome copies/mL were detected in two oral swabs, from pig 56 (on day 7) and pig 51 (on day 11), respectively (Supplementary Table S2).

No anti-ASFV-specific antibodies were detected in the serum obtained at euthanasia from the pigs within groups 1 and 2 (see Supplementary Table S1).

3.4. Variant Analysis of ASFV in the Inoculum

The ASFV DNA in the inoculum was screened for the presence of single nucleotide polymorphisms (SNPs) and insertions and deletions (indels) in the MiSeq reads. All the PCR fragments had read coverage $\geq$ 50 for the entirety of the fragments, except for fragments 05a, 06 and 14, which lacked this degree of coverage at the 562, 124 and 405 genome positions in total, respectively. In the virus inoculum, a total of two silent SNPs were present, G33319A and A44769G, located in the MGF 505-2R and MGF 505-10R genes, respectively, at a frequency of 5.8% and 2.6%, respectively. Over 70 indels were identified, ranging in size from 1 to 3 bp; however, the majority were 1 bp in size (>68% deletions and >23% insertions), and all were located in homopolymeric regions.

3.5. Deletion Screening by PCR and Sequencing

The EDTA-blood samples and the inoculum were screened for the presence of the internal deletion event at pos. 6362–16,849 (10,487 bp), which was previously found in the second passage of the virus in PPAMs [23]. The inoculum (third passage) produced products of both ~500 bp and ~11 kb in the del-PCR, which spans the deletion site, consistent with the presence of genomes that had undergone the deletion event as well as the expected full-length PCR product, respectively (Figure 4, panel A). The product of ~1 kb in the noDel PCR, from within the deletion (Figure 4, panel B), is consistent with the ability to produce full-length (11 kb) products in the del-PCR assay. All the pigs yielded the full-length products (Figure 4 panels C and D); however, pigs 70 and 71 (from group 4) at day 7 also produced products consistent with harboring the deletion variant virus. The blood sample from pig 70 also contained shortened virus genomes at day 4; however, pig 71 was not screened in this assay on day 4 due to the low level of viral DNA present in the EDTA-blood at that time (Cq value above 30). The viral DNA from pig 70 also generated a prominent product of ~3700 bp (see Figure 4, panel C), which was extracted and sequenced using Nanopore technology, which revealed that this other deletion occurred between positions 7115 and 14,504, a total of 7389 bp; this deletion should result in a PCR product of ~3568 bp (consistent with the observed product). This deletion results in the loss of 15 complete genes; the majority are members of the MGF 110 (3L–12L) family, with a 5′ truncation of the MGF 110-2L gene and a 3′ truncation of the MGF 110-13La-13Lb, as well as ASFV G ACD 00120, 00160, 00190, 00240, MGF100-1R and 285L.

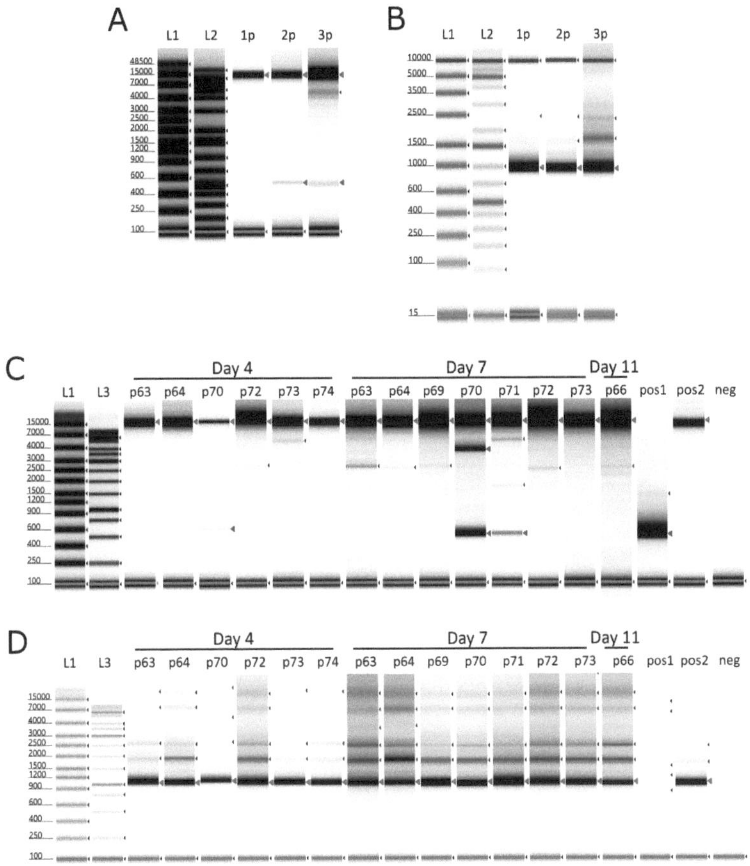

Figure 4. Deletion screening of ASFV POL/2015/Podlaskie inoculums (1p, 2p, 3p) together with infected pigs from group 1 (pig 63, 64, 66) and group 4 from sampling days with Cq values in the ASFV qPCR below 30 in the EDTA-blood. (**A**) PCRs with primers covering nt 6188–17,145 (del-PCR). (**B**) PCR with primers covering nt 6708–7668 (noDel-PCR). (**C**) PCR with primers covering nt 6188–17,145 (del-PCR). (**D**) PCR with primers covering nt 6708–7668 (noDel-PCR). L1: TapeStation ladder, L2: GeneRuler 1 kb plus, L3: GeneRuler 1 kb, 1p: First passage, 2p: Second passage, 3p: Third passage, p63: Pig 63 EDTA-blood, p64: Pig 64 EDTA-blood, p66: Pig 66 EDTA-blood, p69: Pig 69 EDTA-blood, p70: Pig 70 EDTA-blood, p71: Pig 71 EDTA-blood, p72: Pig 72 EDTA-blood, p73: Pig 73 EDTA-blood, p74: Pig 74 EDTA-blood, pos1: Positive control with deletion, pos2: Positive control without deletion, neg: Negative control (H_2O). Red arrowheads indicate the wild type, whereas blue arrowheads indicate the deletion variant. Green and purple bands indicate the lower and upper molecular weight markers, respectively, whereas small black arrowheads indicate bands detected by the TapeStation Analysis software version 5.1 (Agilent Technologies).

4. Discussion

In this study, we were able to demonstrate infection with ASFV following feeding with a high dose (10^5 $TCID_{50}$) of the ASFV POL/2015/Podlaskie. Oral administration of the lower doses of the same virus did not result in infection. In contrast, intranasal inoculation with the low dose (10^3 $TCID_{50}$) of ASFV was very efficient in establishing infection, underlining that infection with ASFV by the intranasal route is much easier to establish than infection via oral uptake. A higher efficiency of intranasal versus oral inoculation was described previously with the highly virulent ASFV-Malawi strain (genotype VIII, [32]) [16].

The relatively high dose needed for the ASFV POL/2015/Podlaskie to establish oral infection in the current study is also consistent with other studies exposing pigs to ASFV via oral uptake of feed spiked with the virus. In one study, the infection of pigs could not be established after 14 consecutive days of being fed with commercial feed spiked with serum from a pig infected with ASFV Georgia 2007/1 at levels of ~10^4 to 10^5 $TCID_{50}$ [14]. In another study, using virus doses ranging from 10^3 to 10^8 $TCID_{50}$, a minimum infectious dose of 10^4 $TCID_{50}$ was reported after one feed with compound feed spiked with spleen material from a pig infected with the ASFV Georgia 2007 [13]. In studies using oral inoculation with different insects, the feeding of 24 pigs with 50 virus-fed larvae, calculated to contain about 10^5 $TCID_{50}$ of the ASFV POL/2015/Podlaskie in total, did not result in infection of any of the pigs [15]. The larvae had either been fed on serum (*T. molitor*) or spleen suspension from ASFV-infected pigs (*H. illucens*). However, in an earlier study using the same virus, oral uptake of 20 blood-feeding flies fed on EDTA-blood containing infectious ASFV (calculated virus load in 20 flies was 10^5 $TCID_{50}$) did result in infection in 50% of the exposed pigs [9].

The reasons for the differences reported in the ability of ASFV to establish infection after the feeding of pigs, even at medium to high doses (10^4 $TCID_{50}$ and up), could reflect other factors than the virus dose itself. The materials used for spiking and the feeding material themselves could also impact the outcome of the virus exposures. For spiking, blood (EDTA-blood or serum), organ material (spleen) and infected-cell supernatants have each been used in various studies. It has previously been demonstrated that serum can have a stabilizing effect on ASFV [15,33], and it seems as if serum can enhance the infectivity of extracellular (but not intracellular) ASFV virions [34]. It should be noted that the studies that did report successful oral infection used spleen suspension [13], cell supernatant diluted in 5% serum (current study) or EDTA-blood [9]. On the other hand, the studies that failed to demonstrate the infection of pigs used either spleen material [15] or serum [14,15]. Hence, a clear indication of the effect of the type of spiking material used cannot be described based on the available evidence.

Another factor affecting the infection efficiency may be the feed material itself, i.e., the material and its preparation (e.g., the adsorption time of the spiking virus to the feed as also discussed previously by others [14]). These processes could have an enhancing or inhibitory effect on either the stability of the virus or the delivery of the virus dose (e.g., for establishing contact between the oropharynx and the virus). Based on the differences in the minimum infectious dose observed for the oral uptake of solid feed (10^4 $TCID_{50}$) versus liquid feed (1 $TCID_{50}$) in one study, it was suggested that the liquid provided a more suitable medium for virus contact to the tonsils or other tissues where primary virus replication can occur following oronasal or intraoropharyngeal virus exposure [13]. In the current study, the soft cakes most likely provided a "sticky matrix" that allowed for prolonged exposure of the inoculation material to the lymphoid tissues in the upper gastrointestinal tract. This would include the tonsils and tissues drained by the medial retropharyngeal lymph nodes. These lymph nodes have recently been identified as key entry points for infection with ASFV in both domestic pigs and wild boar [35]. Even though the soft cakes have proven suitable for achieving oral infection, using either virus suspension in the current study or flies in a previous study [9], future studies could aim at developing more user-friendly, highly palatable and standardized delivery systems for oral exposures with ASFV (e.g., for infection and possible vaccination). From the data available so far, it seems as if a delivery system that, like the quite sticky soft cakes, allows the virus to come into prolonged contact with the upper gastrointestinal tract (e.g., using gels) could be a feasible approach.

In addition to the inoculation material used (i.e., the spiking virus and the feed), other factors can also be of importance for the outcome of this study. In the current study, a wire snare was used for restraining the pigs during sampling, which was performed prior to the inoculations of the pigs. Wire snares can result in microlesions in the oral cavity, which could be hypothesized to increase the risk of (parental) infection via these lesions, even when using lower doses of the virus. In the current study, the use of the wire snare did not, however, seem to increase the incidence of oral infection of pigs using lower doses of ASFV.

From the back-titration of the administered dose, it appears as if pigs in the low dose and medium dose groups could, unintentionally, have been given slightly lower doses than anticipated. Due to logistical reasons, the back-titrations were (non-optimally) performed after an additional two freeze/thaw cycles when compared to the time of inoculation, which could have affected the apparent level of the virus in the back-titration. If so, this effect seems to have been more pronounced in the more diluted virus suspensions (low dose and medium dose) when compared to the high dose suspension. As serum seems to have a stabilizing effect on the virus [15,33,34], a higher concentration of serum could be used for preparation of the inoculum in future studies. In the low dose and medium dose groups, the lack of infection of the pigs following 13 consecutive inoculations with virus doses from ~10^2 to 10^4 $TCID_{50}$ is not consistent with the statistical model predictions from an earlier study [13]. The model applied in that study was based on infection of 40% of the pigs that were exposed to 10^4 $TCID_{50}$ orally, and it predicted that 10 oral exposures to 10^2, 10^3 or 10^4 $TCID_{50}$ would lead to infection of 25%, 50% and 100% of the exposed pigs, respectively. For the high dose used in the current study, 10^5 $TCID_{50}$, the same model was based on 44% of the pigs being exposed to 10^5 $TCID_{50}$ in feed becoming infected with ASFV, leading to a prediction of three exposures resulting in 75% of the exposed pigs becoming infected. In the current study, using a dose of 10^5 $TCID_{50}$, we found that only 33–50% of the exposed pigs were infected with ASFV following up to three doses. Further studies are required to determine the effects of multiple oral inoculations of ASFV. The apparently low ability of ASFV to establish infections via the oral route could be problematic for the development of oral vaccines to combat the disease. Currently, two intramuscular live-attenuated vaccines (LAVs) are licensed for use in domestic pigs in Vietnam [36]; however, in the European scenario, in which the virus is maintained within the wild boar population, oral vaccination using baits deployed in the field would be the most feasible approach [37]. The results obtained in the current study indicate that besides a vaccine candidate virus being effective and safe, a high dose of the vaccine virus could be needed within the bait in order to ensure successful immunization.

Due to the nature of the study design used here, it cannot be readily determined whether two or three pigs (63, 64, and 66) became infected due to feeding with the high dose of ASFV or after which of the three oral inoculations the infection was actually established. However, when comparing the time course of the infection in the three pigs to the course of infection in the intranasally inoculated pigs, it seems most likely that pigs 63 and 64 were already infected from the first inoculation at day 0. This is also in line with earlier results obtained from feeding of pigs with the same virus [9]. The delayed appearance of infection in pig 66 could either indicate that infection was established in this pig after the inoculation on day 4 or that the pig was infected via direct contact to its infected pen mates (pigs 62 and 64). Using the same virus, a four-day delay in the course of infection has previously been observed between intranasally inoculated pigs and "in contact" animals [18]. Interestingly, the lack of virus transmission to the last three pigs, pigs 65, 67 and 68, within the same pen underlines that the transmission of ASFV is not always very efficient, especially when blood is not present in the pen environment [12]. We believe that the low levels of ASFV DNA detected in the oral and nasal swabs obtained from these three pigs at day 7 indicate exposure to virus excreted from their infected pen mates, and it appears as if they were exposed primarily via the less efficient oral route. The low efficiency for establishing oral infection together with the low dose exposures have been linked to the low contagiousness of ASFV observed in field settings in Europe. This low contagiousness along with the high stability of the virus and the high case fatality rate have been linked to the persistence and spread of ASFV on the European continent [38].

The internal deletion event at pos. 6362–16,849 (10,487 bp) originally arose in the second passage in cell culture [23] and was clearly maintained in the third passage, used as inoculum in this study. It was maintained as a minority variant within the viral population of pigs 70 and 71 but does not seem to have affected pathogenicity. Pig 70 contained an additional internal deletion of 7389 bp within the same region of the genome as the 10,847 bp deletion. This smaller deletion likely occurred independently in the viral population. It could be non-essential (i.e., selected due to quicker replication of smaller genomes) or could indicate that this region of the genome is under selective pressure, suggesting a complex evolutionary landscape, with different variants competing or coexisting.

NGS revealed that, at the consensus level, the sequence of the inoculum matched the published sequence of ASFV POL/2015/Podlaskie and that only minority SNPs and 1–3 base-pair indels in homopolymeric regions were detected in the variant analysis. It is difficult to determine the veracity of such short indels in homopolymeric regions by other means, as most sequencing technologies have difficulties with these [39–42].

In conclusion, we have confirmed that infection with ASFV was not easily established following oral uptake of the virus. Only a high dose of a genotype II virus, the ASFV POL/2015/Podlaskie, was sufficient to establish infection of half of the pigs exposed by this route in our experimental setting. The high dose of ASFV needed in order to infect pigs orally could have implications for the dose needed for any future live-attenuated baited vaccine.

Supplementary Materials: The following supporting information can be downloaded at https://www.mdpi.com/article/10.3390/pathogens14020119/s1, Table S1: Summary of data from pigs. Table S2: Raw data from pigs.

Author Contributions: Conceptualization, A.S.O., C.M.L., F.A., T.B.R., A.B., L.L. and G.J.B.; methodology, A.S.O., C.M.L., F.A., C.M.J., T.B.R., A.B., L.L. and G.J.B.; investigation, A.S.O., C.M.L., F.A. and C.M.J.; resources, L.L. and G.J.B.; data curation, A.S.O., C.M.L. and C.M.J.; writing—original draft preparation, A.S.O.; writing—review and editing, C.M.L., F.A., C.M.J., T.B.R., A.B., L.L. and G.J.B. All authors have read and agreed to the published version of the manuscript.

Funding: This research was funded by internal resources at the University of Copenhagen and Statens Serum Institut and by the Danish Veterinary and Food Administration (FVST), the latter as part of the agreement for commissioned work between the Danish Ministry of Food and Agriculture and Fisheries and the University of Copenhagen and Statens Serum Institut.

Institutional Review Board Statement: The pig study was approved by the Ethical and Animal Welfare Committee of the Generalitat de Catalunya (Autonomous Government of Catalonia; permit number: 12187).

Informed Consent Statement: Not applicable.

Data Availability Statement: The sequence data have been deposited as a BioProject at the NCBI (Accession number PRJNA1192983). All other necessary data are contained within the article and the Supplementary Information.

Acknowledgments: We are very grateful to the staff at the IRTA-CReSA and Statens Serum Institut. Especially, we owe great thanks to Guillermo Cantero, Iván Cordon, Marta Muñoz, Xavier Abad, Marie Hornstrup Christensen and Fie Fisker Brønnum Christiansen for their excellent work.

Conflicts of Interest: The authors declare no conflicts of interest.

References

1. Alonso, C.; Borca, M.; Dixon, L.; Revilla, Y.; Rodriguez, F.; Escribano, J.M. ICTV virus taxonomy profile: Asfarviridae. *J. Gen. Virol.* **2018**, *99*, 613–614. [CrossRef] [PubMed]
2. Urbano, A.; Forth, J.H.; Olesen, A.S.; Dixon, L.; Rasmussen, T.B.; Cackett, G.; Werner, F.; Karger, A.; Andrés, G.; Wang, X.; et al. African swine fever virus: Cellular and molecular aspects. In *Understanding and Combatting African Swine Fever*; Iacolina, L., Penrith, M.-L., Bellini, S., Chenais, E., Jori, F., Montoya, M., Ståhl, K., Gavier-Widén, D., Eds.; Wageningen Academic Publishers: Wageningen, The Netherlands, 2021; pp. 25–61. [CrossRef]
3. WOAH. 2024. Available online: https://www.woah.org/en/disease/african-swine-fever/#ui-id-2 (accessed on 8 November 2024).
4. WOAH. 2024. Available online: https://www.woah.org/app/uploads/2021/03/african-swine-fever.pdf (accessed on 2 December 2024).
5. Olesen, A.S.; Belsham, G.J.; Rasmussen, T.B.; Lohse, L.; Bødker, R.; Halasa, T.; Boklund, A.; Bøtner, A. Potential routes for indirect transmission of African swine fever virus into domestic pig herds. *Transbound. Emerg. Dis.* **2020**, *67*, 1472–1484. [CrossRef]
6. McKercher, P.O.; Yedloutschnig, R.J.; Calllis, J.J.; Murphy, R.; Panina, G.F.; Civardi, A.; Bugnetti, M.; Foni, E.; Laddomada, A.; Scarano, C.; et al. Survival of Viruses in "Prosciutto di Parma" (Parma ham). *Can. Inst. Food Sci. Technol. J.* **1987**, *20*, 267–272. [CrossRef]
7. Mebus, C.A.; House, C.; Gonzalvo, F.R.; Pineda, J.M.; Tapiador, J.; Pire, J.J.; Bergada, J.; Yedloutschnig, R.J.; Sahu, S.; Becerra, V.; et al. Survival of foot-and mouth disease, African swine fever, and hog cholera viruses in Spanish serrano cured hams and Iberian cured hams, shoulders and loins. *Food Microbiol.* **1993**, *10*, 133–143. [CrossRef]
8. Petrini, S.; Feliziani, F.; Casciari, C.; Giammarioli, M.; Torresi, C.; Mario De Mia, G. Survival of African swine fever virus (ASFV) in various traditional Italian dry-cured meat products. *Prev. Vet. Med.* **2019**, *162*, 126–130. [CrossRef]
9. Olesen, A.S.; Lohse, L.; Hansen, M.F.; Boklund, A.; Halasa, T.; Belsham, G.J.; Rasmussen, T.B.; Bøtner, A.; Bødker, R. Infection of pigs with African swine fever virus via ingestion of stable flies (*Stomoxys calcitrans*). *Transbound. Emerg. Dis.* **2018**, *65*, 1152–1157. [CrossRef]
10. Montgomery, E. On a form of swine fever occurring in British East Africa (Kenya colony). *J. Comp. Pathol. Ther.* **1921**, *34*, 159–191. [CrossRef]
11. Olesen, A.S.; Lohse, L.; Boklund, A.; Halasa, T.; Belsham, G.J.; Rasmussen, T.B.; Bøtner, A. Short time window for transmissibility of African swine fever virus from a contaminated environment. *Transbound. Emerg. Dis.* **2018**, *65*, 1024–1032. [CrossRef]
12. Olesen, A.S.; Lohse, L.; Accensi, F.; Goldswain, H.; Belsham, G.J.; Bøtner, A.; Netherton, C.L.; Dixon, D.; Portugal, R. Inefficient transmission of African swine fever to sentinel pigs from an environment contaminated by ASFV-infected pigs under experimental conditions. *Transbound. Emerg. Dis.* **2024**, *2024*, 8863641. [CrossRef]
13. Niederwerder, M.C.; Stoian, A.M.M.; Rowland, R.R.R.; Dritz, S.S.; Petrovan, V.; Constance, L.A.; Gebhardt, J.T.; Olcha, M.; Jones, C.K.; Woodworth, J.C.; et al. Infectious dose of African swine fever virus when consumed naturally in liquid. *Emerg. Infect. Dis.* **2019**, *25*, 891–897. [CrossRef]
14. Blázquez, E.; Pujols, J.; Segalè, J.; Rodrìguez, F.; Crenshaw, J.; Rodrìguez, C.; Ròdenas, J.; Polo, J. Commercial feed containing porcine plasma spiked with African swine fever is not infective in pigs when administered for 14 consecutive days. *PLoS ONE* **2020**, *15*, e0235895. [CrossRef]
15. Olesen, A.S.; Lazov, C.M.; Lecocq, A.; Accensi, F.; Jensen, A.B.; Lohse, L.; Rasmussen, T.B.; Belsham, G.J.; Bøtner, A. Uptake and survival of African swine fever virus in mealworm (*Tenebrio molitor*) and black soldier fly (*Hermetia illucens*) larvae. *Pathogens* **2023**, *12*, 47. [CrossRef]
16. Howey, E.B.; O'Donnell, V.; de Carvalho Ferreira, H.C.; Borca, M.V.; Arzt, J. Pathogenesis of highly virulent African swine fever virus in domestic pigs exposed via intraoropharyngeal, intranasopharyngeal, and intramuscular inoculation, and by direct contact with infected pigs. *Virus Res.* **2013**, *178*, 328–339. [CrossRef] [PubMed]
17. Pietschmann, J.; Guinat, C.; Beer, M.; Pronin, V.; Tauscher, K.; Petrov, A.; Keil, G.; Blome, S. Course and transmission characteristics of oral low-dose infection of domestic pigs and European wild boar with a Caucasian African swine fever virus isolate. *Arch. Virol.* **2015**, *160*, 1657–1667. [CrossRef] [PubMed]
18. Olesen, A.S.; Lohse, L.; Boklund, A.; Halasa, T.; Gallardo, C.; Pejsak, Z.; Belsham, G.J.; Rasmussen, T.B.; Bøtner, A. Transmission of African swine fever virus from infected pigs by direct contact and aerosol routes. *Vet. Microbiol.* **2017**, *211*, 92–102. [CrossRef] [PubMed]

19. Olesen, A.S.; Kodama, M.; Lohse, L.; Accensi, F.; Rasmussen, T.B.; Lazov, C.M.; Limborg, M.T.; Gilbert, M.T.P.; Bøtner, A.; Belsham, G.J. Identification of African swine fever virus transcription within peripheral blood mononuclear cells of acutely infected pigs. *Viruses* **2021**, *13*, 2333. [CrossRef]
20. Bøtner, A.; Nielsen, J.; Bille-Hansen, V. Isolation of porcine reproductive and respiratory syndrome (PRRS) virus in a Danish swine herd and experimental infection of pregnant gilts with the virus. *Vet. Microbiol.* **1994**, *40*, 351–360. [CrossRef]
21. Reed, L.J.; Muench, H. A simple method for estimating fifty percent endpoints. *Am. J. Epidemiol.* **1938**, *27*, 493–497. [CrossRef]
22. Tignon, M.; Gallardo, C.; Iscaro, C.; Hutet, E.; Van der Stede, Y.; Kolbasov, D.; De Mia, G.M.; Le Potier, M.-F.; Bishop, R.P.; Arias, M.; et al. Development and inter-laboratory validation study of an improved new real-time PCR assay with internal control for detection and laboratory diagnosis of African swine fever virus. *J. Virol. Methods* **2011**, *178*, 161–170. [CrossRef]
23. Johnston, C.M.; Olesen, A.S.; Lohse, L.; Madsen, A.M.; Bøtner, A.; Belsham, G.J.; Rasmussen, T.B. A deep sequencing strategy for investigation of virus variants within African swine fever virus-infected pigs. *Pathogens* **2024**, *13*, 154. [CrossRef]
24. Schubert, M.; Lindgreen, S.; Orlando, L. AdapterRemoval v2: Rapid Adapter Trimming, Identification, and Read Merging. *BMC Res. Notes* **2016**, *9*, 88. [CrossRef]
25. Li, H. Aligning Sequence Reads, Clone Sequences and Assembly Contigs with BWA-MEM. *arXiv* **2013**, arXiv:1303.3997. [CrossRef]
26. Danecek, P.; Bonfield, J.K.; Liddle, J.; Marshall, J.; Ohan, V.; Pollard, M.O.; Whitwham, A.; Keane, T.; McCarthy, S.A.; Davies, R.M.; et al. Twelve Years of SAMtools and BCFtools. *Gigascience* **2021**, *10*, giab008. [CrossRef] [PubMed]
27. Wilm, A.; Aw, P.P.K.; Bertrand, D.; Yeo, G.H.T.; Ong, S.H.; Wong, C.H.; Khor, C.C.; Petric, R.; Hibberd, M.L.; Nagarajan, N. LoFreq: A Sequence-Quality Aware, Ultra-Sensitive Variant Caller for Uncovering Cell-Population Heterogeneity from High-Throughput Sequencing Datasets. *Nucleic Acids Res.* **2012**, *40*, 11189–11201. [CrossRef] [PubMed]
28. Cingolani, P.; Platts, A.; Wang, L.L.; Coon, M.; Nguyen, T.; Wang, L.; Land, S.J.; Lu, X.; Ruden, D.M. A Program for Annotating and Predicting the Effects of Single Nucleotide Polymorphisms, SnpEff: SNPs in the Genome of *Drosophila melanogaster* Strain W1118; Iso-2; Iso-3. *Fly* **2012**, *6*, 80–92. [CrossRef] [PubMed]
29. Shen, W.; Le, S.; Li, Y.; Hu, F. SeqKit: A Cross-Platform and Ultrafast Toolkit for FASTA/Q File Manipulation. *PLoS ONE* **2016**, *11*, e0163962. [CrossRef]
30. De Coster, W.; Rademakers, R. NanoPack2: Population-Scale Evaluation of Long-Read Sequencing Data. *Bioinformatics* **2023**, *39*, btad311. [CrossRef]
31. Li, H. Minimap2: Pairwise Alignment for Nucleotide Sequences. *Bioinformatics* **2018**, *34*, 3094–3100. [CrossRef]
32. Chapman, D.A.G.; Darby, A.C.; Da Silva, M.; Upton, C.; Radford, A.D.; Dixon, L.K. Genomic analysis of highly virulent isolate of African swine fever virus. *Emerg. Infect. Dis.* **2017**, *17*, 599–605. [CrossRef]
33. Plowright, W.; Parker, J. The stability of African swine fever virus with particular reference to heat and pH inactivation. *Arch. Gesamte. Virusforsch.* **1967**, *21*, 383–402. [CrossRef]
34. Canter, J.A.; Aponte, T.; Ramirez-Medina, E.; Pruitt, S.; Gladue, D.P.; Borca, M.V.; Zhu, J.J. Serum neutralizing and enhancing effects on African swine fever virus infectivity in adherent pig PBMC. *Viruses* **2022**, *14*, 1249. [CrossRef] [PubMed]
35. Sanchez-Cordon, P.J.; Lean, F.Z.X.; Batten, C.; Steinbach, F.; Neimani, A.; Le Potier, M.-F.; Wikstrom-Lassa, E.; Wynne, F.; Strong, R.; McCleary, S.; et al. Comparative evaluation of disease dynamics in wild boar and domestic pigs experimentally inoculated intranasally with the European highly virulent African swine fever virus genotype II strain "Armenia 2007". *Vet. Res.* **2024**, *55*, 89. [CrossRef] [PubMed]
36. Diep, N.V.; Duc, N.V.; Ngoc, N.T.; Dang, V.X.; Tiep, T.N.; Nguyen, V.D.; Than, T.T.; Maydaniuk, D.; Goonewardene, K.; Ambagala, A.; et al. Genotype II Live-Attenuated ASFV Vaccine Strains Unable to Completely Protect Pigs against the Emerging Recombinant ASFV Genotype I/II Strain in Vietnam. *Vaccines* **2024**, *12*, 1114. [CrossRef]
37. Gavier-Widén, D.; Ståhl, K.; Dixon, L. No hasty solutions for African swine fever. *Science* **2020**, *367*, 622–624. [CrossRef] [PubMed]
38. Chenais, E.; Depner, K.; Guberti, V.; Dietze, K.; Viltrop, A.; Ståhl, K. Epidemiological considerations on African swine fever in Europe 2014–2018. *Porc Health Manag.* **2019**, *5*, 6. [CrossRef]
39. Kieleczawa, J. Fundamentals of Sequencing of Difficult Templates—An Overview. *J. Biomol. Tech.* **2006**, *17*, 207–217.
40. Zeng, F.; Jiang, R.; Chen, T. PyroHMMsnp: An SNP Caller for Ion Torrent and 454 Sequencing Data. *Nucleic Acids Res.* **2013**, *41*, e136. [CrossRef]
41. Fuller, C.W.; Middendorf, L.R.; Benner, S.A.; Church, G.M.; Harris, T.; Huang, X.; Jovanovich, S.B.; Nelson, J.R.; Schloss, J.A.; Schwartz, D.C.; et al. The Challenges of Sequencing by Synthesis. *Nat. Biotechnol.* **2009**, *27*, 1013–1023. [CrossRef]
42. Pfeiffer, F.; Gröber, C.; Blank, M.; Händler, K.; Beyer, M.; Schultze, J.L.; Mayer, G. Systematic Evaluation of Error Rates and Causes in Short Samples in Next-Generation Sequencing. *Sci. Rep.* **2018**, *8*, 10950. [CrossRef]

pathogens

MDPI

Article

Pathological Characteristics of Domestic Pigs Orally Infected with the Virus Strain Causing the First Reported African Swine Fever Outbreaks in Vietnam

Thi Thu Huyen Nguyen [1,2], Van Tam Nguyen [3], Phuong Nam Le [3], Nguyen Tuan Anh Mai [3], Van Hieu Dong [1], Tran Anh Dao Bui [1], Thi Lan Nguyen [1], Aruna Ambagala [4] and Van Phan Le [1,3,*]

1 College of Veterinary Medicine, Vietnam National University of Agriculture, Hanoi 100000, Vietnam
2 Faculty of Animal Science and Veterinary Medicine, Bac Giang Agriculture and Forestry University, Bac Giang 230000, Vietnam
3 Institute of Veterinary Science and Technology (IVST), Hanoi 100000, Vietnam
4 National Centre for Foreign Animal Disease, Canadian Food Inspection Agency, Winnipeg, MB R3E 3M4, Canada
* Correspondence: letranphan@vnua.edu.vn

Abstract: African swine fever (ASF) is currently Vietnam's most economically significant swine disease. The first ASF outbreak in Vietnam was reported in February 2019. In this study, VNUA/HY/ASF1 strain isolated from the first ASF outbreak was used to infect 10 eight-week-old pigs orally with 10^3 HAD_{50} per animal. The pigs were observed daily for clinical signs, and whole blood samples were collected from each animal for viremia detection. Dead pigs were subjected to full post-mortem analyses. All 10 pigs displayed acute or subacute clinical signs and succumbed to the infection between 10 to 27 (19.8 ± 4.66) days post-inoculation (dpi). The onset of clinical signs started around 4–14 dpi. Viremia was observed in pigs from 6–16 dpi (11.2 ± 3.55). Enlarged, hyperemic, and hemorrhagic lymph nodes, enlarged spleen, pneumonia, and hydropericardium were observed at post-mortem examinations.

Keywords: ASFV; clinical signs; gross findings; orally experimental infection; pathology

Citation: Nguyen, T.T.H.; Nguyen, V.T.; Le, P.N.; Mai, N.T.A.; Dong, V.H.; Bui, T.A.D.; Nguyen, T.L.; Ambagala, A.; Le, V.P. Pathological Characteristics of Domestic Pigs Orally Infected with the Virus Strain Causing the First Reported African Swine Fever Outbreaks in Vietnam. *Pathogens* **2023**, *12*, 393. https://doi.org/10.3390/pathogens12030393

Academic Editors: Lauro Velazquez-Salinas, Gisselle N. Medina, Elizabeth Ramirez-Medina and Natasha N. Gaudreault

Received: 15 December 2022
Revised: 14 January 2023
Accepted: 24 February 2023
Published: 1 March 2023

1. Introduction

African swine fever (ASF) is a viral disease that is highly contagious and devastating to domestic and wild pig populations. The economic impact of this disease is profound, with significant losses incurred globally due to high mortality rates and trade restrictions. Although ASF is primarily present in African countries, the disease has now spread to other areas such as Asia, Europe, and America [1]. ASF transmission occurs through direct contact between infected and susceptible pigs, but contaminated feed, water, equipment, and clothing can also spread the disease [2]. Once ASF enters a pig population, it spreads rapidly, leading to high fever, loss of appetite, and internal bleeding. Mortality rates can be as high as 100%, and currently, there are no specific treatments or vaccines for ASF [3]. Strict biosecurity measures and culling infected pigs are the only viable means of controlling the spread of the disease [4].

ASF virus (ASFV), the causative agent of African swine fever, is a large, enveloped virus with a double-stranded DNA genome. It belongs to the Asfivirus genus within the *Asfarviridae* family. The viral genome of ASFV ranges from 170 to 190 kbp in length, encoding over 170 proteins [5,6]. ASFV strains are divided into 24 genotypes based on the partial sequence of the ASFV B646L gene encoding p72 protein [7]. ASFV can be further sub-genotyped through various methods, which target different viral genome regions to identify genetic variations among different virus strains. Common sub-genotyping methods for ASFV include the analysis of specific genes, such as the E183L gene encoding p54 protein (genotype I-XXIV), EP402R gene encoding CD2v protein (serotype I-VIII), and

B602L gene (multiple variants), as well as the intergenic region (IGR) between MGF 505-9R and MGF 505-10R, and other genes like K205R, MGF 110-14L, H240R, O174L, E199L, K145R, I9R, and MGF 505-5R [8–16]. Sub-genotyping ASFV can help researchers better understand the virus's epidemiology and evolution, which is essential for developing effective control and prevention measures against this economically important disease.

ASF was first reported in Kenya in 1921 [17]. The first ASF transcontinental spread was reported in 1957 from Africa to Spain. The virus, which belonged to ASFV p72 genotype I, then spread to Portugal (1960) and the neighboring countries. ASF was eliminated from Europe in the mid-1990s, except in Sardinia, which remains an endemic [18,19]. The second ASFV transcontinental spread happened in 2007 [8,18,19]. An ASFV p72 genotype II was detected in Georgia and later spread to Armenia, the Russian Federation, Haiti regions, and eastern European countries mainly through wild boars [19,20].

In Asia, the first ASF outbreak was reported in mainland China in August 2018. ASFV then rapidly spread across China to almost all the neighboring countries, including Vietnam, Mongolia, Cambodia, Laos, Myanmar, Hong Kong, North Korea, South Korea, the Philippines, India, Indonesia, and Thailand. The responsible virus was a highly virulent ASFV strain belonging to ASFV p72 genotype II [21]. In 2021, besides genotype II, genotype I ASFVs were isolated from lymph nodes and spleens in China, which caused chronic forms including intermittent fever, arthroncus, and cutaneous necrosis [22].

Previous studies showed that ASF could manifest in four clinical forms, depending on the virus strain, transmission route, dose, and the host's characteristics. These four clinical forms are peracute, acute, subacute, and chronic. In peracute cases, pigs may die suddenly without showing any prior clinical signs. In acute cases, pigs may exhibit high fever, anorexia, lethargy, and hemorrhages on the skin and internal organs. In subacute and chronic cases, pigs may show milder clinical signs, such as loss of appetite, weight loss, and fever [23]. ASFV strains are classified based on their virulence, which can significantly impact the severity of the disease. Highly virulent strains can cause mortality rates of up to 100%, while low virulent strains may cause only mild or asymptomatic infections. Once infected, pigs shed the virus at high levels in all secretions and excretions, contributing to the rapid spread of the disease. Effective control measures, including strict biosecurity protocols and rapid detection, are essential to prevent the spread of ASFV and limit its impact on the swine industry [1].

In Vietnam, the first two outbreaks of ASF were reported in two adjacent northern provinces, Hung Yen and Thai Binh, in the red river delta in early 2019 [14,21]. After the first two outbreaks, the virus quickly spread to all 63 provinces in Vietnam. The highly contagious nature of ASF and the lack of an effective vaccine or treatment for the disease led to the mass culling of infected and contact pigs as a control measure. The impact of ASF on the pig industry in Vietnam has been significant, with nearly 5.9 million pigs destroyed, accounting for 20% of the total swine population in the country. This has led to a pork shortage and a sharp increase in the prices of pork products. Many small-scale pig farmers, who make up the majority of the pig industry in Vietnam, have been severely affected by the outbreak, with some losing their entire herds and livelihoods (FAO & Ministry of Agriculture and Rural Development, Vietnam). The results of the genetic sequencing analysis indicated that the ASFV strains responsible for the initial outbreaks in Vietnam belonged to genotype II, which is comparable to the ASF Georgia 2007/1 strain [21]. Although there have been many studies on the molecular epidemiology of ASFV strains causing disease in pigs in Vietnam, studies on their pathology are still very limited [24]. A better understanding of the virulence of ASFV strains is crucial for effectively controlling and managing ASF. Therefore this study was conducted to understand better the pathogenicity of the ASFV strain that caused the first reported outbreak in Vietnam.

2. Materials and Methods

2.1. Ethics Statements

The animal experiment was conducted at the large animal biosafety level 2 facility according to the animal use guidelines at the Vietnam National University of Agriculture (VNUA), Hanoi, Vietnam.

2.2. Virus Strain

The virus strain used for the study was VNUA/HY/ASF1 (GenBank Accession no. MK554698), which belonged to p72 genotype II and originated from infected pigs during the first outbreak of ASF in Vietnam in February 2019 [21]. Healthy pigs aged 8 to 10 weeks were used to collect primary porcine alveolar macrophages (PAMs) for ASFV culture. Real-time PCR was performed to confirm the absence of various viruses such as Porcine circovirus type 2 (PCV2), Classical swine fever (CSF), Porcine reproductive and respiratory syndrome (PRRS), and African swine fever (ASF) using a kit from Median Diagnostics Inc. (http://www.mediandiagnostics.com, accessed on 19 May 2021). The PAM cells were cultured in RPMI 1640 medium (Gibco) supplemented with 10% fetal bovine serum (FBS) and 1% antibiotic. The cells were seeded at a density of approximately 4×10^5 cells/cm^2 onto tissue culture plastic plates. After 48 h of ASFV infection, 20 μL of 1% porcine red blood cells in RPMI medium was added to each well of PAM cells. The formation of hemadsorption (HAD) rosettes on ASFV-infected PAM cells was observed daily under an inverted microscope for five days. As previously described, the virus titer was determined using the Hemadsorption (HAD) assay and calculated using the Reed and Muench method [25,26].

2.3. Animals

A total of 15 healthy eight-week-old crossbred (Yorkshire - Landrace) pigs were used in this study. All pigs were previously confirmed to be negative for African swine fever virus, porcine circovirus 2, foot-and-mouth virus; classical swine fever virus, and porcine reproductive and respiratory syndrome virus by conventional PCR and real-time PCR (data not shown). Pigs were also tested negative for antibodies to ASFV by enzyme-linked immunosorbent assay (VDPro® ASFV Ab i-ELISA ver 2.0 Kit, Median Diagnostics, Seoul, Republic of Korea). Pigs were housed at the biosafety level 2 large animal facilities at the Vietnam National University of Agriculture, Hanoi, Vietnam, provided ad libitum feed and water, and observed daily. After one week of acclimatization, pigs were randomly divided into two groups, one with 10 pigs (ASFV-infected group) and the other with 5 pigs (mock-infected group), and moved to two separate pens. Five pigs in the control group were inoculated with one ml of sterile DMEM. Pigs were monitored for clinical signs, and their rectal temperatures were measured using a digital thermometer daily.

2.4. Sample Collection

In this study, blood and oral fluid samples were collected from each pig to assess the ASFV load using a kit from Median Diagnostics Inc. (http://www.mediandiagnostics.com, accessed on 19 May 2021). The cotton rope chewing method was used to collect oral fluid samples, which had previously been used in a similar study [27]. In detail, the pigs were allowed to chew on the rope for 45 min until the rope was sufficiently wet. The wet rope was then compressed into plastic bags to recover the oral fluid. Approximately 2.5 mL of the oral fluid was transferred into a 15 mL tube for analysis. Blood samples were collected from the jugular vein of each pig before feeding in the morning at 0, 2, 4, 6, 8, 10, and 12 dpi or daily when the trial pigs showed clinical signs (e.g., fever) for viremia detection and not causing stress on experimental pigs. The blood samples were then placed into blood collection tubes containing an anticoagulant solution (EDTA). After the pigs died, a necropsy was conducted immediately by designated veterinarians, and organ samples were collected for ASFV genome detection. The results of the DNA load analysis of both oral fluid and blood samples and the organ samples were used to assess the severity of ASFV infection in the pigs.

2.5. DNA Extraction and Real-Time PCR

To detect ASFV, DNA was extracted from the whole blood, oral fluid, and tissue samples obtained from necropsied pigs. The DNeasy Blood & Tissue Kit (Qiagen, Hilden, Germany) was used to extract DNA from whole blood and oral fluid samples, while DNA was extracted from each homogenized tissue sample using the DNA Mini Kit (Qiagen, Hilden, Germany) following the manufacturer's instructions. ASFV was detected using the VDx ASFV qPCR kit (Median Diagnostics). Briefly, 5 μL of extracted DNA was mixed with 10 μL of 2× master mix and 5 μL of 4× oligo mix in a PCR tube. The reaction proceeded under the following conditions: 40 cycles of 95 °C for 15 s and 58 °C for 60 s using a CFX96 Touch Real-Time PCR Detection System (Bio-Rad Laboratories Ltd., Hercules, CA, USA). Samples with a Ct (cycle threshold) value less than 40 were considered positive for ASFV.

2.6. Scoring of ASF Symptoms and Gross Pathology Findings

To evaluate the virulence of the ASFV strain VNUA/HY/ASF1, the clinical scoring system was used as described previously [24,28]. For each criterion, a score was recorded for either normal (score 0), slightly altered (score 1), distinct clinical symptom (score 2), or severe ASF symptom (score 3). The scores of all parameters and individual pigs' total scores were estimated daily. The possible total score was a maximum of 18. Pigs with a total clinical score higher than 3 were defined as pigs having ASFV infection. Any dead pigs were immediately necropsied and through post-mortem. When any pigs had severe clinical signs (fever, anorexia, cough, diarrhea, and so on) for more than two consecutive days, or when the total accumulated clinical sign score exceeded 18 points, which was the humane endpoint for the diseased pigs. These diseased pigs were treated with anesthesia drugs and disposed of humanely [20].

2.7. Statistical Analysis

Student-*t* test was used to evaluate differences between the two groups for statistical significance. Statistical analyses and data visualization were performed with SPSS statistic 20.

3. Results

3.1. Clinical Signs

Although pigs were inoculated with the same dose at the same time using the same route, the onset of clinical signs ranged from 4 to 14 dpi. Fever and lethargy were the first clinical signs observed. Two out of ten ASFV-infected pigs showed increasing rectal temperatures at 4 dpi. The remaining animals started fever later, and most showed intermittent fever (Table 1). None of the animals in the uninfected group developed fever (Figure 1). At 15 dpi, the mean rectal temperature of the pen increased to 40.8 °C ($p < 0.05$). Although the mean temperature during the first 3 days after virus infection differed between the ASFV-infected and un-infected pig groups, this difference was not statistically significant ($p > 0.05$).

Table 1. Dates (shown as dpi) of onset of clinical signs and viremia.

No.	Date of Onset of Clinical Manifestations							Dead	Onset of Viremia
	Anorexia	Recumbency	Diarrhea	Cough	Lethargy	Fever	Skin Hemorrhages		
1	14	15	14	11	5	4	15	18	8
2	14	16	-	-	11	11	-	21	12
3	16	19	18	11	10	15	-	20	14
4	22	23	-	-	12	19	-	25	16
5	9	-	-	5	6	4	8	10	6
6	18	19	-	-	14	15	-	22	12
7	22	23	-	-	12	20	26	27	16
8	15	-	-	7	8	8	-	18	8
9	16	19	19	-	12	15	-	20	8
10	14	15	-	15	12	11	15	17	12

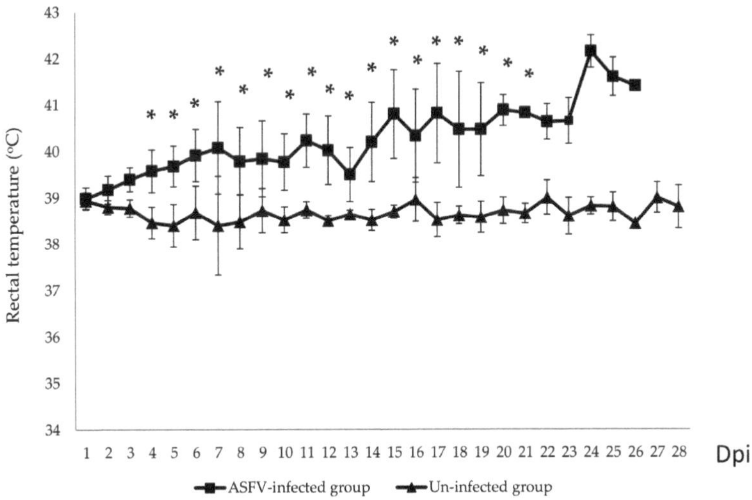

Figure 1. Mean of daily rectal temperatures (°C) of the ASFV- infected and un-infected pig groups. * $p < 0.05$.

Other clinical signs observed in the ASFV-infected group included lethargy, anorexia, fever (10/10), recumbency (8/10), cough (5/10 pigs), diarrhea (3/10 pigs), and skin hemorrhages (4/10 pigs). Lethargy was first observed in three out of ten pigs at 5–8 dpi (Table 1). The pigs succumbed to the infection between 10 to 27 (19.8 ± 4.66) dpi (Table 1 and Figure 2). In addition, the mean clinical scores of the ASFV-infected group changed over time. There was a statistically significant difference between the mean clinical scores of the ASFV-infected and uninfected groups ($p > 0.05$) from 6 to 23 dpi. The pigs of the uninfected group did not show clinical signs (Figure 3). Fever and lethargy were observed in all animals at least 4–6 days before the pigs succumbed to the infection. Skin hemorrhages, diarrhea, and cough were observed in a few animals (Figure 4). Despite the clinical sign majority of the animals stayed active with normal appetite until late in the infection, and two of the 10 infected animals stayed active until they were found dead the next day. Overall clinical sign findings in this study suggested acute and/or subacute ASF infection, as described in a previous report [23,29].

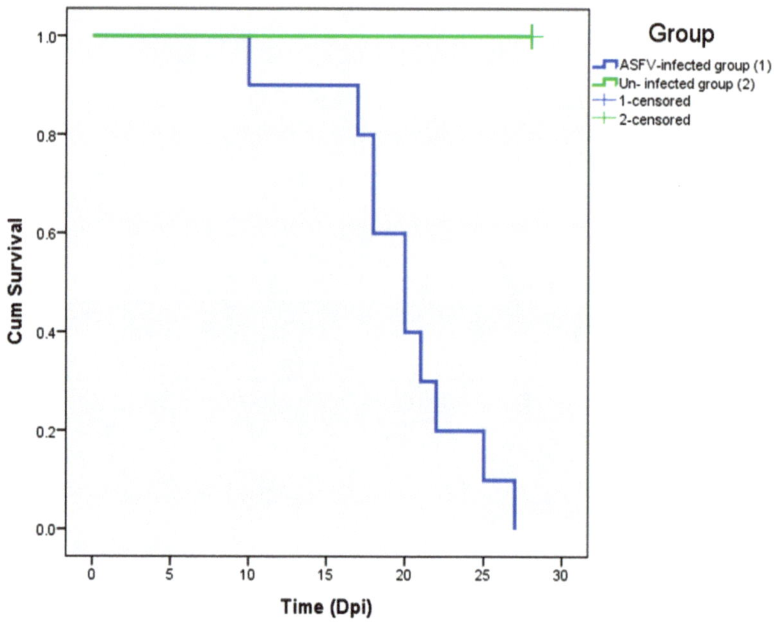

Figure 2. The survival rate of ASFV-infected (blue) and un-infected (green) pigs.

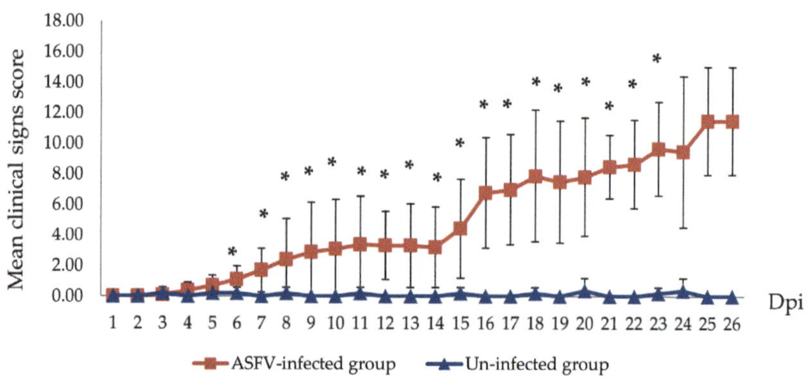

Figure 3. Mean clinical scores of ASFV-infected (red line) and un-infected (blue line) pig groups. The scores were calculated as previously described by I. Galindo-Cardiel and H. S. Lee [24,28]. (* $p < 0.05$).

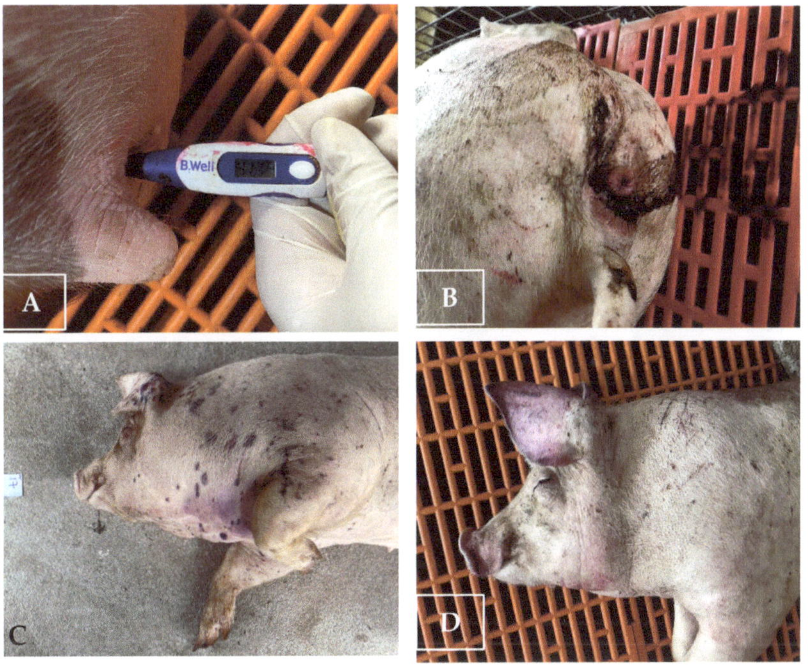

Figure 4. Some of the clinical signs of pigs infected with ASFV. (**A**) Fever; (**B**) diarrhea; (**C**) petechial hemorrhages on the skin; (**D**) erythema and skin hemorrhages.

3.2. Blood and Oral Fluid Sample Analysis

Following oral inoculation, the first pig (#5) showed viremia at 6 dpi (Table 1). ASFV genome was detected in oral fluid starting at 9 dpi, three days after the first detection of viremia in pen (Figure 5). Despite mild fever and lethargy observed by 9 dpi, all animals remained active and had a normal appetite. All the pigs in the control group stayed healthy throughout the experiment and were negative for the ASFV genome by real-time PCR (data not shown).

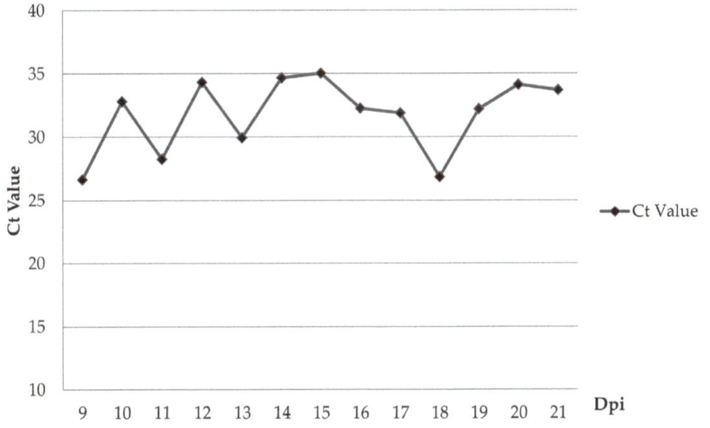

Figure 5. The mean viral load (Ct value) of the ASFV in oral fluids collected from the ASFV-infected pig group.

3.3. Gross Pathological Findings

The spleen was significantly larger and dark red with congestion and hemorrhage (Figure 6A, red arrow) in all pigs who succumbed to the infection. Tonsil, mandibular, mediastinal, and inguinal lymph nodes were enlarged, hemorrhagic, and/or congested (Figure 6G, E). Mesenteric lymph nodes were also markedly red and enlarged (Figure 6A-red circle). The liver was enlarged, congested, and had scattered pale foci (Figure 6A, blue arrow). Petechial hemorrhages were observed on the cortex (Figure 6B, black arrow). Lungs were congested, hemorrhagic, and edematous, with moderate to severe interstitial pneumonia (Figure 6C). Kidneys were congested (Figure 6F). Hydropericardium, hemorrhages in gastric mucosa, and gallbladder wall were observed in some pigs (Figure 6D-black arrow, H and I-black arrow).

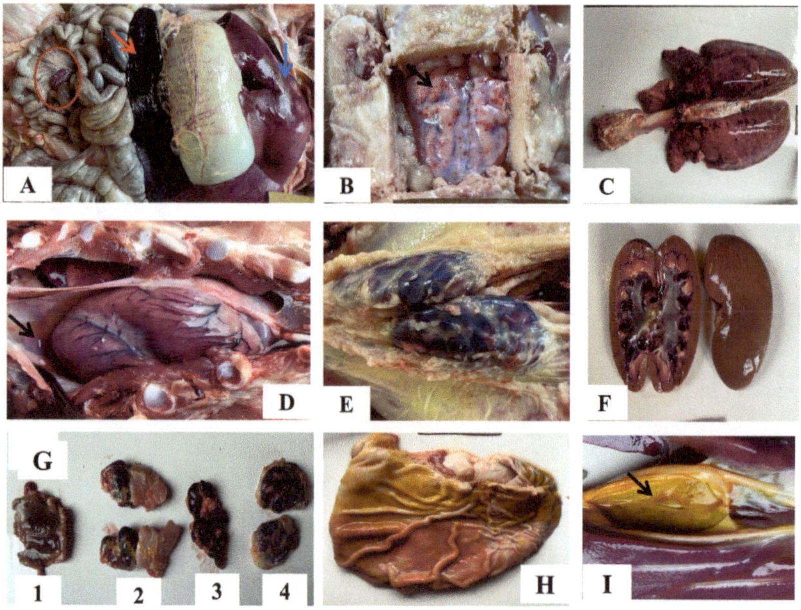

Figure 6. Gross pathological lesions were observed in pigs infected with ASFV. (**A**) Congested and enlarged spleen (red arrow); hemorrhagic mesenteric lymph nodes (red circle); enlarged liver with hemorrhages (blue arrow); (**B**) hemorrhagic meninges; (**C**) congested and interstitial pneumonia; (**D**) hydropericardium (black arrow); (**E**) enlarged and hemorrhagic inguinal lymph nodes; (**F**) petechial hemorrhage in kidneys; (**G1**) hemorrhagic tonsils; (**G2**) enlarged and hemorrhagic mandibular lymph nodes; (**G3**) mediastinal lymph nodes; and (**G4**) inguinal lymph nodes; (**H**) gastric mucosa with superficial lesions and hemorrhage; (**I**) petechiae gallbladder wall (black arrow).

None of the pigs from the uninfected group showed ASF-related lesions (Figures not shown).

4. Discussion

African swine fever (ASF) is a contagious hemorrhagic fever of domestic and wild swine. ASF has become a global veterinary concern since it was first reported in China in 2018. Later it rapidly spread to almost all Asian countries near China, including the most recent outbreaks in India [4]. Until now, there is no safe and effective ASF vaccine available worldwide. In Vietnam, the first ASFV outbreak was reported in Hung Yen province in February 2019. The infected pigs showed typical clinical signs of ASF, such as fever, lack of appetite, skin hemorrhages, bloody diarrhea, cyanosis, and 100% mortality.

However, the clinical signs reported with the ASF outbreaks in Vietnam did not always align with the peracute or acute forms of the disease. The responsible VNUA/HY/ASF1 virus belonged to ASFV p72 genotype II [21]. During the second ASF-reported outbreak in Thai Binh province, pigs quickly showed anorexia and high fever for three days before they were found dead [30]. The ASFV VNUA/HY/ASF1 isolate had been inoculated intramuscularly (titer, $10^{3.5}$ HAD_{50}/mL per pig) into ten 7-to-8-week-old pigs (Yorkshire × Landrace × Duroc), and all of the inoculated pigs died within 5–8 dpi. Fever developed in three pigs at 3 dpi, and viremia was noted in those pigs at around 2 dpi. Oral fluid collected from the pigs tested positive for the ASFV genome as early as 3 dpi and continued to be positive until the end of the study [24]. Previous studies showed that the clinical outcome of ASF depends on many factors, including the pathogenicity of the virus isolates, the dose, the route of infection, and host characteristics [23,31]. Unlike CSF, which mainly affects young pigs, all age groups are considered equally susceptible to ASF. For easy handling, pigs 8 to 12 weeks old are often chosen to evaluate the pathogenicity of the ASFV strains by intramuscular, intranasal inoculations, and direct contact [32–34]. In addition, the oral inoculation dose of 10^3 HAD_{50}/mL per pig has been used in several studies to assess clinical signs of ASFV-infected pigs [35,36]. Therefore, in this study, VNUA/HY/ASF1 strain isolated from the first ASF outbreak was used to infect 10 eight-week-old pigs orally with 10^3 HAD_{50} per animal.

In this study, compared to the previous study that used intramuscular inoculation, oral inoculation of the ASFV VNUA/HY/ASF1 strain resulted in a delayed onset and protracted clinical disease in pigs. Two of the orally inoculated pigs developed a mild fever at 4 dpi; viremia was noted only at 6 dpi in one of the pigs. The first animal dies of ASF at 10 dpi, and the last one at 27 dpi. ASFV genome was detected in oral fluid only after 9 dpi. Out of 10 pigs inoculated orally, only four got infected within 8 days post-inoculation. This suggests that the remaining six animals acquired the infection through direct contact with the four pigs that got infected following experimental inoculation. In a recent study conducted by Niederwerder et al. [35], the infection probability of ASFV Georgia 2007 via oral route following a single dose of 10^3 $TCID_{50}$ was calculated to be around 83.3%. Surprisingly our study only showed a 40% infection rate, which could be due to many reasons, including the method of inoculation, breed, and health of the animals, etc. Despite these changes, clinical signs observed in this experiment were similar to those reported before (fever, anorexia, depression, diarrhea) [24,30]. The appearance of clinical symptoms such as fever and lethargy was expected to appear after or coincide with the induction of viremia [24,37,38]. However, several pigs in this study developed fever and lethargy 2–4 days before the viremia was observed. This could be due to many reasons, including concurrent infections, ambient temperature fluctuation in animal pens, and handling of pigs for blood collection. The pigs used in the experiment were purchased from a commercial farm in Vietnam. They were not specific-pathogen-free animals. Therefore, it is possible that some of the pigs in the study were infected with common bacterial and viral agents found in swine herds in Vietnam. Such infections can induce fever and lethargy irrespective of the ASFV infection [39]. Hence, some non-ASF-specific clinical signs, such as fever and lethargy, could appear early or at the same time as ASF viremia in some animals. Therefore, we believe that the clinical picture observed in this study is closer to what happens in Vietnam and is more aligned with acute and subacute forms of the disease. No pigs died suddenly (per acute), and the mean time to death of pigs after oral inoculation was 19.8 ± 4.66 days. The first animal died of ASF infection at 10 dpi, and almost all animals stayed healthy despite an ongoing ASF infection in the pen until 20 dpi when 60% of the animals in the pen died. Despite protracted pathogenicity, all pigs that died of ASF had similar gross pathological lesions, i.e., splenomegaly and hemorrhagic lymphadenitis. Interestingly, in five animals, we also observed meningeal hemorrhages. In this study, there was a fluctuation of the Ct value obtained from oral fluid samples of pigs during the infection process, which could be affected by many factors, such as the health conditions after ASFV infection and chewing activities on the ropes. It was noted that some

ASFV-infected pigs could not chew on ropes when they had severe clinical symptoms. The variations of the Ct value obtained from the oral fluid were also reported by Goonewardene, K. B. et al. [40].

As observed in this study, the protracted clinical picture following ASF's introduction into the pig pens has significant practical relevance. As the observations from the study point out, ASF infections can quickly spread unnoticed and/or slowly on the farm for two to three weeks once the farm is infected with ASFV. Therefore, any abnormal behavior in a pig farm should be quickly addressed by submitting samples for ASF detection. Such an approach to contain and control ASF spread, especially in ASF-endemic countries such as Vietnam.

5. Conclusions

In this study, we evaluated the pathogenicity of the ASFV VNUA/HY/ASF1 strain following oral inoculation of 8 weeks old commercial pigs. Pigs infected showed acute to subacute clinical form with a mean time of death around 19.8 $\pm$ 4.66 days. The onset of the clinical signs of different pigs was also different, ranging from 4 to 14 dpi. The viremia pigs infected with ASFV were detected only after 6 dpi (11.2 $\pm$ 3.55). This study provides valuable information regarding the pathogenesis of ASF outbreak strain VNUA/HY/ASF1 in Vietnam following oral infection in domestic pigs—a similar model to the natural exposure of ASF in the field.

Author Contributions: Conceptualization, V.P.L., T.L.N., T.A.D.B. and A.A.; Methodology, V.P.L., T.L.N. and A.A.; Software, T.T.H.N., N.T.A.M. and V.H.D.; Validation, V.P.L., T.A.D.B. and A.A.; Formal analysis, V.P.L., T.T.H.N. and A.A.; Investigation, T.T.H.N., P.N.L., N.T.A.M. and V.T.N.; Resources, V.P.L., P.N.L. and T.T.H.N.; Data curation, T.T.H.N., P.N.L., N.T.A.M. and V.T.N.; Writing—original draft preparation, T.T.H.N., N.T.A.M. and V.T.N.; Writing—review and editing V.P.L., V.H.D. and A.A.; Visualization, V.P.L., V.H.D. and T.T.H.N.; Supervision, V.P.L. and A.A.; Project administration, V.P.L.; Funding acquisition, V.P.L. All authors have read and agreed to the published version of the manuscript.

Funding: This work was supported by the Vietnam National Project under project code no. DTDL.CN-53/19.

Institutional Review Board Statement: The animal experiment was conducted in line with the guidelines of good experimental practices and was approved by the Committee on Animal Research and Ethics of the Faculty of Veterinary Medicine, Vietnam National University of Agriculture (Approval number, CARE-2021/05; Approval date, 2 June 2021).

Informed Consent Statement: Not applicable.

Data Availability Statement: The data presented in this study are available on request from the corresponding author.

Conflicts of Interest: The authors declare no conflict of interest. Study design, data collection, analysis, and decision to publish or prepare the manuscript belong to the authors only.

References

1. Liu, Y.; Zhang, X.; Qi, W.; Yang, Y.; Liu, Z.; An, T.; Wu, X.; Chen, J. Prevention and Control Strategies of African Swine Fever and Progress on Pig Farm Repopulation in China. *Viruses* **2021**, *13*, 2552. [CrossRef]
2. Dee, S.A.; Bauermann, F.V.; Niederwerder, M.C.; Singrey, A.; Clement, T.; de Lima, M.; Long, C.; Patterson, G.; Sheahan, M.A.; Stoian, A.M.M.; et al. Survival of Viral Pathogens in Animal Feed Ingredients under Transboundary Shipping Models. *PLoS ONE* **2018**, *13*, e0194509. [CrossRef]
3. Urbano, A.C.; Ferreira, F. African Swine Fever Control and Prevention: An Update on Vaccine Development. *Emerg. Microbes Infect.* **2022**, *11*, 2021–2033. [CrossRef]
4. Kim, Y.J.; Park, B.; Kang, H.E. Control Measures to African Swine Fever Outbreak: Active Response in South Korea, Preparation for the Future, and Cooperation. *J. Vet. Sci.* **2021**, *22*, e13. [CrossRef]
5. Dixon, L.K.; Chapman, D.A.G.; Netherton, C.L.; Upton, C. African Swine Fever Virus Replication and Genomics. *Virus Res.* **2013**, *173*, 3–14. [CrossRef]
6. Galindo, I.; Alonso, C. African Swine Fever Virus: A Review. *Viruses* **2017**, *9*, E103. [CrossRef]

7. Quembo, C.J.; Jori, F.; Vosloo, W.; Heath, L. Genetic Characterization of African Swine Fever Virus Isolates from Soft Ticks at the Wildlife/Domestic Interface in Mozambique and Identification of a Novel Genotype. *Transbound. Emerg. Dis.* **2018**, *65*, 420–431. [CrossRef]
8. Gallardo, C.; Fernández-Pinero, J.; Pelayo, V.; Gazaev, I.; Markowska-Daniel, I.; Pridotkas, G.; Nieto, R.; Fernández-Pacheco, P.; Bokhan, S.; Nevolko, O.; et al. Genetic Variation among African Swine Fever Genotype II Viruses, Eastern and Central Europe. *Emerg. Infect. Dis.* **2014**, *20*, 1544–1547. [CrossRef]
9. Kim, H.-J.; Cho, K.-H.; Ryu, J.-H.; Jang, M.-K.; Chae, H.-G.; Choi, J.-D.; Nah, J.-J.; Kim, Y.-J.; Kang, H.-E. Isolation and Genetic Characterization of African Swine Fever Virus from Domestic Pig Farms in South Korea, 2019. *Viruses* **2020**, *12*, 1237. [CrossRef]
10. Kovalenko, G.; Ducluzeau, A.-L.; Ishchenko, L.; Sushko, M.; Sapachova, M.; Rudova, N.; Solodiankin, O.; Gerilovych, A.; Dagdag, R.; Redlinger, M.; et al. Complete Genome Sequence of a Virulent African Swine Fever Virus from a Domestic Pig in Ukraine. *Microbiol. Resour. Announc.* **2019**, *8*, e00883-19. [CrossRef]
11. Olesen, A.S.; Lohse, L.; Dalgaard, M.D.; Woźniakowski, G.; Belsham, G.J.; Bøtner, A.; Rasmussen, T.B. Complete Genome Sequence of an African Swine Fever Virus (ASFV POL/2015/Podlaskie) Determined Directly from Pig Erythrocyte-Associated Nucleic Acid. *J. Virol. Methods* **2018**, *261*, 14–16. [CrossRef]
12. Sanna, G.; Dei Giudici, S.; Bacciu, D.; Angioi, P.P.; Giammarioli, M.; De Mia, G.M.; Oggiano, A. Improved Strategy for Molecular Characterization of African Swine Fever Viruses from Sardinia, Based on Analysis of P30, CD2V and I73R/I329L Variable Regions. *Transbound. Emerg. Dis.* **2017**, *64*, 1280–1286. [CrossRef]
13. Sauter-Louis, C.; Forth, J.H.; Probst, C.; Staubach, C.; Hlinak, A.; Rudovsky, A.; Holland, D.; Schlieben, P.; Göldner, M.; Schatz, J.; et al. Joining the Club: First Detection of African Swine Fever in Wild Boar in Germany. *Transbound. Emerg. Dis.* **2021**, *68*, 1744–1752. [CrossRef]
14. Mai, N.T.A.; Vu, X.D.; Nguyen, T.T.H.; Nguyen, V.T.; Trinh, T.B.N.; Kim, Y.J.; Kim, H.-J.; Cho, K.-H.; Nguyen, T.L.; Bui, T.T.N.; et al. Molecular Profile of African Swine Fever Virus (ASFV) Circulating in Vietnam during 2019-2020 Outbreaks. *Arch. Virol.* **2021**, *166*, 885–890. [CrossRef]
15. Mazur-Panasiuk, N.; Walczak, M.; Juszkiewicz, M.; Woźniakowski, G. The Spillover of African Swine Fever in Western Poland Revealed Its Estimated Origin on the Basis of O174L, K145R, MGF 505-5R and IGR I73R/I329L Genomic Sequences. *Viruses* **2020**, *12*, 1094. [CrossRef]
16. Mazur-Panasiuk, N.; Woźniakowski, G. The Unique Genetic Variation within the O174L Gene of Polish Strains of African Swine Fever Virus Facilitates Tracking Virus Origin. *Arch. Virol.* **2019**, *164*, 1667–1672. [CrossRef]
17. Eustace Montgomery, R. On A Form of Swine Fever Occurring in British East Africa (Kenya Colony). *J. Comp. Pathol. Ther.* **1921**, *34*, 159–191. [CrossRef]
18. Sánchez-Vizcaíno, J.M.; Mur, L.; Martínez-López, B. African Swine Fever (ASF): Five Years around Europe. *Vet. Microbiol.* **2013**, *165*, 45–50. [CrossRef]
19. Revilla, Y.; Pérez-Núñez, D.; Richt, J.A. African Swine Fever Virus Biology and Vaccine Approaches. *Adv. Virus Res.* **2018**, *100*, 41–74.
20. Ramirez-Medina, E.; O'Donnell, V.; Silva, E.; Espinoza, N.; Velazquez-Salinas, L.; Moran, K.; Daite, D.A.; Barrette, R.; Faburay, B.; Holland, R.; et al. Experimental Infection of Domestic Pigs with an African Swine Fever Virus Field Strain Isolated in 2021 from the Dominican Republic. *Viruses* **2022**, *14*, 1090. [CrossRef]
21. Le, V.P.; Jeong, D.G.; Yoon, S.-W.; Kwon, H.-M.; Trinh, T.B.N.; Nguyen, T.L.; Bui, T.T.N.; Oh, J.; Kim, J.B.; Cheong, K.M.; et al. Outbreak of African Swine Fever, Vietnam, 2019. *Emerg. Infect. Dis.* **2019**, *25*, 1433–1435. [CrossRef] [PubMed]
22. Sun, E.; Huang, L.; Zhang, X.; Zhang, J.; Shen, D.; Zhang, Z.; Wang, Z.; Huo, H.; Wang, W.; Huangfu, H.; et al. Genotype I African Swine Fever Viruses Emerged in Domestic Pigs in China and Caused Chronic Infection. *Emerg. Microbes Infect.* **2021**, *10*, 2183–2193. [CrossRef] [PubMed]
23. Sánchez-Vizcaíno, J.M.; Mur, L.; Gomez-Villamandos, J.C.; Carrasco, L. An Update on the Epidemiology and Pathology of African Swine Fever. *J. Comp. Pathol.* **2015**, *152*, 9–21. [CrossRef] [PubMed]
24. Lee, H.S.; Bui, V.N.; Dao, D.T.; Bui, N.A.; Le, T.D.; Kieu, M.A.; Nguyen, Q.H.; Tran, L.H.; Roh, J.-H.; So, K.-M.; et al. Pathogenicity of an African Swine Fever Virus Strain Isolated in Vietnam and Alternative Diagnostic Specimens for Early Detection of Viral Infection. *Porc. Health Manag.* **2021**, *7*, 36. [CrossRef]
25. Reed, L.J.; Muench, H. A Simple Method Of Estimating Fifty Per Cent Endpoints. *Am. J. Epidemiol.* **1938**, *27*, 493–497. [CrossRef]
26. Malmquist, W.A.; Hay, D. Hemadsorption and Cytopathic Effect Produced by African Swine Fever Virus in Swine Bone Marrow and Buffy Coat Cultures. *Am. J. Vet. Res.* **1960**, *21*, 104–108.
27. Mur, L.; Gallardo, C.; Soler, A.; Zimmermman, J.; Pelayo, V.; Nieto, R.; Sánchez-Vizcaíno, J.M.; Arias, M. Potential Use of Oral Fluid Samples for Serological Diagnosis of African Swine Fever. *Vet. Microbiol.* **2013**, *165*, 135–139. [CrossRef]
28. Galindo-Cardiel, I.; Ballester, M.; Solanes, D.; Nofrarías, M.; López-Soria, S.; Argilaguet, J.M.; Lacasta, A.; Accensi, F.; Rodríguez, F.; Segalés, J. Standardization of Pathological Investigations in the Framework of Experimental ASFV Infections. *Virus Res.* **2013**, *173*, 180–190. [CrossRef]
29. Salguero, F.J. Comparative Pathology and Pathogenesis of African Swine Fever Infection in Swine. *Front. Vet. Sci.* **2020**, *7*, 282. [CrossRef]
30. Nga, B.T.T.; Tran Anh Dao, B.; Nguyen Thi, L.; Osaki, M.; Kawashima, K.; Song, D.; Salguero, F.J.; Le, V.P. Clinical and Pathological Study of the First Outbreak Cases of African Swine Fever in Vietnam, 2019. *Front. Vet. Sci.* **2020**, *7*, 392. [CrossRef]

31. Guinat, C.; Reis, A.L.; Netherton, C.L.; Goatley, L.; Pfeiffer, D.U.; Dixon, L. Dynamics of African Swine Fever Virus Shedding and Excretion in Domestic Pigs Infected by Intramuscular Inoculation and Contact Transmission. *Vet. Res.* **2014**, *45*, 93. [CrossRef]
32. Lohse, L.; Nielsen, J.; Uttenthal, Å.; Olesen, A.S.; Strandbygaard, B.; Rasmussen, T.B.; Belsham, G.J.; Bøtner, A. Experimental Infections of Pigs with African Swine Fever Virus (Genotype II); Studies in Young Animals and Pregnant Sows. *Viruses* **2022**, *14*, 1387. [CrossRef]
33. Tran, X.H.; Le, T.T.P.; Nguyen, Q.H.; Do, T.T.; Nguyen, V.D.; Gay, C.G.; Borca, M.V.; Gladue, D.P. African Swine Fever Virus Vaccine Candidate ASFV-G-ΔI177L Efficiently Protects European and Native Pig Breeds against Circulating Vietnamese Field Strain. *Transbound. Emerg. Dis.* **2021**, *69*, e497–e504.
34. Post, J.; Weesendorp, E.; Montoya, M.; Loeffen, W.L. Influence of Age and Dose of African Swine Fever Virus Infections on Clinical Outcome and Blood Parameters in Pigs. *Viral Immunol.* **2017**, *30*, 58–69. [CrossRef]
35. Niederwerder, M.C.; Stoian, A.M.M.; Rowland, R.R.R.; Dritz, S.S.; Petrovan, V.; Constance, L.A.; Gebhardt, J.T.; Olcha, M.; Jones, C.K.; Woodworth, J.C.; et al. Infectious Dose of African Swine Fever Virus When Consumed Naturally in Liquid or Feed. *Emerg. Infect. Dis.* **2019**, *25*, 891–897. [CrossRef]
36. Howey, E.B.; O'Donnell, V.; de Carvalho Ferreira, H.C.; Borca, M.V.; Arzt, J. Pathogenesis of Highly Virulent African Swine Fever Virus in Domestic Pigs Exposed via Intraoropharyngeal, Intranasopharyngeal, and Intramuscular Inoculation, and by Direct Contact with Infected Pigs. *Virus Res.* **2013**, *178*, 328–339. [CrossRef]
37. Cho, K.-H.; Hong, S.-K.; Jang, M.-K.; Ryu, J.-H.; Kim, H.-J.; Lee, Y.-R.; Roh, I.-S.; Sohn, H.-J.; Kang, H.-E.; Park, J.-Y. Comparison of the Virulence of Korean African Swine Fever Isolates from Pig Farms during 2019–2021. *Viruses* **2022**, *14*, 2512. [CrossRef]
38. Walczak, M.; Żmudzki, J.; Mazur-Panasiuk, N.; Juszkiewicz, M.; Woźniakowski, G. Analysis of the Clinical Course of Experimental Infection with Highly Pathogenic African Swine Fever Strain, Isolated from an Outbreak in Poland. Aspects Related to the Disease Suspicion at the Farm Level. *Pathogens* **2020**, *9*, 237. [CrossRef]
39. Kameyama, K.; Nishi, T.; Yamada, M.; Masujin, K.; Morioka, K.; Kokuho, T.; Fukai, K. Experimental Infection of Pigs with a Classical Swine Fever Virus Isolated in Japan for the First Time in 26 Years. *J. Vet. Med. Sci.* **2019**, *81*, 1277–1284. [CrossRef]
40. Goonewardene, K.B.; Chung, C.J.; Goolia, M.; Blakemore, L.; Fabian, A.; Mohamed, F.; Nfon, C.; Clavijo, A.; Dodd, K.A.; Ambagala, A. Evaluation of Oral Fluid as an Aggregate Sample for Early Detection of African Swine Fever Virus Using Four Independent Pen-Based Experimental Studies. *Transbound. Emerg. Dis.* **2021**, *68*, 2867–2877. [CrossRef]

pathogens

MDPI

Article

Pathology and Clinics of Naturally Occurring Low-Virulence Variants of African Swine Fever Emerged in Domestic Pigs in the South Caucasus

Hranush Avagyan [1,2], Sona Hakobyan [1], Bagrat Baghdasaryan [1], Hranush Arzumanyan [1], Arpine Poghosyan [1], Nane Bayramyan [1], Anna Semerjyan [3], Mariam Sargsyan [4], Henry Voskanyan [1], Tigranuhi Vardanyan [1], Naira Karalyan [5], Lina Hakobyan [1], Liana Abroyan [1], Aida Avetisyan [1,2], Elena Karalova [1,2], Zara Semerjyan [1,2] and Zaven Karalyan [1,3,*]

1 Laboratory of Cell Biology and Virology, Institute of Molecular Biology of NAS RA, Yerevan 0014, Armenia; a.avagian@yahoo.com (H.A.); 777sona7@gmail.com (S.H.); artdrbaghdasaryan@gmail.com (B.B.); arzumanyan0712@gmail.com (H.A.); arpi.poghosyan21@gmail.com (A.P.); naneramaz@mail.ru (N.B.); aleksvoskanyan@mail.ru (H.V.); tika.vardanyan86@mail.ru (T.V.); lina.hakobyan@gmail.com (L.H.); abroyan.liana@gmail.com (L.A.); a.avetis@mail.ru (A.A.); hatussili@yahoo.com (E.K.); zarasem@gmail.com (Z.S.)

2 Experimental Laboratory, Yerevan State Medical University after M. Heratsy, Yerevan 0025, Armenia

3 Department of Medical Biology and Genetics, Yerevan State Medical University after M. Heratsy, Yerevan 0025, Armenia; annasemerjian@hotmail.com

4 Department of Epizootiology and Parasitology, Armenian National Agrarian University, Yerevan 0009, Armenia; mariam.sargsyan.1960@mail.ru

5 Department of Pathological Anatomy and Clinical Morphology, Yerevan State Medical University after M. Heratsi, Yerevan 0025, Armenia; naira_karalyan@yahoo.com

* Correspondence: zkaralyan@yahoo.com

Citation: Avagyan, H.; Hakobyan, S.; Baghdasaryan, B.; Arzumanyan, H.; Poghosyan, A.; Bayramyan, N.; Semerjyan, A.; Sargsyan, M.; Voskanyan, H.; Vardanyan, T.; et al. Pathology and Clinics of Naturally Occurring Low-Virulence Variants of African Swine Fever Emerged in Domestic Pigs in the South Caucasus. *Pathogens* **2024**, *13*, 130. https://doi.org/10.3390/pathogens13020130

Academic Editor: Fun-In Wang

Received: 24 December 2023
Revised: 23 January 2024
Accepted: 24 January 2024
Published: 29 January 2024

Abstract: Shortly after the establishment of African swine fever virus (ASFV) genotype II in 2007, cases of acute fatal infection were observed. However, after several years of circulation in the Eurasian region, the clinical signs of the disease changed. Currently, this disease can occur acutely, subclinically, chronically, or asymptomatically. Cases of the complete recovery of infected pigs, and the disappearance of ASFV from their tissues and secretions have been described. This form of the disease first appeared in Armenia at the end of 2011. This virus was described and identified as the Dilijan2011IMB strain. The goal of our research was to study the main features of clinical, pathological, immunological, virological, and genetic parameters involved in the development of new forms of African swine fever (ASF). Chronic ASF was characterized with low titers of the virus and a decrease in the intensity of hemadsorption. Additionally, a reduced intensity in clinical symptoms and pathoanatomical results was noted. The absolute, but not the relative number of immune cells changes; the neutropenia (in bone marrow and spleen), lymphopenia (in bone marrow), lymphocytosis (only in spleen), lymphoid cell depletion (in bone marrow), and pancytopenia (in bone marrow) observed in the chronic form of ASF were less pronounced compared to in the acute form. When comparing the late stage of chronic ASF to the acute form, the key cytological indicators in the spleen, lymph nodes, and blood were less severe in the chronic stage. Bone marrow failure in the chronic form, expressed in a pronounced decrease in all cell types, generally coincided with the data in the acute form of ASF. The same data were obtained after assessing serum TNF-alpha levels. Thus, we can conclude that the chronic form of ASF occurs due to a less pronounced immune response, as well as a decrease in virus titers in the blood and tissues of infected pigs.

Keywords: African swine fever virus; chronization; immunopathology; histopathology; clinics

1. Introduction

African swine fever virus (ASFV) (genotype II) entered Eurasia in 2007, and at the initial stage of circulation, showed the presence of acute and super-acute forms of the disease with a 100% lethal outcome [1–3]. However, after several years, individual cases of subacute forms of the disease have been recorded. Thus, at the end of 2011, cases of chronic or subacute form of ASF were identified in Dilijan (Armenia) [4].

As African swine fever (ASF) spreads beyond its initial African location to domestic pigs in other regions, the disease's progression typically accelerates. This results in more animals entering subacute and chronic phases, potentially leading to longer lifespans in affected pigs and reduced visible clinical symptoms. Thus, in Spain, two years after the introduction of ASF, domestic virus-carrying pigs were detected [5]. Some studies have identified reduced-virulence isolate ASFV genomes in apparently healthy pigs in Uganda and Kenya [6,7]. In Armenia, similar processes occurred 4 years after the emergence of ASFV.

Previously, the following main characteristics of the chronicity of the disease were recorded: (1) a reduction of the virus in infected organs and blood; (2) a decrease in the hemadsorption activity of the virus; (3) a decrease in the manifestation of the main clinical and laboratory parameters of ASF (as decreased in hemorrhages and immunopathological manifestations); and (4) prolonged viremia [4].

In 2018–2020, several outbreaks of an unknown disease occurred in the South Caucasus in pigs vaccinated against classical swine fever. After field and laboratory examinations, the presence of the African swine fever virus in all the studied samples was confirmed in our laboratory.

In recent years, there have been increasing reports of new cases of ASF virus with an atypical course. The pathology is characterized by a significantly prolonged disease duration, sometimes reduced mortality, and prolonged viremia. Such changes are often associated with the circulation of the virus in the feral pig population [8,9]. In China, there have been many cases of ASF virus with reduced virulence in domestic pigs since 2018 [10,11].

Attenuated viruses cause less severe pathological alterations compared to the more aggressive pathological changes induced by their virulent wild-type origins. Although there is evidence of the presence of mildly pathogenic strains of ASFV belonging to genotype II [10,12–15], no study has provided a definitive description of the pathology development in natural attenuation. The current information on the hemadsorption activity (Leitão et al., 2001) prospects of decreasing activity usually correlate with attenuation and prospects of their use as vaccines [15,16], so it has been shown that naturally attenuated isolates can provide fairly reliable protection in wild boar [17]. In this article, we try to show the differences in the pathogenetic characteristics of naturally attenuated strains of ASFV detected in the South Caucasus in 2018–2020.

We conducted a comparative analysis of the ASFV content in the serums of pigs with chronic and acute forms of ASF. Long viremia was observed, starting from the first clinical manifestations, and ending with the terminal stage of the disease. Viremia levels were 1.5–3.5 log lower than similar values in the acute form of ASF.

A pathological analysis of an autopsy of pig organs was also performed.

2. Materials and Methods

2.1. Animals

All the studied animals were kept at 14 farms located in the South Caucasus (38.960246, 46.595191). The first outbreak (Kovsakan2018) was recorded in the late summer of 2018 (number of animals: 19); the second outbreak (Kovsakan2019), in the summer of 2019 (number of animals: 19); the third outbreak (Kovsakan2020) occurred at the end of 2019, beginning of 2020 (number of animals: 17). This article presents data from 55 pigs with unusual ASF (Arm007) variants (tentatively designated as chronic form and persistent form, which is possible with periodic reactivation).

2.2. Sample Collection

The Institutional Review Board/Independent Ethics Committee of the Institute of Molecular Biology of NAS RA approved the collection of biological samples (reference number IRB00004079). Biological samples were collected from August 2018 to September 2018 for Kovsakan1 2018, from July 2019 to September 2019 for Kovsakan2 2019, and from November 2019 to January 2020 for Kovsakan3 2020. The studied pigs were negative for other known porcine viral diseases and vaccinated against classical swine fever. Samples from the liver, brain, bone marrow, heart, kidney, spleen, lymph nodes, and lung were kept in separate, disposable plastic containers. Clinical signs of infection were recorded daily [4]. Gross anatomical pathology characteristics were observed during routine postmortem examinations.

2.3. DNA Isolation and Quantitative Real-Time PCR (p72)

In order to determine ASFV gene expressions in different organs of pigs, first, total viral RNA/DNA was isolated (HiGene™ Viral RNA/DNA Prep Kit (BIOFACT Daejeon, Republic of Korea)). Quantitative real-time PCR was carried out using SYBR green methods on an Eco Illumina Real-Time PCR System device (Illumina Inc. San Diego, CA, USA) [18,19]. Each reaction mixture (20 μL) composition was as follows: 4 μL 5 × HOT FIREPol® EvaGreen® qPCR Mix Plus (ROX) (Solis BioDyne Tartu, Estonia), 0.2 μL of each specific primer, 4 μL template DNA/cDNA, and 11.6 μL ddH20. The thermal profile was set as follows: Polymerase activation: 95 °C for 12 min, 40 cycles: 95 °C for 15 s, 52 °C for 30 s, and 72 °C for 30 s. Standard curves were generated using serial 10-fold dilutions of viral DNA. The fluorescence threshold (Ct) was calculated using the ECO-Illumina system software v5.0. Primers used for amplification were designed and ordered from Integrated DNA Technology-IDT (https://www.idtdna.com/pagesas) (accessed on 5 February 2019) [20].

The primers and fluorescent-labeled probe used were as follows:
ASFV B646L gene:
Fluorescent probe—6-FAM/TAMRA
Sequence 1—TGC TCA TGG TAT CAA TCT TAT CG
Sequence 2—CCA CTG GGT TGG TAT TCC TC
Sequence 3—/56-FAM/TTC CAT CAA AGT TCT GCA GCT CTT/36-TAMSp/
β-actin gene:
Fluorescent probe—TET/ZEN/IBFQ
Sequence 1—CTC GAT CAT GAA GTG CGA CGT
Sequence 2—GTG ATC TCC TTC TGC ATC CTG TC
Sequence 3—/5TET/AT CAG GAA G/Zen/G ACC TCT ACG CCA ACA CGG/3IABkFQ/
The β-actin gene was used as a housekeeping gene.

To align the cDNA plots and ASFV infection titers, Cq values were rescaled after comparison with viral genome copies and modified in absolute amounts along the y-axis for better visualization. To evaluate the profile of the ASFV replication efficiency, the genes with different temporal expression patterns were identified [21,22].

2.4. Hemadsorption Assay

A hemadsorption assay (HAD) was performed and expressed in log 10 hemadsorption units (HADU50/mL) [23].

2.5. Tissue Samples

Liver, kidney, and lung samples were preserved for 24 h in 10% buffered formalin solution (pH 7.2). Following fixation, the specimens underwent a progressive phase of alcohol dehydration, followed by a xylol wash, and a routine process for embedding in paraffin wax in preparation for light microscopy. Wax-embedded samples were cut (Microm HM 355; 5 μm) and stained with hematoxylin and eosin in accordance with the manufacturer's instructions (Sigma-Aldrich, Steinheim am Albuch, Germany) for structural analysis. A light microscope was used to conduct the histological examination.

2.6. Serum Collection and ELISA

Blood and serum were obtained via puncture of the jugular vein using a vacutainer system. Healthy porcine blood samples were taken to obtain control values. For the detection of TNF-a levels in serum, a commercial ELISA kit (#MBS745775; MyBioSource, San Diego, CA, USA) was used.

2.7. Cytology of Blood and Hemolymphoid Organs

Blood smears, lymph nodes, spleen, and bone marrow were prepared routinely according to [24]. In accordance with the manufacturer's instructions (Sigma-Aldrich), slides were fixed in pure methanol and stained using modified Giemsa solution (aquare B/aquare II, eosin, and methylene blue) for cell examination. Using a light microscope set to 1250 magnification, cells were examined and counted in 100 randomly chosen fields (0.01 mm^2). The microscopic evaluation of cells based on morphological characteristics was performed as described previously [25]. The evaluation of monocytes, monoblasts, and macrophages was based on morphologic characteristics [26].

Each print or smear was reviewed by three independent experts. Average data were provided for a minimum of 1000 cells in each preparation.

The main criteria for describing atypical lymphocytes in ASF have been described previously [27]. For atypical lymphocytes, the main criteria are altered nuclei and/or larger cells with slightly less condensed chromatin and cells with a lower nucleus:cytoplasm ratio. Also, an increased amount of nuclear DNA (hyperdiploid or polyploid nuclei) is often described.

3. Results

3.1. Virus Load

The presence of ASFV DNA in the blood and spleen was confirmed via polymerase chain reaction (PCR) (Figure 1A). We contrasted the nucleotide sequences from the p72-based PCRs with those from representative samples that were previously reported (unpublished data). As predicted, the Dilijan 2011 IMB ASFV clustered within p72 genotype II. It demonstrated 100% nucleotide similarity with all compared ASFVs circulating in the Caucasus regions since 2007.

The content of the ASF virus (Kovsakan1/2018 isolate) in the blood of sick pigs approximately corresponded to 10^5 units of HADU when infected with the Arm07 isolate (Figure 1A). This index is approximately 0.5 decimal logarithms lower than the maximum determined during infection with Arm07 isolate. However, taking into account the fact that it was difficult to determine the maximum titers of the virus (Kovsakan1/2018) in the blood of pigs due to the duration of the disease, it can be assumed that the number of genome copies is approximately the same when infected with Arm07 and Kovsakan1/2018 isolates. There is a discrepancy in the quantitative indicators of the ASF virus between the RT-PCR and HAD data when analyzing the viremia of the Kovkasan1/2018 isolate (a decrease in virus content expressed in HADU is not accompanied by a decrease in copies of the gene encoding p72). This is most likely explained by the reduced HAD activity of all new virus isolates. The number of genome copies of isolates Kovsakan2/2019 and Kovsakan3/2020 was significantly lower and corresponded to 10^2 HADU (Figure 1A). The titer levels of all tested virus isolates in the blood of infected pigs, when analyzed via HAD, were significantly lower than those during infection with the Arm07 isolate (approximately 2.0 decimal logarithms) (Figure 1B). In studied samples, not only the levels of HADU per ml, but also the intensity of hemadsorption were reduced.

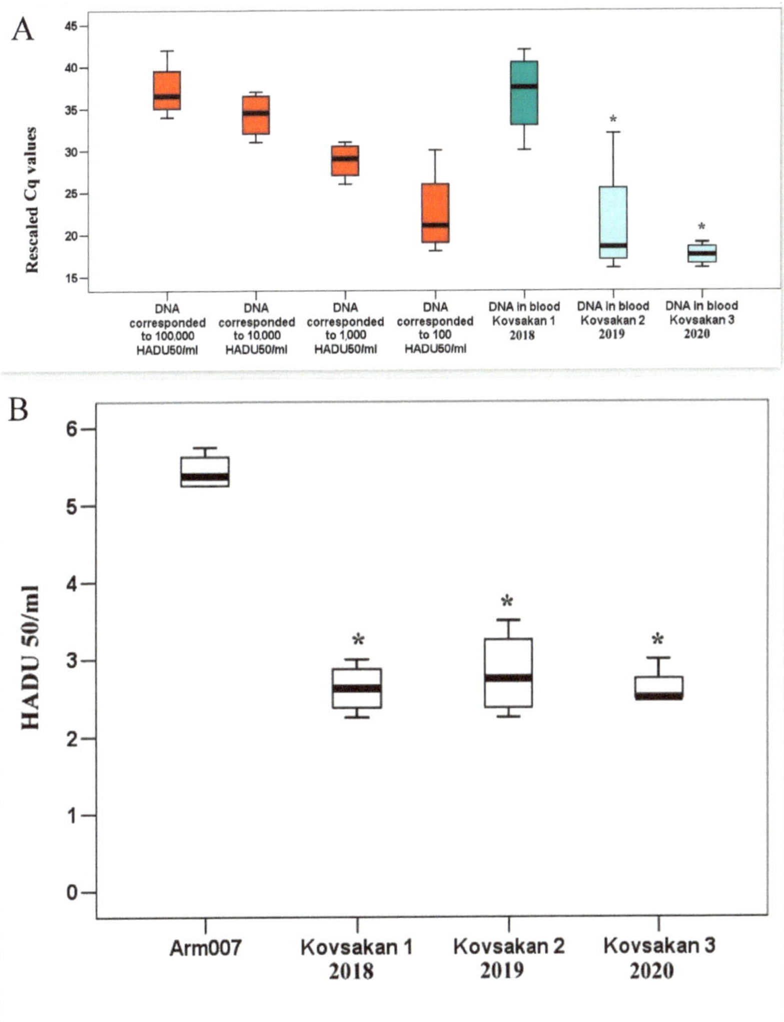

Figure 1. ASFV HADU and DNA levels (rescaled Ct values) in acute and chronic infected pigs' serum. (**A**) ASFV DNA levels p72 (Ct values from real-time PCR) in the blood of infected pigs. * Significant compared to Kovsakan1 ($p < 0.05$) (**B**) HADU levels in the blood of infected pigs. * Significant compared to Arm 007 ($p < 0.05$–0.01).

3.2. Epidemiology

All three outbreaks were identified in the South Caucasus (Kovsakan). The first outbreak (Kovsakan1/2018) was recorded in the late summer of 2018 (August–September); the second outbreak (Kovsakan2/2019), in summer 2019 (July–September); the third outbreak (Kovsakan3/2020), occurred from November 2019 to January 2020.

The sources of ASF disease in all of the three outbreaks were not identified. All cases of the disease were detected in small farms (the number of animals in the farms usually ranged from 2–3 pigs to 15–20). There was a wild boar population in the region, but there were no wild boars or forest areas in the vicinity of the farms. In some of the affected farms (mostly with very few animals), the pigs were kept free-range. Most of the larger farms kept animals in barns without free range.

On the other hand, before the detection of ASF, there was a free exchange of animals and feed between farmers, which explains the duration of the outbreak and the route of infection within the rural community.

3.3. Clinical Manifestation

In the Kovsakan1/2018 cases, the aged sows were affected more often than young piglets (up to 3 months old) [4]. Nineteen of them developed the disease described below (Table 1).

Table 1. Clinical characteristics of naturally attenuated forms of African swine fever.

	Arm07	Kovsakan2018	Kovsakan2019 and Kovsakan2020
	Late Stage	Late Stage	Late Stage
Permanent loss of appetite (%)	100	100	47
Skin hemorrhages	Severe (on ears, abdomen, etc.)	Mild (predominantly on ears)	Mild (predominantly on ears)
Internal bleeding	Severe	Mild	Mild, sometimes absent
Lethality (%)	100	12 out of 19 (63.6%)	4 out of 17 (26.7%) *
Average life expectancy (day)	4–7	25–35 *	48 and more *

* significant compared with Arm07 ($p < 0.05$–$p < 0.01$).

The time of illness was noted from the moment the first symptoms were observed: apathy, sometimes temporary loss of appetite, and sometimes short-term fever. Among the diseased animals, 12 were 8–21 months old, and 7 were piglets aged 2–2.5 months. Weight loss and some apathy were observed in all investigated animals. Arthritis was observed in some pigs (2 animals) and also short-term attacks of fever (14 animals). Fever can be described as moderate and irregular, but is often absent. The disease was characterized by a permanent, but prolonged tendency to worsen. Petechial hemorrhages (sometimes localized only on the ears) occur at a late stage of the disease (1–4 weeks after the onset of the first symptoms), and the intensity of hemorrhages varies greatly in different animals. Clinical disease in animals infected with ASFV Kovsakan1/2018 progressively worsened, with all animals euthanized in extremis by 35–44 days post infection.

The time of illness was noted from the moment the first symptoms were detected: loss of appetite and rise in temperature. Among the investigated animals, 19 were 6–18 months old, and 17 were piglets aged 2–4 months. Clinical manifestations in pigs infected with Kovsakan2/2019 (n = 19) and Kovsakan3/2020 (n = 17) isolates were almost similar but milder. In the latter cases, sometimes clinically detectable symptoms disappeared and reappeared. It was not possible to reliably identify the causes of this phenomenon (recurrence of persistence or cure with a new infection). Some pigs have survived long enough to show signs of recovery, even to the point of gaining weight.

Table 1 represents the average life expectancy in ASFV-infected pigs (days) after the first symptoms arose until the last stage of the disease in both groups. As shown, the average life expectancy increased in a new form of ASF at least three times. In two cases, pigs survived the disease. In this case, ASF symptoms disappeared, and after 3 months, all samples from blood, urine, and stool were free from ASFV (the evaluations were performed via HAD). Other animals were euthanized at the last stage of the disease. In the chronic form of ASF, the temperature curve after the early stage of the disease shows a tendency to decrease down to normal levels, with separate short-term paroxysms. For pigs infected by the chronic strain of ASF, one of the specific characteristics of the disease was a lack or absence of skin hemorrhages, almost total absence of blood in urine and stool.

In investigations on the serum levels of TNF-alpha, pigs infected by all chronic ASFV strains were significantly lower compared with the acute form of the disease (all measurements were performed at a late stage of the disease).

3.4. *Pathology*

3.4.1. Kovsakan1/2018

An autopsy revealed severe damage to internal organs. Severe damage was found in the vast majority of autopsy cases. This manifested presenting with single and limited hemorrhages, hemorrhages in the lungs, intestine, liver, and kidneys. In some cases, bleeding was observed in the abdominal cavity, as well as blood-stained pleural effusion. Splenomegaly with massive hemorrhages was observed in 63.6% of the pigs (12 out of 19).

3.4.2. Kovsakan2/2019 and Kovsakan3/2020

All examined organs were edematous. There were no massive hemorrhages. In the lungs, isolated pinpoint hemorrhages present were. Splenomegaly with massive hemorrhages was observed in 13.3% of the pigs (2 out of 17).

Macroscopically, hemorrhages were detected in some lymph nodes.

3.5. *Histopathology*

Kovsakan1/2018, Kovsakan2/2019, and Kovsakan3/2020

A histological analysis revealed isolated foci of microhemorrhages and diapedesis of erythroid tissue in lymph nodes. In lungs, isolated pinpoint hemorrhages were present.

The histological structure of the kidneys was preserved, with a mild proliferation of mesangial cells (similar to chronic mesangioproliferative glomerulonephritis). The epithelium of the proximal convoluted and straight tubules was preserved with protein-containing urine in the lumen (Figure 2A,B).

The general lobular structure of the liver was preserved, and there was congestion in the vessels of the portal tract and sinusoids, especially in the central part of the lobules and central veins, possibly due to stagnation of blood in the systemic veins, possibly due to heart failure. Portal fields were slightly enlarged. There were activated Kupffer cells, single lymphocytes, and neutrophils (segmented and bend), especially in the central part of the lobules around necrotic hepatocytes. Diapedesis hemorrhages and necrotic changes were detected in all areas (Figure 2A,B).

In the brain tissue, the extravasation of individual cells (mainly erythrocytes and leukocytes) was observed (Figure 2C,D).

There was pronounced venous congestion in the lungs with foci of hemorrhage, interalveolar septa were thickened due to venous congestion and connective tissue growth, and alveoli were mostly empty (Figure 2E,F). Interalveolar septa were thickened via erythrodiapedesis and leukodiapedesis (Figure 2G,H).

Overall, the histopathological findings resemble those of the chronic forms of ASF with mild pathogenesis.

3.6. *Immunopathology*

Cellular Immune Responses Associated with Chronic ASF

It was previously shown that the number of leukocytes in the blood of ASFV-infected pigs significantly decreased at the last stages of acute ASF, and this was also observed in the chronic form of ASF. The total white blood cell count in the Kovsakan1/2018 ASF form was slightly (insignificantly) below the cutoff for the established normal range (Figure 3A), but significantly higher compared with the acute form (Arm007). In Kovsakan2/2019 and particularly in Kovsakan3/2020, the numbers of nucleated cells in blood were significantly higher compared with those in the acute form (Arm007) and healthy pigs; the data of Kovsakan3/2020 isolate correspond to the data of the norm (and Arm007).

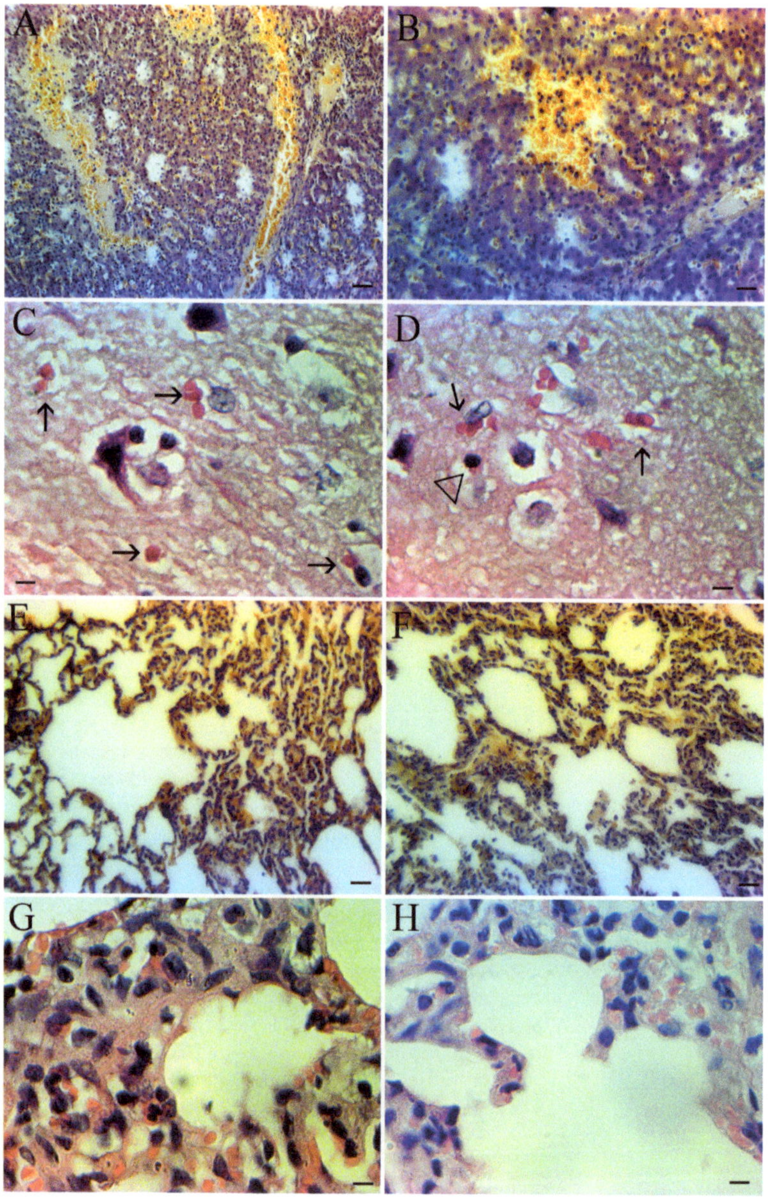

Figure 2. (**A**) Diapedesis and local hemorrhages in the liver at the late stage of the chronic form of ASF. Scale bar 100 µm. (**B**) Local hemorrhage in the liver at the late stage of the chronic form of ASF. Scale bar 50 µm. (**C**) Mild diapedesis in brain tissue. Erythrocytes in brain (arrowed). Scale bar 10 µm. (**D**) Mild diapedesis in brain tissue. Erythrocytes in brain (arrowed), and lymphoid cell (triangle). Scale bar 10 µm. (**E**) Mixed diapedesis in lung tissue at the late stage of the chronic form of ASF Scale bar 50 µm. (**F**) Mixed diapedesis and thickness of the alveolar wall at the late stage of the chronic form of ASF. Scale bar 50 µm. (**G**) Thickening of the interalveolar septa due to erythrodiapedesis and leukodiapedesis. Kovsakan2019. Scale bar 10 µm (**H**) Thickening of the interalveolar septa due to erythrodiapedesis and leukodiapedesis. Kovsakan2020. Scale bar 10 µm.

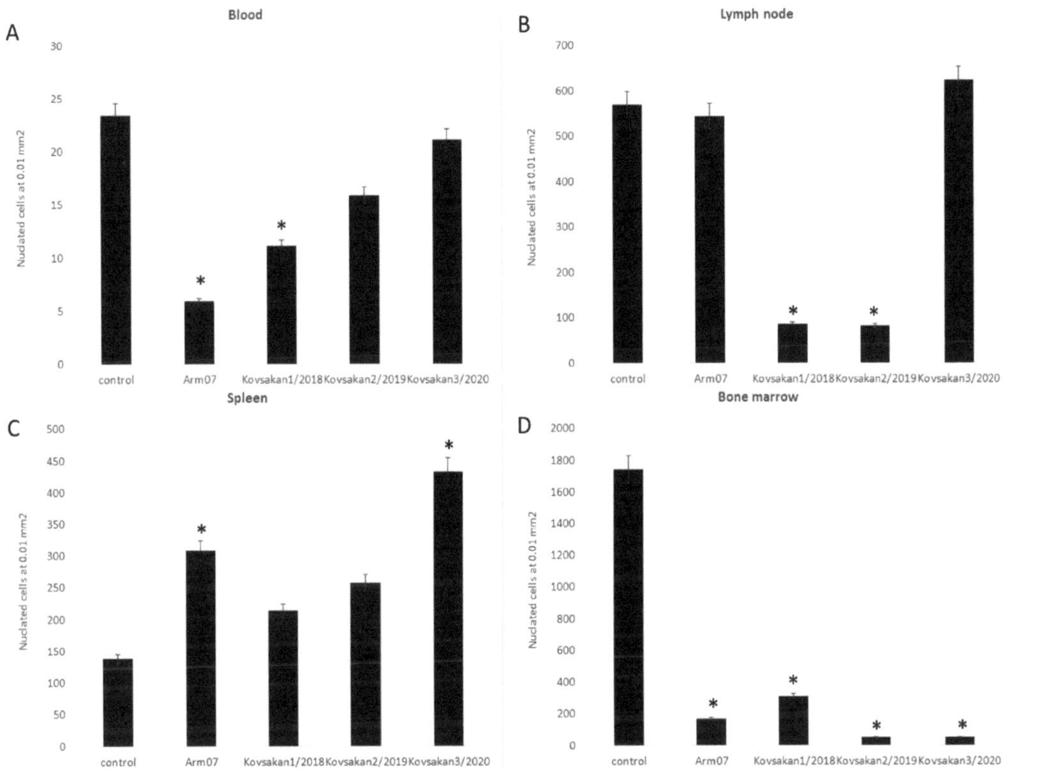

Figure 3. Nucleated cell amount in the blood, lymph nodes, spleen, and bone marrow at late stages of Kovsakan1/2018, Kovsakan2/2019, and Kovsakan3/2020. (**A**) Nucleated cell amount in the blood. (**B**) Nucleated cell amount in lymph nodes. (**C**) Nucleated cell amount in the spleen. (**D**) Nucleated cell amount in the bone marrow. * Significant compared to control (p < 0.05–0.01).

In lymph nodes, the cell counts for Kovsakan1/2018 and Kovsakan2/2019 were significantly lower (Figure 3B) compared with that of the acute form (Arm007). A sphenogram (Figure 3C) in pigs with Kovsakan3/2020 revealed that the number of nucleated cells in blood was significantly higher compared with that of the acute form (Arm007). A myelogram (Figure 3D) reveals that the nucleated cell numbers for Kovsakan2/2019 and Kovsakan3/2020 were significantly lower (Figure 3D) compared with those of the acute form (Arm007).

Marked neutropenia (mature forms) were observed in the late stages of Kovsakan2/2019 and Kovsakan3/2020, but not Kovsakan1/2018 (Table 2). The percentage of lymphoid cells in the blood of pigs with the Kovsakan3/2020 form almost corresponds to the control values; however, at the same time, the content of monocytes sharply decreases.

As follows from Table 2, with Kovsakan1/2018, a more prominent shift is observed to the left in all WBC populations of peripheral blood compared with acute ASF (lymphoid, myeloid populations). This phenomenon occurs in all described forms of the disease, but the most obvious is Kovsakan1/2018. Also, significant amounts of destroyed cells arise. However, the smudge cell numbers at terminal stages of all forms of ASF, except in cases of Kovsakan2/2019 and Kovsakan3/2020, were more prominent.

As follows from Table 3, upon ASF (Arm07) infection, a severe lymphopenia in the BM was detected in the late phase of infection. Isolates Kovsakan1/2018 and Kovsakan2/2020 also showed reductions in lymphocytes, however less prominent. Isolate Kovsakan3/2020 showed a reverse tendency to increase the percentage of lymphoid cells. Very prominent decreases in monocytes and monoblasts occurred in all new isolates of ASFV. The content of nucleated erythroid cells corresponded to the control values and was significantly lower compared to the acute form caused by the Arm007 isolate (Table 3), and Kovsakan3/2020 exhibited a significantly increased amount of macrophages, sometimes with a pathologically activated phenotype.

Table 2. Blood cell populations compared in the late stages of Arm07 and Kovsakan1–3 infections.

Blood		Arm07	Kovsakan1/2018	Kovsakan2/2019	Kovsakan3/2020
Cells	Control	Late Stage	Late Stage		
Erythroblasts	0	3.1 ± 0.3	8 ± 0.9 *	6.2 ± 0.9	4.1 ± 0.5
Lymphoblast	0	2.8 ± 0.3	5 ± 0.9 *	6.3 ± 1.1	2.2 ± 0.1
Lymphocyte	56.0 ± 4.7	44 ± 6.1	24 ± 3.6 *	32.9 ± 4.4	53.7 ± 6.9
Monoblast	0.5 ± 0.1	0.2 ± 0.01	5 ± 0.7 *	1.0 ± 0.3	0.8 ± 0.2
Monocyte	7.7 ± 1.3	7.4 ± 0.8	8 ± 1.1	3.3 ± 0.5	1.5 ± 0.1 *
Myeloid	0	0.1 ± 0.01	2 ± 0.4 *	0.8 ± 0.2	1.3 ± 0.3
Metamielocyte	0.1 ± 0.02	7.3 ± 1.0	12 ± 2.1 **	2.1 ± 0.5	5.8 ± 1.4 **
Band	7.7 ± 1.0	5.6 ± 0.7	13 ± 0.9 *	9.9 ± 1.4	6.8 ± 0.3 *
Segment	23.1 ± 3.3	1.9 ± 0.3	5 ± 0.8 *	16.1 ± 3.4	7.7 ± 1.2 *
Pathol neutrophil	0	2.1 ± 0.5	9 ± 1.1 *	5.6 ± 2.1	2.4 ± 0.6 *
Eosinophil	4.9 ± 2.1	1.9 ± 0.4	2 ± 0.3	2.8 ± 0.8	3.5 ± 0.9
Basophil	0.2 ± 0.1	0.4 ± 0.1	>0.1	>0.1	>0.1
Smudge cells (remnants of cells) without clear cytoplasmic borders	0	22.7 ± 4.7	6 ± 1.2 *	12.9 ± 4.5 *	10.1 ± 3.0 *

* significant compared with Arm07 ($p < 0.05$–$p < 0.01$). ** tendency ($p < 0.1$).

In BM, a sharp decrease in the content of all nuclear cells is noticeable in all forms of ASF. The most pronounced pancytopenia was observed in Kovsakan3/2020 when the BM can be assessed as almost empty. The main feature of the disease caused by isolate Kovsakan3/2020 is a significantly increased number of macrophages, almost all of them with pathologically activated phenotypes. Similar macrophages in infection with other forms of ASF are also found, but much less frequently. In acute ASF, the number of lymphocytes significantly decreased, and the same occurred in Kovsakan1/2018 and 2/2019. The numbers of monocytes and monoblasts, as well as mature neutrophils, decreased in all investigated isolates observed at 3 dpi (Table 3). Compared with those in healthy pigs, immature myeloid cells increased in Kovsakan1/2018 and partially in Kovsakan2/2019, but decreased in Kovsakan3/2020. The number of nucleated erythroid cells significantly decreases with Kovsakan1/2018 infection, is significantly higher than in healthy pigs after infection with Kovsakan2/2019, and approximately corresponds to the norm in Kovsakan3/2020.

As follows from Table 3, as terminal stages of chronic ASF take place, a significant decrease in all forms of elytroid populations in the BM was observed, compared not only with the control but also with acute ASF. Also, BM smears showed a more prominent shift to the left in the myeloid population compared with terminal stages of acute ASF. This also describes the proliferation of mature macrophages in BM tissue with a significant increase in the macrophage number compared with corresponding indices at terminal stages of acute ASF. There were no significant changes in other cell populations.

The general trend of changes in the population composition of spleen cells (Table 4) is reduced to a decrease in the content of early cells (lymphoblasts, monoblasts, myeloid cells) with an increase in the content of lymphocytes.

Table 3. Bone marrow cell populations compared in the late stages of Arm07 and Kovsakan1–3 infections.

Bone Marrow		Arm07	Kovsakan1/2018	Kovsakan2/2019	Kovsakan3/2020
Cells	Control	Late Stage	Late Stage	Late Stage	Late Stage
Nucleated erythroid cells	31.9 ± 4.4	32.1 ± 2.7	7.7 ± 0.9 *	54 ± 9.5 *	28 ± 4.3
Lymphoblasts	6.5 ± 0.8	8.1 ± 1.2	7.1 ± 1.3	2 ± 0.4 *	10 ± 0.5
Lymphocytes	20.6 ± 1.2	14.6 ± 2.2	23.4 ± 3.0 *	26 ± 3.8	41 ± 7.6 *
Monoblasts	4.2 ± 0.5	7 ± 0.9	1.0 ± 0.2	1.0 ± 0.5	2.0 ± 0.4
Monocytes	3.2 ± 0.8	4.1 ± 0.7	2.5 ± 0.7	0.001	0.001
Myeloid cells	2.7 ± 0.6	3 ± 1.0	7.6 ± 1.3 *	5.1 ± 0.7	3.0 ± 0.1
Metamielocytes	6.3 ± 0.9	3 ± 0.5	9.2 ± 1.1	3.0 ± 0.9	3.1 ± 1.1
Band neutrophils	10.6 ± 2.1	7.8 ± 0.9	20.3 ± 2.8 *	0.9 ± 0.3 *	2.8 ± 0.5 *
Segment neutrophils	4.1 ± 0.9	3.8 ± 0.9	1.5 ± 0.2	2.1 ± 0.6	0.9 ± 0.4 *
Pathol neutrophils	0	0.4 ± 0.1	1.0 ± 0.1	0	0
Eosinophils	8.5 ± 0.8	7.1 ± 0.9	6.1 ± 0.6	3.7 ± 0.8	3.9 ± 1.1 *
Basophils	1.4 ± 0.1	2.4 ± 0.2	0.5 ± 0.1 *	0.001	0
Macrophages	0.1 ± 0.04	1.3 ± 0.1	5.1 ± 0.9 *	1.1 ± 0.05	3.9 ± 0.6 *
Atypical lymphocytes	0	4.4 ± 0.3	3.0 ± 0.5	0	0
Smudge cells (remnants of cells) without clear cytoplasmic borders	0	1 ± 0.4	4.1 ± 0.9 *	1.9 ± 0.3	1.1 ± 0.2

* significant compared with Arm07 ($p < 0.05$–$p < 0.01$).

Table 4. Spleen cell populations compared in the late stages of Arm07 and Kovsakan1–3 infections.

Spleen		Arm07	Kovsakan1/2018	Kovsakan2/2019	Kovsakan3/2020
Cells	Control	Late Stage	Late Stage	Late Stage	Late Stage
Nucleated erythroid cells	5.5 ± 1.6	16.5 ± 2.8	5.0 ± 2.0	5.9 ± 1.1	5.8 ± 1.9
Lymphoblasts	5.7 ± 1.3	4.4 ± 1.6	4.1 ± 1.2	4.4 ± 1.0	1.9 ± 0.8 *
Lymphocytes	60.4 ± 6.3	26.1 ± 4.9	44.3 ± 4.8 *	47.3 ± 5.6	82.6 ± 9.1 *
Monoblasts	3.1 ± 0.9	2.4 ± 1.0	0.5 ± 0.1 *	0.5 ± 0.6	0.4 ± 0.3 *
Monocytes	7.0 ± 1.2	2.2 ± 0.3	0.3 ± 0.1 *	0.3 ± 0.1 *	0.2 ± 0.1 *
Myeloid cells	1.0 ± 0.5	10.5 ± 2.6	8.2 ± 2.3	9.2 ± 1.8	1.3 ± 0.4
Metamielocytes	3.1 ± 0.8	4.8 ± 1.6	5.1 ± 1.1	5.1 ± 1.2	1.0 ± 0.1
Band neutrophils	4.3 ± 0.7	7.1 ± 1.8	1.2 ± 0.2 *	1.2 ± 0.5 *	1.3 ± 0.1 *
Segment neutrophils	4.0 ± 0.7	6.5 ± 2.2	0.8 ± 0.2 *	0.8 ± 0.1 *	1.3 ± 0.1 *
Pathol neutrophils	0	0.1 ± 0.1	0.1 ± 0.1	0.1 ± 0.1	0.05 ± 0.1
Eosinophils	4.7 ± 1.0	11.0 ± 2.8	18.1 ± 3.9	3.3 ± 1.0 *	1.3 ± 0.2 *
Basophils	1.2 ± 0.6	1.1 ± 0.3	0.1 ± 0.1	0.0 ± 0.0	0.01 ± 0.01
Macrophages	0.1 ± 0.1	0.2 ± 0.1	0.3 ± 0.1	0.3 ± 0.1	0.8 ± 0.1
Atypical lymphocytes	0	3.0 ± 1.0	0.7 ± 0.2 *	21.6 ± 2.8 *	1.9 ± 0.7
Smudge cells (remnants of cells) without clear cytoplasmic borders	0.1 ± 0.1	4.5 ± 2.2	11.2 ± 3.0	4.4 ± 1.3	5.8 ± 1.7

* significant compared with Arm07 ($p < 0.05$–$p < 0.01$).

During the late stages of the disease, ASF (Arm07) infection in lymph nodes also caused a twofold reduction in lymphocyte counts, which was accompanied by an increase in monocyte counts (lymphopenia) (Table 5). Lymphopenia was most pronounced in Kovsakan1/2018; Kovsakan2/2019 has an intermediate position; and Kovsakan3/2020, according to the content of lymphocytes, corresponded to the norm. According to the content of nucleated erythroid cells, Kovsakan1/2018 and 2/2019 correspond to the pathology caused by Arm, and Kovsakan3/2020 practically corresponds to the norm.

Table 5. Lymph node cell populations compared in the late stages of Arm07 and Kovsakan1–3 infections.

Lymph Node		Arm07	Kovsakan1/2018	Kovsakan2/2019	Kovsakan3/2020
Cells	Control	Late Stage	Late Stage		
Nucleated erythroid cells	1.4 ± 0.1	20.7 ± 3.4	17.9 ± 3.3	17.6 ± 1.7	0.5 ± 0.4 *
Lymphoblasts	1.9 ± 0.4	1.0 ± 0.7	8.1 ± 2.4 *	8.1 ± 1.1 *	1.3 ± 1.0
Lymphocytes	83.0 ± 5.1	38.5 ± 5.9	49.7 ± 3.7 *	60.7 ± 5.2 *	92.4 ± 10.7 *
Monoblasts	1.0 ± 0.2	0.2 ± 0.1	1.4 ± 0.3 *	1.4 ± 0.4 *	0.3 ± 0.1
Monocytes	1.1 ± 0.2	5.0 ± 1.3	0.3 ± 0.1 *	0.3 ± 0.1 *	0.3 ± 0.1 *
Myeloid cells	4.3 ± 1.1	18.5 ± 3.4	3.1 ± 0.4 *	3.1 ± 0.5 *	0.0
Metamielocytes	2.5 ± 1.0	3.7 ± 1.5	5.1 ± 0.9	0.1 ± 0.1 *	0.0
Band neutrophils	1.2 ± 0.5	3.9 ± 2.6	0.1 ± 0.1 *	0.2 ± 0.1 *	0.1 ± 0.1 *
Segment neutrophils	0.8 ± 0.4	1.2 ± 0.8	0.1 ± 0.1 *	0.1 ± 0.1 *	0.9 ± 0.2
Pathol neutrophils	0	0.1 ± 0.1	0.1 ± 0.01	0.1 ± 0.1	0.0
Eosinophils	2.2 ± 0.6	2.0 ± 0.9	1.7 ± 0.4	1.7 ± 0.4	0.3 ± 0.1 *
Basophils	0.6 ± 0.1	0.2 ± 0.1	0	0.0	0.1 ± 0.1
Macrophages	0	0.1 ± 0.1	1.0 ± 0.2	1.0 ± 0.2	0.7 ± 0.2
Atypical lymphocytes	0	1.0 ± 0.5	0.2 ± 0.1	5.7 ± 0.8 *	3.1 ± 1.1 *
Smudge cells (remnants of cells) without clear cytoplasmic borders	0.1 ± 0.1	2.1 ± 0.4	11.2 ± 4.4 *	8.1 ± 1.1 *	1.3 ± 0.6

* significant compared with Arm07 ($p < 0.05$–$p < 0.01$).

In Figure 4, we present the histopathology of immune organs with the Kovsakan3/2020 form of ASFV. In the lymph nodes, with preserved structure and cellularity, there is a slight noticeable erythrocyte infiltration (Figure 4A,B), a diapedesis (Figure 4C); however, usually, hemorrhages are not described (Figure 4D). Spleen follicles, either intact or demonstrated, observed a local reduction in nucleated cells (Figure 4E–H). In bone marrow, sever pancytopenia was detected, with infiltration of macrophages, sometimes with a hyperactive phenotype (Figure 4I,J). Some macrophages, in addition to vacuolization, were sharply increased in size (Figure 4I,K). A population of early unidentifiable cells was also noted (Figure 4L). Erythroblastic islets were preserved, but the number of erythroblasts in them was minimal (Figure 4L).

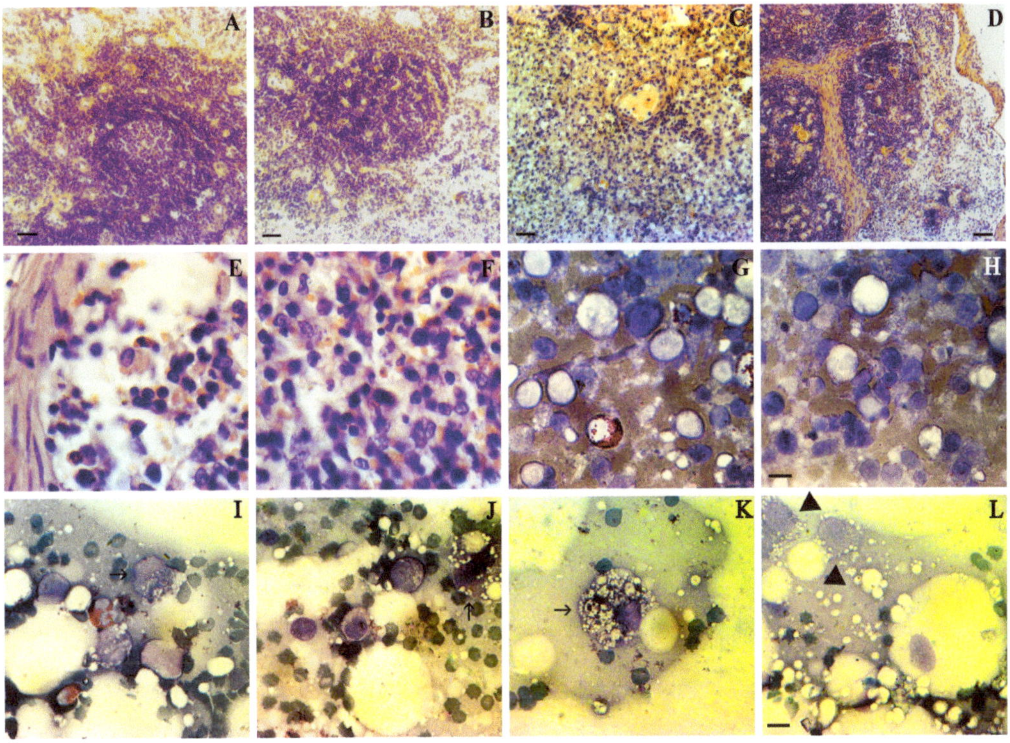

Figure 4. Pathology of immune organs in the late stages of the Kovsakan3/2020 form of ASFV. (**A**) Lymph node. Preserved structure and cellularity. Small but multiple foci of erythrocyte infiltration. Staining by hematoxylin eosin. Scale bar 50 µm. (**B**) Lymph node. Preserved structure and cellularity. Small but multiple foci of erythrocyte infiltration. Staining by hematoxylin eosin. Scale bar 50 µm. (**C**) Lymph node. Severe diapedesis. Staining by hematoxylin eosin. Scale bar 50 µm. (**D**) Lymph node. Preserved structure of the lymph node and absence of hemorrhages. Staining by hematoxylin eosin. Scale bar 100 µm. (**E**) Spleen, follicle local reduction in nucleated cells. Staining by hematoxylin eosin. The scale bar for E-H is 10 µm. (**F**) Spleen, almost intact follicle with mild diapedesis. Staining by hematoxylin eosin. (**G**) Spleen, reduction in nucleated cells. Staining by Giemsa. (**H**) Spleen, lymphocytosis. Staining by Giemsa. (**I**) Bone marrow smear. Pancytopenia. Hyperactive macrophage (arrowed). Staining by Giemsa. The scale bar for I-L is 10 µm. (**J**) Bone marrow smear. Pancytopenia. Hyperactive macrophage (arrowed). Myeloid cells. Staining by Giemsa. (**K**) Bone marrow smear. Pancytopenia. Hyperactive giant macrophage (arrowed). Staining by Giemsa. (**L**) Early blast cells with vacuolized cytoplasm (triangles), erythroblastic islet with minimal erythroblasts. Staining by Giemsa.

4. Discussion

In Eurasia, the emergence of several variants of the ASF virus has been shown to occur within a short period of time, and the virus has been shown to differ in virulence [8]. Sometimes, the change in virulence of a virus can be explained as a result of major genetic changes [28], and sometimes, the explanation of low virulence requires further investigation [29]. Three isolates of the ASF virus with moderate pathological manifestations and reduced mortality have been identified in the South Caucasus in 2018–2020, differing from each other in different degrees of attenuation.

The new ASF isolates cause an unusual clinical picture—described isolates differ from the usual subacute form by a reduced ability of hemadsorption and a reduced frequency and intensity of hemorrhages. A less pronounced lesion of the blood coagulation system is also shown, compared with acute ASF [30]. The causes of subacute and chronic forms may be penetration into the primary ASF zone of spread of new types of viruses with lower virulence, or a change in the virus with the appearance of new, less virulent mutant strains. So, along with the primary highly virulent isolate identified in the Republic of Armenia in 2007, we can conclude that at the end of 2011, a moderately virulent isolate for domestic pigs appeared [4].

The main aspect of the attenuation of the described isolates of the ASF virus is the long-term carrying of the virus with viremia [31]. At the same time, the level of the virus in the blood and tissue of infected animals is significantly lower than that when infected with a highly pathogenic strain (Arm07). For the attenuation of the virus, either increased sensitivity to the antiviral defense of the host is required, or a decrease in the replication activity of the virus, or changes in the virus genome [8].

Generally, a lower virulence of ASFV is associated with genomic changes present in attenuated ASFV strains and differences in the immune response of infected animals [32]; a combination of both factors is also possible. The decrease in virulence can be explained by the fact that pigs may have been previously exposed to ASF virus and survived. On the one hand, this assumption is supported by the fact that several outbreaks of ASF were recorded in a relatively small region over several years. However, most of the affected pigs were between several weeks and several months old and would not have had enough time to become reinfected with ASF. Spontaneous abortions in pregnant pigs were not observed on the farms investigated. Viremia was not detected in healthy pigs from neighboring farms or in the majority of healthy pigs (either via rtPSR or HAD). These data reduce (but do not exclude) the likelihood of reinfection with ASF virus.

The decrease in virus virulence can also be explained by the transfer of antibodies to the ASF virus in the mothers' milk. It is known that when antibodies are transferred via milk, virus titers into the blood of piglets, the duration of viremia decreases, and the clinical manifestation becomes much less pronounced [33]. This is explained by the protective effect of passively acquired antibodies [34]. Therefore, this aspect cannot be ignored, especially as some sows were more severely affected by ASF than the piglets.

From our point of view, the most likely explanation is a change in the genome of the virus, due to the evolution characteristic of viruses. From studying the evolution of viruses, we know that genomic changes will almost inevitably reduce the pathogenicity of viruses given enough time [35]. Although DNA viruses have a significantly lower rate of change in the genome, their variability far exceeds that of cellular organisms. Draft sequences of some genes shown modest mutations compared to ASFV Georgia 2007 (Figure S1 MGF 360-2L gene; Figure S2 G1340L gene; Figure S3 MGF360-11L gene; Figure S4 MGF360-10L gene) [36,37] The emergence of new, less virulent ASF virus mutants undoubtedly has an important role in the survival of the virus, increasing the possibilities of its transmission. It is commonly recognized that DNA viruses frequently cause chronic, invisible, long-lasting illnesses [38]. The changes in the clinical manifestations of the unusual ASF form described in our article support this view. The changes we detected in the clinical manifestations of the disease confirm this point of view.

One of the most impressive manifestations of the pathology of partially attenuated ASFV isolates is a very pronounced pancytopenia accompanied by the devastation of all hematopoietic organs, primarily the bone marrow. Numerous viruses, both human and animal, can result in severe bone marrow depression. These viruses include the human immunodeficiency virus, dengue virus, parvovirus B19, Epstein–Barr virus, cytomegalovirus, and feline leukemia and panleukopenia viruses [39]. The authors state that pancytopenia's clinical manifestation can result from a number of pathogenic pathways. A number of viruses have the potential to directly harm hematopoietic progenitors, expose the bone marrow to immune-mediated destruction, or cause the stromal microenvironment's vital nurturing elements to be lost. We tend to consider the second postulate as the most probable. Previously, we identified the occurrence of hemophagocytic lymphohistiocytosis in the acute form of ASF and provided its criteria in pigs [40], which allows us to assume similar pathogenesis and lethal forms when infected with partially attenuated ASF virus isolates.

It should also be noted that in the blood and lymph nodes, the total white blood cell content does not differ from that of healthy pigs (unlike in the acute form of ASF and Kovsakan2018 infection); therefore, compensatory mechanisms of bone marrow depletion should not be excluded. This is also supported by splenic hyperplasia with the possible development of compensatory extramedullary hematopoiesis [41].

Kovsakan2/2019 and Kovsakan3/2020 isolates are very close to each other in terms of clinical manifestations and the amount of ASFV in the blood of the pigs but differ in terms of pathological and laboratory parameters. We do not rule out that these isolates are varieties of the same virus or that the above differences are a reflection of different stages of the disease. The latter is quite likely, since the disease is characterized by a very blurred nature, and it is difficult to determine whether the animals are in the stage of manifestation or of convalescence of the disease.

5. Conclusions

The identified new virus isolates resemble a strongly prolonged form of persistent ASF with reduced mortality. Therefore, the most important characteristics of the natural attenuation of the ASF virus should be considered a reduction in the infectious titers of the virus in the blood and a change in the pathogenesis of the disease.

The direct mechanisms of virus attenuation have not been reliably identified; however, the pathogenetic reasons are a pronounced decrease in the manifestations of the symptoms of the disease, which allows the bodies of pigs to develop compensatory mechanisms.

The undoubted reason for attenuation from the perspective of the virus is a pronounced prolongation of the viremia.

Supplementary Materials: The following supporting information can be downloaded at: https://www.mdpi.com/article/10.3390/pathogens13020130/s1, Figure S1: ASFV MGF 360-2L gene (Kovsakan 3 2020) sequence compared to Georgia 2007; Figure S2: ASFV G1340L gene (Kovsakan 3 2020) sequence compared to Georgia 2007; Figure S3: ASFV MGF360-11L gene (Kovsakan 3 2020) sequence compared to Georgia 2007; Figure S4: ASFV MGF360-10L gene (Kovsakan 3 2020) sequence compared to Georgia 2007.

Author Contributions: Conceptualization, Z.K., H.A. (Hranush Avagyan) and N.K.; methodology, A.S., M.S., H.V., N.K., L.H., L.A., A.A. and E.K; software, Z.K. and H.A. (Hranush Avagyan); validation, Z.K., H.A. (Hranush Avagyan) and N.K.; formal analysis, Z.K., N.K. and A.S.; investigation, Z.K., H.A. (Hranush Avagyan), S.H., B.B., H.A. (Hranush Arzumanyan), A.P., N.B., A.S., M.S., H.V., T.V., N.K., L.H., L.A., A.A., E.K. and Z.S.; resources, Z.K. and H.A. (Hranush Avagyan); data curation, Z.K., H.A. (Hranush Avagyan) and N.K.; writing—original draft preparation, Z.K.; writing—review and editing, Z.K. and A.P.; visualization, L.H., L.A., A.A. and E.K.; supervision, Z.K. and H.A. (Hranush Avagyan); project administration, Z.K. and H.A. (Hranush Avagyan); funding acquisition, Z.K. and H.A. (Hranush Avagyan). All authors have read and agreed to the published version of the manuscript.

Funding: Under the framework of research project NO 21APP-1F007, the Science Committee of RA provided partial support for the work.

Institutional Review Board Statement: The Institute of Molecular Biology NAS RA's Ethics Committee examined and approved the experiments (IRB 00004079, 2013; Protocol N5 from 25 May 2018). The animal study protocol was approved by the Ethics Committee of the Institute of Molecular Biology NAS RA (IRB 06042021/1, 2021).

Informed Consent Statement: Not applicable.

Data Availability Statement: Data available in a publicly accessible repository.

Conflicts of Interest: The authors declare no competing interests.

References

1. Pikalo, J.; Zani, L.; Hühr, J.; Beer, M.; Blome, S. Pathogenesis of African swine fever in domestic pigs and European wild boar—Lessons learned from recent animal trials. *Virus Res.* **2019**, *271*, 197614. [CrossRef] [PubMed]
2. Rowlands, R.J.; Michaud, V.; Heath, L.; Hutchings, G.; Oura, C.; Vosloo, W.; Dwarka, R.; Onashvili, T.; Albina, E.; Dixon, L.K. African swine fever virus isolate, Georgia, 2007. *Emerg. Infect. Dis.* **2008**, *14*, 1870–1874. [CrossRef] [PubMed]
3. Cwynar, P.; Stojkov, J.; Wlazlak, K. African Swine Fever Status in Europe. *Viruses* **2019**, *11*, 310. [CrossRef] [PubMed]
4. Sargsyan, M.A.; Voskanyan, H.E.; Karalova, E.M.; Hakobyan, L.H.; Karalyan, Z.A. Third wave of African swine fever infection in Armenia: Virus demonstrates the reduction of pathogenicity. *Vet. World* **2018**, *11*, 5–9. [CrossRef] [PubMed]
5. Hess, W.R.; Endris, R.G.; Haslett, T.M.; Monahan, M.J.; McCoy, J.P. Potential arthropod vectors of African swine fever virus in North America and the Caribbean basin. *Vet. Parasitol.* **1987**, *26*, 145–155. [CrossRef]
6. Kalenzi Atuhaire, D.; Ochwo, S.; Afayoa, M.; Norbert Mwiine, F.; Kokas, I.; Arinaitwe, E.; Ademun-Okurut, R.A.; Boniface Okuni, J.; Nanteza, A.; Ayebazibwe, C.; et al. Epidemiological Overview of African Swine Fever in Uganda (2001–2012). *J. Vet. Med.* **2013**, *2013*, 949638. [CrossRef] [PubMed]
7. Thomas, L.F.; Bishop, R.P.; Onzere, C.; Mcintosh, M.T.; Lemire, K.A.; de Glanville, W.A.; Cook, E.A.; Fèvre, E.M. Evidence for the presence of African swine fever virus in an endemic region of Western Kenya in the absence of any reported outbreak. *BMC Vet. Res.* **2016**, *12*, 192. [CrossRef]
8. Sehl-Ewert, J.; Deutschmann, P.; Breithaupt, A.; Blome, S. Pathology of African Swine Fever in Wild Boar Carcasses Naturally Infected with German Virus Variants. *Pathogens* **2022**, *11*, 1386. [CrossRef]
9. Martínez Avilés, M.; Bosch, J.; Ivorra, B.; Ramos, Á.M.; Ito, S.; Barasona, J.Á.; Sánchez-Vizcaíno, J.M. Epidemiological impacts of attenuated African swine fever virus circulating in wild boar populations. *Res. Vet. Sci.* **2023**, *162*, 104964. [CrossRef]
10. Sun, E.; Zhang, Z.; Wang, Z.; He, X.; Zhang, X.; Wang, L.; Wang, W.; Huang, L.; Xi, F.; Huangfu, H.; et al. Emergence and prevalence of naturally occurring lower virulent African swine fever viruses in domestic pigs in China in 2020. *Sci. China Life Sci.* **2021**, *64*, 752–765. [CrossRef]
11. Sun, E.; Huang, L.; Zhang, X.; Zhang, J.; Shen, D.; Zhang, Z.; Wang, Z.; Huo, H.; Wang, W.; Huangfu, H.; et al. Genotype I African swine fever viruses emerged in domestic pigs in China and caused chronic infection. *Emerg. Microbes Infect.* **2021**, *10*, 2183–2193. [CrossRef] [PubMed]
12. Gallardo, C.; Soler, A.; Rodze, I.; Nieto, R.; Cano-Gómez, C.; Fernandez-Pinero, J.; Arias, M. Attenuated and non-haemadsorbing (non-HAD) genotype II African swine fever virus (ASFV) isolated in Europe, Latvia 2017. *Transbound. Emerg. Dis.* **2019**, *66*, 1399–1404. [CrossRef]
13. Souto, R.; Mutowembwa, P.; van Heerden, J.; Fosgate, G.T.; Heath, L.; Vosloo, W. Vaccine Potential of Two Previously Uncharacterized African Swine Fever Virus Isolates from Southern Africa and Heterologous Cross Protection of an Avirulent European Isolate. *Transbound. Emerg. Dis.* **2016**, *63*, 224–231. [CrossRef]
14. Zani, L.; Forth, J.H.; Forth, L.; Nurmoja, I.; Leidenberger, S.; Henke, J.; Carlson, J.; Breidenstein, C.; Viltrop, A.; Höper, D.; et al. Deletion at the 5′-end of Estonian ASFV strains associated with an attenuated phenotype. *Sci. Rep.* **2018**, *8*, 6510. [CrossRef]
15. Wang, Z.; Ai, Q.; Huang, S.; Ou, Y.; Gao, Y.; Tong, T.; Fan, H. Immune Escape Mechanism and Vaccine Research Progress of African Swine Fever Virus. *Vaccines* **2022**, *10*, 344. [CrossRef] [PubMed]
16. Boinas, F.S.; Hutchings, G.H.; Dixon, L.K.; Wilkinson, P.J. Characterization of pathogenic and non-pathogenic African swine fever virus isolates from Ornithodoros erraticus inhabiting pig premises in Portugal. *J. Gen. Virol.* **2004**, *85 Pt 8*, 2177–2187. [CrossRef]
17. Urbano, A.C.; Ferreira, F. African swine fever control and prevention: An update on vaccine development. *Emerg. Microbes Infect.* **2022**, *11*, 2021–2033. [CrossRef]
18. Ginzinger, D.G. Gene quantification using real-time quantitative PCR: An emerging technology hits the mainstream. *Exp. Hematol.* **2002**, *30*, 503–512. [CrossRef] [PubMed]
19. Yin, J.L.; Shackel, N.A.; Zekry, A.; McGuinness, P.H.; Richards, C.; Putten, K.V.; McCaughan, G.W.; Eris, J.M.; Bishop, G.A. Real-time reverse transcriptase-polymerase chain reaction (RT-PCR) for measurement of cytokine and growth factor mRNA expression with fluorogenic probes or SYBR Green I. *Immunol. Cell Biol.* **2001**, *79*, 213–221. [CrossRef]
20. Available online: https://eu.idtdna.com/pagesas (accessed on 5 February 2019).
21. Avagyan, H.R.; Hakobyan, S.A.; Poghosyan, A.A.; Bayramyan, N.V.; Arzumanyan, H.H.; Abroyan, L.O.; Avetisyan, A.S.; Hakobyan, L.A.; Karalova, E.M.; Karalyan, Z.A. African Swine Fever Virus Manipulates the Cell Cycle of G0-Infected Cells to Access Cellular Nucleotides. *Viruses* **2022**, *14*, 1593. [CrossRef]
22. Cackett, G.; Portugal, R.; Matelska, D.; Dixon, L.; Werner, F. African Swine Fever Virus and Host Response: Transcriptome Profiling of the Georgia 2007/1 Strain and Porcine Macrophages. *J. Virol.* **2022**, *96*, e0193921. [CrossRef]
23. Enjuanes, L.; Carrascosa, A.L.; Moreno, M.A.; Viñuela, E. Titration of African swine fever (ASF) virus. *J. Gen. Virol.* **1976**, *32*, 471–477. [CrossRef] [PubMed]
24. Zakaryan, H.; Cholakyans, V.; Simonyan, L.; Misakyan, A.; Karalova, E.; Chavushyan, A.; Karalyan, Z. A study of lymphoid organs and serum proinflammatory cytokines in pigs infected with African swine fever virus genotype II. *Arch. Virol.* **2015**, *160*, 1407–1414. [CrossRef] [PubMed]

25. Karalova, E.M.; Sargsyan, K.hV.; Hampikian, G.K.; Voskanyan, H.E.; Abroyan, L.O.; Avetisyan, A.S.; Hakobyan, L.A.; Arzumanyan, H.H.; Zakaryan, H.S.; Karalyan, Z.A. Phenotypic and cytologic studies of lymphoid cells and monocytes in primary culture of porcine bone marrow during infection of African swine fever virus. *Vitr. Cell. Dev. Biol. Anim.* **2011**, *47*, 200–204. [CrossRef] [PubMed]
26. Harvey, J.W. *Atlas of Veterinary Hematology: Blood and Bone Marrow of Domestic Animals*; W.O. Saunders Company: Philadelphia, PA, USA, 2001; p. 224.
27. Karalyan, Z.A.; Ter-Pogossyan, Z.R.; Abroyan, L.O.; Hakobyan, L.H.; Avetisyan, A.S.; Yu, K.N.; Karalova, E.M. Characterization of the atypical lymphocytes in African swine fever. *Vet. World* **2016**, *9*, 792–800. [CrossRef] [PubMed]
28. Ito, S.; Bosch, J.; Martínez-Avilés, M.; Sánchez-Vizcaíno, J.M. The Evolution of African Swine Fever in China: A Global Threat? *Front. Vet. Sci.* **2022**, *9*, 828498. [CrossRef]
29. Gallardo, C.; Nurmoja, I.; Soler, A.; Delicado, V.; Simón, A.; Martin, E.; Perez, C.; Nieto, R.; Arias, M. Evolution in Europe of African swine fever genotype II viruses from highly to moderately virulent. *Vet. Microbiol.* **2018**, *219*, 70–79.
30. Zakaryan, H.; Karalova, E.; Voskanyan, H.; Ter-Pogossyan, Z.; Nersisyan, N.; Hakobyan, A.; Saroyan, D.; Karalyan, Z. Evaluation of hemostaseological status of pigs experimentally infected with African swine fever virus. *Vet. Microbiol.* **2014**, *174*, 223–228. [CrossRef] [PubMed]
31. Zhenzhong, W.; Chuanxiang, Q.; Shengqiang, G.; Jinming, L.; Yongxin, H.; Xiaoyue, Z.; Yan, L.; Naijun, H.; Xiaodong, W.; Zhiliang, W.; et al. Genetic variation and evolution of attenuated African swine fever virus strain isolated in the field: A review. *Virus Res.* **2022**, *319*, 198874. [CrossRef]
32. Ayanwale, A.; Trapp, S.; Guabiraba, R.; Caballero, I.; Roesch, F. New Insights in the Interplay Between African Swine Fever Virus and Innate Immunity and Its Impact on Viral Pathogenicity. *Front. Microbiol.* **2022**, *13*, 958307. [CrossRef]
33. Atyeo, C.; Alter, G. The multifaceted roles of breast milk antibodies. *Cell* **2021**, *184*, 1486–1499. [CrossRef] [PubMed]
34. Schlafer, D.H.; McVicar, J.W.; Mebus, C.A. African swine fever convalescent sows: Subsequent pregnancy and the effect of colostral antibody on challenge inoculation of their pigs. *Am. J. Vet. Res.* **1984**, *45*, 1361–1366. [PubMed]
35. Armengaud, J.; Delaunay-Moisan, A.; Thuret, J.Y.; van Anken, E.; Acosta-Alvear, D.; Aragón, T.; Arias, C.; Blondel, M.; Braakman, I.; Collet, J.F.; et al. The importance of naturally attenuated SARS-CoV-2in the fight against COVID-19. *Environ. Microbiol.* **2020**, *22*, 1997–2000. [CrossRef] [PubMed]
36. Warr, A.; Newman, C.; Craig, N.; Vendelė, I.; Pilare, R.; Cariazo Cruz, L.; Barangan, T.G.; Morales, R.G.; Opriessnig, T.; Venturina, V.M.; et al. No part gets left behind: Tiled nanopore sequencing of whole ASFV genomes stitched together using Lilo. *bioRxiv* **2021**. [CrossRef]
37. Multiple Sequence Alignment (MSA). Available online: https://www.ebi.ac.uk/Tools/msa/ (accessed on 15 January 2023).
38. Domingo, E.; Verdaguer, N.; Ochoa, W.F.; Ruiz-Jarabo, C.M.; Sevilla, N.; Baranowski, E.; Mateu, M.G.; Fita, I. Biochemical and structural studies with neutralizing antibodies raised against foot-and-mouth disease virus. *Virus Res.* **1999**, *62*, 169–175. [CrossRef] [PubMed]
39. Rosenfeld, S.J.; Young, N.S. Viruses and bone marrow failure. *Blood Rev.* **1991**, *5*, 71–77. [CrossRef]
40. Karalyan, Z.R.; Ter-Pogossyan, Z.R.; Karalyan, N.Y.; Semerjyan, Z.B.; Tatoyan, M.R.; Karapetyan, S.A.; Karalova, E.M. Hemophagocytic lymphohistiocytosis in acute African swine fever clinic. *Vet. Immunol. Immunopathol.* **2017**, *187*, 64–68. [CrossRef]
41. Li, Q.X.; Fan, H. Bone marrow depletion by 89Sr complements preleukemic defect in a long terminal repeat variant of Moloney murine leukemia virus. *J. Virol.* **1991**, *65*, 4442–4448. [CrossRef]

pathogens

MDPI

Article

Pathology of African Swine Fever in Wild Boar Carcasses Naturally Infected with German Virus Variants

Julia Sehl-Ewert [1,*], Paul Deutschmann [2], Angele Breithaupt [1] and Sandra Blome [2]

1 Department of Experimental Animal Facilities and Biorisk Management, Friedrich-Loeffler-Institut, Südufer 10, 17493 Greifswald-Insel Riems, Germany
2 Institute of Diagnostic Virology, Friedrich-Loeffler-Institut, Südufer 10, 17493 Greifswald-Insel Riems, Germany
* Correspondence: julia.sehl-ewert@fli.de

Abstract: In 2020, African swine fever (ASF) was first identified in German wild boar, reaching a case number of about 4400 to date. Upon experimental infection, pathology is well documented; however, data on field infections are scarce in domestic pigs and not available from wild boar, respectively. Although the ASF viral genome is considered exceptionally stable, a total of five lineages with 10 distinct virus variants of genotype II have emerged in Eastern Germany. To investigate the pathology in naturally infected wild boar and to evaluate virus variants II, III and IV for their virulence, wild boar carcasses were obtained from three different outbreak areas. The carcasses underwent virological and pathomorphological investigation. The animals revealed characteristic ASF lesions of the highest severity accompanied by bacterial infections in several cases. In particular, wild boar infected with variant IV from Spree-Neiße (SN) district showed lower viral genome loads and total viral antigen scores, but simultaneously tended to reveal more chronic lesions. Our findings indicate a protracted course of the disease at least after infection with variant IV, but need confirmation under standardized experimental conditions. There is a strong need to monitor differences in the virulence among variants to identify potential attenuation that might complicate diagnosis. In addition, veterinarians, hunters and farmers need to be made aware of less acute courses of ASF to consider this as an important differential to chronic classical swine fever.

Keywords: ASFV; pathology; Germany; virus variant; wild boar; natural infection

Citation: Sehl-Ewert, J.; Deutschmann, P.; Breithaupt, A.; Blome, S. Pathology of African Swine Fever in Wild Boar Carcasses Naturally Infected with German Virus Variants. *Pathogens* **2022**, *11*, 1386. https://doi.org/10.3390/pathogens11111386

Academic Editors: Lauro Velazquez-Salinas, Gisselle N. Medina and Elizabeth Ramirez-Medina

Received: 27 September 2022
Accepted: 17 November 2022
Published: 20 November 2022

1. Introduction

Since its first occurrence in Georgia in 2007, African swine fever (ASF) has continuously spread from the Trans-Caucasian region to Russia, and, in 2014, further to countries of Europe [1]. In September 2020, the disease was confirmed for the first time in a wild boar found in the Spree-Neiße (SN) district in Eastern Germany close to the German–Polish border [2]. To date, more than 4400 cases in German wild boar in the Eastern federal states Brandenburg, Saxony and Mecklenburg-Western Pomerania as well as seven outbreaks in domestic pig holdings located in Brandenburg, Mecklenburg-Western Pomerania, Baden-Wurttemberg and Lower Saxony have been officially identified (https://tsis.fli.de/Reports/Info.aspx, accessed on 9 September 2022).

African swine fever, which is caused by the large, enveloped, double-stranded DNA African swine fever virus (ASFV), can occur as acute, subacute, chronic and subclinical disease courses depending on the virulence of the virus strain as well as on the age and immunological background of the animals [3]. In European countries except Sardinia, highly virulent virus strains of genotype II are prevalent in domestic and wild pigs typically causing acute-lethal disease similar to a hemorrhagic fever [4–6]. Genotype II strains were also identified in the German federal states of Brandenburg and Saxony including the outbreak areas Märkisch-Oderland (MOL), Oder-Spree (LOS), Spree-Neiße (SN) and

Görlitz, in which, surprisingly, five lineages (I-V) including a total of ten viral variants (I, II, II.1, III, III.1, IV, IV.1, IV.2, IV.3, V) have emerged due to single nucleotide variations, insertions and deletions affecting different genes including five multigene families [7]. More specifically, variants III and IV comprise genetic variations in four multigene family (MGF) genes MGF360-10L, MGF360-15R, MGF100-3L and MGF505-4R while variant II shows variation only in the A240L gene coding for the ASFV thymidylate kinase. Whereas the functions of these genes are largely unknown, ASFV MGF360 and MGF505 have been associated with the virulence and pathogenicity of the virus [8,9]. Geographic mapping showed that variant II was predominantly spread in the outbreak area LOS, variant III in MOL and variant IV in the southern part of the outbreak area SN as well as in the federal state of Saxony.

To date, macroscopic pathological records of varying depths of detail largely exist only for experimentally ASFV-infected domestic pigs [10–16] and less frequently for wild boar [4,17–20], which is mainly due to the limited access to wild boar and the associated difficulties to keep them under experimental conditions. Moreover, histopathological data obtained from animal experiments are much less available, but gained importance in the last few years [16,18,21]. Very recently, the first three reports were published concerning naturally ASFV-infected domestic pigs from an outbreak in Vietnam, reporting on the clinical and pathological findings of succumbing and surviving pigs [21,22] and describing ASF-associated age-related lesions [23]. In contrast, descriptions of pathological findings of wild boar that succumb to infection under field conditions are completely missing although this animal species is of great relevance in the maintenance and spread of ASFV in Europe. Hence, the diversity and dimensions of ASFV-associated lesions in the field are only very sparsely represented urging more thorough investigations.

Based on this, we aimed to perform pathological examination of wild boar carcasses infected with ASFV to gain more profound knowledge of the pathology of the animals succumbing to ASF under natural conditions. We took the opportunity to analyze whether three different variations of the emerging virus variants in Germany may have an impact on the virulence of ASFV and the severity and duration of the disease. For this purpose, detailed pathological and molecular virological investigations were performed on wild boar carcasses infected with variants II, III and IV found in LOS, MOL and SN, respectively.

2. Material and Methods

2.1. Study Design

In accordance with the Animal Disease Crisis Unit of the federal states of Brandenburg and Saxony, sixteen wild boar carcasses were obtained from different outbreak areas (n = 7 from Landkreis Oder-Spree (LOS), n = 5 from Märkisch-Oderland (MOL), n = 4 from Spree-Neiße (SN)) between February and March 2021 where ASF virus variants II, III and IV have emerged as published previously [7]. Following legal requirements, the carcasses were tested positive for ASFV by the federal state laboratories of Brandenburg and Saxony. ASF diagnosis was confirmed by the national reference laboratory for ASF. The carcasses were transported to the Friedrich-Loeffler-Institut in compliance with national animal disease and hygiene regulations. The wild boar carcasses were examined in pathological and virological detail. Bacteriologic investigations of secondary bacterial infections were not performed for biosecurity reasons. Details on the cadaver material including location of origin, detection of virus variant, age, sex, weight and preservation status are given in Table 1.

Table 1. Summary presentation of examined wild boar carcasses from three different German outbreak areas LOS, MOL and SN.

No	Origin	Virus Variant	Age (Year)	Sex	Weight (kg)	Stages of Decomposition *	Found Dead/Shot	Anomalies/Comments
1	LOS	II	<1	female	10	fresh stage	dead	Brachygnathia superior
2	LOS	II	<1	female	30	fresh stage	dead	/
3	LOS	II	>2	female	62	bloat stage	dead	/
4	LOS	II	<1	female	40	bloat stage	dead	/
5	LOS	II	<1	female	31	fresh stage	dead	/
6	LOS	II	<1	male	37	bloat stage	dead	/
7	LOS	II	<1	female	27	fresh stage	dead	/
8	MOL	III	<1	female	22	fresh stage	dead	/
9	MOL	III	<1	female	28	fresh stage	dead	/
10	MOL	III	<1	female	36	fresh stage	dead	/
11	MOL	III	<1	female	38	bloat stage	dead	/
12	MOL	III	<1	female	36	bloat stage	shot	Lung not available
13	SN	IV	<1	male	36	fresh stage	dead	/
14	SN	IV	<1	male	30	bloat stage	dead	Scavenger feeding marks (thorax)
15	SN	IV	<1	female	31	fresh stage	dead	/
16	SN	IV	>2	female	75	bloat stage	dead	/

* Stages of decomposition were classified as reviewed by Brooks [24] with further modifications. Fresh stage: no bloating, no discoloration. Bloat stage: bloating, gray to green discoloration of organs.

2.1.1. Pathological Examination

Necropsy

Full necropsies were performed on the wild boar carcasses (n = 16). The organ lesions were scored from 0 to 3 (0 = normal, 1 = mild, 2 = moderate, 3 = severe; unless not otherwise stated) as recently published [25] with the additional modifications shown in Table 2. Tissues samples including the popliteal lymph node, spleen, lung, kidney, liver, heart, brain (cerebellum and cerebrum) and adrenal gland were taken from wild boar and fixed in 10% neutral-buffered formalin for at least 3 weeks.

Table 2. Assessment of gross pathological criteria in ASFV-infected wild boar.

Organ	Macroscopic Finding	Annotations
Lymph node (popliteal)	Enlargement	/
	Hemorrhage	
Lung	Alveolar edema	/
	Interstitial edema	
	Hemorrhage	
	Collapse	
	Consolidation	
	Thoracic effusion	
	Pleuropneumonia	

Table 2. *Cont.*

Organ	Macroscopic Finding	Annotations
Kidney *	Hemorrhage	Assessment of size (petechia, ecchymosis) and distributional pattern (focal (n = 1), oligofocal (n ≤ 20), multifocal (n ≥ 20))
	Pelvic dilation	/
	Pelvic hemorrhage	
Liver and gall bladder *	Congestion	/
	Gall bladder wall hemorrhage/edema	
Spleen *	Determination of relative spleen weight	/
Pancreas	Hemorrhage/edema	/
	Necrosis	
Abdominal cavity *	Peritonitis	/
	Ascitis	
Urinary bladder	Hemorrhage	/
Bone marrow	Hemorrhage	/
Heart	Hemorrhage	Describing localization: endocardial, myocardial, epicardial
	Pericardial effusion	/
	Pericarditis	
Tonsils	Hemorrhage	/
	Necrosis	
Brain	Hemorrhage	/
Adrenal gland		
Genitals		
Skin		
Larynx		

* Further lesions were described.

Histopathology and Immunohistochemistry

The tissue samples were embedded in paraffin wax and cut at 2–3 µm slices. Hematoxylin-eosin (HE) staining was performed to examine the main macroscopic lesions in more histological detail. To visualize viral antigens, anti-ASFV p72 immunohistochemistry was conducted on the respective organs as described earlier [17,18]. In brief, sections were treated with an in-house rabbit polyclonal primary antibody against the major capsid protein p72 of ASFV (diluted in TBS 1:1600, 1 h), followed by incubation with a secondary, biotinylated goat anti-rabbit IgG (Vector Laboratories, Burlingame, CA, USA; diluted in TBS in 1:200, 30 min). Positive antigen detection was visualized by the Avidin–Biotin Complex (ABC) method providing horseradish peroxidase that converted the added chromogen 3-amino-9-ethylcarbazole (AEC) into insoluble red-colored deposits at the reaction site. As negative control, consecutive sections were labeled with an irrelevant antibody (M protein of Influenza A virus, ATCC clone Hb64). An ASF positive control slide was included in each run.

Histopathology including Semiquantitative Antigen Scoring

The slides were scanned using a Hamamatsu S60 scanner and evaluated using ND-Pview.2 plus software (Version 2.8.24, Hamamatsu Photonics, K.K. Japan). While the histopathological lesions obtained on HE-stained sections were described only qualitatively (present/absent) due to autolysis-related limited assessability, the viral antigen content in the respective organ was determined on a semiquantitative scoring scale as previously published [18]. The most affected area (420 × 260 μm) per sample sections was scored with score 0 (no antigen), score 1 (1–3 positive cells), score 2 (4–15 cells) or score 3 (>16 cells). Cells with fine granular cytoplasmic labeling were considered positive whereas chromogen aggregations without cellular association were not counted.

2.1.2. Detection of ASFV Genome

To determine the viral genome load, the tissue samples were homogenized in 1 mL of phosphate buffered saline with a metal bead using a TissueLyzer II (Qiagen GmbH, Hilden, Germany). Viral nucleic acids were extracted from blood and homogenized spleen, lung, liver, kidney, popliteal lymph node and brain with the NucleoMag Vet Kit (Machery-Nagel, Düren, Germany) on the KingFisher extraction platform (Thermo Scientific, Waltham, MA, USA). Quantitative real-time PCR (qPCR) was conducted according to the protocol published by King et al. [26] with an in-house full virus standard for determination of genome loads on a C1000 thermal cycler with the CFX96 Real-Time System (Biorad, Hercules, CA, USA).

2.1.3. Detection of Anti-ASFV Antibodies

For investigation of ASFV-specific antibodies, an accredited indirect immunoperoxidase test (IPT) was applied according to the standard protocol SOP/CISA/ASF/IPT/1 provided by the European Reference laboratory for ASF with modifications regarding cell and virus type (https://asf-referencelab.info/asf/images/ficherosasf/PROTOCOLOS-EN/2021_UPDATE/SOP-ASF-IPT-1_2021.pdf, accessed on 4 April 2022). As sample material, plasma was obtained from EDTA blood by centrifugation at 18.000 g-force for 10 min from German wild boar carcasses and domestic pigs infected with ASFV "Estonia 2014" from a previous trial for comparison. Titers were determined semiquantitatively by endpoint dilution from 1:40 to 1:12,800.

2.1.4. Statistical Analysis

Using GraphPad Prism (Version 8.4.2), statistical analysis was conducted to determine overall group differences in terms of viral genome load, viral antigen amount, macroscopic lesion scores and antibody titers. For this purpose, the non-parametric Kruskal–Wallis test with post hoc Dunn's test was performed. A p value ≤ 0.05 was considered significant.

3. Results

3.1. Pathogen Detection in Blood and Tissues

Full necropsies were performed on all wild boar obtained from the outbreak areas LOS, MOL and SN to determine the amount of viral genome and antigen. The results are shown in Figure 1 and details are given in Supplementary Tables S1 and S2.

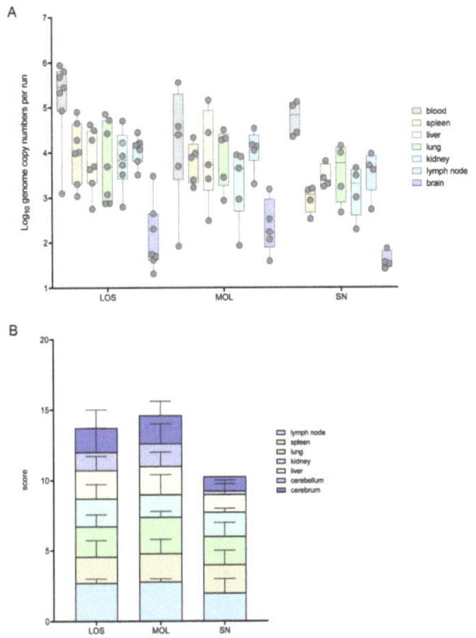

Figure 1. Pathogen detection in blood and tissue samples of ASFV-infected wild boar carcasses from LOS, MOL and SN. (**A**) Box plot presenting the individual viral genome load in blood and organ samples. (**B**) Corresponding stacked bar diagram showing the median viral antigen score with range per organ. Organs were scored on a scale from 0 to 3 based on the number of positively labeled cells in the most-affected tissue area per high power field.

Viral genome could be found in all the samples of the infected wild boar. In general, the highest viral genome loads were detected in blood samples, varying between 1×10^2 and 9×10^5 genome copies (gc)/5 µL nucleic acid. In general, genome loads in most organ samples were roughly one logarithmic step lower than the corresponding blood samples. A lower mean viral genome load was detected in wild boar found in SN when compared to animals from LOS and MOL (Figure 1A).

The viral antigen score of selected tissue sections reflected the results obtained by qPCR. Consistent with the lower number of viral genome copies, wild boar from SN also reached lower viral antigen scores (Figure 1B). Details on immunohistochemistry are included in the histopathological evaluation of organ systems in the following section.

3.2. *Pathological Assessment of Organ Systems*

All carcasses were scored macroscopically based on a standardized scoring system [25] with further modifications as indicated in Table 2. Histopathological alterations were reported only as present/absent due to the reduced number of well-preserved available tissues. A summary of all macroscopical and histopathological ASF-associated [27] and bacterial-induced or background alterations [28–31] including immunohistochemistry results are shown in Table 3.

Table 3. Summary of macroscopical and microscopical lesions in ASF-infected wild boar carcasses. Pathological findings are listed as primary lesions, characteristically associated with ASF [27] and as lesions, usually induced by bacteria or as common background lesions [28–31].

Organs/Tissues	Gross Pathology	Histopathology	Immunohistochemistry
Immune system	**Primary lesions associated with ASF** Lymph nodes: • Hemorrhagic lymph-adenopathy Spleen: • Increased spleen weight Bone marrow: • Hemorrhages	**Primary lesions associated with ASF** Lymph nodes: • Lymphoid depletion • Thrombosis • Necrosis of interfollicular, paracortical areas and medullary chords Spleen: • Lymphoid depletion • Apoptosis/necrosis of myelomonocytic cells Bone marrow: • N/A	Lymph nodes: • Positive, macrophages Spleen: • Positive, macrophages Bone marrow: • N/A
Respiratory system	**Primary lesions associated with ASF** Lung: • Alveolar edema • Hemorrhages • Consolidation • Loss of collapse Nose: • Hemorrhages **Lesions, usually induced by bacteria or common background lesions** Lung: • Fibrous pleuropneumonia	**Primary lesions associated with ASF** Lung: • Alveolar edema • Hemorrhages • Necrotizing interstitial pneumonia Nose: • N/A **Lesions, usually induced by bacteria or common background lesions** Lung: • Fibrino-suppurative/necrotizing bronchopneumonia	Lung: • Positive, alveolar/interstitial macrophages Nose: • N/A
Cardiovascular system	**Primary lesions associated with ASF** Heart: • Hemorrhages (epi-, myo-, endocardial) **Lesions, usually induced by bacteria or common back-ground lesions** Heart: • Fibrous pericarditis	**Primary lesions associated with ASF** Heart: • Hemorrhages Mononuclear infiltration (endo-/subendocardial)	Heart: • Positive, macrophages

Table 3. *Cont.*

Organs/Tissues	Gross Pathology	Histopathology	Immunohistochemistry
Urinary system	**Primary lesions associated with ASF** Kidney: • Hemorrhages (cortical, medullary, pelvic) • Perirenal edema and hemorrhages Urinary bladder: • Hemorrhages (mucosal, serosal, transmural)	**Primary lesions associated with ASF** Kidney: • Hemorrhages (interstitial, glomerular) • Vascular thrombosis Urinary bladder: • N/A **Lesions, usually induced by bacteria or common background lesions** Kidney: • Non-suppurative tubulointerstitial nephritis • Tubular necrosis	Kidney: • Positive, macrophages Urinary bladder: • N/A Urinary bladder: • N/A
Gastrointestinal sys-tem/abdominal cavity	**Primary lesions associated with ASF** Liver: • Congestion • Hemorrhages (subcapsular) Gall bladder: • Wall edema • Wall hemorrhages Stomach: • Hemorrhagic gastritis Small intestine: • Hemorrhages (serosal, mucosal) Large intestine: • Hemorrhages (serosal, mucosal) Abdominal cavity: • Hemorrhagic ascites **Lesions, usually induced by bacteria or common back-ground lesions** Stomach: • Ulcerative gastritis Abdominal cavity: • Fibrous peritonitis	**Primary lesions associated with ASF** Liver: • Apoptosis/necrosis of Kupffer cells and hepatocytes Gall bladder: • N/A Stomach: • N/A Intestine: • N/A **Lesions, usually induced by bacteria or common back-ground lesions** Liver: • Mixed-cellular sinusoidal and periportal infiltration	Liver: • Positive, Kupffer cells Gall blad-der/stomach/intestine: • N/A
Nervous system	**Primary lesions associated with ASF** Brain: • Hemorrhages	**Primary lesions associated with ASF** Brain: • Hemorrhages • Non-suppurative meningitis (cerebral, cerebelar) • Non-suppurative encephalitis (cerebral, cerebellar) • Non-suppurative plexus choroiditis	Brain: • Positive, macrophages

Table 3. *Cont.*

Organs/Tissues	Gross Pathology	Histopathology	Immunohistochemistry
Endocrine system	**Primary lesions associated with ASF** Adrenal gland: • Hemorrhages Pancreas: • Hemorrhages • Edema	**Primary lesions associated with ASF** Adrenal gland: • Hemorrhages (cortical, medullary) • Sinusoidal thrombosis • Mixed-cellular infiltration (medullary Pancreas: • N/A	Adrenal gland: • Positive, macrophages Pancreas: • N/A
Reproductive system	**Primary lesions associated with ASF** Testicle (spermatic chord): • Hemorrhages Vestibulum: • Hemorrhages	• N/A	• N/A

The overall score obtained upon macroscopical evaluation turned out to be the opposite when compared to the viral genome load and antigen score. Therefore, wild boar from SN tended to show a higher total score when compared to wild boar from LOS and MOL (Figure 2). Individual animal scores given for macroscopical findings are included in Supplementary Table S3.

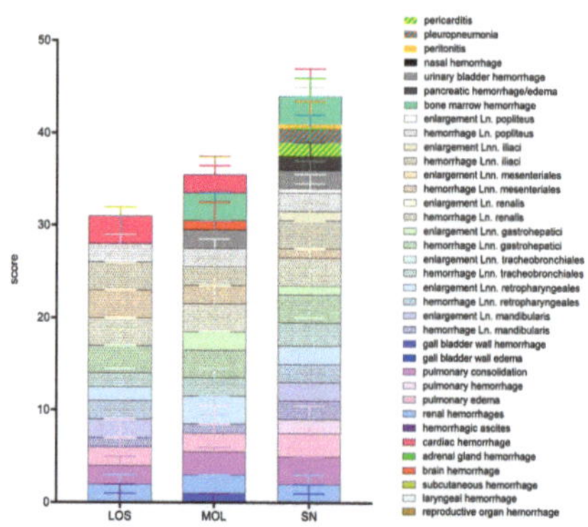

Figure 2. Summary of scoring results following macroscopical investigation of ASFV-infected wild boar carcasses from LOS, MOL and SN. Stacked bar diagram showing the total gross lesion score, which is composed of individual scores given for macroscopical findings shown on the right. Lesions were scored on a scale from 0 to 3. Bars indicate the median with range.

In the following gross and histopathological findings, the different organ systems will be described.

3.3. Immune System

3.3.1. Lymph Nodes

Gross Pathology

In general, hemorrhagic lymphadenopathy was present in all animals irrespective of the outbreak area (Figure 3). Details are given in Supplementary File S1 and Table S3.

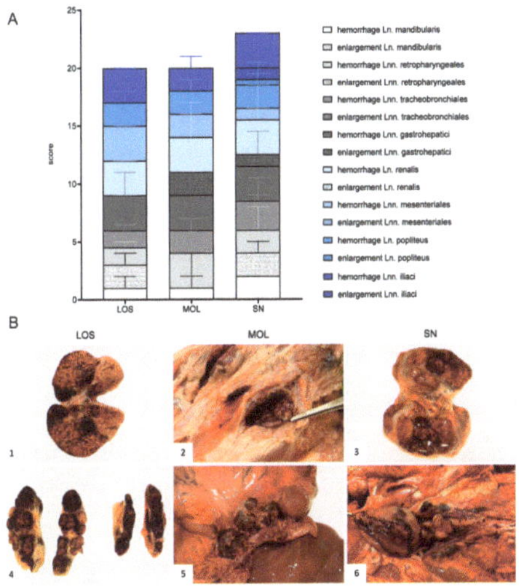

Figure 3. Representative macroscopical findings of lymph nodes in ASFV-infected wild boar carcasses from German outbreak areas. (**A**) Stacked bar diagram showing the total gross lesion score given for enlargement and hemorrhages of various lymph nodes evaluated on a scale from 0 to 3. Bars indicate the median with range per finding. (**B**) Lymph nodes (Ln. mandibularis ((**B1**)–(**B3**)), Ln. renalis (**B4**), Lnn. gastrohepatici (**B5**), Lnn. iliaci (**B6**)) revealed hemorrhages of varying degree.

Histopathology

The popliteal lymph node was examined in more histological detail as demonstrated in Figure 4. The findings were characterized by lymphoid depletion, hemorrhages (Figure 4A,B), necrosis (Figure 4B) and vascular thrombosis (Figure 4C). Animals showed p72 positively labeled cells morphologically consistent with macrophages (Figure 4D). Details for individual animals are given in Table S2.

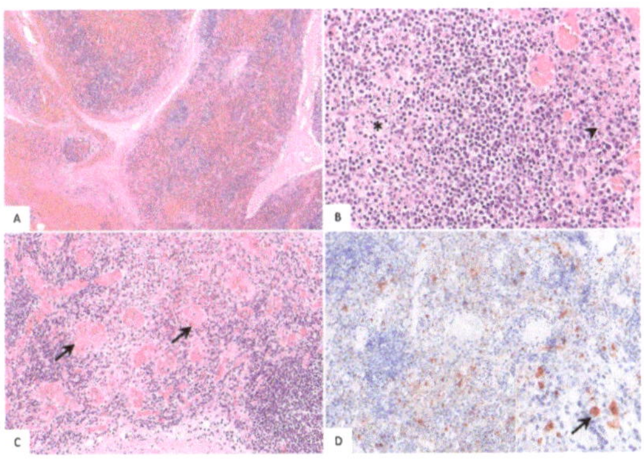

Figure 4. Pathohistological findings of the popliteal lymph node in German ASFV-infected wild boar carcasses. (**A**) Diffuse lymphoid depletion and hemorrhage affected the follicles, paracortex and medullary chords thereby effacing the physiological lymph node architecture, HE stain. (**B**) A lymphoid follicle with lymphoid depletion (asterisk) was surrounded by necrosis (arrowhead), HE. (**C**) Numerous vessels were occluded by fibrin thrombi (arrows) throughout the lymph node, HE. (**D**) A large number of viral-antigen-positive cells are shown, which were morphologically consistent with macrophages (inlay), anti-p72 immunohistochemistry, ABC method.

3.3.2. Spleen

Gross Pathology

Macroscopic assessment of the spleen was limited due to poor preservation. Therefore, the spleen was evaluated by determination of the relative spleen weight based on a recent publication in domestic pigs [32] shown in Figure S1.

High relative spleen weight values were observed in all wild boar irrespective of the district of origin. Median values reached 0.81 (LOS), 0.69 (MOL) and 0.97 (SN).

Histopathology

Briefly, histological examination of the spleen revealed congestion and hemorrhage with lymphoid depletion in all wild boar (Figure S2A,B). Immunopositive cells were detected, phenotypically consistent with macrophages (Figure S2C,D). Individual histopathological results are summarized in Table S2.

3.3.3. Bone Marrow

Gross Pathology

Pathological changes in the femoral bone marrow included hemorrhages in all groups (Figure 5). All scores given for each individual animal can be found in Table S3.

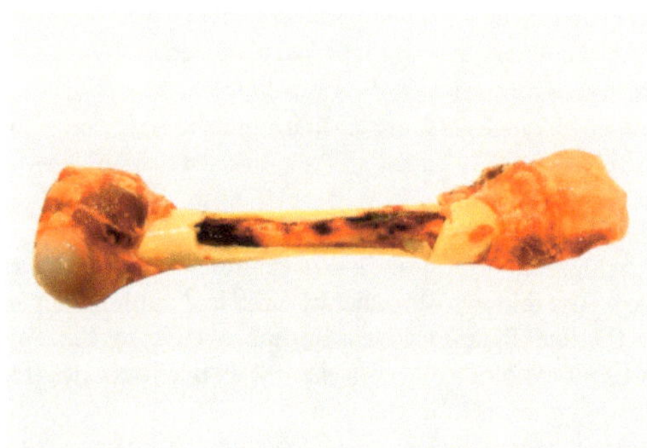

Figure 5. Gross pathology of the bone marrow of naturally ASFV-infected wild boar carcasses from German outbreak areas. Bone marrow hemorrhages, if present, were severe throughout.

Histopathological examination was not performed, because in the majority of animals, progression from red to yellow marrow had already occurred.

3.4. Respiratory System

3.4.1. Lung

Gross Pathology

ASF-associated macroscopic findings of the lung were up to severe pulmonary edema, incomplete collapse with foci of consolidation and congestion as well as hemorrhages (Figure 6B, 1–4). In individual wild boar, fibrous pleuropneumonia, likely as a consequence of bacterial infection, was detected (Figure 6B, 5–6). Furthermore, isolated verminous pneumonia was present in wild boar from all districts. Details on macroscopic lung scores of all animals are summarized in Table S3 and described in Supplementary File S1.

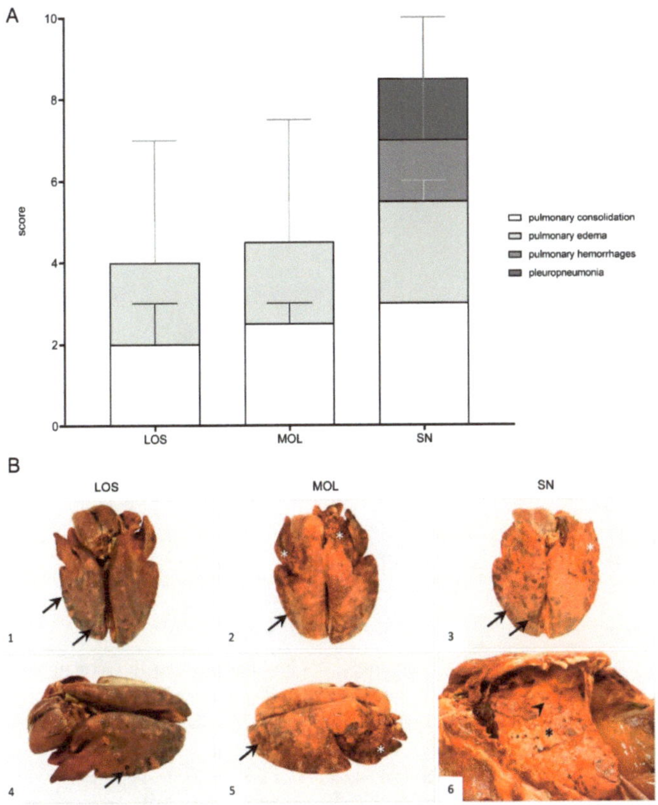

Figure 6. Macroscopical lung lesions of ASFV-infected wild boar carcasses found in German districts. (**A**) Stacked bar diagram demonstrating the median with range of individual scores given for each pathological criterion shown on the right legend. The presence and severity of each finding was scored from 0 to 3. (**B**) All lungs showed consolidated areas of different size (asterisk) and loss of pulmonary collapse ((**B1**)–(**B6**)). Pulmonary hemorrhages of varying severity are demonstrated by arrows ((**B1**)–(**B5**)). Chronic pleuropneumonia, likely due to bacterial infection, is shown in B6 with extensive fibrous pleural adhesions (arrowhead).

Histopathology

Histopathological findings are shown in Figure 7 and Table S2. Pulmonary inflammation either presented as fibrino-suppurative to necrotizing bronchopneumonias, probably due to bacterial infections (Figure 7A–C), or interstitial pneumonia (Figure 7D). Positively labeled cells consistent with intravascular, -alveolar and interstitial macrophages were detected by immunohistochemistry (Figure 7E,F).

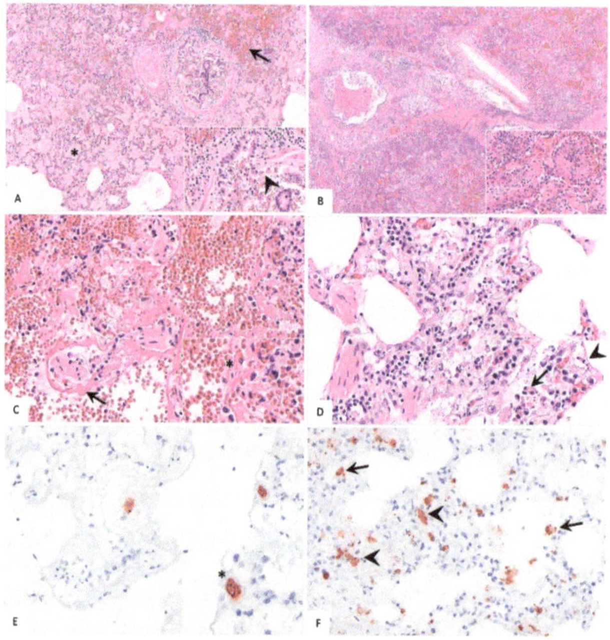

Figure 7. Histopathological findings of lungs in German naturally ASFV-infected wild boar carcasses. (**A**) Alveoli were filled with protein-rich edema fluid (asterisk), erythrocytes (arrow) and fibrin strands. The bronchiolus revealed epithelial necrosis (inlay, arrowhead) and contained cellular debris and erythrocytes. A distended pulmonary vein with fibrin thrombi was present left from the bronchiole, HE. (**B**) In a few animals, severe fibrino-suppurative to necrotizing bronchopneumonia was detected. Alveoli were densely filled with cellular debris, fibrin, viable and degenerate neutrophils, plasma cells, macrophages and lymphocytes as well as erythrocytes (inlay), HE. (**C**) A low number of wild boar showed loss of alveolar epithelium and hyaline membranes (arrow). An intravascular macrophage is indicated by asterisk (inlay), HE. (**D**) Alveolar septa showed epithelial necrosis (inlay, arrowhead), infiltration by necrotic macrophages (inlay, arrow), neutrophils, lymphocytes and plasma cells, HE. (**E**) and (**F**) Immunohistochemistry showed viral antigen-positive cells morphologically consistent with intravascular ((**E**), asterisk, consecutive section of (**C**)) intraalveolar ((**F**), arrow) and interstitial ((**F**), arrowhead) macrophages, anti p72-immunohistochemistry, ABC method.

3.5. Cardiovascular System

3.5.1. Heart

Gross Pathology

Hemorrhages affected wild boar of each group (Figure 8). Bacterial infection led to pericarditis with fibrous adhesions in one animal from SN. Details for individual animals are given in Supplementary File S1 and Table S3.

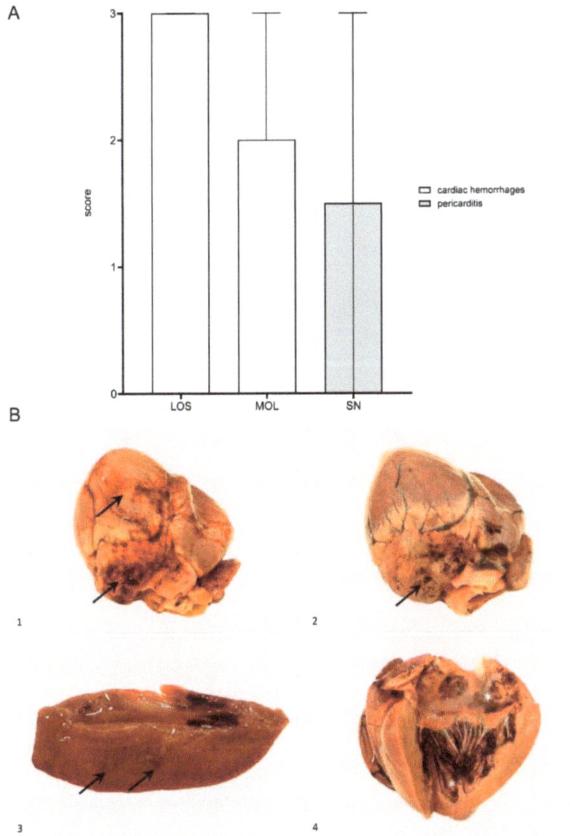

Figure 8. Heart lesions in naturally ASFV-infected wild boar carcasses from German outbreak areas. (**A**) Scoring of the heart included the presence and severity of hemorrhages as well as pericarditis, which were evaluated on a scale from 0 to 3. Bars indicate the median with range. (**B**) Hemorrhagic lesions of different locations and severity of ASFV-infected wild boar are shown. Multifocal paintbrush to coalescing hemorrhages were found in the epicardium (arrow) to a variable extent ((**B1**),(**B2**)). Scant myocardial hemorrhages (arrow) are indicated in (**B3**). Multifocal endocardial hemorrhages (asterisk) are present in (**B4**). The darker blood coagulum had to be differentiated from hemorrhages.

Histopathology

In addition to hemorrhages (Figure 9A,B), in a few animals, there was endocardial and subendocardial infiltration by mononuclear cells (Figure 9C,D). Viral antigen was found in cells mostly morphologically consistent with macrophages (Figure 9E,F). Detailed histopathological evaluation of the heart can be found in Table S2.

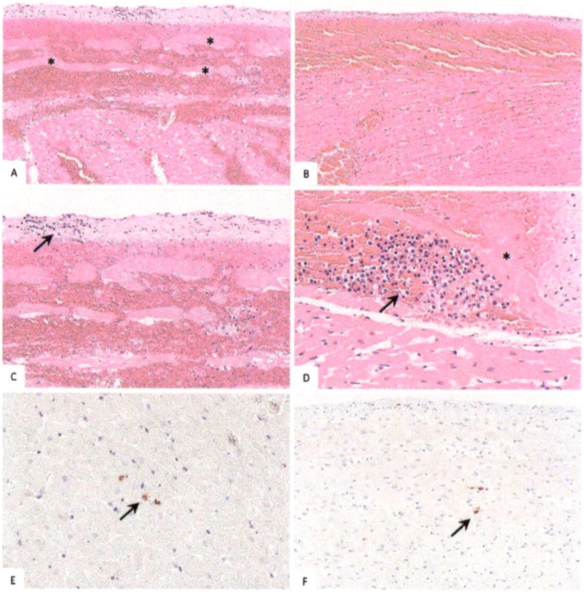

Figure 9. Histopathology of the heart in naturally ASFV-infected wild boar carcasses from Germany. (**A**) Massive hemorrhage involved the endocardium as well as the myocardium, displacing subendocardial Purkinje fibers (asterisk), HE. (**B**) The epicardium was also affected by diffuse hemorrhage radiating into the myocardium, HE. (**C**) Higher magnification from (**A**) shows minimal accumulation of infiltrating mononuclear cells in the endocardium (arrow), HE. (**D**) Subendocardial infiltrates (arrow) were also present between Purkinje fibers (asterisk), HE. (**E**,**F**) Immunohistochemistry of the heart showed only few positive macrophages (arrow), anti-p72 immunohistochemistry, ABC method.

3.6. *Urinary System*

3.6.1. Kidney

Gross Pathology

Renal and perirenal hemorrhages were present in all wild boar irrespective of the district (Figure 10). Details can be found in Supplementary File S1 and Table S3.

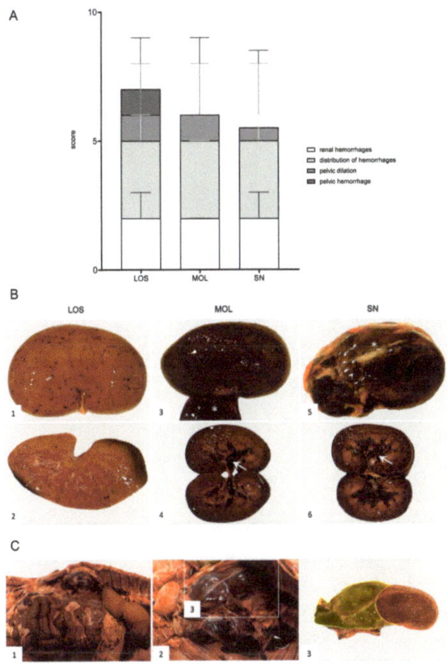

Figure 10. Pathologic changes in kidneys of ASFV-infected German wild boar carcasses from different outbreak areas. (**A**) Stacked bar diagram of gross lesion scoring of pathological criteria listed on the right. Scoring was conducted on a scale from 0 to 3 or from 0 to 4 (distribution pattern of hemorrhages). Individual scores are given as median values with range. (**B**) Hemorrhagic lesions of various size and severity affecting different parts of the organ are shown in ((**B1**)–(**B6**)). Multifocal petechiae with fewer ecchymoses primarily located to the renal cortex are depicted in ((**B1**),(**B2**)). Gray discoloration of the kidney periphery was due to beginning autolysis (**B2**). Mainly affecting the renal cortex (cortico-medullar pattern), diffuse ecchymoses are present in ((**B3**),(**B4**)). Marked dilation and diffuse bleeding into the renal pelvis are depicted in ((**B4**),(**B6**)) (arrows). To a lesser extent, oligofocal petechiae (arrowhead) could be found in the medulla (**B6**). Edema of the perirenal tissue is represented in ((**B3**),(**B5**)) (asterisk). (**C**) Massive hemorrhage resulted in expansion and bulging of the renal capsule ((**C1**),(**C2**)). The hemorrhage further extended into the perirenal and retroperitoneal tissue including the ureter (**C2**). To better distinguish the kidney and the extent of hemorrhage from (**C2**), the kidney was shaded red and the hemorrhage was highlighted in yellow (**C3**).

Histopathology

The histopathological findings included hemorrhages (Figure 11A), glomerular alterations (Figure 11B), non-suppurative tubulointerstitial nephritis, (Figure 11C), tubular epithelial necrosis (Figure 11D) and renal vein thrombosis (Figure 11E). Immunohistochemistry revealed positive cells morphologically consistent with macrophages (Figure 11F). Details on the histopathological findings are shown in Table S2.

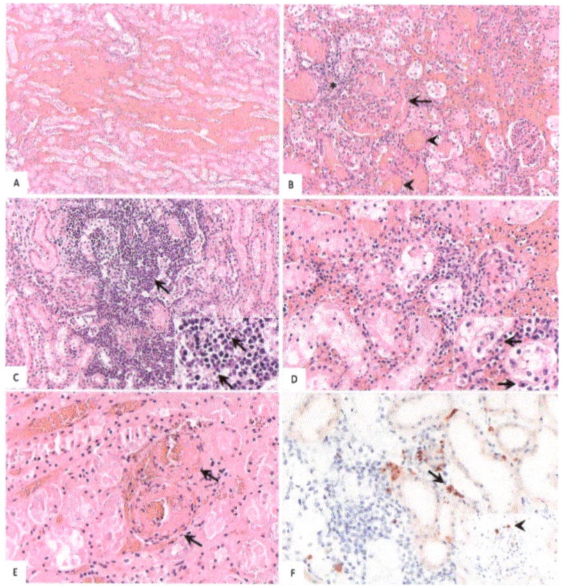

Figure 11. Pathohistological findings of the kidney in naturally ASFV-infected wild boar carcasses. (**A**) Diffuse hemorrhages were present expanding the renal medullary interstitium, HE. (**B**) A glomerulus showed extravasation of fibrin admixed with erythrocytes into the Bowman's space (arrow). There was periglomerular infiltration of partly degenerated mononuclear cells (asterisk). Red blood cell casts were present in several tubules surrounding the glomerulus (arrowhead), HE. (**C**) Extensive mononuclear cell infiltrates accumulated around tubules and glomeruli (arrow) and revealed multiple foci of apoptosis/necrosis (inlay, arrow), HE. (**D**) In some areas, tubulointerstitial nephritis was associated with tubular epithelial apoptosis/necrosis (inlay, arrow), HE. (**E**) Fibrinoid vascular necrosis could be found in varying amounts of renal veins (arrow), HE. (**F**) Representative immunohistochemical image showing moderate numbers of positively labeled macrophages in the renal interstitium (arrow) or glomerular capillaries (inlay, arrowhead), anti-p72 immunohistochemistry, ABC method.

3.6.2. Urinary Bladder

Gross Pathology

The urinary bladder presented with hemorrhages in wild boar of all three groups (Figure 12). Details on the lesions found in the animals as well as individual scores can be found in Supplementary File S1 and Table S3.

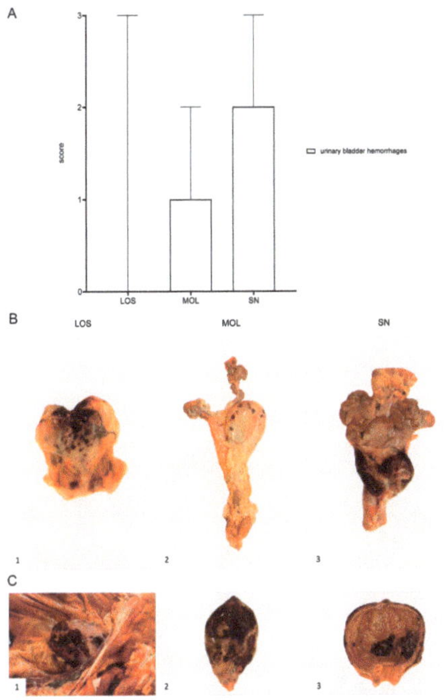

Figure 12. Pathology of the urinary bladder in naturally ASFV-infected wild boar carcasses from German outbreak districts. (**A**) Bar diagram showing hemorrhagic changes of the urinary bladder scored on scale from 0 to 3. Bars indicate the median with range. (**B**) Hemorrhages of varying severity were observed during necropsy. Multifocal-to-coalescing hemorrhages (**B1**) and multiple ecchymoses (**B2**) or severe, diffuse hemorrhage of the urinary bladder radiating into surrounding connective tissue (**B3**) were found. (**C**) Severe hemorrhages were located to the serosa ((**C1**),(**C2**)) as well as to the mucosal surface of the urinary bladder (**C3**).

Histopathological examination was not performed due to poor preservation.

3.7. Gastrointestinal System

3.7.1. Liver and Gall Bladder

Gross Pathology

Due to poor preservation, not all livers could be examined. Hepatic congestion and hemorrhages as well as edema affecting the gall bladder wall were present (Figure S3). Details on lesions are given in Supplementary File S1 and Table S3.

Histopathology

Microscopical lesions of well-preserved livers included apoptosis/necrosis of Kupffer cells (Figure S4A) and hepatocytes (Figure S4B), and sinusoidal and periportal infiltrates (Figure S4C). Immunohistochemistry revealed positive immunolabeling of cells phenotypically consistent with Kupffer cells (Figure S4D). A summary of histopathological observations is included in Table S2.

3.7.2. Stomach and Intestine

Gross Pathology

Due to progressive autolysis, the gastrointestinal tract could be evaluated only in individual animals. Macroscopic findings included hemorrhagic gastritis and hemorrhages in the small and large intestine as indicated in Figure S5. Hemorrhagic ascites was further detected. Occasionally, gastric ulcers as well as fibrous peritonitis, likely associated with bacterial infection, were also found in animals from SN. Supplementary File S1 and Table S3 provide detailed results.

Histopathological examination was not carried out due to advanced autolysis of the gastrointestinal tract.

3.8. Nervous System

3.8.1. Brain

Gross Pathology

The brain was affected by hemorrhages only occasionally in some animals from MOL as shown in Figure 13A. Both the cerebellum and cerebrum were further evaluated by histopathology since data on respective lesions are sparse.

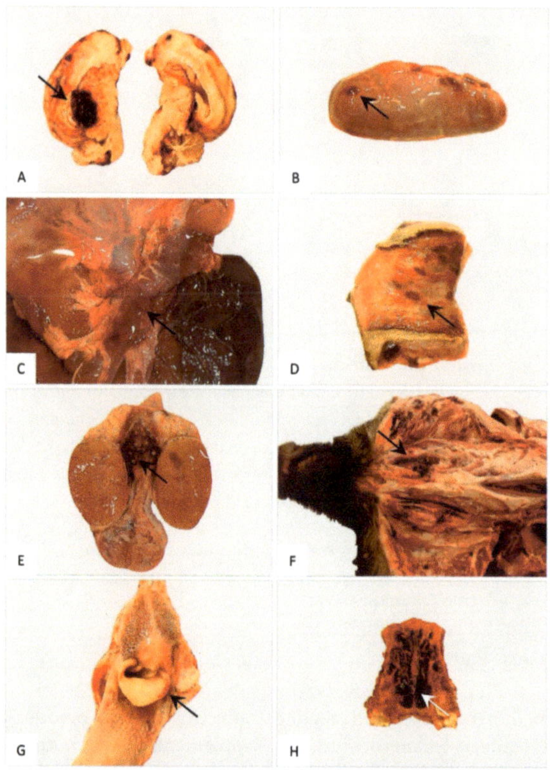

Figure 13. Gross pathology of the nervous, endocrine and reproductive organ systems and other findings in naturally ASFV-infected wild boar carcasses from Germany. Representative lesions included hemorrhages in the cerebrum (**A**), adrenal gland (**B**), pancreas (**C**), vestibulum vaginae (**D**), testis (**E**), subcutaneous tissue (**F**), larynx (**G**) and nasal mucosa (**H**). Arrows indicate hemorrhagic changes in the respective organs.

Histopathology

Microscopical findings of the cerebellum and cerebrum included meningitis, encephalitis and plexus choroiditis as depicted in Figures 14 and 15, respectively. Occasionally, hemorrhage as well as satellitosis and microgliosis were detected. Detailed histopathological results are described in Supplementary File S1 and Table S2.

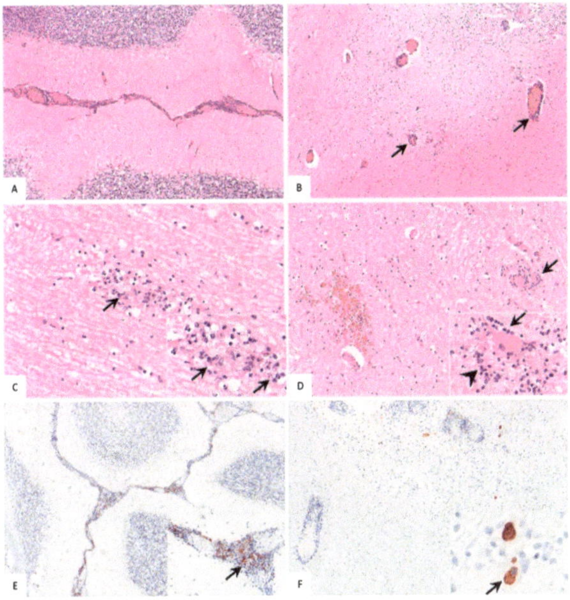

Figure 14. Histopathological findings in the cerebellum of ASFV-infected wild boar carcasses. (**A**) Meningitis was present in affected animals. (**B**) Cerebellar encephalitis was characterized by multifocal perivascular cuffs consisting of mononuclear cell infiltrates. (**C**) Parenchymal mononuclear infiltrates (arrow) showed multifocal apoptosis/necrosis (inlay, arrow). (**D**) Hemorrhage (left), perineural satellitosis (arrow, also see inlay) and microgliosis (inlay, arrowhead) were recognized. (**E**) and (**F**) Cerebellar meninges as well as brain parenchyma revealed positively labeled macrophages of differing amounts (inlays, arrow).

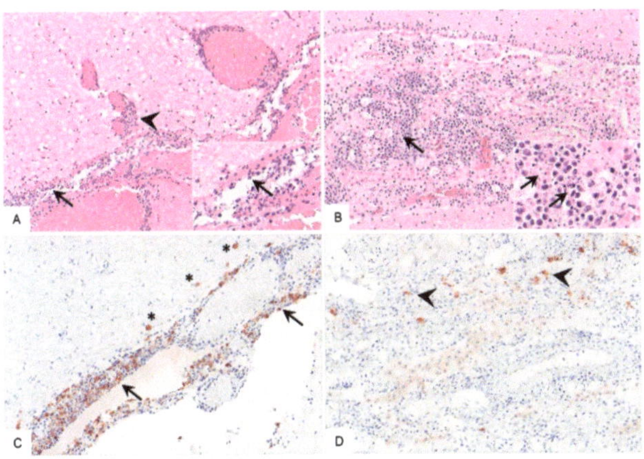

Figure 15. Histopathology of the cerebrum of ASFV-infected wild boar carcasses. (**A**) The meninges (arrow) and adjacent brain parenchyma (arrowhead) were infiltrated by mononuclear cells via Virchow Robin spaces. Mononuclear cells showed multifocal apoptosis/necrosis (inlay, arrow). Meningeal vessels were prominently dilated. (**B**) Mononuclear inflammation was limited to the choroid plexus within ventricles (arrow) with multifocal apoptosis/necrosis of infiltrating cells (inlay, arrow). There was degeneration of only a few plexus epithelial cells. (**C**,**D**) Immunopositive cells were present to variable extents in the meninges (arrow) and brain parenchyma (asterisk) as well as in the choroid plexus epithelium (arrowhead), phenotypically consistent with macrophages.

Immunohistochemical results showed viral antigen-positive cells with macrophage morphology.

3.9. Endocrine System

3.9.1. Adrenal Gland

Gross Pathology

Hemorrhages were observed in the adrenal glands of animals from LOS and SN (Figure 13B).

Histopathology

Histopathology revealed hemorrhages (Figure 16A), sinusoidal thrombosis and necrosis (Figure 16B,C) as well as inflammation (Figure 16D,E). Positively labeled macrophages were detected by immunohistochemistry (Figure 16F). Individual histopathological results are listed in Table S2.

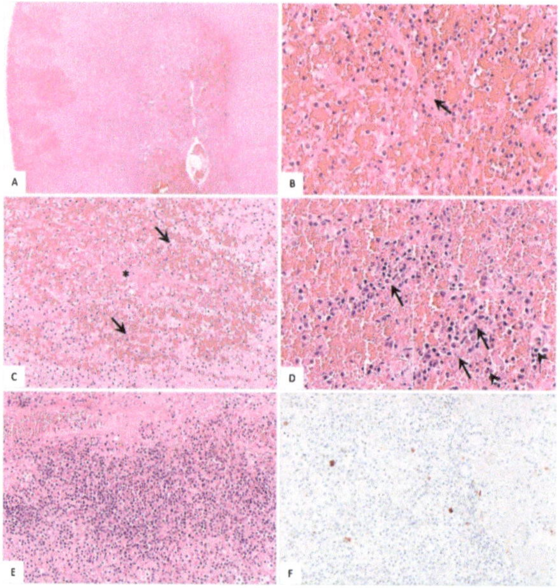

Figure 16. Histopathological findings of the adrenal gland in ASFV-infected wild boar carcasses. (**A**) Overview of the adrenal gland of a deceased wild boar. The adrenal gland showed extensive cortical and medullary hemorrhages. (**B**) Multifocally, fibrin thrombi were visible in the sinusoids (arrow). (**C**) Occasionally, areas of necrosis were present in the cortex (arrow). There was fibrin deposition (asterisk) and massive hemorrhage in the affected location. (**D**) The medulla was markedly expanded by hemorrhage. Infiltrating mononuclear cells as well as a few neutrophilic granulocytes (inlay, arrowhead) accumulated around degenerated cells (inlay, arrow). (**E**) The adrenal medulla was severely infiltrated by mononuclear cells admixed with fewer neutrophils. (**F**) Moderate amounts of antigen-positive macrophages were found in the majority of animals.

3.9.2. Pancreas

Gross Pathology

Pancreatic edema and hemorrhage were detected in animals from SN (Figure 13C). Histopathological examination was not carried out due to advanced autolysis.

3.10. Reproductive System

Occasionally, hemorrhages were found in the vaginal vestibulum in one wild boar from LOS (Figure 13D) and in the spermatic cord in a wild boar from MOL (Figure 13E).

3.11. Occasional Findings

Further hemorrhages were found in the subcutis in animals from LOS (Figure 13F), and in the epiglottis (Figure 13G) and nasal cavity (Figure 13H) in wild boar from SN.

3.12. Antibody Detection against African Swine Fever Virus

All animals were tested for anti-ASF antibodies by IPT as shown in Figure S6. Except for one animal from LOS, all wild boar developed antibodies of different titers between 200 and 800. Higher titers tended to be found in the animals from MOL, titers ranging from 200 to 1600, and in two wild boar from SN having titers of 800 and 3200, respectively. One animal from SN showed a titer of 40. In the fourth wild boar from SN, no test could be performed due to limited sample material.

For comparison, three domestic pigs from a previous study inoculated with the moderately virulent ASFV strain "Estonia 2014" were analyzed for anti-ASFV-specific antibodies. Starting at day 14 pi, all pigs developed antibody titers between 200 and 400. Since one pig had died at day 14 pi, only two animals could be analyzed in the following days. On day 21 pi, titers increased to 800 and 1600. On day 28 pi, titers further increased to 3200 or even remained at the same level of 1600 while on day 35 pi antibodies dropped in one animal to 800, but increased in the other pig to 3200. On day 41 pi, a second increase in the titer to 1600 was noted in one pig whereas in the other one antibodies remained constantly high at 3600.

4. Discussion

Filling the documentation gap on the pathology after ASF field infection, the aim of the present study was to examine ASFV-infected wild boar that succumbed to the disease under natural conditions in both virological and pathomorphological detail. Furthermore, the impact on the virulence of emerging virus variants II, III and IV in the ASF outbreak areas of Eastern Germany was analyzed.

A total of 16 wild boar aged between 0 and 2 years of different sexes were investigated. Despite the different preservation status, the organs of each animal could be examined for ASFV genome load and revealed consistently positive results. While it has to be noted that a direct comparison has to be conducted with great care due to many unknown factors, all animals were found as carcasses in affected regions and that would allow us to assume they reached a similar point of infection, i.e., the terminal phase. At that point, significant differences were not found between animals of different outbreak areas and the three variants, but wild boar from SN tended to show both lower viral genome loads and viral antigen scores compared to animals from LOS and MOL. However, the viral genome load has limited informative value at this point since viral genome can be detected up to 100 days after infection [33] and the time at which the genome load decreases varies greatly between experiments [11,18,19].

In addition to organ-wide detection of viral genome, all wild boar irrespective of the outbreak area and virus variant were diagnosed with characteristic and severe ASF lesions resembling a systemic hemorrhagic disease [6]. While no data exist for wild boar that died of ASF under natural conditions, pathology in domestic pigs has recently been described [21,34,35]. Typically, domestic pigs show comparable lesions such as hemorrhagic lymphadenopathy, splenomegaly, pulmonary consolidation and edema, hemorrhages in the heart and kidneys and hepatomegaly with edema of the gallbladder wall, as well as edematous, hemorrhagic meninges.

While most of the macroscopic findings in this study have been described after experimental infection in wild boar [36], they do not reflect the severity and diversity seen under field conditions. Comparing the three different virus variants, striking, but not significant, differences were evident. Interestingly, the highest total score for gross pathological changes was given for wild boar from SN infected with variant IV, followed by animals from MOL infected with variant III, which showed an intermediated total score, and wild boar from LOS infected with variant II, had the lowest macroscopical score.

For ASF, four different courses of the disease have been described and include peracute, acute, subacute and chronic stages, which are associated with typical lesions [6]. Petrov et al. [33] moreover specified the subacute stage as chronic-like and differentiated into lethal and transient course after infection with moderately virulent ASFV. Gross pathomorphological changes of the subacute/chronic-like stage include multifocal hemorrhages, edema, lymphadenitis, interstitial pneumonia and ascites [6,27,33] whereas bacterial secondary infections inducing fibrinous polyserositis, chronic pneumonia and necrosis of tonsils, however, without vascular changes, predominate in chronic courses [6]. Lesions in acutely and chronically ASFV-infected domestic pigs were also already presented in detail decades ago [12]. The animals with chronic disease showed comparable lesions as observed in the acutely infected pigs, but additionally revealed chronic changes particularly including pericarditis, pneumonia and lymphadenitis. In the present study, in contrast to the animals from LOS and MOL, although without statistical significance, wild boar from SN more frequently showed lesions most likely associated with bacterial infections indicative for a lethal subacute protracted disease course.

More specifically, chronic inflammatory processes such as fibrous pericarditis, pleuropneumonia and peritonitis were more frequently detected in SN animals. At the same time, wild boar from SN, and to a lesser extent also animals from MOL, tended to show more severe hemorrhages in the urinary bladder and bone marrow, but fewer acute hemorrhages as detected in the hearts of animals from LOS. Detailed pathomorphological investigation of experimentally infected wild boar that succumbed to highly virulent ASFV "Armenia07" infection revealed only mild petechiae of the urinary bladder, variable hemorrhages of the heart and congestion of the bone marrow while extensive hemorrhages or lesions induced by other circulating pathogens were absent [36].

Based on this, and in line with virological and immunohistochemical data, this may indicate that at least wild boar infected with the SN variant experienced a more protracted disease course than pigs from LOS suggesting a slightly decreased virulence of the SN virus variant IV to wild boar that still led to the death of the respective animals. This demonstrates that veterinarians, hunters and farmers need to be aware of less acute courses of ASF, usually attributed to classical swine fever, in order to consider this as important differential diagnosis in each case. However, considering the small number of carcasses and the indefinite sample material, this should be interpreted with caution and must be confirmed experimentally under standardized conditions in any case.

Although the majority of organs could be assessed macroscopically, we had to refrain from a detailed semiquantitative histopathological analysis because autolysis had already progressed too far in some cases, which would have considerably reduced the number of samples for investigation. However, in line with macroscopic findings, histopathology confirmed the severe course of disease in all animals regardless of the outbreak area and the virus variant. Since most wild boar studies focus only on macroscopic pathology, it is even more important to study the histopathology of natural ASF infection in more depth [17,18,36,37].

Most of the histopathological findings obtained in this study are fully comparable with those observed in domestic animals investigated upon outbreaks [21,34]. However, some of the observed lesions have already been described, but are not associated with ASF in the first line. For example, adrenal hemorrhages, which have been described to occur in wild boar upon experimental infection [36], were examined in more histopathological detail and revealed interesting results in the present study. Our findings mirror a condition known as Waterhouse Friderichsen syndrome [38]. It has been correlated with several bacterial and viral diseases and is characterized by severe hemorrhage, necrosis and microvascular thrombosis. Although the pathophysiology is not fully understood, hemorrhages are explained by a stress-induced release of adrenaline, vasculitis and coagulation disorders including disseminated intravascular coagulation. In line with the latter, microvascular thrombosis could be shown in multiple organs as signs of acute organ injury in wild boar investigated in this study [12].

Of note, histopathology further highlighted the unique finding of localized inflammation of the cerebral choroid plexus, which occurred in wild boar irrespective of the outbreak district, but mainly affected the majority of animals from MOL and SN. So far, there are only minor reports on ASF lesions in the central nervous system [12,34,39], which can occur at all stages of the disease as demonstrated by Moulton and Coggins [12] in acutely and chronically succumbing as well as in surviving pigs after experimental and natural infection. In addition to mononuclear infiltration of meningeal and cerebral vessels, perivascular hemorrhage, occasional vascular thrombosis and neuronal degeneration, necrosis of the choroid plexus epithelium has been described only once in a few acutely infected animals [12]. The naturally infected wild boar presented in this study showed pronounced mononuclear inflammation with massive cell deaths in addition to occasional necrosis of the plexus epithelium, again suggesting a longer disease course, at least in animals from SN.

To further extrapolate how long naturally infected wild boar might have lived with the disease, antibody titers were determined and compared to those of surviving ASFV "Estonia 2014" experimentally infected domestic pigs from a previous trial. In domestic pigs, low antibody titers were detectable from day 14 to a maximum titer of 400, then increased to a max of 3200 by day 28, and remained constantly high until 41 days post infection, at least in one domestic pig. However, the other pig showed a drop from 3200 to 800 on day 35 pi and a second subsequent increase. While it cannot be excluded that a consumption or decay of antibodies occurred, one should also consider measurement inaccuracies of the semiquantitative test when targeting the fluctuant antibody titers. When comparing this to wild boar, which showed titers of at least 200, the majority of animals independent of the outbreak area might have lived with ASF for more than 14 days.

As suspected, based on the pathological data in animals from SN, but also MOL, the course of the disease was probably longer since they tended to show higher titers of max 3200 and 1600, respectively, while wild boar from LOS reached titers of only max 800. Surprisingly, antibody titers showed no clear correlation to the chronicity of lesions observed in several wild boar since one animal from SN with obviously chronic lesions produced only minimal antibody titers. On the one hand, the chronic lesions in this animal could have already existed before and might not necessarily be associated with ASFV infection. On the other hand, as hypothesized above, the antibodies may have declined over time. To date, little is known about the host's immune response against ASFV, but it is of general acceptance that antibodies directed against ASFV are not sufficient for protection against the disease [40]. However, experiments to investigate the dynamics of antibody development in ASF could be useful to draw conclusions on the disease in wildlife.

5. Conclusions

In summary, this is the first study describing the lesion spectrum in wild boar succumbing to ASF after infection with the different virus variants that have emerged within one year in Germany. Virological and pathomorphological data suggest possible differences in the virulence of the variants. At least, wild boar infected with the SN variant IV tended to experience a more protracted but nevertheless lethal disease course compared to animals infected with LOS variant II or the MOL variant III, which is more likely to be classified as intermediate. These findings are particularly important with regard to the spread and continued occurrence of the ASFV in endemic areas. To elucidate the pathogenicity and differences in the virulence and disease dynamics of the emerging virus variants more thoroughly, further experimental studies in wild boar as well as comparative investigations in domestic pigs under late human endpoint conditions are urgently needed. These studies should also address the impact of protracted disease courses on shedding and thus transmission characteristics.

Supplementary Materials: The following supporting information can be downloaded at: https://www.mdpi.com/article/10.3390/pathogens11111386/s1, Table S1: Summary of individual organ genome copy numbers in wild boar; Table S2: Summary list of histopathological changes and immunohistochemistry results in wild boar; Table S3: Summary list of gross lesions scored on a semiquantitative scale in wild boar; Figure S1: Relative spleen weights of naturally ASFV-infected wild boar, Figure S2: Histopathology of the spleen of naturally ASFV-infected wild boar carcasses, Figure S3: Macroscopical findings of the liver in German ASFV-infected wild boar carcasses, Figure S4: Histopathological results detected in the liver of naturally ASFV-infected wild boar carcasses from Germany, Figure S5: Gross pathology of the gastrointestinal tract in naturally ASF-infected wild boar carcasses from German outbreak areas, Figure S6: Antibody titers determined by immunoperoxidase test in German wild boar carcasses compared to experimentally infected domestic pigs on different days pi, File S1: Detailed analysis.

Author Contributions: Conceptualization, J.S.-E. and S.B.; Methodology, J.S.-E., S.B., P.D. and A.B.; Validation, J.S.-E. and S.B.; Formal Analysis, J.S.-E. and P.D.; Investigation, J.S.-E., S.B., P.D. and A.B.; Resources, S.B. and J.S.-E.; Data Curation, J.S.-E. and P.D.; Writing—Original Draft Preparation, J.S.-E. and P.D.; Writing—Review and Editing, J.S.-E., S.B., P.D. and A.B.; Visualization, J.S.-E. and P.D.; Supervision, J.S.-E. and S.B. All authors have read and agreed to the published version of the manuscript.

Funding: This research received no external funding.

Informed Consent Statement: Not applicable.

Data Availability Statement: All data available can be obtained on request from the corresponding authors.

Acknowledgments: We would like to thank the Brandenburg Crisis Center and the veterinary authorities in the districts Märkisch-Oderland, Oder-Spree and Spree-Neiße for their help and the submission of the examined wild boar carcasses. Further, we would like to thank Christian Loth, Silvia Schuparis, Ulrike Kleinert and Robin Brandt for their valuable support during necropsies and in the laboratory.

Conflicts of Interest: The authors declare no conflict of interest.

References

1. Blome, S.; Franzke, K.; Beer, M. African swine fever—A review of current knowledge. *Virus Res.* **2020**, *287*, 198099. [CrossRef] [PubMed]
2. Sauter-Louis, C.; Forth, J.H.; Probst, C.; Staubach, C.; Hlinak, A.; Rudovsky, A.; Holland, D.; Schlieben, P.; Göldner, M.; Schatz, J.; et al. Joining the club: First detection of African swine fever in wild boar in Germany. *Transbound Emerg. Dis.* **2021**, *68*, 1744–1752. [CrossRef]
3. Sánchez-Vizcaíno, J.M.; Mur, L.; Gomez-Villamandos, J.C.; Carrasco, L. An update on the epidemiology and pathology of African swine fever. *J. Comp. Pathol.* **2015**, *152*, 9–21. [CrossRef] [PubMed]
4. Gabriel, C.; Blome, S.; Malogolovkin, A.; Parilov, S.; Kolbasov, D.; Teifke, J.P.; Beer, M. Characterization of African swine fever virus Caucasus isolate in European wild boars. *Emerg. Infect. Dis.* **2011**, *17*, 2342–2345. [CrossRef] [PubMed]
5. Sauter-Louis, C.; Conraths, F.J.; Probst, C.; Blohm, U.; Schulz, K.; Sehl, J.; Fischer, M.; Forth, J.H.; Zani, L.; Depner, K.; et al. African Swine Fever in Wild Boar in Europe-A Review. *Viruses* **2021**, *13*, 1717. [CrossRef]
6. Salguero, F.J. Comparative Pathology and Pathogenesis of African Swine Fever Infection in Swine. *Front. Vet. Sci.* **2020**, *7*, 282. [CrossRef] [PubMed]
7. Forth, J.H.; Calvelage, S.; Fischer, M.; Hellert, J.; Sehl-Ewert, J.; Roszyk, H.; Deutschmann, P.; Reichold, A.; Lange, M.; Thulke, H.-H.; et al. African swine fever virus—Variants on the rise. *Emerg. Microbes Infect.* **2022**. [CrossRef] [PubMed]
8. Afonso, C.L.; Piccone, M.E.; Zaffuto, K.M.; Neilan, J.; Kutish, G.F.; Lu, Z.; Balinsky, C.A.; Gibb, T.R.; Bean, T.J.; Zsak, L.; et al. African swine fever virus multigene family 360 and 530 genes affect host interferon response. *J. Virol.* **2004**, *78*, 1858–1864. [CrossRef] [PubMed]
9. O'Donnell, V.; Holinka, L.G.; Gladue, D.P.; Sanford, B.; Krug, P.W.; Lu, X.; Arzt, J.; Reese, B.; Carrillo, C.; Risatti, G.R.; et al. African Swine Fever Virus Georgia Isolate Harboring Deletions of MGF360 and MGF505 Genes Is Attenuated in Swine and Confers Protection against Challenge with Virulent Parental Virus. *J. Virol.* **2015**, *89*, 6048–6056. [CrossRef]

10. Gallardo, C.; Soler, A.; Nieto, R.; Cano, C.; Pelayo, V.; Sánchez, M.A.; Pridotkas, G.; Fernandez-Pinero, J.; Briones, V.; Arias, M. Experimental Infection of Domestic Pigs with African Swine Fever Virus Lithuania 2014 Genotype II Field Isolate. *Transbound Emerg. Dis.* **2017**, *64*, 300–304. [CrossRef]
11. Zani, L.; Forth, J.H.; Forth, L.; Nurmoja, I.; Leidenberger, S.; Henke, J.; Carlson, J.; Breidenstein, C.; Viltrop, A.; Hoper, D.; et al. Deletion at the 5′-end of Estonian ASFV strains associated with an attenuated phenotype. *Sci. Rep.* **2018**, *8*, 6510. [CrossRef] [PubMed]
12. Moulton, J.; Coggins, L. Comparison of lesions in acute and chronic African swine fever. *Cornell Vet.* **1968**, *58*, 364–388. [PubMed]
13. Moulton, J.E.; Pan, I.C.; Hess, W.R.; DeBoer, C.J.; Tessler, J. Pathologic features of chronic pneumonia in pigs with experimentally induced African swine fever. *Am. J. Vet. Res.* **1975**, *36*, 27–32. [PubMed]
14. Hervás, J.; Gómez-Villamandos, J.C.; Méndez, A.; Carrasco, L.; Sierra, M.A. The lesional changes and pathogenesis in the kidney in African swine fever. *Vet. Res. Commun.* **1996**, *20*, 285–299. [CrossRef]
15. Gómez-Villamandos, J.C.; Hervás, J.; Méndez, A.; Carrasco, L.; Villeda, C.J.; Wilkinson, P.J.; Sierra, M.A. Pathological changes in the renal interstitial capillaries of pigs inoculated with two different strains of African swine fever virus. *J. Comp. Pathol.* **1995**, *112*, 283–298. [CrossRef]
16. Pikalo, J.; Schoder, M.-E.; Sehl-Ewert, J.; Breithaupt, A.; Cay, A.B.; Lhoëst, C.; van Campe, W.; Mostin, L.; Deutschmann, P.; Roszyk, H.; et al. Towards Efficient Early Warning: Pathobiology of African Swine Fever Virus "Belgium 2018/1" in Domestic Pigs of Different Age Classes. *Animals* **2021**, *11*, 2602. [CrossRef] [PubMed]
17. Pikalo, J.; Schoder, M.E.; Sehl, J.; Breithaupt, A.; Tignon, M.; Cay, A.B.; Gager, A.M.; Fischer, M.; Beer, M.; Blome, S. The African swine fever virus isolate Belgium 2018/1 shows high virulence in European wild boar. *Transbound Emerg. Dis.* **2020**, *67*, 1654–1659. [CrossRef] [PubMed]
18. Sehl, J.; Pikalo, J.; Schäfer, A.; Franzke, K.; Pannhorst, K.; Elnagar, A.; Blohm, U.; Blome, S.; Breithaupt, A. Comparative Pathology of Domestic Pigs and Wild Boar Infected with the Moderately Virulent African Swine Fever Virus Strain "Estonia 2014". *Pathogens* **2020**, *9*, 662. [CrossRef]
19. Nurmoja, I.; Petrov, A.; Breidenstein, C.; Zani, L.; Forth, J.H.; Beer, M.; Kristian, M.; Viltrop, A.; Blome, S. Biological characterization of African swine fever virus genotype II strains from north-eastern Estonia in European wild boar. *Transbound Emerg. Dis.* **2017**, *64*, 2034–2041. [CrossRef] [PubMed]
20. Pietschmann, J.; Guinat, C.; Beer, M.; Pronin, V.; Tauscher, K.; Petrov, A.; Keil, G.; Blome, S. Course and transmission characteristics of oral low-dose infection of domestic pigs and European wild boar with a Caucasian African swine fever virus isolate. *Arch. Virol.* **2015**, *160*, 1657–1667. [CrossRef]
21. Nga, B.T.T.; Tran Anh Dao, B.; Nguyen Thi, L.; Osaki, M.; Kawashima, K.; Song, D.; Salguero, F.J.; Le, V.P. Clinical and Pathological Study of the First Outbreak Cases of African Swine Fever in Vietnam, 2019. *Front. Vet. Sci.* **2020**, *7*, 392. [CrossRef] [PubMed]
22. Pornthummawat, A.; Truong, Q.L.; Hoa, N.T.; Lan, N.T.; Izzati, U.Z.; Suwanruengsri, M.; Nueangphuet, P.; Hirai, T.; Yamaguchi, R. Pathological lesions and presence of viral antigens in four surviving pigs in African swine fever outbreak farms in Vietnam. *J. Vet. Med. Sci.* **2021**, *83*, 1653–1660. [CrossRef]
23. Oh, T.; Do, D.T.; Lai, D.C.; Nguyen, T.C.; Vo, H.V.; Chae, C. Age-related viral load and severity of systemic pathological lesions in acute naturally occurring African swine fever virus genotype II infections. *Comp. Immunol. Microbiol. Infect. Dis.* **2021**, *79*, 101709. [CrossRef]
24. Brooks, J.W. Postmortem Changes in Animal Carcasses and Estimation of the Postmortem Interval. *Vet. Pathol.* **2016**, *53*, 929–940. [CrossRef] [PubMed]
25. Galindo-Cardiel, I.; Ballester, M.; Solanes, D.; Nofrarias, M.; Lopez-Soria, S.; Argilaguet, J.M.; Lacasta, A.; Accensi, F.; Rodriguez, F.; Segales, J. Standardization of pathological investigations in the framework of experimental ASFV infections. *Virus Res.* **2013**, *173*, 180–190. [CrossRef]
26. King, D.P.; Reid, S.M.; Hutchings, G.H.; Grierson, S.S.; Wilkinson, P.J.; Dixon, L.K.; Bastos, A.D.; Drew, T.W. Development of a TaqMan PCR assay with internal amplification control for the detection of African swine fever virus. *J. Virol. Methods* **2003**, *107*, 53–61. [CrossRef]
27. Sánchez-Cordón, P.J.; Vidaña, B.; Neimanis, A.; Núñez, A.; Wikström, E.; Gavier-Widén, D. 4—Pathology of African swine fever. In *Understanding and Combatting African Swine Fever*; Wageningen Academic Publishers: Wageningen, The Netherlands, 2021; pp. 87–139.
28. Spiekermeier, I.; Freitag, M.; Baumgärtner, W. Interstitial nephritis in pigs. *Tierarztl. Prax. Ausg. G Grosstiere/Nutztiere* **2017**, *45*, 350–356.
29. Paz-Sánchez, Y.; Herráez, P.; Quesada-Canales, Ó.; Poveda, C.G.; Díaz-Delgado, J.; Quintana-Montesdeoca, M.D.P.; Plamenova Stefanova, E.; Andrada, M. Assessment of Lung Disease in Finishing Pigs at Slaughter: Pulmonary Lesions and Implications on Productivity Parameters. *Animals* **2021**, *11*, 3604. [CrossRef] [PubMed]
30. Nakagawa, M.; Yoshihara, S.; Suda, H.; Ikeda, K. Pathological studies on white spots of the liver in fattening pigs. *Natl. Inst. Anim. Health Q.* **1983**, *23*, 138–149.
31. Rosell, C.; Segalés, J.; Domingo, M. Hepatitis and Staging of Hepatic Damage in Pigs Naturally Infected with Porcine Circovirus Type 2. *Vet. Pathol.* **2000**, *37*, 687–692. [CrossRef]

32. Lin, Z.; Li, M.; Wang, Y.S.; Tell, L.A.; Baynes, R.E.; Davis, J.L.; Vickroy, T.W.; Riviere, J.E. Physiological parameter values for physiologically based pharmacokinetic models in food-producing animals. Part I: Cattle and swine. *J. Vet. Pharmacol. Ther.* **2020**, *43*, 385–420. [CrossRef]
33. Petrov, A.; Forth, J.H.; Zani, L.; Beer, M.; Blome, S. No evidence for long-term carrier status of pigs after African swine fever virus infection. *Transbound Emerg. Dis.* **2018**, *65*, 1318–1328. [CrossRef]
34. Izzati, U.Z.; Inanaga, M.; Hoa, N.T.; Nueangphuet, P.; Myint, O.; Truong, Q.L.; Lan, N.T.; Norimine, J.; Hirai, T.; Yamaguchi, R. Pathological investigation and viral antigen distribution of emerging African swine fever in Vietnam. *Transbound Emerg. Dis.* **2021**, *68*, 2039–2050. [CrossRef]
35. Truong, A.D.; Ly, D.V.; Vu, T.H.; Hoang, V.T.; Nguyen, T.C.; Chu, T.N.; Nguyen, H.T.; Nguyen, T.V.; Pham, N.T.; Tran, H.T.T.; et al. Unexpected cases in field diagnosis of African swine fever virus in Vietnam: The needs consideration when performing molecular diagnostic tests. *Open Vet. J.* **2020**, *10*, 189–197. [CrossRef]
36. Rodríguez-Bertos, A.; Cadenas-Fernández, E.; Rebollada-Merino, A.; Porras-González, N.; Mayoral-Alegre, F.J.; Barreno, L.; Kosowska, A.; Tomé-Sánchez, I.; Barasona, J.A.; Sánchez-Vizcaíno, J.M. Clinical Course and Gross Pathological Findings in Wild Boar Infected with a Highly Virulent Strain of African Swine Fever Virus Genotype II. *Pathogens* **2020**, *9*, 688. [CrossRef]
37. Roszyk, H.; Franzke, K.; Breithaupt, A.; Deutschmann, P.; Pikalo, J.; Carrau, T.; Blome, S.; Sehl-Ewert, J. The Role of Male Reproductive Organs in the Transmission of African Swine Fever—Implications for Transmission. *Viruses* **2022**, *14*, 31.
38. Mascolo, P.; Feola, A.; Zangani, P.; Famularo, D.; Liguori, B.; Mansueto, G.; Campobasso, C.P. Waterhouse Friderichsen Syndrome: Medico-legal issues. *Forensic Sci. Int. Rep.* **2021**, *4*, 100235. [CrossRef]
39. Karalyan, Z.A.; Sargsyan, M.A.; Arzumanyan, H.H.; Kotsinyan, A.A.; Hakobyan, L.H.; Karalova, E.M.; Voskanyan, H.E. Pathomorphology of the brain in the acute form of African swine fever. *Ann. Parasitol.* **2017**, *63*, 347–352. [PubMed]
40. Neilan, J.G.; Zsak, L.; Lu, Z.; Burrage, T.G.; Kutish, G.F.; Rock, D.L. Neutralizing antibodies to African swine fever virus proteins p30, p54, and p72 are not sufficient for antibody-mediated protection. *Virology* **2004**, *319*, 337–342. [CrossRef] [PubMed]

pathogens

Article

Transcriptome Profiling in Swine Macrophages Infected with African Swine Fever Virus (ASFV) Uncovers the Complex and Close Relationship with Host

Zhaoyao Li [1,2,†], Wenxian Chen [1,2,†], Xiaowen Li [1,2], Keke Wu [1,2], Xinyan Wang [1], Weijun Wang [1], Yuwan Li [1,3], Lin Yi [1,3], Mingqiu Zhao [1,2,3], Hongxing Ding [1,2,3], Shuangqi Fan [1,2,*] and Jinding Chen [1,2,3,*]

1 College of Veterinary Medicine, South China Agricultural University, Guangzhou 510642, China
2 Guangdong Laboratory for Lingnan Modern Agriculture, Guangzhou 510642, China
3 Key Laboratory of Zoonosis Prevention and Control of Guangdong Province, Guangzhou 510642, China
* Correspondence: hnndfsq@126.com (S.F.); jdchen@scau.edu.cn (J.C.); Fax: +86-20-8528-0245 (J.C.)
† These authors contributed equally to this work.

Abstract: African swine fever virus (ASFV) is a pathogen to cause devastating and economically significant diseases in domestic and feral swine. ASFV mainly infects macrophages and monocytes and regulates its replication process by affecting the content of cytokines in the infected cells. There is a limited understanding of host gene expression and differential profiles before and after ASFV infection in susceptible cells. In this study, RNA-seq technology was used to analyze the transcriptomic change in PAMs infected with ASFV at different time points (0 h, 12 h, 24 h). As a result, a total of 2748, 1570, and 560 genes were enriched in group V12 h vs. MOCK, V24 h vs. MOCK, and V24 h vs. V12 h, respectively. These DEGs (differentially expressed genes) in each group were mainly concentrated in the KEGG (Kyoto Encyclopedia of Genes and Genomes) pathways related to innate immunization and inflammation, including the NF-κB signaling pathway, Toll-like receptor signaling pathway, TNF signaling pathway, IL-17 signaling pathway, cytokine-cytokine receptor interaction, and chemokine signaling pathway. Furthermore, the increased levels of IL-1β, TNF-α, IKKβ, CXCL2, and TRAF2 and decreased level of IκBα were validated through the qPCR method. These results suggested that ASFV infection can activate the NF-κB signaling pathway in the early stage. In general, this study provides a theoretical basis for further understanding the pathogenesis and immune escape mechanism of ASFV.

Keywords: African swine fever virus (ASFV); RNA-seq; innate immunity; inflammation; NF-κB signaling pathway

Citation: Li, Z.; Chen, W.; Li, X.; Wu, K.; Wang, X.; Wang, W.; Li, Y.; Yi, L.; Zhao, M.; Ding, H.; et al. Transcriptome Profiling in Swine Macrophages Infected with African Swine Fever Virus (ASFV) Uncovers the Complex and Close Relationship with Host. *Pathogens* **2022**, *11*, 1411. https://doi.org/10.3390/pathogens11121411

Academic Editors: Lauro Velazquez-Salinas, Gisselle N. Medina, Elizabeth Ramirez-Medina and Natasha N. Gaudreault

Received: 30 August 2022
Accepted: 19 November 2022
Published: 24 November 2022

Publisher's Note: MDPI stays neutral with regard to jurisdictional claims in published maps and institutional affiliations.

1. Introduction

African swine fever (ASF), a devastating disease for the livestock industry, seriously threatens global pork production and food security. African swine fever virus (ASFV) causes a virulent, hemorrhagic disease in infected domestic and wild pigs with a mortality rate of up to 100% [1]. Infected pigs show signs of high fever, systemic hemorrhage, vomiting, blood in stools, as well as respiratory disturbances. Splenomegaly and lymphatic hemorrhage could also be observed in diseased pigs. ASF was first reported in East Africa in the early 1900s, then spread to different countries in Africa, Europe, America, and Asia [2,3]. In August 2018, ASF entered China causing great economic losses [4]. Ma et al. collected and analyzed the related data about ASF cases throughout the entire China mainland and discovered that ASF transmission showed a northeast-southwest directional trend [4]. It was considered necessary to restrict the transportation of live pigs and pork products from ASF-affected areas and reduce the pig breeding density [4]. The pig industry in China contracts for almost half of global pork production; therefore, the outbreak of the ASF epidemic in August 2018 has had a great impact on the Chinese economy. It was estimated that China experienced direct economic losses of US $141 billion in September 2019 [5]. In addition, the rise in pork prices has had a negative impact on consumers. According to the OIE (2022), ASF has been reported in 35 countries since January 2020, involving more than 1,100,000 pigs with more than 1,800,000 animal losses [6]. The total consumption of animal feeds such as soy has also been affected by the declining number of pigs [5].

ASFV has a sizeable linear dsDNA genome of 170 to 193 kb, containing 150–167 open reading frames (ORF) [7]. The viral genome encodes more than 170 proteins, many of which have been confirmed to help viruses escape the host immune system by utilizing different mechanisms, including interferon (IFN) response inhibition, inflammatory response, apoptosis, and autophagy. Several members of the multigene family 360 (MGF360) and MGF505 strongly inhibited IL-1β maturation and IFN-β promoter activation in porcine alveolar macrophages (PAMs) [8]. Anyway, I239L, A276R, DP96R, and E120R were identified as having an essential role in the negative regulation of type I interferon [9–12]. Some ASFV proteins, such as A179Lp, A224Lp, and EP402R, regulate programmed cell death pathways in the early stage of viral infection. ASFV E199L protein induced a complete autophagy process in Vero and HEK-293T cells [13]. In addition, ASFV A137R protein inhibited IFN-β production through the autophagy-mediated lysosomal degradation of TANK-binding kinase 1 (TBK1) [14]. ASFV mainly infects macrophages and monocytes and regulates its replication process by affecting the content of cytokines in the infected cells. Although the immune functions of many ASFV proteins have been characterized, the mechanism by which ASFV interacts with the host remains unclear, given its large and complex genome structure.

In this study, as the primary target cells for ASFV, PAMs were used to construct a cell model of ASFV infection. RNA sequencing (RNA-seq) of the transcriptome is now a common method to analyze the gene expression difference in cells or tissues [15]. RNA-seq technology was utilized to analyze the gene expression patterns in PAMs infected with ASFV at different time points (0 h, 12 h, 24 h). This study aimed to understand the ASFV pathogenic characteristics from the perspective of immune pathway changes in PAMs before and after infection to provide a theoretical basis for further study of the pathogenesis and immune escape mechanism of ASFV.

2. Materials and Methods

2.1. Cell Culture and ASFV Infection of PAMs

The isolated PAMs were maintained in RPMI 1640 medium supplemented with 10% FBS at 37 °C with 5% CO_2. When the number of viable cells in the T75 cell culture flask reached 7×10^6, they were infected with ASFV pig/HLJ/2018 strain (MOI = 3) and incubated for 1 h at 37 °C before replacing the cell culture medium. PAMs infected for different time points (0 h, 12 h, and 24 h) were collected and used for RNA/DNA extraction.

2.2. Real-Time Quantitative PCR (qPCR)

Total RNA and ASFV genomic DNA were extracted from PAMs using Trizol reagent and E.Z.N.A. Viral DNA kit (OMEGA, New York, NY, USA), respectively. Genomic RNA was reverse transcribed with the PrimeScript RT kit (TaKaRa, San Jose, CA, USA). In order to verify the viral load and the results obtained by the RNA-seq analysis in ASFV infected PAMs, qPCR was performed with HiScript II One Step qRT-PCR SYBR Green Kit (Vazyme, Nanjing, China) for target genes amplification in Bio-rad equipment using 96-well plates. qPCR amplification conditions were as follows: one cycle of 95 °C for 30 s, 40 cycles of 95 °C for 10 s and 60 °C for 30 s, and one cycle of 95 °C for 15 s, 60 °C for 1 min and 95 °C for 15 s. All samples were analyzed in triplicate. The relative mRNA levels of ASFV and host genes were normalized to the swine β-actin mRNA level. Relative expression levels of the target genes were calculated using the comparative cycle threshold ($2^{-\Delta\Delta CT}$) method [16]. The information about qPCR primers related to ASFV detection and other host genes is listed in Table 1.

Table 1. qPCR primers used in this study.

Primers	Sequence (5′-3′)	References or Genbank
ASFV-B646L-F	CTGCTCATGGTATCAATCTTATCGA	[17]
ASFV-B646L-R	GATACCACAAGATCRGCCGT	
CXCL8-F	AGC CCG TGT CAA CAT GAC TT	[18]
CXCL8-R	TGG AAA GGT GTG GAA TGC GT	
LDHA-F	CGTCAGCAAGAGGGAGA	[19]
LDHA-R	AAGCACTGGATTGGAAGCAACAA	
ENO1-F	AAGCCCTGGAGCTGCTGA	XM_021095279.1
ENO1-R	CGTACTTGCCCGACCTGTAGAA	
SQSTM1-F	GCTGCTCTTCCGACCCT	XM_003123639.4
SQSTM1-R	GCGATCTTATTCATTTGCTCC	
CD163-F	ATTCATCATCCTCGGACCCAT	[20]
CD163-R	CCCAGCACAACGACCACCT	
CD9-F	CCAGGATTTCTACAGGGACA	NM_214006.1
CD9-R	GCATAGTGGATGGCTTTCAG	
CD74-F	AGAGCAAGTGCAGCCGTGGAG	[21]
CD74-R	GGTACAGGAAGTAGGCGGTGGTG	
EEF1A1-F	AGTGCTAATATGCCTTGGTT	NM_001097418.2
EEF1A1-R	TTGTCAGTTGGACGAGTTG	
CTSD-F	ACGTGAAGAACGGCACCACC	NM_001037721.1
CTSD-R	GCCCAACAACGCAGAATTACA	
APOE-F	TGGGAGGAGTCCAAGTGGCA	NM_214308.1
APOE-R	GCTCCGTCAGTTCCTGGGTGA	
IL-1β-F	ACCTGGACCTTGGTTCTCTG	[22]
IL-1β-R	CATCTGCCTGATGCTCTTG	
TNF-α-F	TGGCCCAAGGACTCAGATCAT	[23]
TNF-α-R	TCGGCTTTGACATTGGCTACA	
IκBα-F	ACCAACCAGCCAGAAATCG	NM_001005150.1
IκBα-R	CACAGGCAAGGTGTAGAGGG	
IKKβ-F	AGAGGATCTTCTGCGAGTA	NM_001244129.1
IKKβ-R	CTTTGGGTGCGTAACTG	
CXCL2-F	GTGGAAACAGCAACTGCTCA	[18]
CXCL2-R	AGGGCTTGGTAGTTGTCAGG	
TRAF2-F	CCACCGCTACTGCTCCTACTGC	XM_005652719.2
TRAF2-R	CGCCTTCTTCATAAATGCCCTC	
GAPDH-F	TGGAGTCCACTGGTGTCTTCAC	[23]
GAPDH-R	TTCACGCCCATCACAAACA	

2.3. Transcriptome Sequencing and Data Analysis

The total RNA of PAMs was extracted using Trizol Reagent following the manufacturer's protocol. mRNA was enriched by magnetic beads with Oligo(dT) and fragmented by fragmentation buffer. The cDNA, which utilized the above RNA as a template, was synthesized in the presence of random hexamers, DNA polymerase I, dNTPs, and RNase H. The amplified cDNA was subjected to end repair and adaptor ligation, and PCR. The PCR products were purified by AMPure XP beads to obtain the final cDNA library. The cDNA library was sequenced using Illumina high-throughput sequencing platform and converted to Raw Reads. After quality control and filtration by fastp software and Bowtie2 software, the filtering readings were aligned against with reference genome using HISAT2, and the differential expression conditions of genes among different samples were analyzed using RSEM and edgeR. The false discovery rate (FDR) ≤ 0.05 and the criteria of a fold difference $|\log2FC| \geq 1$ were considered for DEGs. GO (Gene Ontology) analysis and KEGG analysis were performed using cluster Profiler software. All data were obtained from at least three replicates. Values of $p \leq 0.05$ were assumed to be statistically significant.

2.4. Statistical Analysis of Alternative Splicing Events

Alternative Splicing (AS) is a common gene expression mode in most eukaryotic cells. In eukaryotes, different mature mRNA and proteins can be produced by the same gene encoding mRNA through different splicing modes. The rMATS is a software developed for RNA-seq data, which can classify not only alternative splicing events but also perform differential analysis of alternative splicing events between different samples. In this study, rMATS was used to quantify the expression of alternative splicing events in different samples.

3. Results

3.1. ASFV Infection Conditions in PAMs

To validate whether PAMs were successfully infected with ASFV, the viral load in PAMs infected with ASFV (MOI = 3) for 0 h,12 h, and 24 h were determined utilizing qPCR methods. As shown in Figure 1, ASFV was detected from ASFV-infected PAMs, and the viral load in cells increased over time. The results indicated that PAMs were successfully infected with ASFV.

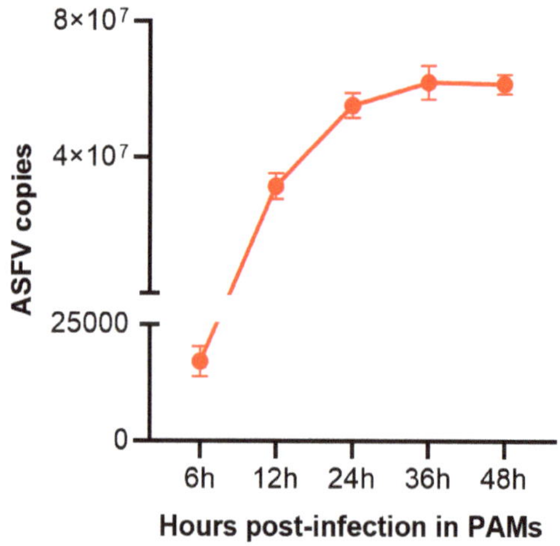

Figure 1. The viral copies number in PAMs infected with ASFV for different times (6 h, 12 h, 24 h,36 h, 48 h). Data are presented as mean ± SD of three independent experiments.

3.2. Gene Expression Statistics and Differential Analysis among Different Samples

The correlation of transcriptome gene expression among infected and mock-infected samples was analyzed, and the Pearson correlation coefficient was obtained to make a heatmap. In addition, PCA (Principal Component Analysis) analysis was also performed. These results showed that the sample biological repeatability in each group as well. In addition, the V12 h and V24 h groups were significantly different from the control group. The transcriptome profiles of each group (V12 h vs. MOCK, V24 h vs. MOCK, V24 h vs. V12 h) were compared using edgeR software, and the DEGs quantities were identified based on the screening conditions of DEGs with FDR $\leq$ 0.05 and $|\log2FC| \geq 1$. In addition, compared with the V12 h group, the number of DEGs in the V24 h group was 560, among which the level of 309 genes was upregulated (Figure 2c). The results of the Venn diagram revealed that the number of common upregulated and downregulated DEGs in the three groups was 12 and 10, independently (Figure 2a,c).

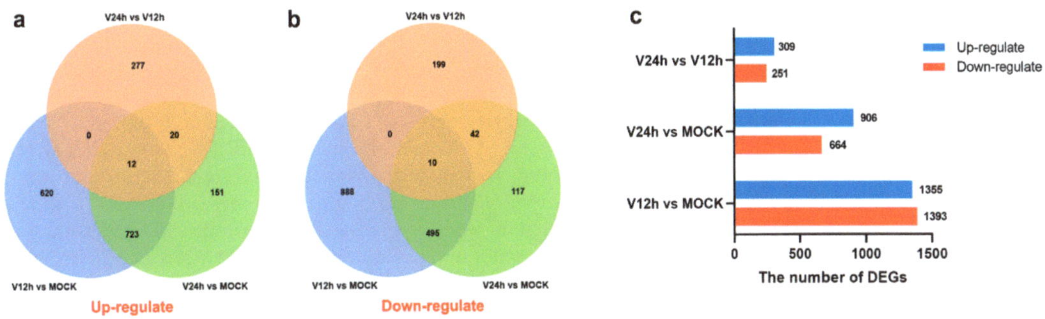

Figure 2. (**a**) Venn diagram of significantly upregulated genes in different groups. (**b**) Venn diagram of significantly downregulated genes in different groups. (**c**) The total number of up/down-regulated DEGs in different groups (FDR ≤ 0.05 and $|\log2FC| \geq 1$).

Hierarchical clustering analysis of the relationship between samples and genes was performed based on gene expression, and the clustering results of the top 30 expressed genes are shown in Figure 3a. Furthermore, 10 genes (CXCL8, LDHA, ENO1, SQSTM1, CD163, CD9, CD74, EEF1A1, CTSD, APOE) were picked out for RT-qPCR to verify the accuracy of the RNA-seq data. As shown in Figure 3b–k, the RT-qPCR results of these genes were generally consistent with their transcriptomic data, which indicated the accuracy and validity of the RNA-seq data.

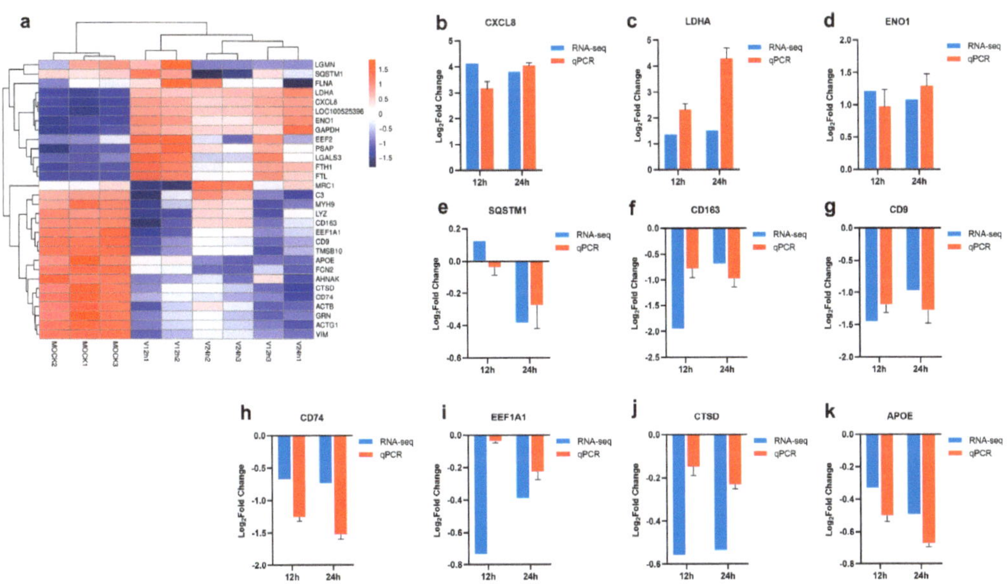

Figure 3. Validation of RNA-seq data by RT-qPCR. (**a**) Cluster heatmap of the top 30 expressed genes. (**b**–**k**) RT-qPCR validation of representative 10 genes selected from the top 30 expressed genes.

3.3. KEGG Pathway Analysis of DEGs

To further clarify the relevant pathways and potential biological functions involved in differentially expressed genes, the KEGG database was used for enrichment analysis of the DEGs in each group. The enrichment results of the top 30 KEGG pathways with small P-values are shown in Figure 4. Compared with the MOCK group, the DEGs in the V12h group (Figure 4a) were mainly concentrated in the KEGG pathways related to innate immunization and inflammation, including the NF-κB signaling pathway, Toll-like receptor signaling pathway, TNF signaling pathway, IL-17 signaling pathway, cytokine-cytokine receptor interaction, and chemokine signaling pathway, which were similar to the enrichment results of the other two groups. In addition, the DEGs were also involved in the metabolism regulation containing glutathione metabolism, HIF-1 signaling pathway, central carbon metabolism in cancer, choline metabolism in cancery, and fructose and mannose metabolism. The DEGs in the V24 h group (Figure 4b) compared to MOCK were also abundant in the MAPK signaling pathway, glycolysis/gluconeogenesis, starch and sucrose metabolism, and pyruvate metabolism. In addition, the DEGs of V24 h vs. V12 h (Figure 4c) were also involved in pantothenate and CoA biosynthesis, glycerolipid metabolism, fatty acid degradation, riboflavin metabolism, sphingolipid metabolism, primary bile acid biosynthesis, etc. The above results showed the profound changes in different signaling pathways and cytokines in host cells after ASFV invasion. The host cells resist virus invasion by regulating sundry innate immunity and metabolism pathways, while viruses may achieve the purpose of persistent infection by influencing host cell responses.

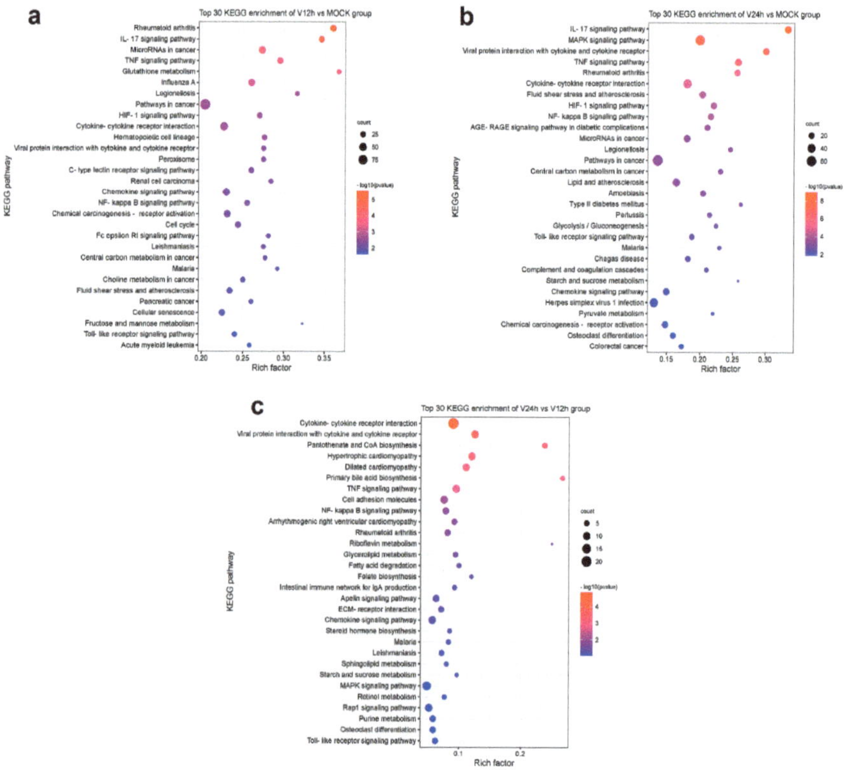

Figure 4. Top 30 KEGG enrichment pathways in each group. (**a**) V12 h vs. MOCK. (**b**) V24 h vs. MOCK. (**c**) V24 h vs. V12 h.

3.4. The Activation of the NF-κB Signaling Pathway in ASFV-Infected PAMs

KEGG pathway enrichment analysis of the transcriptomic data above indicated that DEGs in three groups were enriched in the NF-κB signaling pathway. After ASFV infection for 12 h, the transcriptional level of IL-1β, CXCL8, CXCL2, LOC100525396, PRKCB, TRIM25, IL-1β2, TNF-α, RELB, NFκB2, IKKβ, CD14, CFLAR, IRAK1, TRAF2, and LOC100739325 was increased, and the down-regulated genes including GADD45A, PLAU, GADD45B, TN-

FSF13B, PTGS2, BTK, PARP1, TRAF5, TNFRSF11A, GADD45G were enriched (Figure 5a). There are 12 upregulated genes (IL-1β, CXCL8, LOC100525396, IL-1β2, CXCL2, PRKCB, CD14, TRIM25, TNF-α, CFLAR, LOC100739325, LCK) and 10 down-regulated genes (GADD45B, GADD45A, PLAU, GADD45G, PTGS2, TNFSF14, MAP3K14, TRAF6, TNFAIP3, TRAF5) related to NF-κB signaling pathway in PAMs infected with ASFV for 24 h. NFκB2 (p100) and RELB associated with NF-κB noncanonical pathway were enriched in the V12 h group instead of the V24 h group compared to MOCK, which suggested that ASFV seems to induce both classical and nonclassical NF-κB pathways in the early stage, thus triggering a series of inflammatory processes in the infected host cell (Figure 5c,d). Furthermore, the increased levels of IL-1β, TNF-α, IKKβ, CXCL2, and TRAF2 and decreased level of IκBα were validated through the qPCR method (Figure 5e–j). In general, these results suggested that ASFV infection may activate the NF-κB signaling pathway.

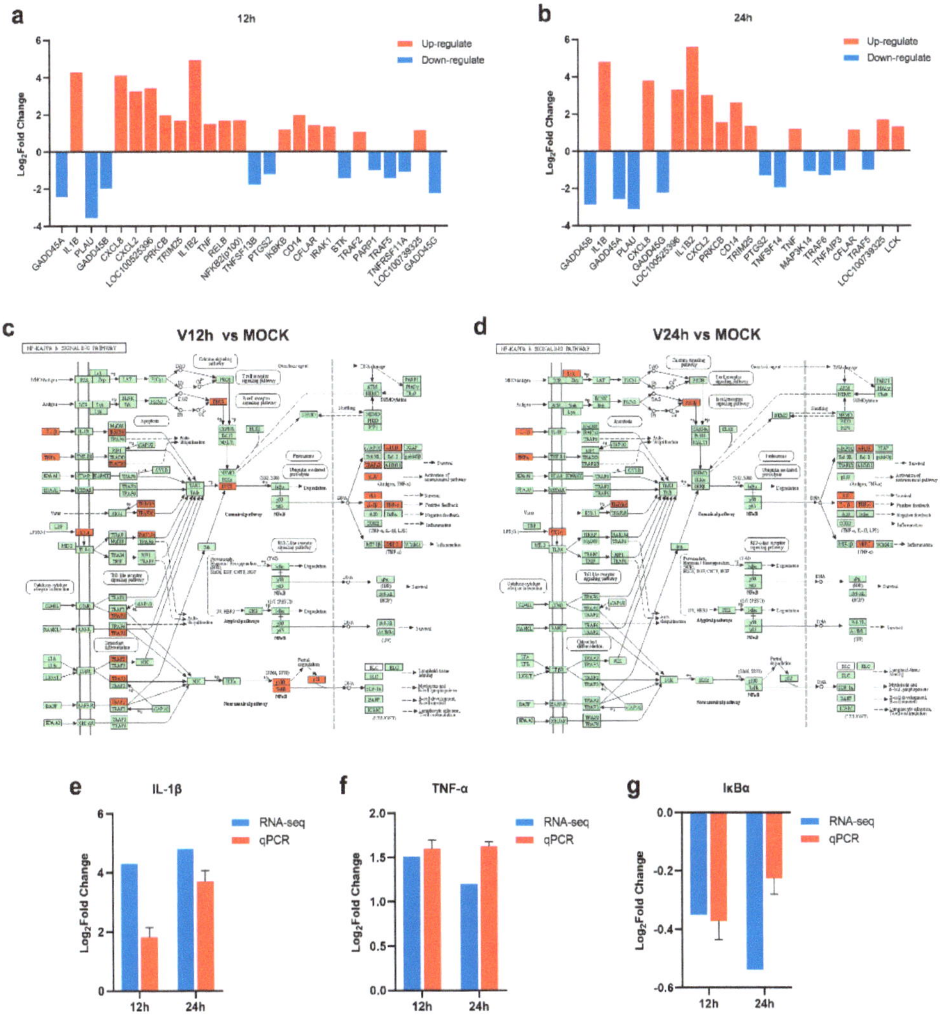

Figure 5. *Cont.*

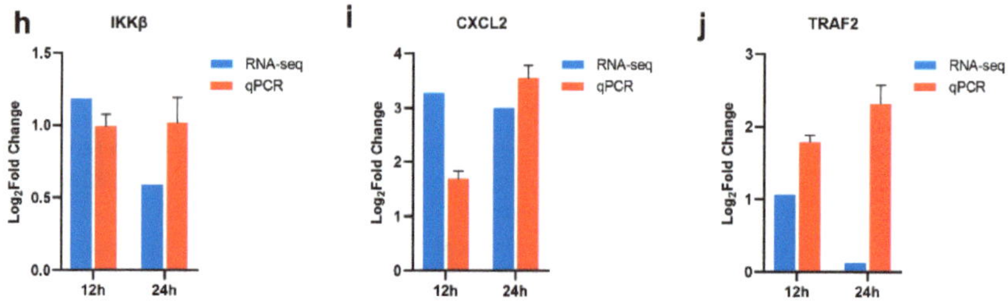

Figure 5. The activation of NF-κB signaling pathway in ASFV infected PAMs (**a**,**b**) The DEGs enriched in the NF-κB signaling pathway in PAMs infected with ASFV for different times ((**a**), 12 h; (**b**), 24 h). (**c**,**d**) The upregulated genes related to the NF-κB signaling pathway in PAMs infected with ASFV for 12 h (**c**)/24 h (**d**) (Upregulated DEGs were marked in red). (**e**–**j**) RT-qPCR validation of representative genes related to NF-κB signaling pathway.

4. Discussion

ASF vaccine development is difficult because of the large genome and complex structure of ASFV. Before the complete control and eradication of ASF, it is necessary to fully understand the infection and immune mechanisms and identify ASFV's major immunogenic genes. ASFV mainly infects porcine monocytes and macrophages, and the early studies on ASFV were primarily in Vero cells. Macrophages are the primary target cells of ASFV and critical immune cells of the host [24]. Transcriptome differential analysis of ASFV-infected macrophages may help to understand the mechanism of host-pathogen interaction. Infectious progeny virus could be produced in cells after ASFV infection for 16 h. In this study, PAMs were used as an in vitro infection model, and the cell samples were harvested at 0 h, 12 h, and 24 h after ASFV infection. Furthermore, RNA-seq technology was used to analyze the transcriptional levels change of cells infected with ASFV at different time points. In the V12 h vs. MOCK group, a total of 2748 DEGs were identified, of which 1355 genes were upregulated and 1393 genes were downregulated (Figure 2c).

The RNA-seq results found that the level of the SQSTM1 gene decreased with time (Figure 3e), which was also confirmed through the qPCR method. The results suggested that ASFV may mediate the autophagy pathway. SQSTM1, also named P62, is one of the well-known autophagy proteins. In autophagy, autophagy receptor P62 binds to ubiquitinated proteins, then forms a complex with LC3-II proteins localized in the autophagic membrane, which are degraded in the autophagic lysosomes [25]. One study has described the enhancement of the autophagy process through the interaction between ASFV E199L and PYCR2 [11]. In addition, ASFV A137R and pI215L proteins were proven to inhibit type I interferon production by regulating the autophagy pathway [14,26]. However, other research indicated that ASFV A179L could interact with the Beclin-1 BH3 motif to inhibit autophagy [27,28]. So far, research on the relationship between autophagy and ASFV infection is still limited. ENO1 is a multifunctional protein involved in several biological and pathophysiological processes, including cell glycolysis, proliferation, migration, invasion, and tumorigenesis [29]. LDHA is a crucial component of glycolysis, promoting the expression of effector T cytokines [30]. It has been reported that many viruses can reprogram glucose metabolism in the host cells [31–34]. H1N1 infection can activate the glycolytic pathway of glucose metabolism to support efficient viral replication [35]. The increased levels of ENO1 and LDHA suggested the vital role of glycolysis in ASFV infection (Figure 3c,d). Similar to other viruses which could regulate glucose metabolism, ASFV may also maintain infection in host cells by reprogramming the glycolytic process. However, Xue et al. confirmed that ASFV infection did not significantly affect the glycolysis pathway, but the produced pyruvate in PAMs after ASFV infection enhanced the lactate production and TCA cycle, which further promoted ASFV replication and immune escape [36]. CXCL8 is the primary mediator of an inflammatory response, attracting neutrophils, basophils, NK cells, and T cells [37]. The gene level of CXCL8 in PAMs infected with ASFV for different time points was higher than MOCK group (Figure 3a). Some studies have reported that low-virulent ASFV strains (OURT88/3) can produce higher levels of CXCL8 and CXCL10 than virulent strains [37]. In addition, the enhanced transcriptional levels of other inflammation-related cytokines, including CCL4, CCL5, CXCL13, IL-1 and TNF-α were also described after ASFV infection [18,22,38]. It proves the close correlation between ASFV and inflammatory response. Infected host cells may clear the virus by inducing inflammation-related cytokines, and the virus may regulate the levels of these chemokines to maintain its replication. The relationship between ASFV and inflammation requires further investigation. As shown in Figure 3f–k, decreased levels of CD163, CD9, CD74, EEF1A1, CTSD, and APOE were observed in this study. Some studies demonstrated that genetically edited pigs lacking CD163 were non-permissive for PRRSV infection but could still be infected with the Georgia 2007/1 ASFV isolates [39,40]. CD163 may not be necessary for ASFV infection. CD9 is involved in cell adhesion, movement, activation, and differentiation. CD74 is an integral transmembrane molecule playing a role in the intracellular sorting of MHC class II molecules, T-cell and B-cell developments, dendritic cell (DC) motility, macrophage inflammation, and thymic selection [41]. Inactivation of eEF1A proteins leads to immunodeficiency, and neural and muscular defects and favors apoptosis [42]. CTSD is an aspartate lysosomal enzyme involved in the degradation of endocytosed and cellular proteins, apoptosis, and brain development [43]. APOE is involved in lipid metabolism and cholesterol metabolism [44]. These DEGs may play essential roles in opposing ASFV, which needs further study.

In this study, many DEGs in ASFV-infected PAMs were enriched in the pathways related to innate immunization and inflammation, including cytokine-cytokine receptor interaction, chemokine signaling pathway, Toll-like receptor signaling pathway, NF-κB signaling pathway, TNF signaling pathway and IL-17 signaling pathway (Figure 4). Different kinds of cytokines and chemokines produced by PAM cells can induce robust immune and inflammatory responses and play an important role in the host antiviral response. Some studies have shown diverse cytokines induced in ASFV-infected cells [18,22,38]. In addition, upon ASFV infection, a wide range of pro-inflammatory factors was secreted in the spleen and renal and gastrohepatic lymph nodes [45]. In vivo studies found that virulent ASFV can evade the host immune system and promote viral replication by delaying the inflammatory response in animals [46]. Nuclear factor kappa B (NF-κB) is an important family of transcription factors consisting of RelA (p65), RelB, c-Rel, p50/p105 (NF-κB1), and p52/p100 (NF-κB2), and can regulate the expression of multiple genes implicated in immunity, inflammation, stress and cell activity [47,48]. The NF-κB pathway is widely regarded as a typical pro-inflammatory signal transduction pathway, the activation of which can promote the secretion of inflammatory cytokines, chemokines, and adhesion molecules [22]. Analysis of transcriptome data from this study revealed that ASFV appeared to induce both classical and nonclassical NF-κB pathways in the early stage and inhibited them in the late stage (Figure 5c,d). NF-κB activated genes were found to be strongly upregulated in Porcine Macrophages infected with the highly virulent ASFV Georgia 2007/1 strain (GRG) [49]. ASFV infection activates the NF-κB signaling pathway, and the inhibitor of this pathway could restrain viral replication [22]. Additionally, several ASFV proteins have been confirmed that they have a strong ability to evade or subvert the antiviral innate immune response and can maintain their own replication by regulating various cytokines in the host [50,51]. The interplay between ASFV and NF-κB signaling is complex and two-sided. On the one hand, ASFV promotes early viral replication using the anti-apoptotic function of NF-κB. On the other hand, ASFV avoids NF-κB-mediated antiviral cytokine responses through different mechanisms of regulation [52]. Further studies are needed to understand how the host immune system interacts with the virus to determine cell survival or death in ASFV infection. It is worth noting that the DEGs were also enriched in metabolism pathways (Figure 4). Multiple viral infections trigger intracellular metabolic reprogramming to support viral replication or rapid cell growth. Wild-type adenovirus 5 (ADWT) could regulate glucose and glutamine metabolism to encourage viral genome replication [53,54]. Dengue virus (DENV)-induced autophagy regulates cellular lipid metabolism and releases free fatty acids conducive to efficient viral replication [55]. A large number of amino acids were discovered to be significantly upregulated in PAMs at the early stages of ASFV infection, and the aspartate and glutamate could promote ASFV replication [36]. In addition, the TCA cycle was critical for the replication of ASFV. TCA cycle may increase ATP and amino acid production to facilitate viral replication [36].

5. Conclusions

In summary, we compared and analyzed the DEGs of PAMs infected with ASFV at different time points (0 h, 12 h, 24 h) and found the significant enrichment and change of transcriptomic factors in diverse pathways, including immunization, inflammation, and metabolism. In addition, ASFV appeared to induce both classical and nonclassical NF-κB pathways in the early stage and inhibit them in the late stage. This study helps understand the interaction between ASFV and the host and lays a foundation for further exploration of the pathogenic mechanism of ASFV.

Author Contributions: Conceptualization, J.C. and S.F.; methodology, validation, and analysis, Z.L.; writing—original draft preparation, W.C.; writing—review and editing, X.L. and K.W.; supervision, and project administration, X.W., W.W., Y.L., L.Y., M.Z. and H.D. All authors contributed to both the conception and design of the work. All authors have read and agreed to the published version of the manuscript.

Funding: This research was funded by the Guangdong Major Project of Basic and Applied Basic Research (No. 2020B0301030007), the Science and Technology Program of Guangzhou, China (No. 202206010161), the Key Research Projects of Universities in Guangdong Province (No. 2019KZDXM026) and the Quality and Efficiency Improvement Project of South China Agricultural University (No. C18).

Institutional Review Board Statement: Not applicable.

Informed Consent Statement: Not applicable.

Data Availability Statement: The data presented in this study are available in the article.

Conflicts of Interest: The authors declare no conflict of interest.

References

1. Dixon, L.K.; Islam, M.; Nash, R.; Reis, A.L. African swine fever virus evasion of host defences. *Virus Res.* **2019**, *266*, 25–33. [CrossRef]
2. Boklund, A.; Cay, B.; Depner, K.; Földi, Z.; Guberti, V.; Masiulis, M.; Miteva, A.; More, S.; Olsevskis, E.; Šatrán, P.; et al. Epidemiological analyses of African swine fever in the European Union (November 2017 until November 2018). *EFSA J. Eur. Food Saf. Auth.* **2018**, *16*, e05494.
3. Wu, K.; Liu, J.; Wang, L.; Fan, S.; Li, Z.; Li, Y.; Yi, L.; Ding, H.; Zhao, M.; Chen, J. Current State of Global African Swine Fever Vaccine Development under the Prevalence and Transmission of ASF in China. *Vaccines* **2020**, *8*, 531. [CrossRef]
4. Ma, J.; Chen, H.; Gao, X.; Xiao, J.; Wang, H. African swine fever emerging in China: Distribution characteristics and high-risk areas. *Prev. Vet. Med.* **2020**, *175*, 104861. [CrossRef] [PubMed]
5. The Global Economic Impact of ASF—WOAH Bulletin. Available online: https://bulletin.woah.org/?panorama=02-2-2-2020-1-economic (accessed on 29 August 2022).
6. ASF Distribution and the Situation in 2020–2022 (Based on INs, FURs and SMRs). Available online: https://www.woah.org/app/uploads/2022/02/asf-report6.pdf (accessed on 20 July 2022).
7. Galindo, I.; Alonso, C. African Swine Fever Virus: A Review. *Viruses* **2017**, *9*, 103. [CrossRef]
8. Li, J.; Song, J.; Kang, L.; Huang, L.; Zhou, S.; Hu, L.; Zheng, J.; Li, C.; Zhang, X.; He, X.; et al. pMGF505-7R determines pathogenicity of African swine fever virus infection by inhibiting IL-1beta and type I IFN production. *PLoS Pathog.* **2021**, *17*, e1009733. [CrossRef]
9. Correia, S.; Ventura, S.; Parkhouse, R.M. Identification and utility of innate immune system evasion mechanisms of ASFV. *Virus Res.* **2013**, *173*, 87–100. [CrossRef]
10. Liu, H.; Zhu, Z.; Feng, T.; Ma, Z.; Xue, Q.; Wu, P.; Li, P.; Li, S.; Yang, F.; Cao, W.; et al. African Swine Fever Virus E120R Protein Inhibits Interferon Beta Production by Interacting with IRF3 To Block Its Activation. *J. Virol.* **2021**, *95*, e82421. [CrossRef]
11. Whittall, J.T.; Parkhouse, R.M. Changes in swine macrophage phenotype after infection with African swine fever virus: Cytokine production and responsiveness to interferon-gamma and lipopolysaccharide. *Immunology* **1997**, *91*, 444–449. [CrossRef] [PubMed]

12. Wang, X.; Wu, J.; Wu, Y.; Chen, H.; Zhang, S.; Li, J.; Xin, T.; Jia, H.; Hou, S.; Jiang, Y.; et al. Inhibition of cGAS-STING-TBK1 signaling pathway by DP96R of ASFV China 2018/1. *Biochem. Biophys. Res. Commun.* **2018**, *506*, 437–443. [CrossRef]
13. Chen, S.; Zhang, X.; Nie, Y.; Li, H.; Chen, W.; Lin, W.; Chen, F.; Xie, Q. African Swine Fever Virus Protein E199L Promotes Cell Autophagy through the Interaction of PYCR2. *Virol. Sin.* **2021**, *36*, 196–206. [CrossRef]
14. Sun, M.; Yu, S.; Ge, H.; Wang, T.; Li, Y.; Zhou, P.; Pan, L.; Han, Y.; Yang, Y.; Sun, Y.; et al. The A137R Protein of African Swine Fever Virus Inhibits Type I Interferon Production via the Autophagy-Mediated Lysosomal Degradation of TBK1. *J. Virol.* **2022**, *96*, e195721. [CrossRef] [PubMed]
15. Hrdlickova, R.; Toloue, M.; Tian, B. RNA-Seq methods for transcriptome analysis. *Wiley Interdiscip. Rev. RNA* **2017**, *8*, e1364. [CrossRef]
16. Ju, X.; Li, F.; Li, J.; Wu, C.; Xiang, G.; Zhao, X.; Nan, Y.; Zhao, D.; Ding, Q. Genome-wide transcriptomic analysis of highly virulent African swine fever virus infection reveals complex and unique virus host interaction. *Vet. Microbiol.* **2021**, *261*, 109211. [CrossRef] [PubMed]
17. African Swine Fever—WOAH—World Organisation for Animal Health. Available online: https://www.woah.org/en/disease/african-swine-fever/ (accessed on 30 August 2022).
18. Yang, B.; Shen, C.; Zhang, D.; Zhang, T.; Shi, X.; Yang, J.; Hao, Y.; Zhao, D.; Cui, H.; Yuan, X.; et al. Mechanism of interaction between virus and host is inferred from the changes of gene expression in macrophages infected with African swine fever virus CN/GS/2018 strain. *Virol. J.* **2021**, *18*, 170. [CrossRef]
19. Bao, Z.Q.; Liao, T.T.; Yang, W.R.; Wang, Y.; Luo, H.Y.; Wang, X.Z. Heat stress-induced autophagy promotes lactate secretion in cultured immature boar Sertoli cells by inhibiting apoptosis and driving SLC2A3, LDHA, and SLC16A1 expression. *Theriogenology* **2017**, *87*, 339–348. [CrossRef]
20. Zhu, Z.; Zhang, X.; Dong, W.; Wang, X.; He, S.; Zhang, H.; Wang, X.; Wei, R.; Chen, Y.; Liu, X.; et al. TREM2 suppresses the proinflammatory response to facilitate PRRSV infection via PI3K/NF-kappaB signaling. *PLoS Pathog.* **2020**, *16*, e1008543. [CrossRef]
21. Zhang, H.; Liu, C.; Cheng, S.; Wang, X.; Li, W.; Charreyre, C.; Audonnet, J.C.; He, Q. Porcine CD74 is involved in the inflammatory response activated by nuclear factor kappa B during porcine circovirus type 2 (PCV-2) infection. *Arch. Virol.* **2013**, *158*, 2285–2295. [CrossRef] [PubMed]
22. Gao, Q.; Yang, Y.; Feng, Y.; Quan, W.; Luo, Y.; Wang, H.; Zheng, J.; Chen, X.; Huang, Z.; Chen, X.; et al. Effects of the NF-kappaB Signaling Pathway Inhibitor BAY11-7082 in the Replication of ASFV. *Viruses* **2022**, *14*, 297. [CrossRef]
23. Zhu, E.; Wu, H.; Chen, W.; Qin, Y.; Liu, J.; Fan, S.; Ma, S.; Wu, K.; Mao, Q.; Luo, C.; et al. Classical swine fever virus employs the PERK- and IRE1-dependent autophagy for viral replication in cultured cells. *Virulence* **2021**, *12*, 130–149. [CrossRef] [PubMed]
24. Gomez-Villamandos, J.C.; Bautista, M.J.; Sanchez-Cordon, P.J.; Carrasco, L. Pathology of African swine fever: The role of monocyte-macrophage. *Virus Res.* **2013**, *173*, 140–149. [CrossRef]
25. Chen, C.; Gao, H.; Su, X. Autophagy-related signaling pathways are involved in cancer (Review). *Exp. Ther. Med.* **2021**, *22*, 710. [CrossRef]
26. Li, L.; Fu, J.; Li, J.; Guo, S.; Chen, Q.; Zhang, Y.; Liu, Z.; Tan, C.; Chen, H.; Wang, X. African Swine Fever Virus pI215L Inhibits Type I Interferon Signaling by Targeting Interferon Regulatory Factor 9 for Autophagic Degradation. *J. Virol.* **2022**, *96*, e94422. [CrossRef]
27. Hernaez, B.; Cabezas, M.; Munoz-Moreno, R.; Galindo, I.; Cuesta-Geijo, M.A.; Alonso, C. A179L, a new viral Bcl2 homolog targeting Beclin 1 autophagy related protein. *Curr. Mol. Med.* **2013**, *13*, 305–316. [CrossRef] [PubMed]
28. Banjara, S.; Shimmon, G.L.; Dixon, L.K.; Netherton, C.L.; Hinds, M.G.; Kvansakul, M. Crystal Structure of African Swine Fever Virus A179L with the Autophagy Regulator Beclin. *Viruses* **2019**, *11*, 789. [CrossRef]
29. Fu, Q.F.; Liu, Y.; Fan, Y.; Hua, S.N.; Qu, H.Y.; Dong, S.W.; Li, R.L.; Zhao, M.Y.; Zhen, Y.; Yu, X.L.; et al. Alpha-enolase promotes cell glycolysis, growth, migration, and invasion in non-small cell lung cancer through FAK-mediated PI3K/AKT pathway. *J. Hematol. Oncol.* **2015**, *8*, 22. [CrossRef]
30. Xu, K.; Yin, N.; Peng, M.; Stamatiades, E.G.; Shyu, A.; Li, P.; Zhang, X.; Do, M.H.; Wang, Z.; Capistrano, K.J.; et al. Glycolysis fuels phosphoinositide 3-kinase signaling to bolster T cell immunity. *Science* **2021**, *371*, 405–410. [CrossRef]
31. Zhao, Y.; Chahar, H.S.; Komaravelli, N.; Dossumbekova, A.; Casola, A. Human metapneumovirus infection of airway epithelial cells is associated with changes in core metabolic pathways. *Virology* **2019**, *531*, 183–191. [CrossRef]
32. Fontaine, K.A.; Sanchez, E.L.; Camarda, R.; Lagunoff, M. Dengue virus induces and requires glycolysis for optimal replication. *J. Virol.* **2015**, *89*, 2358–2366. [CrossRef] [PubMed]
33. Passalacqua, K.D.; Lu, J.; Goodfellow, I.; Kolawole, A.O.; Arche, J.R.; Maddox, R.J.; Carnahan, K.E.; O'Riordan, M.; Wobus, C.E. Glycolysis Is an Intrinsic Factor for Optimal Replication of a Norovirus. *mbio* **2019**, *10*, e02175-18. [CrossRef]
34. Fan, S.; Wu, K.; Zhao, M.; Yuan, J.; Ma, S.; Zhu, E.; Chen, Y.; Ding, H.; Yi, L.; Chen, J. LDHB inhibition induces mitophagy and facilitates the progression of CSFV infection. *Autophagy* **2021**, *17*, 2305–2324. [CrossRef]
35. Ren, L.; Zhang, W.; Zhang, J.; Zhang, J.; Zhang, H.; Zhu, Y.; Meng, X.; Yi, Z.; Wang, R. Influenza A Virus (H1N1) Infection Induces Glycolysis to Facilitate Viral Replication. *Virol. Sin.* **2021**, *36*, 1532–1542. [CrossRef] [PubMed]
36. Xue, Q.; Liu, H.; Zhu, Z.; Yang, F.; Song, Y.; Li, Z.; Xue, Z.; Cao, W.; Liu, X.; Zheng, H. African Swine Fever Virus Regulates Host Energy and Amino Acid Metabolism To Promote Viral Replication. *J. Virol.* **2022**, *96*, e191921. [CrossRef]

37. Fishbourne, E.; Abrams, C.C.; Takamatsu, H.H.; Dixon, L.K. Modulation of chemokine and chemokine receptor expression following infection of porcine macrophages with African swine fever virus. *Vet. Microbiol.* **2013**, *162*, 937–943. [CrossRef]
38. Sun, H.; Niu, Q.; Yang, J.; Zhao, Y.; Tian, Z.; Fan, J.; Zhang, Z.; Wang, Y.; Geng, S.; Zhang, Y.; et al. Transcriptome Profiling Reveals Features of Immune Response and Metabolism of Acutely Infected, Dead and Asymptomatic Infection of African Swine Fever Virus in Pigs. *Front. Immunol.* **2021**, *12*, 808545. [CrossRef] [PubMed]
39. Whitworth, K.M.; Rowland, R.R.; Ewen, C.L.; Trible, B.R.; Kerrigan, M.A.; Cino-Ozuna, A.G.; Samuel, M.S.; Lightner, J.E.; McLaren, D.G.; Mileham, A.J.; et al. Gene-edited pigs are protected from porcine reproductive and respiratory syndrome virus. *Nat. Biotechnol.* **2016**, *34*, 20–22. [CrossRef] [PubMed]
40. Popescu, L.; Gaudreault, N.N.; Whitworth, K.M.; Murgia, M.V.; Nietfeld, J.C.; Mileham, A.; Samuel, M.; Wells, K.D.; Prather, R.S.; Rowland, R. Genetically edited pigs lacking CD163 show no resistance following infection with the African swine fever virus isolate, Georgia 2007/1. *Virology* **2017**, *501*, 102–106. [CrossRef]
41. Su, H.; Na, N.; Zhang, X.; Zhao, Y. The biological function and significance of CD74 in immune diseases. *Inflamm. Res.* **2017**, *66*, 209–216. [CrossRef]
42. Abbas, W.; Kumar, A.; Herbein, G. The eEF1A Proteins: At the Crossroads of Oncogenesis, Apoptosis, and Viral Infections. *Front. Oncol.* **2015**, *5*, 75. [CrossRef]
43. Payton, A.; van den Boogerd, E.; Davidson, Y.; Gibbons, L.; Ollier, W.; Rabbitt, P.; Worthington, J.; Horan, M.; Pendleton, N. Influence and interactions of cathepsin D, HLA-DRB1 and APOE on cognitive abilities in an older non-demented population. *Genes Brain Behav.* **2006**, *5* (Suppl. 1), 23–31. [CrossRef]
44. Huang, Y.; Mahley, R.W. Apolipoprotein E: Structure and function in lipid metabolism, neurobiology, and Alzheimer's diseases. *Neurobiol. Dis.* **2014**, *72 Pt A*, 3–12. [CrossRef]
45. Salguero, F.J.; Ruiz-Villamor, E.; Bautista, M.J.; Sanchez-Cordon, P.J.; Carrasco, L.; Gomez-Villamandos, J.C. Changes in macrophages in spleen and lymph nodes during acute African swine fever: Expression of cytokines. *Vet. Immunol. Immunopathol.* **2002**, *90*, 11–22. [CrossRef]
46. Wang, S.; Zhang, J.; Zhang, Y.; Yang, J.; Wang, L.; Qi, Y.; Han, X.; Zhou, X.; Miao, F.; Chen, T.; et al. Cytokine Storm in Domestic Pigs Induced by Infection of Virulent African Swine Fever Virus. *Front. Vet. Sci.* **2020**, *7*, 601641. [CrossRef] [PubMed]
47. Dorrington, M.G.; Fraser, I. NF-kappaB Signaling in Macrophages: Dynamics, Crosstalk, and Signal Integration. *Front. Immunol.* **2019**, *10*, 705. [CrossRef] [PubMed]
48. Zhang, Q.; Lenardo, M.J.; Baltimore, D. 30 Years of NF-kappaB: A Blossoming of Relevance to Human Pathobiology. *Cell* **2017**, *168*, 37–57. [CrossRef] [PubMed]
49. Cackett, G.; Portugal, R.; Matelska, D.; Dixon, L.; Werner, F. African Swine Fever Virus and Host Response: Transcriptome Profiling of the Georgia 2007/1 Strain and Porcine Macrophages. *J. Virol.* **2022**, *96*, e193921. [CrossRef]
50. Wang, Q.; Zhou, L.; Wang, J.; Su, D.; Li, D.; Du, Y.; Yang, G.; Zhang, G.; Chu, B. African Swine Fever Virus K205R Induces ER Stress and Consequently Activates Autophagy and the NF-kappaB Signaling Pathway. *Viruses* **2022**, *14*, 394. [CrossRef]
51. Powell, P.P.; Dixon, L.K.; Parkhouse, R.M. An IkappaB homolog encoded by African swine fever virus provides a novel mechanism for downregulation of proinflammatory cytokine responses in host macrophages. *J. Virol.* **1996**, *70*, 8527–8533. [CrossRef]
52. Ayanwale, A.; Trapp, S.; Guabiraba, R.; Caballero, I.; Roesch, F. New Insights in the Interplay Between African Swine Fever Virus and Innate Immunity and Its Impact on Viral Pathogenicity. *Front. Microbiol.* **2022**, *13*, 958307. [CrossRef] [PubMed]
53. Thai, M.; Graham, N.A.; Braas, D.; Nehil, M.; Komisopoulou, E.; Kurdistani, S.K.; McCormick, F.; Graeber, T.G.; Christofk, H.R. Adenovirus E4ORF1-induced MYC activation promotes host cell anabolic glucose metabolism and virus replication. *Cell Metab.* **2014**, *19*, 694–701. [CrossRef]
54. Thai, M.; Thaker, S.K.; Feng, J.; Du, Y.; Hu, H.; Ting, W.T.; Graeber, T.G.; Braas, D.; Christofk, H.R. MYC-induced reprogramming of glutamine catabolism supports optimal virus replication. *Nat. Commun.* **2015**, *6*, 8873. [CrossRef]
55. Heaton, N.S.; Randall, G. Dengue virus-induced autophagy regulates lipid metabolism. *Cell Host Microbe* **2010**, *8*, 422–432. [CrossRef]

pathogens

MDPI

Article

Establishment of a Suitable Diagnostic Workflow to Ensure Sensitive Detection of African Swine Fever Virus Genome in Porcine Semen

Virginia Friedrichs [1], Darwin Reicks [2], Jeffrey J. Zimmerman [3], Eric A. Nelson [4], Carola Sauter-Louis [1], Martin Beer [1], Jane Christopher-Hennings [4,*] and Sandra Blome [1,*]

[1] Friedrich-Loeffler-Institut, Suedufer 10, 17493 Greifswald-Insel Riems, Germany; virginia.friedrichs@fli.de (V.F.); carola.sauter-louis@fli.de (C.S.-L.); martin.beer@fli.de (M.B.)

[2] Reicks Veterinary Research and Consulting, Saint Peter, MN 56082, USA; darwin@rvrcmn.com

[3] Veterinary Diagnostic & Production Animal Medicine, Iowa State University, Ames, IA 50011, USA; jjzimm@iastate.edu

[4] Animal Disease Research & Diagnostic Laboratory, South Dakota State University, Brookings, SD 57007, USA; eric.nelson@sdstate.edu

* Correspondence: jane.hennings@sdstate.edu (J.C.-H.); sandra.blome@fli.de (S.B.); Tel.: +1-605-688-5171 (J.C.-H.); +49-38351-7-1144 (S.B.)

Abstract: The rapid spread of African swine fever virus (ASFV), causing severe and often lethal disease in domestic pigs and Eurasian wild boar, continues to be a threat to pig populations and dependent industries. Despite scientific achievements that have deepened our understanding of ASFV pathogenesis, alternative transmission routes for ASFV remain to be elucidated. We previously demonstrated the efficient transmission of ASFV from infected boars to naïve recipient gilts via artificial insemination, thereby highlighting the importance of surveillance of boar semen prior to its shipment. Since the accurate and reliable detection of even low amounts of ASFV in boar semen is key to disease prevention and control, we established a suitable diagnostic workflow to efficiently detect the ASFV genome in boar semen. Here, we assessed the sensitivity of various routine nucleic acid extraction kits as well as qPCR protocols in detecting the ASFV genome in the blood and semen of infected boars. The feasibility of the respective kits and methods for future use in boar studs was also considered. Variability in sensitivity mostly concerned samples with low to very low amounts of the ASFV genome. Ultimately, we defined a well-suited workflow for precisely detecting the ASFV genome in boar semen as early as 2 days post ASFV infection.

Keywords: African swine fever virus; virus diagnostics; boar semen; commercial qPCR kits; comparison; nucleic acid extraction; sensitivity; performance

Citation: Friedrichs, V.; Reicks, D.; Zimmerman, J.J.; Nelson, E.A.; Sauter-Louis, C.; Beer, M.; Christopher-Hennings, J.; Blome, S. Establishment of a Suitable Diagnostic Workflow to Ensure Sensitive Detection of African Swine Fever Virus Genome in Porcine Semen. *Pathogens* **2024**, *13*, 537. https://doi.org/10.3390/pathogens13070537

Academic Editor: Natasha N. Gaudreault

Received: 7 May 2024
Revised: 19 June 2024
Accepted: 20 June 2024
Published: 25 June 2024

1. Introduction

The rapid spread of African swine fever virus (ASFV), an enveloped, double-stranded DNA virus with a genome of 170–190 kpb, remains a threat to pig populations and economies worldwide [1,2]. Originally, ASFV circulated in an ancient sylvatic cycle among asymptomatic warthog and soft tick (genus: *Ornithodoros*) populations in sub-Saharan Africa [3]. Its introduction into domestic pigs or Eurasian wild boar, however, leads to severe yet rather unspecific clinical signs, resulting in high case fatality rates [4,5]. The disease is notifiable to the World Organization for Animal Health (WOAH). The current African swine fever (ASF) panzootic started in 2007, when ASFV was introduced into Georgia and subsequently into the Russian Federation and many Trans-Caucasian countries. In 2014, ASF entered the European Union [6], in 2018, China [7], and in 2021, the Caribbean [8].

At present, 24 different genotypes, characterized based on variations within the p72 capsid protein encoded by the B464L gene, have been defined [9–12]. However, only genotype I and II strains have been found outside of Africa, and the current panzootic involves genotype II strains only. Despite efforts to understand and restrict the disease, its ongoing spread emphasizes the need to evaluate alternative transmission routes and strengthen early warning systems. In a previous study, we demonstrated that artificial insemination is an efficient route to transmit ASFV from infected boars to naïve recipient gilts. Usually, domestic boar semen originates, with the exception of rural backyard farming, from boar studs [13]. In these facilities, boars are kept individually, and semen is collected regularly on demand. The collected semen has to pass mandatory quality control checks, e.g., count of spermatozoa, amount of abnormalities, and mobility. Since we showed that none of those criteria were affected by early ASFV infection or even acute viraemia, risk-based surveillance for the presence of ASFV in boar semen is of the utmost importance and opens up the only possibility for very early detection. Around the globe, real-time polymerase chain reaction (qPCR) is widely used as a reliable, sensitive, and specific tool for detecting animal diseases such as ASFV [14–17]. In addition to a robust qPCR system, a highly effective extraction method is key to the correct laboratory diagnosis [18]. Recommended extraction systems are listed in the WOAH guidelines [19].

However, the extraction methods and qPCR assays need to be constantly adapted to and meet putative evolutionary changes in the virus and/or the demands of the diagnostic field. This is especially true for such a difficult matrix as semen. In recent years, several commercially available qPCR kits for ASFV genome detection have become available, and performance comparisons have been carried out, e.g., Schoder et al. (2020) [20] and Pikalo et al. (2022) [19]. To standardize the conditions and facilitate data interpretation, regional reference laboratories should follow the guidelines provided by the WOAH and regional (e.g., the European Union) and national reference laboratories (NRLs). These guidelines include a list of registered or licensed diagnostic methods suited to and permitted for routine diagnostics [21].

As semen preparations for artificial insemination contain potentially qPCR-inhibitory components [22], such as sucrose [23], they represent complex and highly challenging matrices for routine diagnostic testing, also affecting ASFV diagnostic workflows. Hence, it is of the utmost importance to define a suitable diagnostic workflow to accurately detect the ASFV genome in boar semen.

Here, we compared three methods for nucleic acid extraction, as well as five qPCR protocols, to establish a suitable workflow for the efficient and early detection of the ASFV genome in boar semen using a standard methodology without special semen treatment. The criteria applied for definition were (I) reliable amplification, (II) handling and time requirements, and (III) efficient detection of the ASFV genome in critical samples, e.g., samples obtained early after inoculation, especially those presumed to contain only very few ASFV genome copies.

2. Materials and Methods

2.1. Samples

All samples used in this study were derived from breeding boars, which were intramuscularly inoculated with 10^4 HAD_{50} of the ASFV strain "Estonia 2014" [24]. This trial included four adult breeding boars, two Large White and two Pietrain boars [25]. Semen samples from all boars were collected regularly after experimental ASFV infection to enable early ASFV screening in porcine semen. Since EDTA blood, already tested for accurate and early detection of the ASFV genome [26], was previously defined as the "gold standard" matrix, blood samples were compared in terms of analytical performance with the semen samples. Positive EDTA blood (n = 18) and semen samples (n = 17) obtained at days 2, 3, 4, 5, 14, and 20 post inoculation were included. Table 1 provides details on the semen samples included in the study. The EDTA blood samples corresponded to these semen samples.

Table 1. Semen samples included in the study. The ASF status as true positive (POS) or negative (NEG) was based on manual extraction using the QIAamp Viral RNA Mini kit (QIAGEN) in triplicate, followed by the VetAlert™ PCR as a reference (three independent runs on all replicates). dpi = days post inoculation; NT = not tested.

dpi	Boar #1	Boar #2	Boar #3	Boar #4	POS/Total
2	**POS**	NEG	NEG	**POS**	**2/4**
3	**POS**	NEG	**POS**	**POS**	**3/4**
4	**POS**	**POS**	**POS**	**POS**	**4/4**
5	**POS**	**POS**	**POS**	NT	**3/3**
14	NT	NT	**POS**	NT	**1/1**
20	NT	NT	**POS**	NT	**1/1**
					n = **14**/17

2.2. Extraction of Viral DNA

All the samples were frozen at −80 °C upon collection to ensure full availability of the cell-bound viral genome [27]. For accurate detection of low ASFV genome loads in the boar semen, the performance of three commercially available, routinely used nucleic acid extraction kits was compared (Table 2): (I) the NucleoMag® VET kit (Macherey-Nagel, Düren, Germany); (II) the MagMAX™ Pathogen RNA/DNA kit (Thermo Fisher, Waltham, MA, USA); and (III) the MagMAX™ 96 Viral RNA isolation kit (Thermo Fisher, Waltham, MA, USA). All the samples were extracted in triplicate (n = 3), and all kits were used according to the manufacturer's instructions unless stated otherwise. Extraction was performed on the automated extraction platform KingFisher™ 96 flex (Thermo Fisher, Darmstadt, Germany) upon utilization of the extraction protocols provided by the manufacturers.

Table 2. Specifications of all nucleic acid extraction kits used to extract DNA from blood and semen.

Extraction Kit Name	Input Volume [µL]	Output Volume [µL]	Steps
NucleoMag® VET	100	100	8
MagMAX™ Pathogen RNA/DNA Kit	100 bl/115 se	90	7 bl/8 se
MagMAX™-96 Viral RNA Isolation Kit	50	90	9

Legend: bl = blood samples; se = semen samples.

Three criteria were used to define the sensitivity of the assay and ultimately the most appropriate extraction method: (I) sensitivity across all "true positive" samples, (II) sensitivity based on the sample set taken early after inoculation, i.e., 2–3 dpi, and (III) preference for lower Cq values.

Because early detection of the ASFV genome in semen is crucial, the semen samples were divided into "early" (2–3 dpi, n = 5 positive samples) and "late" samples (>4 dpi, n = 9 positive samples) prior to performance assessment. True positive samples among the "early" samples were defined according to manual extraction and increasing the number of replicates, which was beneficial for the detection of samples with low amounts of the ASFV genome. For manual extraction, DNA from 85 µL of semen was extracted using the QIAamp Viral RNA Mini kit (QIAGEN). Manual extraction was performed in triplicate (n = 3), and the samples were analyzed in three independent qPCR runs (using the VetAlert™ ASFV DNA Test Kit from Tetracore, Rockville, MD, USA), where each sample was measured in triplicate (n = 3).

2.3. Molecular Assays: qPCR

Following extraction, all the samples were compared using the VetAlert ASFV DNA assay (Tetracore). This assay is accredited in the German national reference laboratory

(NRL) and was among the best commercial qPCR kits in previous studies when using various other sample matrices from ASFV-infected pigs [19]. The best-performing extraction method was subsequently used to assess the analytical performance of the five qPCR assays in ASFV detection (all certified for use in ASFV diagnostics).

The five qPCR protocols included in this study were (Table 3) the (I) VetAlert™ ASFV DNA Test Kit; (II) the virotype ASFV 2.0 PCR Kit (Indical); (III) the VetMax™ ASFV Detection Kit (Thermo Fisher), (IV) the WOAH-recommended protocol published by King et al. (2003) [15] with slight modifications (accredited ASF System 1); and (V) the RealPCR ASFV DNA Test (IDEXX). All protocols were utilized according to the manufacturer's/authors' instructions.

Table 3. Specifications of all ASFV detection (qPCR) protocols and kits used in this study.

qPCR Kit Name	Input Volume [μL]	Internal Control	Cycles	Pipetting Steps/Time
VetAlert™ ASFV DNA Test Kit	5	exogenous	45	3/1 h 36 min
virotype ASFV 2.0 PCR Kit	5	exo-/endogenous	40	2/1 h 2 min
VetMax™ ASFV Detection Kit	5	exogenous	45	2/1 h 18 min
ASF System 1 (WOAH, King et al. [15])	5	endogenous	45	5/2 h 25 min
RealPCR ASFV DNA Test	5	endogenous	45	4/1 h 30 min

To facilitate the detection of low ASFV genome copy numbers in the semen samples, the above-mentioned extracted DNA from the semen samples (n = 3 per boar and time point) was evaluated in triplicate (a total of n = 9 per semen sample) for each qPCR assay. For the qPCR assays, the same sensitivity criteria as used for the extraction methods were employed.

2.4. Data Analyses

All the data generated by qPCR were visualized and analyzed for statistical relevance using GraphPad Prism 9 (GraphPad Software Inc.). Statistical analyses were performed using One-Way ANOVA with Tukey's post hoc testing, and significant differences are depicted as follows: ** $p < 0.01$, *** $p < 0.001$, **** $p < 0.0001$. Subsequently, all the qPCR test results were compared to the best-performing protocol using the Bland–Altman test [28]. The optimal performance was defined as the number of wells with successful detection of the ASFV genome in semen samples 2–3 dpi. The Limit of Agreement (LoA) interval for each comparison was defined as the mean difference ±1.96 standard deviation (SD) of the Cq values. Furthermore, the percentage of positive results among the infected boars (true positive and false negative) was calculated for each qPCR method/matrix. Additionally, using samples from 2 and 3 dpi, a repeated-measures ANOVA, correcting for replicates, was performed to test for statistical differences in the performance of the five qPCR tests. However, the power of the statistical calculations was limited because correlated samples were used (repeated sampling of a few individuals). Due to many samples not yielding Cq values upon analysis, the number of cycles minus Cq values was used for calculations. However, the power of the statistical calculations was limited since data for only a few animals were available.

3. Results

3.1. Assessment of Preparation Time, Handling, and Time Requirements

3.1.1. Nucleic Acid Extraction Kits

The sample preparation time, which correlates directly with the number of mandatory pipetting steps, varied significantly between manufacturers. For example, the NucleoMag® VET kit only required the sample to be vortexed prior to adding the ready-to-use lysis buffer, while the MagMAX™ Pathogen RNA/DNA Kit called for lysis buffer and bead mix preparation (2 and 3 steps, respectively) before DNA extraction. Differences were also found between matrices: EDTA blood could be added to the prepared solutions, while the semen samples required preparation of the lysate in a separate plate. The subsequent semen lysate was then added to the plate containing all the necessary washing/elution solutions. An additional difference was notable for the MagMAX™-96 Viral RNA Isolation Kit, which included the preparation of the lysis/binding buffer and bead mix; however, the EDTA blood and semen samples did not require varying preparation steps.

However, the steps required for nucleic acid extraction were comparable between all kits, as listed in Table 1.

In terms of storage, all the kits contained components that needed to be stored at different temperatures (−20 °C/4 °C/RT for kits by Thermo Fisher, −20 °C/RT for the NucleoMag® VET kit), and all kits allowed for the use of automated nucleic acid extraction platforms.

3.1.2. Differences Amongst Tested ASFV qPCR Assays

The required pipetting steps during the preparation of the qPCR reactions overall ranged from two to five steps (Table 2). Two out of five kits included a ready-to-use mix that only needed mixing with the respective sample in the plate: the virotype ASFV 2.0 PCR Kit and the VetMax™ ASFV Detection Kit. Furthermore, the VetAlert™ ASFV DNA Test Kit needed one extra step: adding the enzyme to the mix. Two methods, the WOAH King et al. protocol and the RealPCR ASFV DNA Test kit, required step-by-step mixing of all reagents, with five and three steps, respectively [15].

In terms of the qPCR cycles for target amplification, all assays included similar cycles, i.e., 45, except for the virotype ASFV 2.0 PCR assay, which runs with 40 cycles. The duration of the qPCR runs was also heterogenous, ranging from 1 h and 2 min (virotype ASFV 2.0 PCR Kit) to 2 h and 25 min (WOAH King et al. [15]).

3.2. *Extraction Efficiency Evaluation of the ASFV Genome from Boar Semen*

As shown in Figure 1, sample extraction using the NucleoMag® VET Kit resulted in 100% detection of positive EDTA blood samples (n = 18/18) and 78.6% detection (n = 11/14) of the infected semen samples. Furthermore, the MagMAX™ Pathogen RNA/DNA Kit achieved detection of up to 88.9% (n = 16/18) and 50.0% (n = 7/14) of the positive EDTA blood and semen samples, respectively. Similar results were obtained for the EDTA blood samples (88.9%; n = 16/18) extracted with the MagMAX™-96 Viral RNA Isolation Kit; however, no semen samples extracted with this kit yielded Cq values when analyzed with the VetAlert™ ASFV DNA assay (Figure 1). The extraction of semen samples using the MagMAX™-96 Viral RNA Isolation Kit was carried out three times, further verifying these results.

Out of fourteen positive semen samples, five were positive at an early time point (2–3 dpi). Extraction with the NucleoMag® VET Kit detected 40.0% (n = 2/5) of the true positive samples, followed by extraction with the MagMAX™ Pathogen RNA/DNA Kit (n = 1/5).

Differences in the subsequent Cq values under the same cycling conditions were also observed between the extraction kits. Extraction with the NucleoMag® VET Kit rendered the lowest Cq values (EDTA blood: 20.3 ± 7 SD; Semen: 34.1 ± 3.3 SD), followed by the MagMAX™ Pathogen RNA/DNA Kit (Blood: 23.6 ± 4.8 SD; Semen: 36.7 ± 3.1 SD) and the MagMAX™-96 Viral RNA Isolation Kit (Blood: 24.6 ± 4.4 SD), as shown in Figure 1. Samples with low amounts of the ASFV genome, e.g., the blood samples of boars #2 (Cq 35.2 ± 0.2 SD) and #3 (Cq 38.1 ± 4.5 SD) at 2 dpi, were not accurately detected after extraction with the MagMAX™ kits.

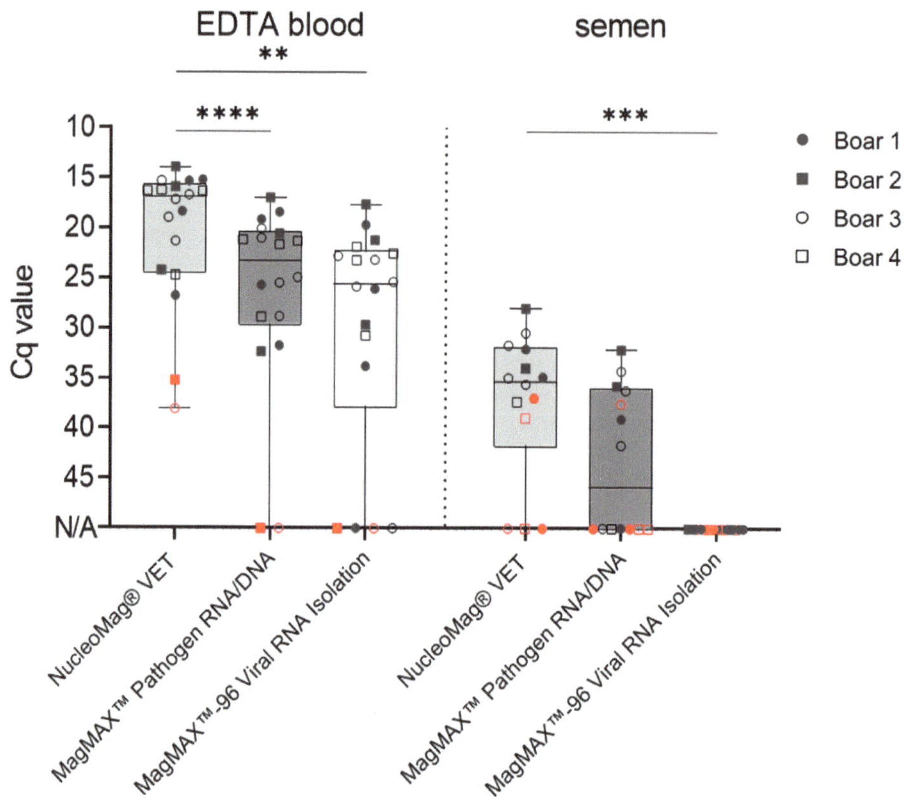

Figure 1. Comparison of ASFV genome extraction efficiency from boar blood and semen. Evaluation of performance using 18 ASFV-positive blood and 14 positive semen samples. Genome loads were evaluated by qPCR with the VetAlert™ ASFV DNA Test Kit. Boxes represent 25/75 percentiles, including the group median with min. and max. values; each individual is represented by a symbol. Critical EDTA blood (2 dpi) and semen (2–3 dpi) samples are indicated in red. All samples were evaluated in triplicate ($n = 3$). N/A = no detection occurred within 45 cycles. Significant differences were assessed by One-Way ANOVA, ** $p < 0.01$, *** $p < 0.001$, **** $p < 0.0001$.

3.3. *Assessment of qPCR Performance in Detecting the ASFV Genome in Boar Semen*

Based on these results, the most suitable routine kit for the extraction of viral genome copies from the semen samples with an automated platform was the NucleoMag® VET Kit. Therefore, a comparison of the analytical performance of the five qPCR assays was carried out using nucleic acids extracted via this method.

Differences amongst matrices were observed in all the included assays. All positive EDTA blood samples ($n = 18$) were detected by all assays (100% sensitivity); however, differences in the Cq values were noted. The VetAlert™ ASFV DNA assay yielded similar values (20.3 ± 7 SD) to ASF System 1, recommended by WOAH, King et al. [15] (20.4 ± 7.3 SD). Following this, the virotype ASFV 2.0 (21.5 ± 6.3 SD) and VetMax™ ASFV Detection assays (21.3 ± 6.6 SD) rendered similar Cq values. Finally, the RealPCR ASFV DNA Test (23.8 ± 6.6 SD) showed higher Cq values for analogous samples (Figure 2).

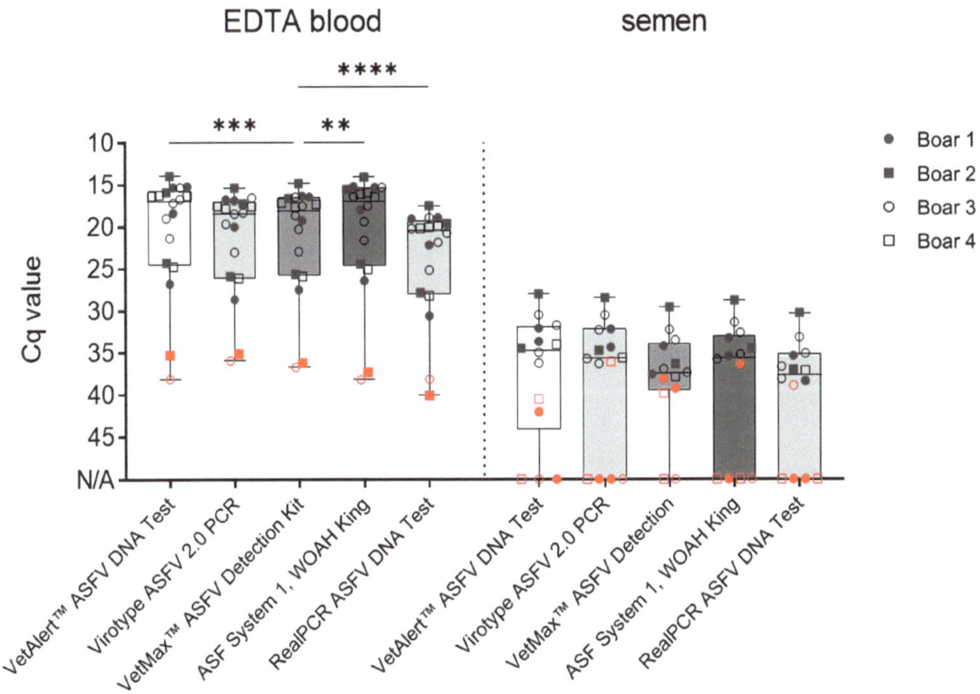

Figure 2. Analytical sensitivity for boar blood and semen in qPCR. Evaluation of qPCR protocol performance, utilizing 18 ASFV-positive blood and 14 positive semen samples. Boxes represent 25/75 percentiles, including the group median with min. and max. values; each individual is represented by a symbol. Critical EDTA blood (boar #2 and #3 at 2 dpi) and semen (2–3 dpi) samples are indicated in red. Samples were evaluated in triplicate (n = 3). N/A = no detection occurred within 45 cycles. Significant differences were assessed by One-Way ANOVA, ** $p < 0.01$, *** $p < 0.001$, **** $p < 0.0001$.

For the second matrix, semen, all the mean Cq values were comparable and ranged from 33.3 ± 2.6 SD (virotype ASFV 2.0 PCR Kit) to 35.7 ± 2.6 SD (RealPCR ASFV DNA Test), as depicted in Figure 2.

However, when the samples were divided into "early" and "late" collected samples, the results were more heterogenous (Figure 2, Table 4). All the qPCR assays successfully detected the ASFV genome in "late" boar semen samples obtained at 4–20 dpi (n = 9). Nonetheless, differences were noted for the "early" semen samples (2–3 dpi; n = 5 true positive). Based on this sample set alone, the VetMax™ ASFV Detection Kit was able to detect 60% of all the true positive early samples (n = 3 with Cq 38.1 ± 2.6 SD), followed by the VetAlert™ ASFV DNA Test Kit, which detected 40% (n = 2 with Cq 39.6 ± 1.8 SD). Finally, ASF System 1 (Cq 33.6 ± 0.8 SD), the virotype ASFV 2.0 PCR Kit (Cq 36), and the RealPCR ASFV DNA Test (Cq 38.9 ± 1.2 SD) were able to detect 20% (n = 1) of the early semen samples. If sensitivity is calculated across all semen samples, the VetMax™ ASFV Detection Kit shows a sensitivity of 85.7% (n = 12/14), the VetAlert™ ASFV DNA Kit a sensitivity of 78.6% (n = 11/14), and the other kits a sensitivity of 71.4% (ASF-System 1, virotype ASFV 2.0 PCR, and RealPCR ASFV DNA Test, all n = 10/14).

Table 4. Overview of successfully detected true positive samples among critical semen samples (2–3 dpi). To account for statistical effects, samples were tested in nine replicates in total (three extraction replicates in three PCR runs).

Kit Name	dpi	No. of Positive Replicates			
		Boar 1	Boar 2	Boar 3	Boar 4
VetAlert™ ASFV DNA	2	**3/9**	—	—	**2/9**
	3	0/9	—	0/9	0/9
virotype ASFV 2.0 PCR	2	0/9	—	—	0/9
	3	0/9	—	0/9	**1/9**
VetMax™ ASFV Detection	2	**4/9**	—	—	**3/9**
	3	**1/9**	—	0/9	0/9
ASF System 1 (WOAH King et al. [15])	2	**2/9**	—	—	0/9
	3	0/9	—	0/9	0/9
RealPCR ASFV DNA	2	0/9	—	—	0/9
	3	0/9	—	**2/9**	0/9

Legend: — = true negative sample.

The analytical performance of the qPCR kits using all the true positive EDTA blood and semen samples (n = 32) was assessed using Bland–Altman plots (Figure 3A) and point-by-point evaluations (Figure 3B). The best-performing assay, the VetMax™ ASFV Detection Kit, was used as a reference for the point-by-point evaluations.

The VetAlert™ ASFV DNA Test Kit (bias 0.97) showed a narrow LoA, indicating a high level of agreement. However, three samples were found outside the LoA, which means that the results of these three samples varied beyond the calculated standard deviation. Two assays, ASF System 1 and the virotype ASFV 2.0 PCR Kit, showed a similar LoA, whereas one and two samples, respectively, were underestimated with these systems when compared with the VetMax™ ASFV Detection Kit. Finally, the RealPCR ASFV DNA Test kit showed samples with Cq values that were both under- and overestimated, indicating higher variability and therefore disagreement in the results between these kits. A manual in-detail comparison revealed once more that these semen samples were detected with a distinct shift in Cq values for the extraction of the ASFV genome. This indicates that out of the tested assays, only the VetAlert™ ASFV DNA Test Kit can be used interchangeably with the best-performing kit, the VetMax™ ASFV Detection Kit. The point-by-point comparison revealed that the VetAlert™ ASFV DNA Test Kit overall had lower Cq values up to a Cq of ~37, where the detection of weakly positive samples was superior using the VetMax™ ASFV Detection Kit (Figure 3B). The virotype ASFV 2.0 PCR Kit and ASF System 1 had comparable Cq values to the VetMax™ ASFV Detection Kit for the EDTA blood samples but failed to detect most of the weakly positive semen samples. The RealPCR ASFV DNA Test had overall higher Cq values but also failed to detect the weakly positive semen samples.

Furthermore, repeated-measures ANOVA resulted in a statistically significant difference in the detection efficiency of the five different PCRs using semen samples from 2 and 3 dpi (p = 0.0068, Figure 4). While the calculations revealed statistical significance between 2 and 3 dpi (p = 0.0042, Figure 4A), no significant variations were observed among the replicates of each sample (n = 8 samples, each 9 replicates, Figure 4B). Of the two-way interactions, significance was confirmed between the results of the five qPCR kits (p = 0.0087), indicating their differing efficiency in detecting samples with low amounts of the ASFV genome correctly.

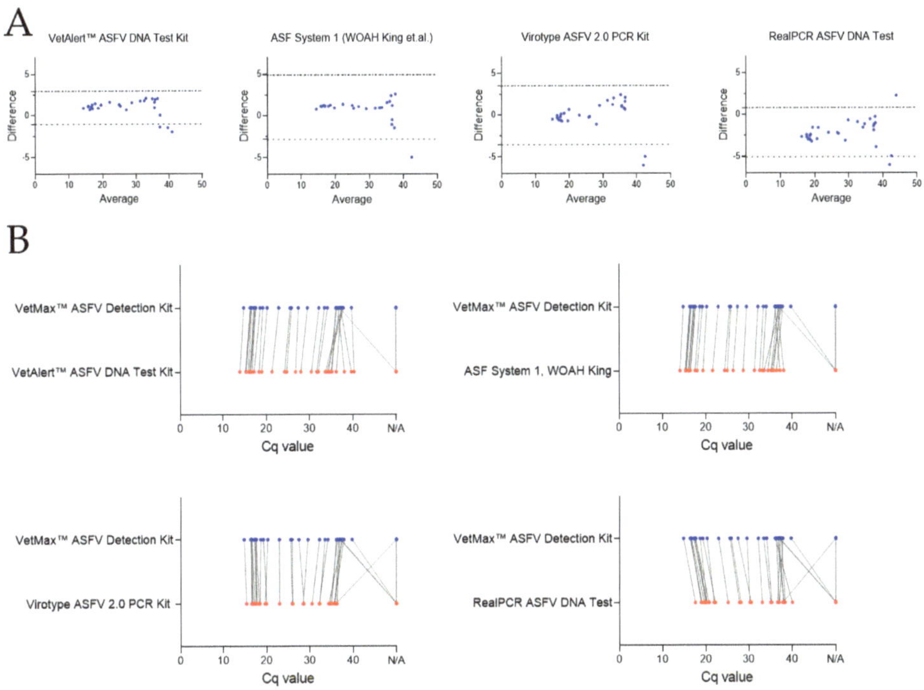

Figure 3. Comparison of five qPCR assays in ASFV genome detection efficiency from boar blood and semen. Evaluation of qPCR method performance, utilizing 18 ASFV-positive blood and 14 positive semen samples. All qPCR kits were compared to the VetMax™ ASFV Detection Kit. (**A**) Bland–Altman plots. Gray areas represent the lower and upper limits of agreement. Blue points display differences between the VetMax™ ASFV Detection Kit and any other qPCR method tested (WOAH King et al. [15]). (**B**) Point-by-point result comparison of all samples tested in relation to the VetMax™ ASFV Detection Kit. N/A = no detection occurred within 40–45 cycles. Lines connect results of corresponding samples.

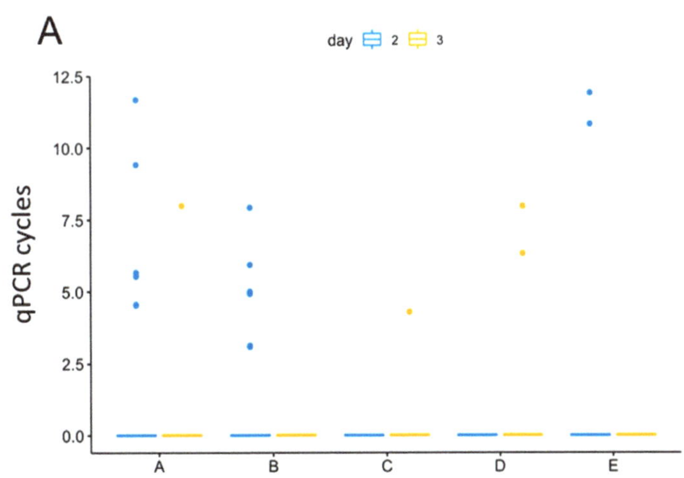

Figure 4. *Cont.*

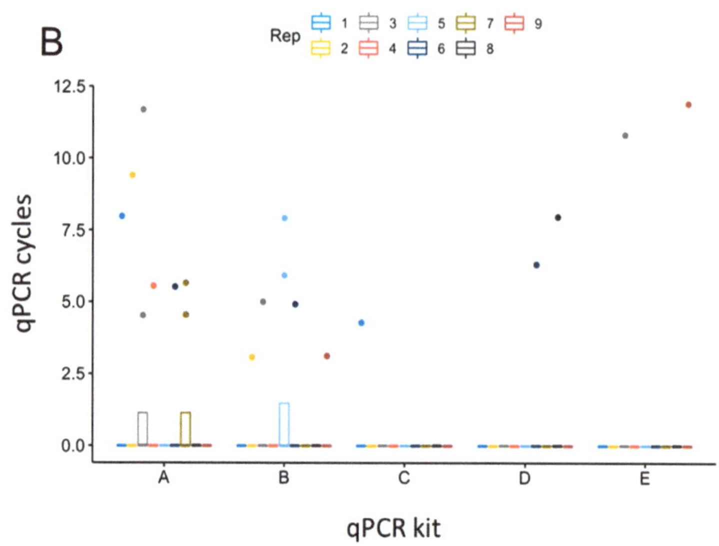

Figure 4. Evaluation of the detection accuracy of five qPCR kits using semen samples of infected boars. Repeated-measures ANOVA was performed to calculate significant differences in detection performance of the five qPCR kits. Variance between 2 and 3 dpi (**A**), as well as among replicates of each sample (**B**) (n = 8 with 9 replicates each), was calculated. X-axis labeling corresponds to qPCR kits: A = VetMax™ ASFV Detection Kit, B = VetAlert™ ASFV DNA assay, C = virotype ASFV 2.0 PCR assay, D = RealPCR ASFV DNA Test, E = ASF System 1 (WOAH King et al. [15]).

4. Discussion

As broadly protective vaccinations or reliable treatment options for ASF are still not available, the accurate and early identification of infected individuals is key to prevent further spread of the disease. The modern pork industry mainly relies on artificial insemination, with the boar semen acquired from boar studs. The semen is collected, diluted with nutrient-containing extenders, and shipped on demand, often nationwide or even across borders. To prevent the spread of ASFV-containing semen, surveillance of semen upon collection is needed. However, fast processing during quality management is essential to ensure the high quality and viability of the spermatozoa. Therefore, we compared various nucleic acid extraction kits, as well as established and validated ASFV-specific qPCR assays, to define a practicable diagnostic workflow without the need for special treatments or protocols for the early detection of even low amounts of the ASFV genome in boar semen in a high-throughput diagnostic laboratory. Similar studies were carried out comparing the performance of WOAH-approved qPCR assays on wild boar samples of varying degrees of decay [19]. Here, the best-suited kits were defined according to the following criteria: (I) sensitivity across the full sample set, (II) sensitivity based on semen samples obtained on days 2 and 3 after inoculation, and (III) yield of low Cq values. Since fresh boar semen typically is shipped less than 24 h after collection, the sample preparation and qPCR run duration served as additional factors for efficiency determination.

Generally, no combination of nucleic acid extraction and qPCR kit was able to detect all the true positive semen samples, as defined by manual DNA extraction. However, regarding the nucleic acid extraction kits, considerable differences in their efficiency in extracting the ASFV genome from boar blood and semen were noted. While extraction with the NucleoMag® VET Kit resulted in the detection of all the true positive blood samples, MagMAX™ kits gave two false negative results for the two samples with the lowest ASFV genome loads (boar #2 and #3 at 2 dpi). Furthermore, the differences were even much more striking regarding semen. Here, extraction with the NucleoMag® VET Kit resulted in

11/14 true positive results, while extraction with the MagMAX™ Pathogen RNA/DNA Kit obtained 7/14 true positive results. However, repeated extraction with the MagMAX™-96 Viral RNA Isolation Kit did not yield any positive results for semen. Conclusively, while their performance on blood samples was largely comparable, the extraction efficiency varied between the kits. Overall, reduced extraction efficiency for semen was noted for all the extraction kits included in this study, indicating the presence of inhibitory components in this matrix, e.g., polysaccharides [23,29]. Additionally, low amounts of the genome in the semen early after infection might result in false negative results, considering that only a tiny fraction of the whole ejaculate is sampled. Hence, detection variability is likely to occur in samples with low amounts of the target genome. Based on the detection efficiency in our study and the short preparation time, the NucleoMag® VET Kit seems well suited to facilitate ASFV monitoring in boar studs. Although the manual extraction of DNA is considered the most sensitive method for ASFV diagnostics, an automated platform with a similar detection efficiency is likely more suitable for monitoring large pig herds. Given that we focused on routine diagnostic workflows with a high throughput, we did not include special extraction protocols for semen (as can be found in the WOAH manual for, e.g., Bovine herpesvirus 1 infection: https://www.woah.org/fileadmin/Home/eng/Health_stanards/tahm/3.04.11_IBR_IPV.pdf (accessed on 6 May 2024)).

Additionally, similar observations were made for the studied qPCR assays. By extracting nucleic acids via the NucleoMag® VET Kit, all the qPCR methods successfully detected all the true positive EDTA blood samples, even samples containing low amounts of the ASFV genome, e.g., boars #2 and #3 at 2 dpi. This further indicated the unmatched suitability of blood samples for the accurate and early detection of the ASFV genome in pigs, as described previously [26]. It is of note that the boars tolerated blood sampling through their saphenous veins without becoming agitated during semen collection. Hence, the collection of small amounts of blood during the procedure might be feasible for obtaining blood for ASFV surveillance, as already routinely applied for PRRSV [30,31]. However, although the Cq values of the EDTA blood samples were largely comparable, they were significantly higher using the RealPCR ASFV DNA Test, suggesting that there could be an impact on the detection of low-positivity samples. Furthermore, although the VetAlert™ ASFV DNA Test Kit and ASF System 1 rendered lower Cq values in general, the Cq values of the weakly positive samples attained with the VetMax™ ASFV Detection Kit were lower, indicating increased performance with weakly/very weakly positive samples as the input matrix. Detection in semen benefited from increasing the number of replicates in qPCR, as described previously [32].

Although no qPCR assay was able to detect all the true positive samples (EDTA blood and semen taken together), the VetMax™ ASFV Detection Kit rendered (I) the most positive results (3/5) and (II) overall lower Cq values, which reached statistical significance for the EDTA blood samples, further enhancing the suitability of this kit for accurate ASFV detection. It is of note that in-depth analyses of the kit performances uncovered that the VetAlert™ ASFV DNA Test Kit can be used interchangeably with the VetMax™ ASFV Detection Kit due to their similar performance. However, in terms of the handling/preparation time and complexity, the VetMax™ ASFV Detection Kit provided a ready-to-use mix, while the VetAlert™ ASFV DNA Test Kit required the preparation of said mix.

In addition, the virotype ASFV 2.0 PCR Kit only detected 1/5 true positive semen samples, which possibly resulted from the number of amplification cycles recommended for this kit. Considering that the Cq values of true positive semen samples can exceed 40, the virotype ASFV 2.0 PCR Kit was likely at a disadvantage due to being optimized for time (with the shortest run among all tested), not sensitivity.

In summary, based on our dataset, we identified the NucleoMag® VET Kit as the most suitable kit for nucleic acid extraction to enable the detection of even low amounts of the ASFV genome in porcine blood and semen samples. Furthermore, the VetMax™ ASFV Detection Kit and, although to a lesser extent, the VetAlert™ ASFV DNA Test Kit provided paramount detection of weakly positive blood and semen samples among all the kits tested. However, the suitability for diagnostic workflows of each kit must be carefully assessed, e.g., handling, the possibility of combining assays, the time needed, and the range of detectable pathogens with kits from one manufacturer. This is especially true when samples with expectedly high Cq values (obtained during the early stages of infection without apparent clinical signs) are handled.

5. Conclusions

With this study, we present a suitable workflow that enables the efficient detection of the ASFV genome in boar semen using routine protocols. We compared the performance of three widely used magnetic-bead-based methods for nucleic acid extraction, as well as WOAH-recommended qPCR methods and commercial qPCR kits. Among the tested options and based on our dataset, the NucleoMag® VET Kit performed best, in combination with the VetMax™ ASFV Detection Kit. Nevertheless, the limitations of this study must be considered, as the sample sizes were small (blood $n = 18$, semen $n = 17$) and the samples were correlated, as they were derived from four individuals rather than being independent.

Author Contributions: Conceptualization, S.B., M.B., J.C.-H., D.R. and J.J.Z.; methodology, V.F. and S.B.; formal analysis, V.F., C.S.-L., J.C.-H., D.R., J.J.Z. and S.B.; investigation, V.F.; data curation, V.F. and C.S.-L.; writing—original draft preparation, V.F. and S.B.; writing—review and editing, V.F., S.B., J.C.-H., D.R., E.A.N. and C.S.-L.; visualization, V.F.; supervision, S.B., J.C.-H., D.R., J.J.Z., E.A.N. and M.B.; project administration, S.B., J.C.-H. and D.R.; funding acquisition, S.B., J.C.-H. and M.B. All authors have read and agreed to the published version of the manuscript.

Funding: This research received funding through the Horizon 2020 ERA-NET Cofund International Coordination of Research on Infectious Animal Diseases (ICRAD), project "ASF-RASH", project ID: 82; from the FLI ASF research network; and in part from the National Pork Board (NPB), a program sponsored by the United States Department of Agriculture. The project ID of the NPB project associated with this work is #21-134.

Institutional Review Board Statement: The study was conducted in accordance with animal welfare regulations, including EU Directive 2010/63/EC, and institutional guidelines. The study was approved by the State Office for Agriculture, Food Safety and Fishery in Mecklenburg—Western Pomerania (LALFF M—V) and is filed under reference number 7221.3-1-071/21.

Informed Consent Statement: Not applicable.

Data Availability Statement: The data are available on request from the corresponding author.

Acknowledgments: We would like to thank all the animal caretakers, Steffen Brenz, Matthias Jahn, Domenique Lux, and Patrice Mary, for their excellence in animal husbandry and during the sampling procedures. Further, we would like to thank Ralf Henkel for preparing and assisting with all the necropsies. For excellent technical assistance during the sample processing and experiments, we thank Ulrike Kleinert.

Conflicts of Interest: The authors declare no conflicts of interest.

References

1. Andres, G.; Charro, D.; Matamoros, T.; Dillard, R.S.; Abrescia, N.G.A. The cryo-EM structure of African swine fever virus unravels a unique architecture comprising two icosahedral protein capsids and two lipoprotein membranes. *J. Biol. Chem.* **2020**, *295*, 1–12. [CrossRef]
2. Alejo, A.; Matamoros, T.; Guerra, M.; Andres, G. A Proteomic Atlas of the African Swine Fever Virus Particle. *J. Virol.* **2018**, *92*, e01293-01218. [CrossRef]
3. Penrith, M.L.; Vosloo, W.; Jori, F.; Bastos, A.D. African swine fever virus eradication in Africa. *Virus Res.* **2013**, *173*, 228–246. [CrossRef]
4. Gallardo, M.C.; Reoyo, A.T.; Fernandez-Pinero, J.; Iglesias, I.; Munoz, M.J.; Arias, M.L. African swine fever: A global view of the current challenge. *Porc. Health Manag.* **2015**, *1*, 21. [CrossRef]
5. Gallardo, C.; Nieto, R.; Soler, A.; Pelayo, V.; Fernandez-Pinero, J.; Markowska-Daniel, I.; Pridotkas, G.; Nurmoja, I.; Granta, R.; Simon, A.; et al. Assessment of African Swine Fever Diagnostic Techniques as a Response to the Epidemic Outbreaks in

Eastern European Union Countries: How To Improve Surveillance and Control Programs. *J. Clin. Microbiol.* **2015**, *53*, 2555–2565. [CrossRef] [PubMed]
6. Chenais, E.; Depner, K.; Guberti, V.; Dietze, K.; Viltrop, A.; Stahl, K. Epidemiological considerations on African swine fever in Europe 2014-2018. *Porc. Health Manag.* **2019**, *5*, 6. [CrossRef]
7. Zhou, X.; Li, N.; Luo, Y.; Liu, Y.; Miao, F.; Chen, T.; Zhang, S.; Cao, P.; Li, X.; Tian, K.; et al. Emergence of African Swine Fever in China, **2018**. *Transbound Emerg. Dis.* **2018**, *65*, 1482–1484. [CrossRef]
8. Gonzales, W.; Moreno, C.; Duran, U.; Henao, N.; Bencosme, M.; Lora, P.; Reyes, R.; Nunez, R.; De Gracia, A.; Perez, A.M. African swine fever in the Dominican Republic. *Transbound Emerg. Dis.* **2021**, *68*, 3018–3019. [CrossRef] [PubMed]
9. Galindo, I.; Alonso, C. African Swine Fever Virus: A Review. *Viruses* **2017**, *9*, 103. [CrossRef] [PubMed]
10. Achenbach, J.E.; Gallardo, C.; Nieto-Pelegrin, E.; Rivera-Arroyo, B.; Degefa-Negi, T.; Arias, M.; Jenberie, S.; Mulisa, D.D.; Gizaw, D.; Gelaye, E.; et al. Identification of a New Genotype of African Swine Fever Virus in Domestic Pigs from Ethiopia. *Transbound Emerg. Dis.* **2017**, *64*, 1393–1404. [CrossRef] [PubMed]
11. Bastos, A.D.; Penrith, M.L.; Cruciere, C.; Edrich, J.L.; Hutchings, G.; Roger, F.; Couacy-Hymann, E.; Thomson, G.R. Genotyping field strains of African swine fever virus by partial p72 gene characterisation. *Arch. Virol.* **2003**, *148*, 693–706. [CrossRef] [PubMed]
12. Blome, S.; Franzke, K.; Beer, M. African swine fever—A review of current knowledge. *Virus Res.* **2020**, *287*, 198099. [CrossRef] [PubMed]
13. Broekhuijse, M.L.; Gaustad, A.H.; Bolarin Guillen, A.; Knol, E.F. Efficient Boar Semen Production and Genetic Contribution: The Impact of Low-Dose Artificial Insemination on Fertility. *Reprod. Domest. Anim.* **2015**, *50* (Suppl. S2), 103–109. [CrossRef] [PubMed]
14. Tignon, M.; Gallardo, C.; Iscaro, C.; Hutet, E.; Van der Stede, Y.; Kolbasov, D.; De Mia, G.M.; Le Potier, M.F.; Bishop, R.P.; Arias, M.; et al. Development and inter-laboratory validation study of an improved new real-time PCR assay with internal control for detection and laboratory diagnosis of African swine fever virus. *J. Virol. Methods* **2011**, *178*, 161–170. [CrossRef] [PubMed]
15. King, D.P.; Reid, S.M.; Hutchings, G.H.; Grierson, S.S.; Wilkinson, P.J.; Dixon, L.K.; Bastos, A.D.; Drew, T.W. Development of a TaqMan PCR assay with internal amplification control for the detection of African swine fever virus. *J. Virol. Methods* **2003**, *107*, 53–61. [CrossRef]
16. Fernandez-Pinero, J.; Gallardo, C.; Elizalde, M.; Robles, A.; Gomez, C.; Bishop, R.; Heath, L.; Couacy-Hymann, E.; Fasina, F.O.; Pelayo, V.; et al. Molecular diagnosis of African Swine Fever by a new real-time PCR using universal probe library. *Transbound Emerg. Dis.* **2013**, *60*, 48–58. [CrossRef]
17. Aguero, M.; Fernandez, J.; Romero, L.; Sanchez Mascaraque, C.; Arias, M.; Sanchez-Vizcaino, J.M. Highly sensitive PCR assay for routine diagnosis of African swine fever virus in clinical samples. *J. Clin. Microbiol.* **2003**, *41*, 4431–4434. [CrossRef] [PubMed]
18. Schulz, C.; van der Poel, W.H.; Ponsart, C.; Cay, A.B.; Steinbach, F.; Zientara, S.; Beer, M.; Hoffmann, B. European interlaboratory comparison of Schmallenberg virus (SBV) real-time RT-PCR detection in experimental and field samples: The method of extraction is critical for SBV RNA detection in semen. *J. Vet. Diagn. Investig.* **2015**, *27*, 422–430. [CrossRef]
19. Pikalo, J.; Carrau, T.; Deutschmann, P.; Fischer, M.; Schlottau, K.; Beer, M.; Blome, S. Performance Characteristics of Real-Time PCRs for African Swine Fever Virus Genome Detection-Comparison of Twelve Kits to an OIE-Recommended Method. *Viruses* **2022**, *14*, 220. [CrossRef]
20. Schoder, M.E.; Tignon, M.; Linden, A.; Vervaeke, M.; Cay, A.B. Evaluation of seven commercial African swine fever virus detection kits and three Taq polymerases on 300 well-characterized field samples. *J. Virol. Methods* **2020**, *280*, 113874. [CrossRef]
21. WOAH. Register of Diagnostic Kits Certified by the OIE as Validated as Fit for Purpose. Available online: https://www.woah.org/en/what-we-offer/veterinary-products/diagnostic-kits/the-register-of-diagnostic-kits/ (accessed on 16 May 2024).
22. Hoffmann, B.; Schulz, C.; Beer, M. First detection of Schmallenberg virus RNA in bovine semen, Germany, 2012. *Vet. Microbiol.* **2013**, *167*, 289–295. [CrossRef] [PubMed]
23. Louwrier, A.; van der Valk, A. Can sucrose affect polymerase chain reaction product formation? *Biotechnol. Lett.* **2001**, *23*, 175–178. [CrossRef]
24. Zani, L.; Forth, J.H.; Forth, L.; Nurmoja, I.; Leidenberger, S.; Henke, J.; Carlson, J.; Breidenstein, C.; Viltrop, A.; Hoper, D.; et al. Deletion at the 5′-end of Estonian ASFV strains associated with an attenuated phenotype. *Sci. Rep.* **2018**, *8*, 6510. [CrossRef] [PubMed]
25. Friedrichs, V.; Reicks, D.; Hasenfuss, T.; Gerstenkorn, E.; Zimmerman, J.J.; Nelson, E.A.; Carrau, T.; Deutschmann, P.; Sehl-Ewert, J.; Roszyk, H.; et al. Artificial Insemination as an Alternative Transmission Route for African Swine Fever Virus. *Pathogens* **2022**, *11*, 1539. [CrossRef] [PubMed]
26. Pikalo, J.; Deutschmann, P.; Fischer, M.; Roszyk, H.; Beer, M.; Blome, S. African Swine Fever Laboratory Diagnosis-Lessons Learned from Recent Animal Trials. *Pathogens* **2021**, *10*, 177. [CrossRef] [PubMed]
27. Mallik, S. Identification Methods | Multilocus Enzyme Electrophoresis. In *Encyclopedia of Food Microbiology*; Elsevier: Amsterdam, The Netherlands, 2014; Volume 2, pp. 336–343.
28. Bland, J.M.; Altman, D.G. Statistical methods for assessing agreement between two methods of clinical measurement. *Lancet* **1986**, *1*, 307–310. [CrossRef] [PubMed]
29. Da Silva, N.; Zardoya, R.; Santurde, G.; Solana, A.; Castro, J.M. Rapid and sensitive detection of the bovine viral diarrhea virus genome in semen. *J. Virol. Methods* **1995**, *55*, 209–218. [CrossRef] [PubMed]

30. Reicks, D.L.; Muñoz-Zanzi, C.; Rossow, K. Sampling of adult boars during early infection with porcine reproductive and respiratory syndrome virus for testing by polymerase chain reaction using a new blood collection technique (blood-swab method). *JSHAP* **2006**, *14*, 258–264.
31. Pepin, B.J.; Kittawornrat, A.; Liu, F.; Gauger, P.C.; Harmon, K.; Abate, S.; Main, R.; Garton, C.; Hargrove, J.; Rademacher, C.; et al. Comparison of specimens for detection of porcine reproductive and respiratory syndrome virus infection in boar studs. *Transbound Emerg. Dis.* **2015**, *62*, 295–304. [CrossRef]
32. Nybo, K. qPCR: Technical Replicate Variation. *Biotechniques* **2011**, *50*, 23. [CrossRef]

pathogens

MDPI

Article

Triplex Crystal Digital PCR for the Detection and Differentiation of the Wild-Type Strain and the MGF505-2R and I177L Gene-Deleted Strain of African Swine Fever Virus

Kaichuang Shi [1,2,*,†], Kang Zhao [1,†], Haina Wei [2], Qingan Zhou [2], Yuwen Shi [1], Shenglan Mo [2], Feng Long [2], Liping Hu [2], Shuping Feng [2] and Meilan Mo [1,*]

1 College of Animal Science and Technology, Guangxi University, Nanning 530005, China; zhaokang0519@163.com (K.Z.); shiyuwen2@126.com (Y.S.)
2 Guangxi Center for Animal Disease Control and Prevention, Nanning 530001, China; weihaina@sina.cn (H.W.); zhouqingan1@163.com (Q.Z.); moshl_2015@126.com (S.M.); longfeng1136@163.com (F.L.); hu.liping@foxmail.com (L.H.); fsp166@163.com (S.F.)
* Correspondence: shikaichuang@126.com (K.S.); momeilan@gxu.edu.cn (M.M.)
† These authors contributed equally to this work.

Abstract: African swine fever (ASF) is a severe and highly contagious viral disease that affects domestic pigs and wild boars, characterized by a high fever and internal bleeding. The disease is caused by African swine fever virus (ASFV), which is prevalent worldwide and has led to significant economic losses in the global pig industry. In this study, three pairs of specific primers and TaqMan probes were designed for the ASFV B646L, MGF505-2R and I177L genes. After optimizing the reaction conditions of the annealing temperature, primer concentration and probe concentration, triplex crystal digital PCR (cdPCR) and triplex real-time quantitative PCR (qPCR) were developed for the detection and differentiation of the wild-type ASFV strain and the MGF505-2R and/or I177L gene-deleted ASFV strains. The results indicate that both triplex cdPCR and triplex qPCR were highly specific, sensitive and repeatable. The assays could detect only the B646L, MGF505-2R and I177L genes, without cross-reaction with other swine viruses (i.e., PRRSV, CSFV, PCV2, PCV3, PEDV, PDCoV and PRV). The limit of detection (LOD) of triplex cdPCR was 12 copies/reaction, and the LOD of triplex qPCR was 500 copies/reaction. The intra-assay and inter-assay coefficients of variation (CVs) for repeatability and reproducibility were less than 2.7% for triplex cdPCR and less than 1.8% for triplex qPCR. A total of 1510 clinical tissue samples were tested with both methods, and the positivity rates of ASFV were 14.17% (214/1510) with triplex cdPCR and 12.98% (196/1510) with triplex qPCR, with a coincidence rate of 98.81% between the two methods. The positivity rate for the MGF505-2R gene-deleted ASFV strains was 0.33% (5/1510), and no I177L gene-deleted ASFV strain was found. The results indicate that triplex cdPCR and triplex qPCR developed in this study can provide rapid, sensitive and accurate methods for the detection and differentiation of the ASFV B646L, MGF505-2R and I177L genes.

Keywords: African swine fever virus (ASFV); multiplex crystal digital PCR (multiplex cdPCR); multiplex real-time quantitative PCR (multiplex qPCR); gene-deleted strain; wild-type strain

Citation: Shi, K.; Zhao, K.; Wei, H.; Zhou, Q.; Shi, Y.; Mo, S.; Long, F.; Hu, L.; Feng, S.; Mo, M. Triplex Crystal Digital PCR for the Detection and Differentiation of the Wild-Type Strain and the MGF505-2R and I177L Gene-Deleted Strain of African Swine Fever Virus. *Pathogens* **2023**, *12*, 1092. https://doi.org/10.3390/pathogens12091092

Academic Editor: Natasha N. Gaudreault

Received: 29 June 2023
Revised: 24 August 2023
Accepted: 25 August 2023
Published: 28 August 2023

1. Introduction

African swine fever (ASF) is an acute, highly contagious infectious disease that causes hemorrhagic fevers in infected pigs, and the mortality rate might reach almost 100% [1]. The etiological agent, African swine fever virus (ASFV), is an enveloped, double-stranded DNA virus that belongs to the family *Asfarviridae*, and it is the only member of the genus *Asfivirus* [2]. The ASFV viral particle size ranges from 260 to 300 nm in diameter and contains a 170–194 kb genome encoding over 160 open reading frames (ORFs) with conserved central regions and variable ends [3,4]. ASF was first discovered in Kenya in 1921, and it subsequently spread to Europe, where it was first detected in Portugal in 1957. Subsequently, ASF was found in the Caucasus and southern Russia in 2007 [5,6]. Currently, outbreaks of ASF are ongoing in various regions of the world, including Africa, the Caucasus, Eastern Europe, the Russian Federation, Asia and Latin America [7,8]. ASF was first reported in China in August 2018 [9] and quickly spread to most provinces within a very short time [10,11]. Thereafter, other Asian countries, including Mongolia, South Korea, Vietnam, Laos, Cambodia, the Philippines and Indonesia, have also reported ASF [12,13]. ASF has caused significant economic losses in the global pig industry.

To date, 24 genotypes of ASFV based on the B646L gene have been discovered around the world, and most genotypes have spread within the African continent. Only genotype I and genotype II strains of ASFV have been reported outside Africa. The genotype I strain of ASFV was first identified in Portugal in 1957, and the genotype II strain of ASFV was first discovered in Georgia in 2007 [5,14,15]. Today, genotype II ASFV is the dominant genotype that is prevalent in various countries around the world [15,16]. Since the first report of ASF in China in 2018, most prevalent strains of ASFV have been found to be genotype II [17,18], but genotype I strains of ASFV have been found in several provinces (Anhui, Shandong and Guangxi provinces) of China [17,19,20]. As the predominant strains of ASFV, the genotype II ASFV strains have posed a serious threat to the healthy development of the Chinese pig industry. In addition, several provinces in China have identified naturally variant ASFV strains that lack the MGF360-505R and EP402R genes [21]. These strains demonstrated reduced virulence, and the infected pigs exhibited low morbidity and mortality rates. Genotype I strains that lack the MGF360 and MGF505 gene families have been discovered in Guangxi province, southern China [17]. Currently, the emergence of gene-deleted strains of ASFV has introduced new challenges to the prevention and control of ASF [21,22]. In addition, many scientists have reported some ASFV vaccine candidates that lack some of the virulence genes. By deleting one or several genes of MGF505-1R, MGF505-2R, MGF505-3R, MGF360-12L, MGF360-13L, MGF360-14L, CD2v, 9GL, DP148R and UK, scientists generated six novel gene-deleted vaccine strains, and these strains have demonstrated good safety and efficacy [23]. Several studies have confirmed that strains lacking the MGF360-505R and EP402R genes can be used as ideal candidates for developing live attenuated vaccines [24–26]. It is noteworthy that an ASF vaccine candidate, which has been genetically modified to have a deletion in part of the I177L gene, has been demonstrated to provide protection to pigs against highly pathogenic strains of ASFV that are presently circulating in Europe and Asia [27–30]. The attenuated vaccine that utilized the I177L gene-deleted strain was officially approved for commercial use in Vietnam in June 2022 (http://link.gov.vn/vtUM759t, accessed on 9 March 2023). This vaccine is the first commercial vaccine approved for clinical use for preventing ASF in the world [31]. Especially, the MGF505-2R and/or I177L gene-deleted strains of ASFV have aroused great interest and become the focus of attention in the pig industry. Therefore, it is very necessary to develop a rapid, sensitive and accurate method to detect and differentiate the wild-type strain and the MGF505-2R and/or I177L gene-deleted strains of ASFV.

Real-time quantitative PCR (qPCR) is a convenient, efficient, accurate and high-throughput technology and has been widely used to test viral nucleic acids [32]. Digital PCR (dPCR) is a powerful technology that has been developed in recent years, particularly in the field of microbiological research. Compared to conventional PCR and qPCR, dPCR does not require a standard curve for absolute quantitative analysis and is more accurate for the detection of the low number of DNA or RNA molecules [32–34]. To date, dPCR has been developed as an accurate method for the detection of ASFV, including droplet digital PCR (ddPCR) [35,36], crystal digital PCR (cdPCR) [37] and other similar technologies [38]. qPCR has been developed for the detection of ASFV [39–41], and several multiplex qPCRs based on the E296R, B646L or E183L genes of ASFV have been reported for the detection of genotypes I and II of ASFV [42–44]. In addition, several assays have been developed for the differential detection of the wild-type strains of ASFV and for the MGF505-2R, MGF-360-14L, CD2v, EP402R and/or I177L gene-deleted strains of ASFV [35,45–48]. In this study, triplex cdPCR and triplex qPCR, which target the ASFV B646L gene, MGF505-2R gene and I177L gene, were developed to detect and differentiate the wild-type strain and the MGF505-2R and I177L gene-deleted strains of the ASFV. The developed assays were used to test 1510 clinical tissue samples collected from Guangxi province, southern China, to validate the application of these assays.

2. Materials and Methods

2.1. Viral Strains

The vaccine strains of classical swine fever virus (CSFV, C strain), porcine circovirus type 2 (PCV2, SX07 strain), porcine reproductive and respiratory syndrome virus (PRRSV, TJM-F92 strain), porcine epidemic diarrhea virus (PEDV, SCJY-1 strain) and pseudorabies virus (PRV, Bartha-K61 strain) were purchased from Huapai Bioengineering Group Co., Ltd. (Chengdu, China). The positive clinical tissue samples of ASFV, PCV3 and porcine deltacoronavirus (PDCoV) were provided by the Guangxi Center for Animal Disease Control and Prevention (CADC), China. These viruses and samples were stored at −80 °C until use.

2.2. Clinical Samples

From January 2022 to December 2022, a total of 1510 clinical tissue samples from 1510 dead pigs were collected from different pig farms, slaughterhouses, farmers' markets and harmless disposal sites in Guangxi province, southern China. The tissue samples from each pig included livers, spleens, kidneys and lymph nodes and were pooled and homogenized before being tested with the developed triplex cdPCR and triplex qPCR. The samples were transported to our laboratory at ≤4 °C within 12 h (from the death of the pig to the arrival of the samples at the laboratory) and were stored at −80 °C until use.

2.3. Design of Primers and Probes

Three pairs of specific primers and TaqMan probes were designed using the Primer Express Software v3.0 (ABI, Los Angeles, CA, USA), which targeted the conserved regions of the B646L, MGF505-2R and I177L genes, respectively. If the ASFV strain is positive for these three genes, it is the wild-type strain; if the ASFV strain is negative for the MGF505-2R and/or I177L genes, it is the MGF505-2R and/or I177L gene-deleted ASFV strain. The primers and probes are shown in Table 1.

Table 1. Primer and probe sequences.

Targeted Gene	Name	Sequences (5′→3′)	Amplicon (bp)
B646L	ASFV-B646L-F	GGCGTATAAAAAGTCCAGGAAATTC	79
	ASFV-B646L-R	TTCGGCGAGCGCTTTATC	
	ASFV-B646L-P	FAM-TCACCAAATCCTTTTGCGATGCAAGCT-BHQ1	
MGF505-2R	ASFV-MGF505-F	AGTCATGCACGGCATATACAA	153
	ASFV-MGF505-R	GGTTTAAACCGTGCCACATCC	
	ASFV-MGF505-P	VIC-ACGCGGCCACCCAATTCAGAGAC-BHQ1	
I177L	ASFV-I177L-F	GGCATAATTATCAAATGCGAAGGG	122
	ASFV-I177L-R	TGGAAAGTTAATGATCAGGGCTT	
	ASFV-I177L-P	Cy5-AATCCTAGCTTGCCGGTAATGGCT-BHQ2	

2.4. Extraction of Nucleic Acids

Phosphate-buffered saline (PBS, pH 7.2) was added to the vaccines and the clinical tissue homogenates (20%, W/V), and the solutions were vortexed for 5 min and centrifuged at 12,000× *g* at 4 °C for 10 min. The total nucleic acids were extracted from the homogenized tissues or vaccine solutions using the GeneRotex 96 Automatic Nucleic Acid Extractor (TIANLONG Scientific, Xi'an, China) with Viral DNA/RNA Isolation Kit 4.0 (TIANLONG Scientific, Xi'an, China) according to the manufacturer's instructions. The obtained nucleic acids were stored at −80 °C until use.

2.5. Construction of the Standard Plasmids

The DNA of ASFV was used as a template to amplify the targeted fragments of the B646L gene, the MGF505-2R gene and the I177L gene via PCR using the specific primers, respectively. The purified amplicons were ligated into the pMD18-T vector (TaKaRa, Dalian, China) and were subsequently transformed into *E. coli* DH5α competent cells (TaKaRa, Dalian, China). The positive clones were cultured at 37 °C for 20–24 h, and the plasmid constructs were extracted using MiniBEST Plasmid Extraction Kit Ver.5.0 (TaKaRa, Dalian, China), confirmed by sequencing. The recombinant plasmids were named pASFV-B646L-1, pASFV-MGF505-2R-1 and pASFV-I177L-1, respectively, and were used as standard plasmids in this study. They were stored at −80 °C until use.

The standard plasmids were quantified using a NanoDrop spectrophotometer (Thermo Fisher, Waltham, MA, USA) to measure their ultraviolet absorbance at 260 nm and 280 nm. Their concentrations were determined using the following formula: plasmid copy number (copies/μL) = (plasmid concentration $\times 10^{-9} \times 6.02 \times 10^{23}$)/(660 Dalton/bases × DNA length).

The sequences located at two ends of the ASFV MGF505-2R and I177L genes were artificially synthesized, respectively, by TaKaRa Biomedical Technology Co. Ltd. (TaKaRa, Dalian, China) and were then inserted into the pMD18-T vector (TaKaRa, Dalian, China) to construct the plasmids without the MGF505-2R gene (named pASFV-ΔMGF505-2R) and the I177L gene (named pASFV-ΔI177L), respectively. They were used as the MGF505-2R and I177L gene-deleted controls. The sequences of pASFV-ΔMGF505-2R and pASFV-ΔI177L are shown in Supplementary Table S1.

2.6. Optimization of the Reaction Conditions

The optimal conditions for triplex qPCR, including the annealing temperature and the concentrations of primers and probes, were determined using the QuantStudio 6 qPCR detection system (ABI, Carlsbad, CA, USA). The parameters used for amplification were as follows: pre-denaturation at 95 °C for 2 min, followed by 40 cycles of denaturation at 95 °C for 5 s and annealing and extension at 59 °C for 30 s. At the end of each cycle, the fluorescent signals were determined. In order to determine the optimal reaction conditions for triplex qPCR, the basic components were as follows: 12.5 μL of Premix Ex Taq (Probe qPCR) (TaKaRa, Dalian, China); 2.5 μL of the mixture of three standard plasmids with a final reaction concentration of 10^8 copies/μL for each plasmid; 3 pairs of primers and corresponding TaqMan probes with different final concentrations (from 100 nM to 500 nM); and nuclease-free distilled water to a total volume of 25 μL. The ultimate reaction conditions were optimized to achieve the highest ΔRn and the lowest threshold cycle (Ct).

The Naica™ sapphire crystal system (Stilla Technologies™, Villejuif, France) was used to perform triplex cdPCR. The entire process was carried out within the Sapphire chip (Stilla Technologies™, Villejuif, France). After thermocycling, the chips were transferred to Naica™ Prism3 (Stilla Technologies™, Villejuif, France) to image the FAM, VIC and CY5 detection channels. The concentrations of the templates were then determined using the Crystal Miner software (Stilla Technologies™, Villejuif, France). The parameters used for amplification were as follows: 95 °C for 2 min; 45 cycles of 95 °C for 5 s; and 59 °C for 30 s. A total volume of 25 μL was used to determine the optimal reaction conditions of triplex cdPCR using the following basic systems: 12.5 μL of PerfeCTa Multiplex qPCR ToughMix (Quanta Biosciences, Gaithersburg, MD, USA); 2.5 μL of Fluorescein Sodium Salt (1 μM) (Apexbio Biotechnology, Beijing, China); a mixture of three pairs of primers and three probes of different final concentrations of 0.1–0.9 μL and 2.5 μL of the mixture of three standard plasmids with concentrations of 10^4 copies/μL for each plasmid; and nuclease-free distilled water to a final volume of 25 μL.

2.7. Analytical Specificity Analysis

The DNA of ASFV, PCV2, PCV3 and PRV was extracted from the vaccine solutions or homogenized positive samples. The RNA of PRRSV, CSFV, PEDV and PDCoV was extracted from the vaccine solutions or homogenized positive samples, and it was then reverse transcribed to cDNA using the PrimeScript II 1st Strand cDNA Synthesis Kit (TaKaRa, Dalian, China).

The DNA or cDNA of ASFV, PCV2, PCV3, PRV, PRRSV, CSFV, PEDV and PDCoV, as well as the recombinant gene-deleted plasmids pASFV-ΔMGF505-2R and pASFV-ΔI177L, was used to evaluate the specificity of the developed assays. The mixture of three standard plasmids was used as the positive control, and nuclease-free distilled water was used as the negative control.

2.8. Analytical Sensitivity Analysis

The standard plasmids pASFV-B646L-1, pASFV-MGF505-2R-1 and pASFV-I177L-1 were mixed at a ratio of 1:1:1 and were 10-fold serially diluted. The sensitivity of triplex cdPCR was analyzed using the mixed plasmids at concentrations from 2.0×10^5 copies/µL to 2.0×10^{-1} copies/µL (final reaction concentration). The sensitivity of triplex qPCR was evaluated using the mixed plasmid at concentrations from 2.0×10^6 copies/µL to 2.0×10^0 copies/µL (final reaction concentration).

2.9. Repeatability Analysis

The standard plasmids pASFV-B646L-1, pASFV-MGF505-2R-1 and pASFV-I177L-1 were mixed at a ratio of 1:1:1 and were 10-fold serially diluted. The concentrations of 2.0×10^4, 2.0×10^3 and 2.0×10^2 copies/µL (final reaction concentration) of each plasmid were used as templates to perform in triplicate for the intra-assay test and on three different days for the inter-assay test. The coefficients of variation (CVs) were obtained.

2.10. Detection of the Clinical Samples

A total of 1510 clinical tissue samples collected in Guangxi province in China from January 2022 to December 2022 were tested with the developed triplex cdPCR and triplex qPCR, and the coincidence rate and Kappa value between the two methods were calculated using SPSS version 26.0 software (https://www.ibm.com/cn-zh/spss, accession on 13 June 2023).

3. Results

3.1. Construction of the Standard Plasmids

The targeted fragments of the ASFV B646L, MGF505-2R and I177L genes were amplified via PCR, purified, ligated into the pMD18-T vector and transformed into DH5α competent cells. The positive clones were cultured, the recombinant plasmids were extracted, and they were confirmed by sequencing. The initial concentrations of the standard plasmids, which were named pASFV-B646L-1, pASFV-MGF505-2R-1 and pASFV-I177L-1, were determined to be 2.16×10^{10}, 2.0×10^{10} and 3.45×10^{10} copies/µL, respectively. All standard plasmids were diluted to 2.0×10^{10} copies/µL and stored at −80 °C until use.

3.2. Determination of the Optimal Parameters

After optimization, the optimal annealing temperature, concentrations of the primers and probes, and amplification cycles were determined for the triplex qPCR assay. The reaction system in a total volume of 25 µL is shown in Table 2. The amplification parameters were as follows: 95 °C for 10 s; 40 cycles of 95 °C for 5 s; and 59 °C for 30 s. Samples with a Ct value of ≤36 were considered to be positive samples, and samples with a Ct value of >36 were considered to be negative samples.

After optimization, the optimal annealing temperature, concentrations of the primers and probes, and amplification cycles were determined for the triplex cdPCR assay (Figure 1). The optimal reaction system in a total volume of 25 µL was obtained (Table 2). The amplification parameters were as follows: 95 °C for 10 s; 45 cycles of 95 °C for 5 s; and 59 °C for 30 s. After amplification, the Naica™ system (Stilla Technologies™, Villejuif, France) automatically reported the absolute concentration of each sample.

Table 2. Optimal reaction system of triplex cdPCR and triplex qPCR.

Reagent	Triplex cdPCR		Triplex qPCR	
	Volume (μL)	Final Concentration (nM)	Volume (μL)	Final Concentration (nM)
PerfeCta Multiplex qPCR ToughMix (2×)	12.5	1×	/	/
Fluorescein Sodium Salt (1 μM)	2.5	100	/	/
Premix Ex Taq (Probe qPCR) (2×)	/	/	12.5	1×
ASFV-B646L-F (25 μM)	0.8	800	0.3	300
ASFV-B646L-R (25 μM)	0.8	800	0.3	300
ASFV-B646L-P (25 μM)	0.3	300	0.3	300
ASFV-MGF505-2R-F (25 μM)	0.8	800	0.4	400
ASFV-MGF505-2R-R (25 μM)	0.8	800	0.4	400
ASFV-MGF505-2R-P (25 μM)	0.3	300	0.4	400
ASFV-I177L-F (25 μM)	0.8	800	0.4	400
ASFV-I177L-R (25 μM)	0.8	800	0.4	400
ASFV-I177L-P (25 μM)	0.3	300	0.4	400
Total nucleic acids	2.5	/	2.5	/
Nuclease-free distilled water	Up to 25	/	Up to 25	/

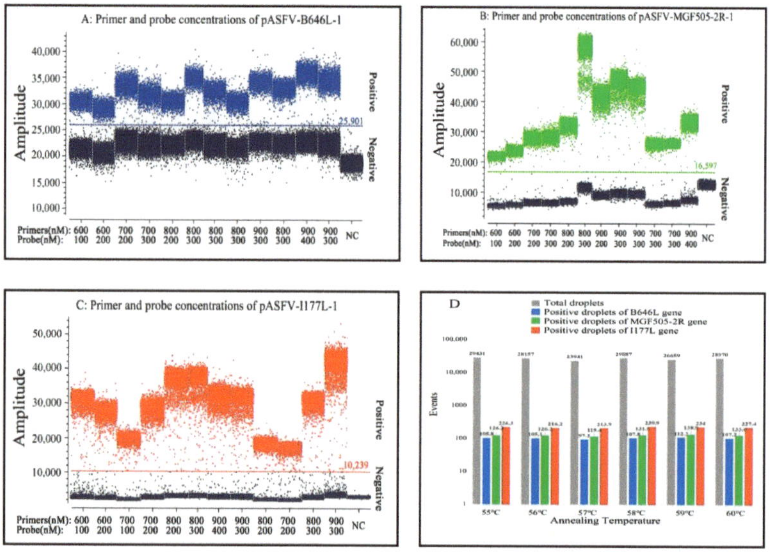

Figure 1. Optimization of the primer and probe concentrations (**A–C**) and the annealing temperature (**D**) for triplex cdPCR. (**A–C**) Amplification results of pASFV-B646L-1, pASFV-MGF505-2R-1 and pASFV-I177L-1 plasmids (all at final reaction concentrations of 2.0 × 10^4 copies/μL) with different probe and primer concentrations. NC: Negative control. (**D**) Amplification results of pASFV-B646L-1, pASFV-MGF505-2R-1 and pASFV-I177L-1 plasmids (all at final reaction concentrations of 2.0 × 10^4 copies/μL) with different annealing temperatures.

3.3. Generation of the Standard Curves

Mixtures of three standard plasmids, pASFV-B646L-1, pASFV-MGF505-2R-1 and pASFV-I177L-1, ranging from 2.0 × 10^7 copies/μL to 2.0 × 10^2 copies/μL (final reaction concentration: 2.0 × 10^6 copies/μL to 2.0 × 10^1 copies/μL), were used as templates to generate the standard curves of the developed qPCR. The results indicate that the slope, R^2 and Eff% were -3.477, 0.9998 and 93.9%, respectively, for the B646L gene; -3.4883, 0.9996 and 93.5%, respectively, for the MGF505-2R gene; and -3.5459, 0.9994 and 91.4%, respectively, for the I177L gene, indicating excellent correlation coefficients (R^2) between the initial concentrations of the template and the Ct values (Figure 2).

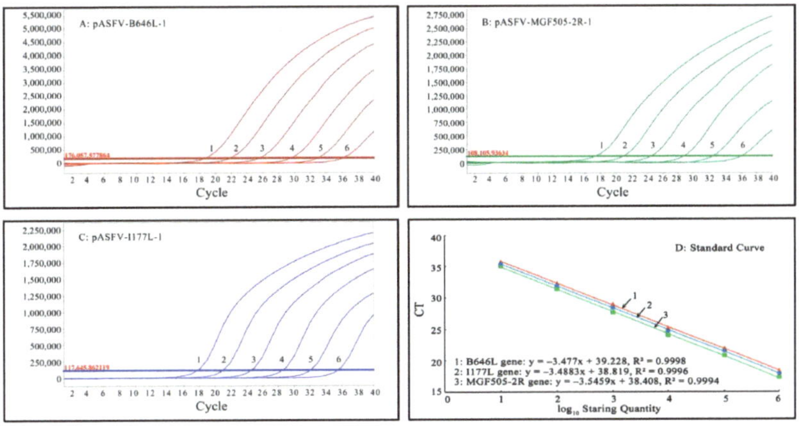

Figure 2. Generation of the standard curves of triplex qPCR. Amplification curves (**A–C**) and standard curves (**D**) of triplex qPCR. (**A–C**) Final reaction concentrations of curves 1 to 6 range from 2.0×10^6 to 2.0×10^1 copies/μL; 8: Negative control.

Mixtures of three standard plasmids, pASFV-B646L-1, pASFV-MGF505-2R-1 and pASFV-I177L-1, ranging from 2.0×10^5 copies/μL to 2.0×10^1 copies/μL (final reaction concentration: 2.0×10^4 copies/μL to 2.0×10^0 copies/μL), were used as templates to generate the standard curves of the developed triplex cdPCR. The results indicate that the slope and R^2 were 1.0603 and 0.9987, respectively, for the B646L gene; 1.0477 and 0.9996, respectively, for the MGF505-2R gene; and 1.0315 and 0.9982, respectively, for the I177L gene, indicating excellent correlation coefficients (R^2) between the initial concentrations of the template and the positive droplets (Figure 3).

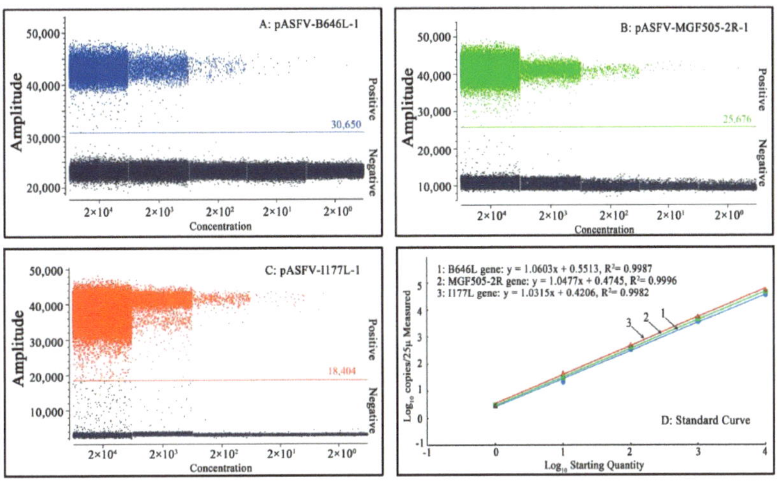

Figure 3. Generation of the standard curves of triplex cdPCR. (**A–C**) Final reaction concentrations of the plasmids range from 2.0×10^4 to 2.0×10^0 copies/μL. (**D**) shows the standard curves.

3.4. Sensitivity Analysis

The limit of detection (LOD) of triplex qPCR was evaluated using a mixture of three standard plasmids: pASFV-B646L-1, pASFV-MGF505-2R-1 and pASFV-I177L-1, ranging from 2.0×10^7 copies/μL to 2.0×10^1 copies/μL (final reaction concentration: 2.0×10^6 copies/μL to 2.0×10^0 copies/μL). The results indicate that the LOD of pASFV-B646L-1, pASFV-MGF505-

2R-1 and pASFV-I177L-1 was 20 copies/μL (Figure 4), which was equal to 500 copies/reaction.

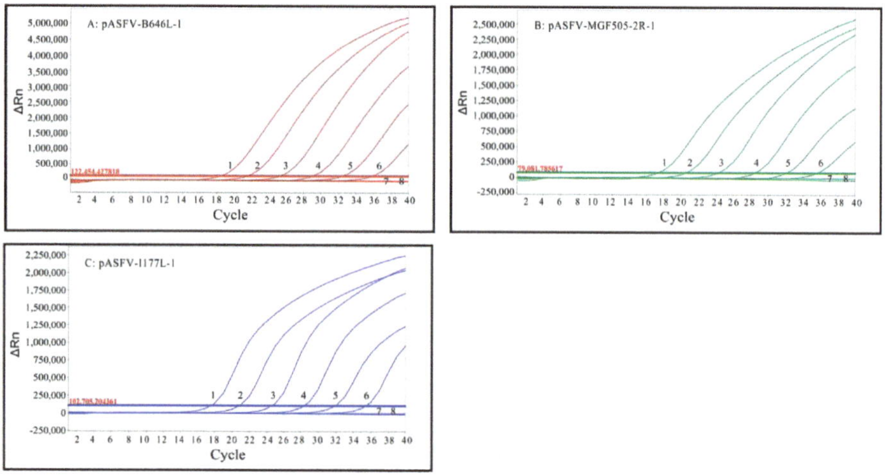

Figure 4. Sensitivity analysis of triplex qPCR. (**A–C**) Final reaction concentrations of curves 1 to 7 range from 2.0×10^6 to 2.0×10^0 copies/μL; 8: Negative control.

The LOD of triplex cdPCR was evaluated using a mixture of three standard plasmids: pASFV-B646L-1, pASFV-MGF505-2R-1 and pASFV-I177L-1, ranging from 2.0×10^6 copies/μL to 2.0×10^0 copies/μL (final reaction concentration: 2.0×10^5 copies/μL to 2.0×10^{-1} copies/μL). The results revealed a gradual decline in the number of positive droplets with decreasing plasmid concentrations. The LOD of pASFV-B646L-1, pASFV-MGF505-2R-1 and pASFV-I177L-1 was 2 copies/μL (Figure 5), and the actual concentration value was revised as 12 copies/reaction via the Poisson distribution. The results demonstrated that triplex cdPCR was 42 times more sensitive than triplex qPCR.

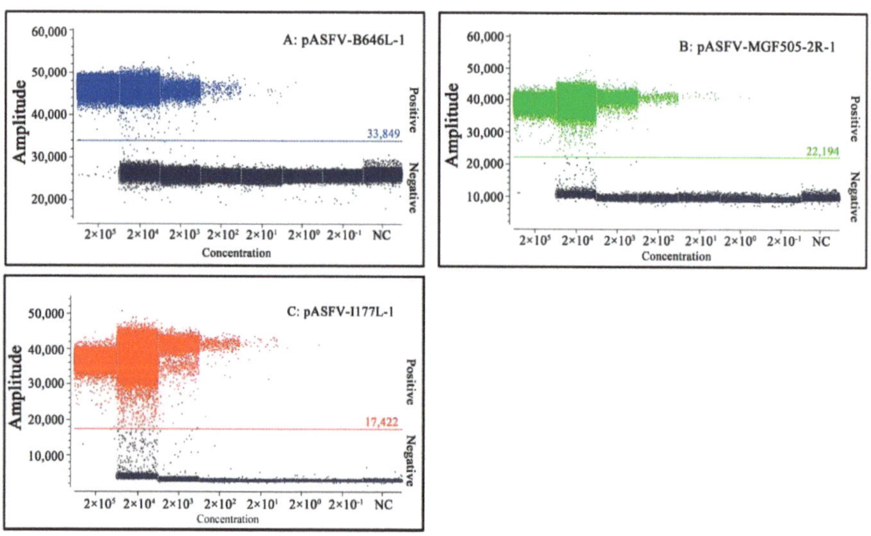

Figure 5. Sensitivity analysis of triplex cdPCR. (**A–C**) Final reaction concentrations of the plasmids range from 2.0×10^5 to 2.0×10^{-1} copies/μL. NC: Negative control.

3.5. Specificity Analysis

The specificity of triplex cdPCR and triplex qPCR was validated using the DNA/cDNA of ASFV, PCV2, PCV3, PRV, PRRSV, CSFV, PEDV and PDCoV, as well as the recombinant plasmids pASFV-ΔEP402R and pASFV-ΔI177L. The results indicate that the assays only generated positive droplets from ASFV, without cross-reaction with other swine viruses, demonstrating the strong specificity of the two assays (Figure 6).

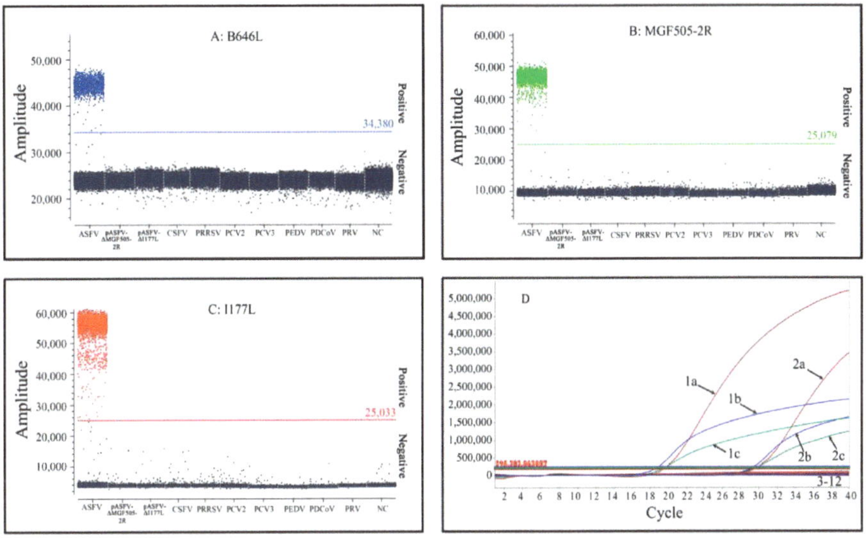

Figure 6. Specificity analysis of triplex cdPCR (**A–C**) and triplex qPCR (**D**). (**A–C**) Results of the specificity tests of the B646L gene (**A**), the MGF505-2R gene (**B**) and the I177L gene (**C**). NC: Negative control. (**D**) Results of the specificity test of triplex qPCR. 1a: pASFV-B646L-1; 1b: pASFV-I177L-1; 1c: pASFV-MGF505-2R-1; 2a: ASFV B646L gene; 2b: ASFV I177L gene; 2c: ASFV MGF505-2R gene; 3: pASFV-ΔI177L; 4: pASFV-ΔMGF505-2R; 5: CSFV; 6: PRRSV; 7: PCV2; 8: PCV3; 9: PEDV; 10: PDCoV; 11: PRV; 12: Negative control.

3.6. Repeatability and Reproducibility Analysis

The repeatability and reproducibility of triplex cdPCR and triplex qPCR was evaluated using mixtures of three standard plasmids with final reaction concentrations of 2.0×10^4 copies/μL, 2.0×10^3 copies/μL and 2.0×10^2 copies/μL. The results indicate that the intra-assay and inter-assay CVs ranged from 1.0% to 2.7% for triplex cdPCR and 0.6% to 1.8% for triplex qPCR (Table 3), indicating excellent repeatability for both developed assays.

Table 3. Evaluation of repeatability and reproducibility.

Plasmid	Concentration (Copies/μL)	Intra-Assay for Repeatability						Inter-Assay for Reproducibility					
		Triplex cdPCR (Copies/Reaction)			Triplex qPCR (Ct)			Triplex cdPCR (Copies/Reaction)			Triplex qPCR (Ct)		
		$\overline{X}$	SD	CV (%)	$\overline{X}$	SD	CV (%)	$\overline{X}$	SD	CV (%)	$\overline{X}$	SD	CV (%)
pASFV-B646L-1	2.0×10^4	33,983.33	500.83	1.47	25.33	0.19	0.75	34,083.33	780.36	2.29	25.46	0.27	1.06
	2.0×10^3	3463.33	41.26	1.19	28.68	0.23	0.80	3455.00	56.35	1.63	28.48	0.40	1.40
	2.0×10^2	331.67	6.29	1.90	32.20	0.36	1.12	335.00	9.01	2.69	32.07	0.59	1.84
pASFV-MGF505-2R-1	2.0×10^4	43,225.00	650.00	1.50	24.44	0.17	0.70	43,418.33	698.75	1.61	24.38	0.23	0.94
	2.0×10^3	4389.17	68.66	1.56	28.38	0.20	0.70	4370.83	91.66	2.10	28.68	0.30	1.05
	2.0×10^2	390.83	5.20	1.33	31.53	0.31	0.98	395.00	10.00	2.53	32.03	0.52	1.62
pASFV-I177L-1	2.0×10^4	54,550.00	912.41	1.67	23.66	0.21	0.89	54,041.67	1179.07	2.18	23.56	0.30	1.27
	2.0×10^3	5607.50	90.93	1.62	28.28	0.24	0.85	5612.50	103.11	1.84	28.28	0.24	0.85
	2.0×10^2	515.00	5.00	0.97	31.49	0.18	0.57	509.17	11.27	2.21	31.69	0.20	0.63

3.7. Application for the Detection of Clinical Samples

A total of 1510 clinical tissue samples collected in Guangxi province, southern China, from January 2022 to December 2022 were tested via triplex cdPCR and triplex qPCR. The results show that the positivity rates of ASFV and the MGF505-2R gene-deleted ASFV strain were 14.17% (214/1510) and 0.33% (5/1510) with triplex cdPCR and 12.98% (196/1510) and 0.33% (5/1510) with triplex qPCR, respectively. No positive sample of the I177L gene-deleted ASFV strain was found with the two methods (Table 4). The coincidence rate between the two methods was 98.81% (Table 5).

Table 4. Detection results of the clinical samples with triplex cdPCR and triplex qPCR.

Pathogen	Number	Triplex qPCR		Triplex cdPCR		Coincidence Rate (%)	Kappa
		Positive	Positive Rate (%)	Positive	Positive Rate (%)		
ASFV	1510	196	12.98	214	14.17	98.81	0.95
ASFV-ΔMGF505-2R	1510	5	0.33	5	0.33	100	/
ASFV-ΔI177L	1510	0	0	0	0	100	/

Table 5. Comparison of the results using triplex cdPCR and triplex qPCR.

Triplex qPCR	Triplex cdPCR			Coincidence Rate (%)	Kappa
	Positive	Negative	Total		
Positive	196	0	196	98.81	0.95
Negative	18	1296	1314		
Total	214	1296	1510		

The 1510 tissue samples showed a 14.17% positivity rate, of which the positivity rates of slaughterhouses, famers' markets and harmless disposal sites were 12.12% (115/949), 16.56% (25/151) and 18.05% (74/410), respectively. The chi-squared test showed that the positivity rate of harmless disposal sites (18.05%) was significantly higher than that of slaughterhouses (12.12%) ($p < 0.05$), and the positivity rate of famers' markets (16.56%) was between those of harmless disposal sites (18.05%) and slaughterhouses (12.12%) but showed no significant difference from them ($p > 0.05$).

4. Discussion

ASF has caused huge economic losses in the pig industry around the world. Since ASFV was first discovered in China in August 2018, virulent genotype II ASFV strains have rapidly spread to most provinces of China and have become the dominant epidemic strains [9,17,49]. The highly virulent genotype II ASFV strain has a nearly 100% fatality rate in infected domestic pigs and wild boars, underscoring the significant threat that the genotype II strain poses to the pig industry [50]. Naturally gene-deleted strains of ASFV, of which the MGF505 gene was the common deleted gene, have also been discovered in pig herds and have caused economic losses in the pig industry [17,21,22]. Today, the development of effective vaccines is one of the most urgent measures for the prevention and control of ASF. Gene-deleted strains of ASFV are promising vaccine candidates. Several reports have demonstrated that candidate vaccine strains, which were developed through the deletion of seven genes from the MGF360 to MGF505 families (ASFV-ΔMGF) or the I177L gene (ASFV-ΔI177L), exhibit protective effects against highly virulent ASFV strains [23–29]. The wild-type strains and the gene-deleted strains of ASFV have become the focus of attention. Therefore, it is necessary to monitor the wild-type and gene-deleted strains of ASFV, especially the MGF505-2R and the I177L gene-deleted strains. To date, several qPCR methods have been developed for the detection and differentiation of the wild-type strains and the gene-deleted strains of ASFV [45–48], and dPCR has also been developed for the detection of ASFV [34–36]. However, no multiplex cdPCR has been developed for the detection and differentiation of the wild-type strain and the MGF505-2R and/or I177L gene-deleted strains of ASFV until now.

cdPCR is a unique droplet-based technology that enables the partitioning of PCR reaction liquids into a single-layer droplet 2D array for PCR amplification, and the sample is read by a three-color fluorescence scanning device to obtain accurate quantitative data [51,52]. In comparison to qPCR, cdPCR offers several advantages, such as absolute quantification that is independent of standard curves, increased precision, low sensitivity to PCR inhibitors and the ability to detect low target concentrations [33,53]. In this study, after the optimization of the reaction system and conditions, triplex cdPCR and triplex qPCR were successfully developed for the detection of the ASFV B646L, MGF505-2R and I177L genes. Both methods demonstrated excellent specificity, high sensitivity and good repeatability. They specifically detected only the ASFV B646L, MGF505-2R and I177L genes, without cross-reaction with other swine viruses. The LOD of triplex cdPCR was 12 copies/reaction, whereas the LOD of triplex qPCR was 500 copies/reaction, indicating that the former was 42 times more sensitive than the latter. The intra-assay and inter-assay CVs ranged from 1.0% to 2.7% for triplex cdPCR and from 0.6% to 1.8% for triplex qPCR. Compared to triplex qPCR, triplex cdPCR is a more sensitive method for the accurate detection of very low concentrations of the wild-type strains and the MGF505-2R and I177L gene-deleted strains of ASFV. ddPCR targeting the ASFV K205R gene, established by Wu et al. [34], showed an LOD of 10 copies/reaction. Nanofluidic chip dPCR targeting the ASFV B646L gene, established by Jia et al. [37], showed an LOD of 30.1995 copies/reaction. Duplex ddPCR based on the ASFV B646L and EP402R genes, established by Zhu et al. [35], showed an LOD of 52 copies/reaction and 8.6 copies/reaction for B646L and EP402R, respectively. The triplex cdPCR developed in this study showed similar or superior sensitivity to the previous results. The results demonstrate that sensitive, efficient and specific triplex cdPCR and triplex qPCR were developed for the detection and differentiation of the wild-type strain and the MGF505-2R and I177L gene-deleted strains of ASFV.

A total of 1510 clinical samples collected from Guangxi province from January 2022 to December 2022 were tested with the developed triplex cdPCR. The results show that the positivity rates of the wild-type strain and the MGF505-2R gene-deleted strain of ASFV were 14.17% and 0.33%, and no positive sample of the I177L gene-deleted strain was found. These findings suggest that both the wild-type strains and the MGF505-2R gene-deleted strains of ASFV were prevalent in China, and no I177L gene-deleted ASFV strain was discovered. Compared with the results of our previous reports [17,20,36,46,54], the positivity rate of ASFV in pig herds has gradually decreased in recent years. The results indicate that important measures, such as early detection and diagnosis, the eradication of pathogenic-positive pigs and strict biosafety, which have been implemented in China in recent years [55,56], are very effective and successful measures for the prevention and control of ASF in China. However, the wild-type strains and naturally gene-deleted strains of ASFV are still circulating in some pig herds, and more and more man-made gene-deleted vaccine strains will be generated in the future. Therefore, new challenges will be confronted for the prevention and control of the disease. The developed assays provide rapid, sensitive and accurate methods to continuously monitor the circulating strains in the field in order to effectively prevent and control ASF and decrease the economic losses of ASF.

Fortunately, the I177L gene-deleted ASFV strain has not been found in the field in Guangxi province of China until now. Guangxi province is adjacent to Vietnam. Every year, there are a large number of travelers and the trade of goods between China and Vietnam, and animals and animal products are often smuggled across the border. The risk of the I177L gene-deleted ASFV strain being introduced into China from Vietnam is very high. The I177L gene-deleted vaccine can be legally used in Vietnam, but its use in China is illegal and strictly prohibited. Therefore, it is necessary to strengthen the monitoring of the I177L gene-deleted strain so that, if this strain appears in China, it can be found in a timely manner, and effective response measures can be taken. The triplex cdPCR and triplex qPCR established in this study can provide sensitive, specific and accurate methods for the detection of this strain.

5. Conclusions

Triplex cdPCR and triplex qPCR were successfully developed for the detection and differentiation of the wild-type strain and the MGF-505-2R and I177L gene-deleted strains of ASFV. The assays demonstrated high specificity, sensitivity and repeatability and could offer reliable methods for evaluating ASFV in clinical samples. In addition, both wild-type strains and MGF505-2R gene-deleted strains of ASFV were found in Guangxi province, southern China.

Supplementary Materials: The following supporting information can be downloaded at: https://www.mdpi.com/article/10.3390/pathogens12091092/s1, Table S1: The sequences of recombinant plasmids pASFV-ΔMGF505-2R and pASFV-ΔI177L.

Author Contributions: Methodology, investigation and writing—original draft preparation: K.Z., H.W., Y.S. and Q.Z.; software and validation: S.F. and F.L.; data curation and supervision: L.H. and S.M.; funding acquisition and writing—review and editing: K.S. and M.M. All authors have read and agreed to the published version of the manuscript.

Funding: This study was supported by the Key Research and Development Program (No. AB21238003) of Guangxi Science and Technology Bureau, China, and the Agricultural Science and Technology Program (No. Z202031, No. Z2022194) of Guangxi Agricultural and Rural Bureau, China.

Institutional Review Board Statement: This study was approved by the Guangxi Center for Animal Disease Control and Prevention (CADC) (No. 2020-A-01). Guangxi CADC was approved by the Ministry of Agriculture and Rural Affairs of the People's Republic of China for the collection and detection of ASFV in clinical samples (approval number: 2018-154-25).

Informed Consent Statement: Written informed consent was obtained to use the clinical samples in this study from the owners of the animals.

Data Availability Statement: Not applicable.

Conflicts of Interest: The authors declare no conflict of interest.

References

1. Li, Z.; Chen, W.; Qiu, Z.; Li, Y.; Fan, J.; Wu, K.; Li, X.; Zhao, M.; Ding, H.; Fan, S.; et al. African Swine Fever Virus: A Review. *Life* **2022**, *12*, 1255. [CrossRef] [PubMed]
2. Alonso, C.; Borca, M.; Dixon, L.; Revilla, Y.; Rodriguez, F.; Escribano, J.M.; ICTV Report Consortium. ICTV Virus Taxonomy Profile Asfarviridae. *J. Gen. Virol.* **2018**, *99*, 613–614. [CrossRef]
3. Karger, A.; Pérez-Núñez, D.; Urquiza, J.; Hinojar, P.; Alonso, C.; Freitas, F.B.; Revilla, Y.; Le Potier, M.-F.; Montoya, M. An Update on African Swine Fever Virology. *Viruses* **2019**, *11*, 864. [CrossRef] [PubMed]
4. Wang, G.; Xie, M.; Wu, W.; Chen, Z. Structures and Functional Diversities of ASFV Proteins. *Viruses* **2021**, *13*, 2124. [CrossRef]
5. Cwynar, P.; Stojkov, J.; Wlazlak, K. African Swine Fever Status in Europe. *Viruses* **2019**, *11*, 310. [CrossRef] [PubMed]
6. Sánchez-Vizcaíno, J.M.; Mur, L.; Martínez-López, B. African Swine Fever: An Epidemiological Update. *Transbound. Emerg. Dis.* **2012**, *59* (Suppl. S1), 27–35. [CrossRef]
7. Danzetta, M.L.; Marenzoni, M.L.; Iannetti, S.; Tizzani, P.; Calistri, P.; Feliziani, F. African Swine Fever: Lessons to Learn From Past Eradication Experiences. A Systematic Review. *Front. Veter-Sci.* **2020**, *7*, 296. [CrossRef]
8. Gaudreault, N.N.; Madden, D.W.; Wilson, W.C.; Trujillo, J.D.; Richt, J.A. African Swine Fever Virus: An Emerging DNA Arbovirus. *Front. Vet. Sci.* **2020**, *7*, 215. [CrossRef]
9. Zhou, X.; Li, N.; Luo, Y.; Liu, Y.; Miao, F.; Chen, T.; Zhang, S.; Cao, P.; Li, X.; Tian, K.; et al. Emergence of African Swine Fever in China, 2018. *Transbound. Emerg. Dis.* **2018**, *65*, 1482–1484. [CrossRef]
10. Tao, D.; Sun, D.; Liu, Y.; Wei, S.; Yang, Z.; An, T.; Shan, F.; Chen, Z.; Liu, J. One year of African swine fever outbreak in China. *Acta Trop.* **2020**, *211*, 105602. [CrossRef]
11. Ito, S.; Bosch, J.; Martínez-Avilés, M.; Sánchez-Vizcaíno, J.M. The evolution of African swine fever in China: A global threat? *Front. Vet. Sci.* **2022**, *9*, 828498. [CrossRef]
12. Mighell, E.; Ward, M.P. African Swine Fever spread across Asia, 2018–2019. *Transbound. Emerg. Dis.* **2021**, *68*, 2722–2732. [CrossRef] [PubMed]
13. Cadenas-Fernández, E.; Ito, S.; Aguilar-Vega, C.; Sánchez-Vizcaíno, J.M.; Bosch, J. The Role of the Wild Boar Spreading African Swine Fever Virus in Asia: Another Underestimated Problem. *Front. Vet. Sci.* **2022**, *9*, 844209. [CrossRef] [PubMed]

14. Blome, S.; Franzke, K.; Beer, M. African swine fever—A review of current knowledge. *Virus Res.* **2020**, *287*, 198099. [CrossRef] [PubMed]
15. Penrith, M.-L.; Van Heerden, J.; Heath, L.; Abworo, E.O.; Bastos, A.D.S. Review of the Pig-Adapted African Swine Fever Viruses in and Outside Africa. *Pathogens* **2022**, *11*, 1190. [CrossRef]
16. Njau, E.P.; Domelevo Entfellner, J.-B.; Machuka, E.M.; Bochere, E.N.; Cleaveland, S.; Shirima, G.M.; Kusiluka, L.J.; Upton, C.; Bishop, R.P.; Pelle, R.; et al. The first genotype II African swine fever virus isolated in Africa provides insight into the current Eurasian pandemic. *Sci. Rep.* **2021**, *11*, 1–13. [CrossRef]
17. Shi, K.; Liu, H.; Yin, Y.; Si, H.; Long, F.; Feng, S. Molecular Characterization of African Swine Fever Virus From 2019-2020 Outbreaks in Guangxi Province, Southern China. *Front. Vet. Sci.* **2022**, *9*, 912224. [CrossRef]
18. Qu, H.; Ge, S.; Zhang, Y.; Wu, X.; Wang, Z. A systematic review of genotypes and serogroups of African swine fever virus. *Virus Genes* **2022**, *58*, 77–87. [CrossRef]
19. Sun, E.; Huang, L.; Zhang, X.; Zhang, J.; Shen, D.; Zhang, Z.; Wang, Z.; Huo, H.; Wang, W.; Huangfu, H.; et al. Genotype I African swine fever viruses emerged in domestic pigs in China and caused chronic infection. *Emerg. Microbes Infect.* **2021**, *10*, 2183–2193. [CrossRef]
20. Liu, H.; Shi, K.; Zhao, J.; Yin, Y.; Chen, Y.; Si, H.; Qu, S.; Long, F.; Lu, W. Development of a one-step multiplex qRT-PCR assay for the detection of African swine fever virus, classical swine fever virus and atypical porcine pestivirus. *BMC Vet. Res.* **2022**, *18*, 43. [CrossRef]
21. Sun, E.; Zhang, Z.; Wang, Z.; He, X.; Zhang, X.; Wang, L.; Wang, W.; Huang, L.; Xi, F.; Huangfu, H.; et al. Emergence and prevalence of naturally occurring lower virulent African swine fever viruses in domestic pigs in China in 2020. *Sci. China Life Sci.* **2021**, *64*, 752–765. [CrossRef] [PubMed]
22. Wang, Z.; Qi, C.; Ge, S.; Li, J.; Hu, Y.; Zhang, X.; Lv, Y.; Han, N.; Wu, X.; Wang, Z.; et al. Genetic variation and evolution of attenuated African swine fever virus strain isolated in the field: A review. *Virus Res.* **2022**, *319*, 198874.
23. Chen, W.; Zhao, D.; He, X.; Liu, R.; Wang, Z.; Zhang, X.; Li, F.; Shan, D.; Chen, H.; Zhang, J.; et al. A seven-gene-deleted African swine fever virus is safe and effective as a live attenuated vaccine in pigs. *Sci. China Life Sci.* **2020**, *63*, 623–634. [CrossRef]
24. O'Donnell, V.; Holinka, L.G.; Gladue, D.P.; Sanford, B.; Krug, P.W.; Lu, X.; Arzt, J.; Reese, B.; Carrillo, C.; Risatti, G.R.; et al. African Swine Fever Virus Georgia Isolate Harboring Deletions of MGF360 and MGF505 Genes Is Attenuated in Swine and Confers Protection against Challenge with Virulent Parental Virus. *J. Virol.* **2015**, *89*, 6048–6056. [CrossRef]
25. Monteagudo, P.L.; Lacasta, A.; López, E.; Bosch, L.; Collado, J.; Pina-Pedrero, S.; Correa-Fiz, F.; Accensi, F.; Navas, M.J.; Vidal, E.; et al. BA71ΔCD2: A New Recombinant Live Attenuated African Swine Fever Virus with Cross-Protective Capabilities. *J. Virol.* **2017**, *91*, e01058-17. [CrossRef]
26. Lopez, E.; Bosch-Camós, L.; Ramirez-Medina, E.; Vuono, E.; Navas, M.J.; Muñoz, M.; Accensi, F.; Zhang, J.; Alonso, U.; Argilaguet, J.; et al. Deletion Mutants of the Attenuated Recombinant ASF Virus, BA71ΔCD2, Show Decreased Vaccine Efficacy. *Viruses* **2021**, *13*, 1678. [CrossRef] [PubMed]
27. Borca, M.V.; Ramirez-Medina, E.; Silva, E.; Vuono, E.; Rai, A.; Pruitt, S.; Holinka, L.G.; Velazquez-Salinas, L.; Zhu, J.; Gladue, D.P. Development of a Highly Effective African Swine Fever Virus Vaccine by Deletion of the I177L Gene Results in Sterile Immunity against the Current Epidemic Eurasia Strain. *J. Virol.* **2020**, *94*, e02017-19. [CrossRef] [PubMed]
28. Borca, M.V.; Ramirez-Medina, E.; Silva, E.; Vuono, E.; Rai, A.; Pruitt, S.; Espinoza, N.; Velazquez-Salinas, L.; Gay, C.G.; Gladue, D.P. ASFV-G-ΔI177L as an Effective Oral Nasal Vaccine against the Eurasia Strain of Africa Swine Fever. *Viruses* **2021**, *13*, 765. [CrossRef]
29. Tran, X.H.; Le, T.T.P.; Nguyen, Q.H.; Do, T.T.; Nguyen, V.D.; Gay, C.G.; Borca, M.V.; Gladue, D.P. African swine fever virus vaccine candidate ASFV-G-DeltaI177L efficiently protects European and native pig breeds against circulating Vietnamese field strain. *Transbound. Emerg. Dis.* **2022**, *69*, e497–e504. [CrossRef]
30. Liu, Y.; Xie, Z.; Li, Y.; Song, Y.; Di, D.; Liu, J.; Gong, L.; Chen, Z.; Wu, J.; Ye, Z.; et al. Evaluation of an I177L gene-based five-gene-deleted African swine fever virus as a live attenuated vaccine in pigs. *Emerg. Microbes Infect.* **2023**, *12*, 2148560. [CrossRef]
31. Zhang, H.; Zhao, S.; Zhang, H.; Qin, Z.; Shan, H.; Cai, X. Vaccines for African swine fever: An update. *Front. Microbiol.* **2023**, *14*, 1139494. [CrossRef]
32. Hawkins, S.F.C.; Guest, P.C. Multiplex Analyses Using Real-Time Quantitative PCR. *Methods Mol. Biol.* **2017**, *1546*, 125–133. [CrossRef]
33. Kuypers, J.; Jerome, K.R. Applications of Digital PCR for Clinical Microbiology. *J. Clin. Microbiol.* **2017**, *55*, 1621–1628. [CrossRef] [PubMed]
34. Hindson, C.M.; Chevillet, J.R.; Briggs, H.A.; Gallichotte, E.N.; Ruf, I.K.; Hindson, B.J.; Vessella, R.L.; Tewari, M. Absolute quantification by droplet digital PCR versus analog real-time PCR. *Nat. Methods* **2013**, *10*, 1003–1005. [CrossRef]
35. Wu, X.; Xiao, L.; Lin, H.; Chen, S.; Yang, M.; An, W.; Wang, Y.; Yang, Z.; Yao, X.; Tang, Z. Development and application of a droplet digital polymerase chain reaction (ddPCR) for detection and investigation of African swine fever virus. *Can. J. Vet. Res.* **2018**, *82*, 70–74. [PubMed]
36. Zhu, J.; Jian, W.; Huang, Y.; Gao, Q.; Gao, F.; Chen, H.; Zhang, G.; Liao, M.; Qi, W. Development and Application of a Duplex Droplet Digital Polymerase Chain Reaction Assay for Detection and Differentiation of EP402R-Deleted and Wild-Type African Swine Fever Virus. *Front. Vet. Sci.* **2022**, *9*, 905706. [CrossRef]

37. Shi, K.; Chen, Y.; Yin, Y.; Long, F.; Feng, S.; Liu, H.; Qu, S.; Si, H. A Multiplex Crystal Digital PCR for Detection of African Swine Fever Virus, Classical Swine Fever Virus, and Porcine Reproductive and Respiratory Syndrome Virus. *Front. Vet. Sci.* **2022**, *9*, 926881. [CrossRef]
38. Jia, R.; Zhang, G.; Liu, H.; Chen, Y.; Zhou, J.; Liu, Y.; Ding, P.; Wang, Y.; Zang, W.; Wang, A. Novel Application of Nanofluidic Chip Digital PCR for Detection of African Swine Fever Virus. *Front. Vet. Sci.* **2020**, *7*, 621840. [CrossRef]
39. Wang, A.; Jia, R.; Liu, Y.; Zhou, J.; Qi, Y.; Chen, Y.; Liu, D.; Zhao, J.; Shi, H.; Zhang, J.; et al. Development of a novel quantitative real-time PCR assay with lyophilized powder reagent to detect African swine fever virus in blood samples of domestic pigs in China. *Transbound. Emerg. Dis.* **2020**, *67*, 284–297. [CrossRef]
40. Wang, Y.; Xu, L.; Noll, L.; Stoy, C.; Porter, E.; Fu, J.; Feng, Y.; Peddireddi, L.; Liu, X.; Dodd, K.A.; et al. Development of a real-time PCR assay for detection of African swine fever virus with an endogenous internal control. *Transbound. Emerg. Dis.* **2020**, *67*, 2446–2454. [CrossRef]
41. Zhan, Y.; Zhang, L.-H.; Lin, Y.; Cai, Y.-F.; Zou, Y.-W.; Hao, Z.-Y.; Luo, Z.-H.; Wang, N.-D.; Deng, Z.-B.; Yang, Y.; et al. Development and preliminary testing of a probe-based duplex real-time PCR assay for the detection of African swine fever virus. *Mol. Cell. Probes* **2021**, *59*, 101764. [CrossRef] [PubMed]
42. Li, X.; Hu, Y.; Liu, P.; Zhu, Z.; Liu, P.; Chen, C.; Wu, X. Development and application of a duplex real-time PCR assay for differentiation of genotypes I and II African swine fever viruses. *Transbound. Emerg. Dis.* **2022**, *69*, 2971–2979. [CrossRef] [PubMed]
43. Cao, S.; Lu, H.; Wu, Z.; Zhu, S. A duplex fluorescent quantitative PCR assay to distinguish the genotype I and II strains of African swine fever virus in Chinese epidemic strains. *Front. Vet. Sci.* **2022**, *9*, 998874. [CrossRef] [PubMed]
44. Gao, Q.; Feng, Y.; Yang, Y.; Luo, Y.; Gong, T.; Wang, H.; Gong, L.; Zhang, G.; Zheng, Z. Establishment of a Dual Real-Time PCR Assay for the Identification of African Swine Fever Virus Genotypes I and II in China. *Front. Vet. Sci.* **2022**, *9*, 882824. [CrossRef]
45. Yang, H.; Peng, Z.; Song, W.; Zhang, C.; Fan, J.; Chen, H.; Hua, L.; Pei, J.; Tang, X.; Chen, H.; et al. A triplex real-time PCR method to detect African swine fever virus gene-deleted and wild type strains. *Front. Vet. Sci.* **2022**, *9*, 943099. [CrossRef]
46. Zhao, K.; Shi, K.; Zhou, Q.; Xiong, C.; Mo, S.; Zhou, H.; Long, F.; Wei, H.; Hu, L.; Mo, M. The Development of a Multiplex Real-Time Quantitative PCR Assay for the Differential Detection of the Wild-Type Strain and the MGF505-2R, EP402R and I177L Gene-Deleted Strain of the African Swine Fever Virus. *Animals* **2022**, *12*, 1754. [CrossRef]
47. Lin, Y.; Cao, C.; Shi, W.; Huang, C.; Zeng, S.; Sun, J.; Wu, J.; Hua, Q. Development of a triplex real-time PCR assay for detection and differentiation of gene-deleted and wild-type African swine fever virus. *J. Virol. Methods* **2020**, *280*, 113875. [CrossRef]
48. Guo, Z.; Li, K.; Qiao, S.; Chen, X.-X.; Deng, R.; Zhang, G. Development and evaluation of duplex TaqMan real-time PCR assay for detection and differentiation of wide-type and MGF505-2R gene-deleted African swine fever viruses. *BMC Vet. Res.* **2020**, *16*, 428. [CrossRef]
49. Ge, S.; Li, J.; Fan, X.; Liu, F.; Li, L.; Wang, Q.; Ren, W.; Bao, J.; Liu, C.; Wang, H.; et al. Molecular Characterization of African Swine Fever Virus, China, 2018. *Emerg. Infect. Dis.* **2018**, *24*, 2131–2133. [CrossRef]
50. Ata, E.B.; Li, Z.-J.; Shi, C.-W.; Yang, G.-L.; Yang, W.-T.; Wang, C.-F. African swine fever virus: A raised global upsurge and a continuous threaten to pig husbandry. *Microb. Pathog.* **2022**, *167*, 105561. [CrossRef]
51. Madic, J.; Zocevic, A.; Senlis, V.; Fradet, E.; Andre, B.; Muller, S.; Dangla, R.; Droniou, M.E. Three-color crystal digital PCR. *Biomol. Detect. Quantif.* **2016**, *10*, 34–46. [CrossRef] [PubMed]
52. Tan, L.L.; Loganathan, N.; Agarwalla, S.; Yang, C.; Yuan, W.; Zeng, J.; Wu, R.; Wang, W.; Duraiswamy, S. Current commercial dPCR platforms: Technology and market review. *Crit. Rev. Biotechnol.* **2023**, *43*, 433–464. [CrossRef]
53. Whale, A.S.; Cowen, S.; Foy, C.A.; Huggett, J.F. Methods for Applying Accurate Digital PCR Analysis on Low Copy DNA Samples. *PLoS ONE* **2013**, *8*, e58177. [CrossRef] [PubMed]
54. Chen, Y.; Shi, K.; Liu, H.; Yin, Y.; Zhao, J.; Long, F.; Lu, W.; Si, H. Development of a multiplex qRT-PCR assay for detection of African swine fever virus, classical swine fever virus and porcine reproductive and respiratory syndrome virus. *J. Vet. Sci.* **2021**, *22*, e87. [CrossRef]
55. Liu, Y.; Zhang, X.; Qi, W.; Yang, Y.; Liu, Z.; An, T.; Wu, X.; Chen, J. Prevention and Control Strategies of African Swine Fever and Progress on Pig Farm Repopulation in China. *Viruses* **2021**, *13*, 2552. [CrossRef] [PubMed]
56. Dixon, L.K.; Stahl, K.; Jori, F.; Vial, L.; Pfeiffer, D.U. African Swine Fever Epidemiology and Control. *Annu. Rev. Anim. Biosci.* **2020**, *8*, 221–246. [CrossRef]

pathogens

MDPI

Article

Development of a Novel Indirect ELISA for the Serological Diagnosis of African Swine Fever Using p11.5 Protein as a Target Antigen

Mizuki Watanabe [1,2], Tomoya Kitamura [2], Koji Nagata [3], Mitsutaka Ikezawa [2], Ken-ichiro Kameyama [2], Kentaro Masujin [2] and Takehiro Kokuho [2,*]

[1] Nippon Institute for Biological Science, Tokyo 198-0024, Japan
[2] Division of Transboundary Animal Disease Research, National Institute of Animal Health (NIAH), National Agriculture and Food Research Organization (NARO), Tokyo 187-0022, Japan
[3] Department of Applied Biological Chemistry, Graduate School of Agricultural and Life Sciences, The University of Tokyo, Tokyo 113-8657, Japan
* Correspondence: takehiro@affrc.go.jp; Tel.: +81-42-321-1440

Abstract: African swine fever is a hemorrhagic viral disease with a mortality rate of nearly 100% in pigs. Hence, it is classified as a notifiable disease by the World Organization for Animal Health. Because no field-available vaccine exists, African swine fever virus (ASFV) control and eradication solely depend on good farm biosecurity management and rapid and accurate diagnosis. In this study, we developed a new indirect serological enzyme-linked immunosorbent assay (ELISA) using recombinant p11.5 protein from ASFV as a solid-phase target antigen. The cutoffs were determined by receiver operating curve analysis performed with serum samples obtained from naïve and infected pigs. Based on the results of a commercially available serological ELISA, the relative sensitivity and specificity of our assay were 93.4% and 94.4% (N = 166; area under the curve = 0.991; 95% confidence interval = 0.982–0.999), respectively. Furthermore, to compare the performance of the serological ELISAs, we conducted the assays on a panel of sera collected from pigs and boars experimentally infected with different ASFV isolates. The results indicated the greater sensitivity of the newly developed assay and its ability to detect anti-ASFV antibodies earlier after virus inoculation.

Keywords: African swine fever virus (ASFV); A137R; p11.5; serological ELISA

Citation: Watanabe, M.; Kitamura, T.; Nagata, K.; Ikezawa, M.; Kameyama, K.-i.; Masujin, K.; Kokuho, T. Development of a Novel Indirect ELISA for the Serological Diagnosis of African Swine Fever Using p11.5 Protein as a Target Antigen. *Pathogens* **2023**, *12*, 774. https://doi.org/10.3390/pathogens12060774

Academic Editors: Lauro Velazquez-Salinas, Gisselle N. Medina and Elizabeth Ramirez-Medina

Received: 11 May 2023
Revised: 26 May 2023
Accepted: 27 May 2023
Published: 29 May 2023

1. Introduction

African swine fever (ASF) is a hemorrhagic viral disease with a maximum mortality rate of 100% in domestic pigs. Therefore, it is listed as a notifiable disease by the World Organization for Animal Health. The etiological agent, African swine fever virus (ASFV), is a large double-stranded DNA virus with icosahedral morphology [1,2].

ASF was first reported by Montgomery in Kenya in 1921 [3]. After a long period of circulation in Africa the virus suddenly spread to Europe and South America in 1957 and 1960, respectively [4]. In 2007, ASF emerged in Georgia and then spread to Armenia, Azerbaijan, and other Caucasian countries [5].

From 2014 to 2018, several European countries including Estonia, Lithuania, Latvia, Poland, Czech Republic, Bulgaria, Belgium, Romania, and Hungary experienced severe outbreaks of ASF. Since then, the disease has gradually spread to additional countries such as Germany, North Macedonia, and Italy. In Asia, ASF was first confirmed in China in 2018, and has since spread to nearly all neighboring countries with the exception of Japan and Chinese Taipei (as of April 2023). Moreover, ASF emerged in Papua New Guinea in Oceania in 2020, and in 2021, it affected the Dominican Republic and Haiti in Central America. As a result, the disease has had a significant socio-economic impact on global pork industries and poses threats to biosafety, biosecurity, and the sustainable supply of food for human consumption [4,6,7].

There is currently a high demand for effective vaccines against ASF, and research and development efforts are underway worldwide. Among these, live attenuated vaccines with genetic mutations or deletions in genes such as 9GL, EP402R (CD2v), DP148R, I177L, I226R, A137R, and E184L, as well as mutations in MGF and LVR gene loci, have been published. Additionally, recombinant viral vector vaccines such as rAd5 + MVA have been designed to express multiple antigenic proteins (B602L, B636L, CP204L, E183L, E199L, EP153R, F317L, and MGF505-5R), either individually or in combination [6,7]. However, despite these advancements, a vaccine that fully meets both safety and efficacy requirements has not yet become available. Therefore, its control and eradication solely depend on the implementation of appropriate biosecurity measures and rapid detection of infected animals [8].

Domestic pigs (*Sus scrofa domesticus*) are affected by ASFV, but manifestations of the disease can vary from highly lethal hemorrhagic, acute disease to subclinical illness depending on the virus strain [9,10]. By contrast, African wild suids such as warthogs (*Phacochoerus aethiopicus*) and bush pigs (*Potamochoerus larvatus*) exhibit milder symptoms or remain asymptomatic during infection. Wild boars (*Sus scrofa*) are also highly susceptible to ASFV, presenting symptoms similar to those of domestic pigs. Thus, they are major vectors of disease spread in many countries without vector ticks [11].

The rapid detection of ASFV-infected animals is imperative in outbreak areas. In cases of infestations by less virulent strains of ASFV or long-term persistence of the virus in endemic areas, serological diagnosis and surveillance are highly demanded. Serologic surveillance aims to detect antibodies against ASFV. Positive ASFV antibody test results can indicate ongoing or previous outbreaks. This is because some animals may recover and remain seropositive for a significant period, possibly life. This could include carrier animals [8].

This study aimed to develop a novel indirect serological enzyme-linked immunosorbent assay (ELISA) to detect immunological responses to avirulent or less virulent ASFVs with higher sensitivity than currently available commercial assays. In our preliminary experiments, to identify an appropriate antigen for serological ELISA, mouse monoclonal antibodies (mAbs) were generated against the pathogenic genotype II strain AQS-C-1-22 (AQS) [12], and the antigenic proteins were identified by screening using a panel of expression vectors encoding a complete set of 192 predicted open-reading frames of Georgia 2007/1, a highly pathogenic genotype II strain. p11.5, the protein encoded by the A137R gene, was identified as a major viral protein with particular antigenic activity.

Hence, we examined the suitability of this molecule as a target antigen in a solid-phase indirect ELISA for the serological detection of ASFV-specific antibodies in infected animals and established a sensitive serological assay system.

2. Materials and Methods

2.1. p11.5 Recombinant DNA Construction

The glutathione S-transferase (GST) tag sequence was inserted into pCold I DNA (Takara Bio Inc., Shiga, Japan) between the *Xho*I and *Eco*RI sites to create pCold-GST-DNA. The p11.5/A137R gene of ASFV Georgia 2007/1 strain (accession No. NC_044959.2 at DDBJ/EMBL/GenBank) was amplified using forward (5′-CCAGGGGCCCGAATTCATGGAAGCAGTTCTTACCAAAC-3′) and reverse primers (5′-TAGACTGCAGGTCGACTTAGCCTTCTTTGATATTCATC-3′). Three different expression plasmids featuring insertions of single, double, and triple p11.5 domain-encoding sequences were prepared by modifying the bacterial expression plasmid pCold-GST using the NEBuilder HiFi DNA assembly cloning kit (New England Biolabs, Ipswich, MA, USA). Briefly, linearized pCold-GST plasmid DNA was ligated to PCR-amplified fragments of p11.5 or to p11.5 with a 3×GGGGS linker sequence at the 3′-end in combination to yield GST-p11.5×1 and GST-p11.5×2 expression vectors, respectively. Similarly, another vector encoding three repetitive sequences of p11.5 with 3×GGGGS and 3×GS linker sequences (GST-p11.5×3) was prepared. The schematic diagram of the inserted sequences in the pCold-GST plasmids are indicated in Figure 1a. These plasmids were then introduced into *Escherichia coli* DH5α (Takara Bio) and incubated overnight at 37 °C on ampicillin-loaded agar plates. We confirmed the inserted sequences by Sanger sequencing (Applied Biosystems 3500 or Applied Biosystems SeqStudio Genetic Analyzer, Thermo Fisher Scientific Inc., Waltham, MA, USA).

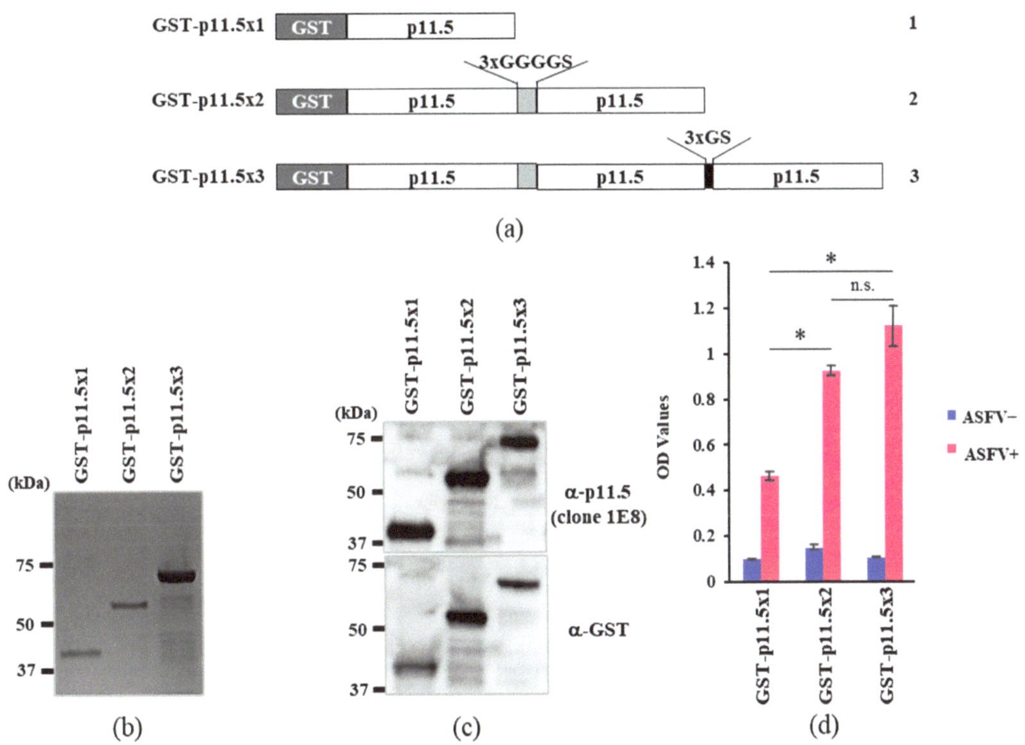

Figure 1. p11.5 antigen preparation and analysis. (**a**) Schematic presentation of the inserted sequences in the pCold-GST plasmids. Two different linkers, namely 3×GGGGS and 3×GS, are also shown. A GST tag was added to the N-terminal domain for purification and identification. Line 1, GST-p11.5×1; line 2, GST-p11.5×2; line 3, GST-p11.5×3. (**b**,**c**) GST-p11.5×1, GST-p11.5×2, and GST-p11.5×3 fusion proteins were expressed in *E. coli* BL21, solubilized, and purified on a GSTrap FF column. Half of the eluted proteins were analyzed by electrophoresis in a denaturing gel and stained with Coomassie brilliant blue G-250 (**b**), and the remaining proteins were analyzed by electrophoresis in a denaturing gel and immunoblotted with the indicated antibodies (**c**). The image is representative of three independent experiments. α, anti-. (**d**) Comparison of the OD_{450} values of ELISA using three different proteins as the target antigen. ASFV+, ASFV-positive serum samples (from pigs infected with ASFV OUR T88/3 at 21 dpi); ASFV−, ASFV-negative serum samples (from uninfected pigs). The antigen proteins were coated at a concentration of 5 μg/mL. The serum samples and secondary antibody were used at dilutions of 1:20 and 1:5000, respectively. Each data point is indicated as the mean ± standard error of three independent experiments. *, $p < 0.01$ (Tukey's test); n.s., not significant.

2.2. Recombinant Protein Generation

Synthetic recombinant plasmids were introduced into *E. coli* BL21 (Takara Bio). Transformed BL21 cells were cultivated with a small amount of 2×YT broth (containing 50 μg/mL ampicillin) in a shaker at 200 rpm and 37 °C and then inoculated into a large bottle of 2×YT at a dilution of 1:100. *E. coli* was grown at 160 rpm and 19 °C when the optical density at 600 nm (OD_{600}) reached 0.4–0.5. Then, 300 μM isopropyl-β-D-thiogalactopyranoside (Nacalai Tesque) was added to induce protein synthesis with overnight incubation at 160 rpm and 19 °C. Following centrifugation, bacteria were collected and suspended in lysis buffer (1 M NaCl, 50 mM Tris pH 8.0, 1% Triton X) containing protease inhibitor cocktails (Merck KGaA, Darmstadt, Germany) and sonicated. After centrifugation at 15,000 rpm for 10 min, the supernatants were collected and filtered through a 0.45 μm syringe filter, and the proteins were purified on a GSTrap FF column using an AKTA start liquid chromatography system (Cytiva, Tokyo, Japan) following standard protocols. Eluted proteins were fractionized by sodium dodecyl sulfate-polyacrylamide gel electrophoresis (SDS-PAGE) and stained for visualization with Coomassie brilliant blue G-250 (Bio-Safe CBB G-250 Stain, Nacalai Tesque).

2.3. Antibodies

Anti-p11.5 antibody (clone 1E8, IgG2b, κ) is a mAb against ASFV p11.5 previously generated in our laboratory. Goat polyclonal antibody against GST was purchased from Cytiva. A horseradish peroxidase (HRP)-conjugated antibody for indirect ELISA, namely Peroxidase AffiniPure Goat Anti-Swine IgG (H + L), was purchased from Jackson ImmunoResearch Inc. (West Grove, PA, USA). Goat anti-Mouse IgG (H + L) Cross-Adsorbed Secondary Antibody, Alexa Fluor 488 (Thermo Fisher Scientific Inc.) was obtained as a secondary antibody for immunofluorescence staining.

2.4. Serum Samples

2.4.1. Animal Experiment and Virus Strains

Animal experiments were performed in compliance with the regulations outlined in the Guide for the Care and Use of Laboratory Animals of the National Institute of Animal Health (NIAH), National Agriculture and Food Research Organization (NARO), Guidelines for Proper Conduct of Animal Experiments of the Science Council of Japan [13], and ARRIVE guidelines [14]. The animal study was reviewed and approved by the Institutional Animal Care and Use Committee at NIAH, NARO (approval numbers: 21-023, 21-052, 22-026, 22-058, 22-059). During the study period, all pigs received weaning/growing diets, and they had free access to water. The animals were observed daily for clinical signs and/or welfare disturbances. Every effort was made to minimize animal distress and reduce the number of animals used. All pigs were checked for the absence of ASFV by quantitative PCR [15–17] after transduction and for the absence of anti-ASFV antibodies using an indirect ELISA kit (ID Screen African Swine Fever Indirect ELISA, IDvet, Grabels, France) [17]. Pigs were then randomly divided into groups. The virus was inoculated intramuscularly into all pigs. Clinical signs and body temperature were monitored daily, and blood samples were obtained at regular intervals. All pigs were then necropsied before the end of the experimental period. Euthanasia was considered a humane endpoint when pigs exhibited significant depression, and it was justified for welfare reasons. For each animal experiment, LWD pigs and wild boars were used. Five infection tests in animals were conducted using ASFV genotypes I, II, and X.

All positive serum samples (N = 76) were obtained from pigs experimentally infected with ASFV. OUR T88/3 (genotype I) [18], Lisbon60V (genotype I) [19], and Kenya05-Tk1 (genotype X) were obtained from the OIE reference laboratory for ASF (Universidad Complutense de Madrid, Spain). AQS-C-1-22 (AQS, genotype II) was described previously [16]. An attenuated AQS strain and Armenia 2007 (Arm07, genotype II) ΔMGF [17] were cloned in our previous studies. Details of the virus inoculation protocol are presented in Supplementary Table S1. Pig serum samples were obtained at regular intervals after inoculation. Anti-ASFV antibody positivity was determined by IDvet indirect ELISA. For negative control serum samples, we collected serum samples from experimental pigs before inoculation (N = 60).

2.4.2. Serum Samples from the Field

ASFV-negative serum samples were collected from healthy pigs (N = 30) from commercial pig farms in Japan.

2.5. Immunoblotting

Western blotting was performed to confirm the expression of purified GST-tagged p11.5×1, p11.5×2, and p11.5×3 recombinant proteins by transformed bacteria. The bacterial lysates containing recombinant proteins were heated at 100 °C for 5 min in SDS sample buffer (Sample Buffer Solution with Reducing Reagent (6×) for SDS-PAGE; Nacalai Tesque), subjected to electrophoresis in denaturing gels (Mini-PROTEAN TGX Gels 4%–20%; Bio-Rad Laboratories, Inc., Hercules, CA, USA), and transferred to PVDF membranes. The membranes were blocked with Bullet Blocking One for Western Blotting (Nacalai Tesque) for 20 min and reacted with the indicated antibodies for at least 1 h at room temperature or overnight at 4 °C. These membranes were then reacted with secondary antibody conjugated with peroxidase (Jackson) and visualized using Chemi-Lumi One L or Chemi-Lumi One Super (Nacalai Tesque) with the chemiluminescence imaging system (ChemiDoc Touch; Bio-Rad Laboratories).

2.6. p11.5-Indirect ELISA

2.6.1. Procedure for p11.5-Indirect ELISA

ELISA plates coated with recombinant proteins diluted in carbonate buffer (34 mM Na_2CO_3, 100 mM $NaHCO_3$, pH 9.5) were incubated at 4 °C overnight or at room temperature for 2 h. Pig serum samples were diluted in ChonBlock Blocking/Sample Dilution Buffer (Chondrex, Inc., Woodinville, WA, USA) and added to each well of the plate after removing the coating antigen. Following incubation at 37 °C for 1 h, the plates were washed three times with 0.1% Tween in PBS (*v*/*v*, PBST), and Peroxidase AffiniPure Goat Anti-Swine IgG (H + L) at an indicated dilution was added to each well. After incubating at 37 °C for 1 h following three washes with PBST, 50 µL of the chromogenic substrate solution (TMB, Sera Care, Milford, MA, USA) was added. Color reaction was developed for 5 min and then stopped by adding 50 µL of 3 M sulfuric acid. OD_{450} of the reactions was measured using a Nivo ELISA plate reader (PerkinElmer Co., Ltd., Waltham, MA, USA). Antigen and antibody concentrations were optimized via checkerboard titration. Antigens were diluted to concentrations of 0.5–20 µg/mL, pig serum samples were diluted to 1:10–1:500, and HRP anti-swine IgG was diluted to 1:1000–1:10,000 to identify the optimal conditions. The optimal conditions were those for which OD_{450} of the positive serum samples exceeded 1, and the OD450 ratio (P/N value) of positive to negative serum samples was highest. Negative serum samples from 90 pigs in commercial pig farms and experimental pigs in the negative control groups in animal experiments and 76 positive serum samples from experimentally infected pigs confirmed positive for ASFV by IDvet indirect ELISA was used to calculate the cutoffs for the newly developed indirect ELISA. Cutoffs were indicated by OD_{450} of 0.386.

2.6.2. Reproducibility of p11.5-Indirect ELISA

The intra- and inter-assay variations of the results of p11.5-indirect ELISA were evaluated using the coefficient of variation (CV). Five negative and positive serum samples each were randomly selected, tested in three replicates in one batch to evaluate intra-assay variation, and analyzed by three independent assays to evaluate inter-assay variation.

2.6.3. Comparison of p11.5-Indirect ELISA and IDvet Indirect ELISA

In total, 166 pig serum samples were tested in duplicate by p11.5-indirect ELISA and IDvet indirect ELISA. The performance of the assays was evaluated by relative sensitivity [(true positive/true positive + false positive) × 100%] and relative specificity [(true negative/true negative + false positive) × 100%].

2.7. Statistical Analysis

We used EZR (The R Foundation for Statistical Computing, Vienna, Austria), a modified version of R commander that adds statistical functions commonly used in biostatistics, to draw receiver operating characteristic (ROC) curves [20]. For multiple comparisons, we used Tukey's test. Significance was indicated by $p < 0.05$.

3. Results

3.1. Expression of p11.5 Recombinant Proteins

The A137R gene of the highly virulent genotype II ASFV isolate AQS was expressed as a GST-tagged protein in *E. coli* BL21. The nucleotide sequence of the A137R gene of AQS was 100% identical to that of Georgia 2007/1, which was considered to be the original strain of the present global epidemic. As shown in Figure 1a, GST-p11.5×2 and GST-p11.5×3 are GST-tagged proteins carrying two and three domains of p11.5 with a linker peptide insertion (3×GGGGS/3×GS), respectively. On SDS-PAGE, three major bands were observed at approximately 43, 59, and 75 kDa, respectively (Figure 1b). To confirm the expression of the expected proteins, Western blotting of the same gel was performed with anti-p11.5 mAb (clone 1E8) and anti-GST antibody (Figure 1c).

3.2. Assessment of the Recombinant Proteins as an Indirect ELISA Target Antigen

We compared the reactivity of recombinant p11.5 proteins in different forms in a solid-phase indirect ELISA format. Recombinant GST-p11.5×1, GST-p11.5×2, and GST-p11.5×3 were used for coating an assay plate at a concentration of 5 μg/mL, and the OD_{450} values of uninfected pig sera (negative) and pig sera infected with the avirulent strain OUR T88/3 (genotype I) was determined by colorimetric ELISA. The highest OD_{450} was achieved when recombinant GST-p11.5×3 was used as a coating antigen (Figure 1d). Hence, we selected recombinant GST-p11.5×3 as the target antigen of indirect ELISA (were in termed "p11.5-indirect ELISA").

3.3. Optimization of the Working Conditions of p11.5-Indirect ELISA

Next, we attempted to optimize p11.5-indirect ELISA using different amounts of the coating antigen and different dilutions of the secondary antibody. Figure 2a,b shows OD_{450} for ASFV-positive and ASFV-negative sera under different assay conditions, respectively. The highest P/N ratios of the reactions were observed when the secondary antibody was used at dilutions of 1:3000 and 1:5000 (Figure 2c). Hence, the secondary antibody was fixed at a dilution of 1:5000 for further investigation. Finally, we used recombinant p11.5 at a concentration of 5 μg/mL to coat the wells of ELISA plates, and the test sera were applied at a dilution of 1:20 for the assay, as these conditions yielded higher OD_{450} values and the highest P/N ratio for ASFV-positive and ASFV-negative sera (Table 1).

3.4. Standardization and Determination of the Cutoffs in p11.5-Indirect ELISA

To determine the cutoff of the assay, 166 porcine sera (90 intact and 76 ASFV-infected) were tested in duplicate by p11.5-indirect ELISA using the optimized protocol. As shown in Figure 3a, the area under the ROC curve (AUC) with a threshold value of 0.386 was determined to be 0.991 [95% confidence interval (CI) = 0.982–0.999], indicating that the assay is "highly accurate." Therefore, using this threshold (0.386) as the cutoff, we achieved a reliable serological assay with relative sensitivity of 93.4% (95% CI = 85.3–97.8) and relative specificity of 94.4% (95% CI = 87.5–98.2; Figure 3, Table 2).

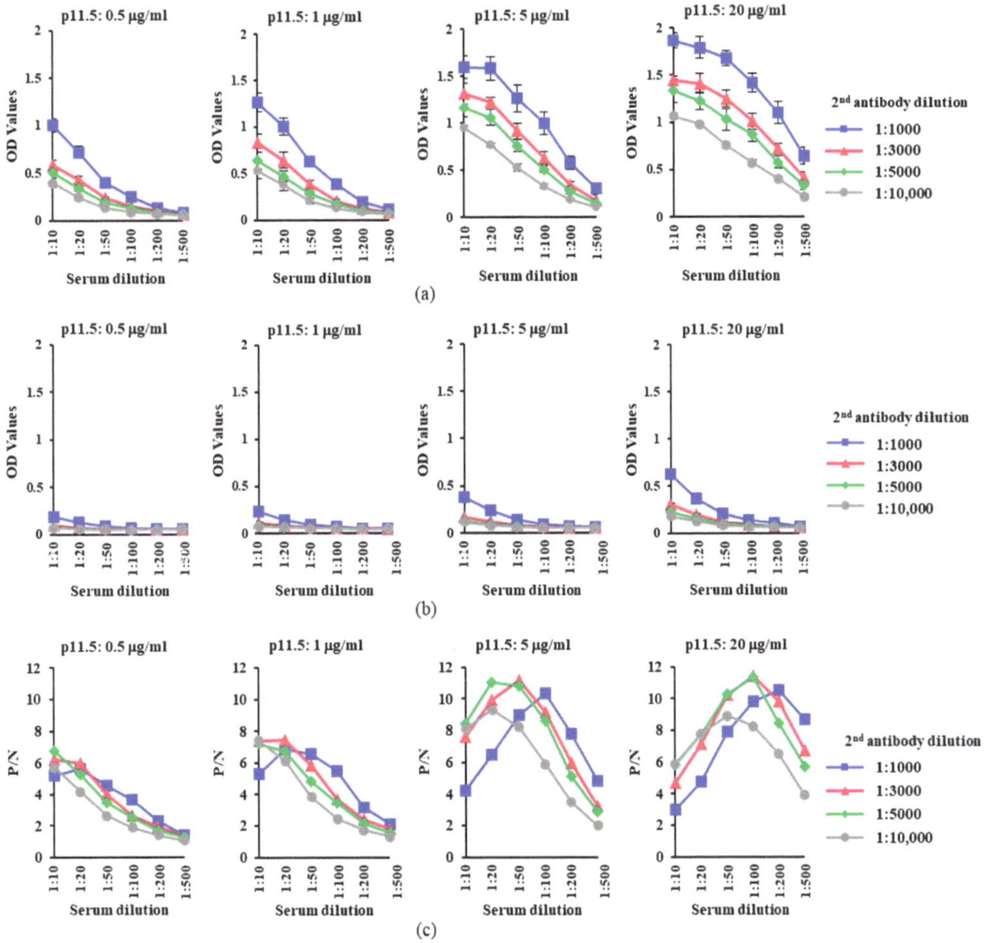

Figure 2. Determination of an appropriate amount of p11.5 protein for coating and optimal concentrations for test samples and the secondary antibody. (**a**,**b**) The p11.5 antigen at concentrations of 0.5, 1, 5, and 20 µg/mL; positive (OUR T88/3 infected pig sera) (**a**) or negative sera (normal pig sera) (**b**) at dilutions of 1:10, 1:20, 1:50, 1:100, 1:200, and 1:500; and anti-swine secondary antibody at dilutions of 1:1000, 1:3000, 1:5000, and 1:10,000 were used. An optimal reaction was obtained with a secondary antibody dilution of 1:5000. Each data point represents the mean ± standard error of three independent experiments; (**c**) The P/N ratios were calculated.

Table 1. Determination of the optimal dilutions of coating antigens and test samples.

Antigen Concentration		Serum Dilution					
		1:10	1:20	1:50	1:100	1:200	1:500
20 μg/mL	P	1.336	1.226	1.028	0.876	0.577	0.331
	N	0.230	0.159	0.100	0.077	0.068	0.058
	P/N	5.81	7.73	10.31	11.38	8.44	5.74
5 μg/mL	P	1.171	1.062	0.761	0.510	0.277	0.152
	N	0.139	0.096	0.070	0.059	0.054	0.053
	P/N	8.42	11.06	10.82	8.60	5.10	2.88
1 μg/mL	P	0.637	0.470	0.285	0.177	0.107	0.071
	N	0.089	0.070	0.059	0.051	0.049	0.046
	P/N	7.19	6.75	4.85	3.48	2.17	1.54
0.5 μg/mL	P	0.506	0.341	0.188	0.128	0.081	0.059
	N	0.075	0.064	0.054	0.050	0.048	0.047
	P/N	6.75	5.30	3.48	2.54	1.71	1.26

α-swine secondary antibody, 1:5000; P, OD450 of infected animals (OUR T88/3-infected pig sera); N, OD450 of uninfected animals (normal pig sera); P/N, P/N ratio. The optimized condition is denoted by a red box. Each well of the plates was coated with 5 μg/mL p11.5, and for testing samples, a dilution of 1:20 was used for further analysis.

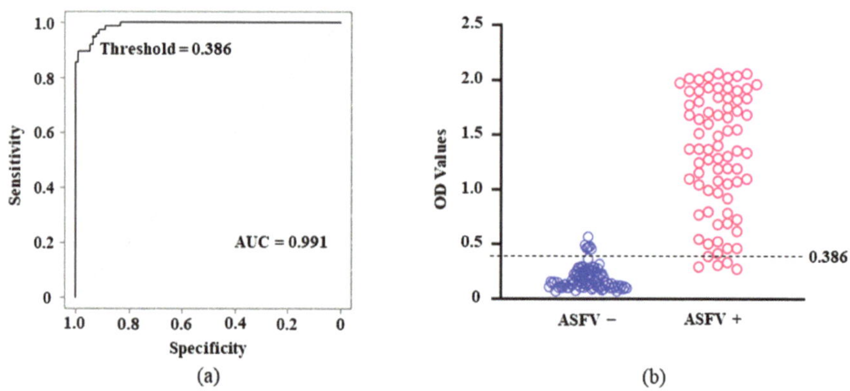

Figure 3. Standardization and determination of the cutoff for p11.5-indirect ELISA ROC analysis. In total, 166 pig sera (90 negative and 76 positive sera) were assayed by p11.5-indirect ELISA and IDvet indirect ELISA. (**a**) The AUC of p11.5-indirect ELISA was determined by ROC analysis (AUC = 0.991; 95% CI = 0.982–0.999). The threshold value (0.386) was calculated by EZR. (**b**) Each circle represents one serum sample of an individual pig. Values above and below the dashed line with an OD cutoff of 0.386 were considered positive and negative, respectively.

Table 2. Comparison of p11.5-indirect ELISA and the commercial kits.

p11.5-Indirect ELISA	IDvet-Indirect ELISA		Total
	+	−	
+	71	5	76
−	5	85	90
Total	76	90	166

Relative sensitivity = 93.4% (71/76), relative specificity = 94.4% (85/90).

3.5. Reproducibility of p11.5-Indirect ELISA

To evaluate the reproducibility of the developed assay, we examined the intra- and inter-assay CVs using uninfected and ASFV-infected porcine sera (one sample each from Groups 1–5; see Supplemental Table S1). The intra- and inter-assay CVs were both lower than 10%, clearly indicating the high repeatability of p11.5-indirect ELISA (Table 3).

Table 3. Reproducibility of p11.5-indirect ELISA.

Samples		Intra-Assay			Inter-Assay		
		Mean OD Value	SD	CV%	Mean OD Value	SD	CV%
Negative	No.1	0.141	0.009	6.39	0.142	0.003	2.01
	No.2	0.107	0.002	1.58	0.103	0.009	8.7
	No.3	0.109	0.003	2.64	0.120	0.011	8.79
	No.4	0.084	0	0	0.098	0.002	2.09
	No.5	0.087	0.002	2.48	0.092	0.004	4.57
Positive	Group 1	1.440	0.045	3.09	1.440	0.028	1.95
	Group 2	1.390	0.068	4.86	1.630	0.09	5.54
	Group 3	0.919	0.081	8.81	0.908	0.024	2.59
	Group 4	1.430	0.073	5.06	1.460	0.081	5.55
	Group 5	0.665	0.014	2.09	0.780	0.068	8.75

The assay was conducted in three replicates for each sample in one assay to determine the intra-assay variation or performed three times separately to determine the inter-assay variation.

3.6. Comparison of Serological ELISAs for Early Detection of Specific Antibodies in ASFV-Infected Animals

In the present study, we compared the performance of serological ELISAs in early detection of specific antibodies in pigs after the inoculation of genotypically varied attenuated ASFVs (Figure 4, Supplemental Table S1). In Group 1, four and three pigs were inoculated with the OUR T88/3 strain at doses of 1×10^6 (Groups 1a and 1b) and 1×10^5 $TCID_{50}$/head (Group 1c), respectively. In Group 1, all pigs became seropositive by 21 days post-inoculation (dpi), whereas one pig was deemed positive at 7 dpi by p11.5-indirect ELISA but not by IDvet indirect ELISA. In Group 2, three pigs were inoculated with another attenuated genotype I ASFV strain (Lisbon60V) at a dose of 1×10^6 $TCID_{50}$/head. All pigs showed seroconversion at 7 dpi by p11.5-indirect ELISA and at 14 dpi by both ELISAs. In Group 3, three pigs each were inoculated with an attenuated derivative of AQS (unpublished) with one dose at 1×10^3 $TCID_{50}$/head (Group 3a), one dose at 1×10^5 $TCID_{50}$/head (Group 3b), two doses at 1×10^3 $TCID_{50}$/head (Group 3c), or two doses at 1×10^5 $TCID_{50}$/head (Group 3d). All of these pigs were then challenged with 1×10^2 $TCID_{50}$/head wild-type AQS at 21 (Groups 3a and 3b) or 28 dpi (Groups 3c and 3d). One pig each in Groups 3a, 3c, and 3d became positive at 35, 41, and 41 dpi, respectively, by both ELISAs, and seroconversion in one pig in Group 3d was only detected by p11.5-indirect ELISA at 35 dpi. In Group 4, three and five pigs (Groups 4a and 4b) were inoculated with another attenuated genotype II ASFV strain (Arm07ΔMGF [17]) at doses of 1×10^5 and 1×10^7 $TCID_{50}$/head, respectively. Another three pigs each were inoculated with a single dose of the same virus at doses of 1×10^3 (Group 4c) and 1×10^5 $TCID_{50}$/head (Group 4d), respectively. Additionally, three pigs each were inoculated twice with the same virus at doses of 1×10^3 (Group 4e) and 1×10^5 $TCID_{50}$/head (Group 4f), respectively. Group 4c–f pigs were injected with 1×10^2 $TCID_{50}$/head wild-type AQS at 21 (4c and 4d) or 28 dpi (4e and 4f). One pig each in Groups 4b and 4f tested positive for the virus earlier by IDvet indirect ELISA (15 and 21 dpi, respectively) than by p11.5-indirect ELISA (21 and 35 dpi, respectively). Conversely, according to p11.5-indirect ELISA, six pigs tested positive at 7 and 15 dpi (Group 4b), 28 dpi (Group 4c), and 14 dpi (Group 4e). For the remaining 12 pigs in this group, similar results were obtained by both ELISAs.

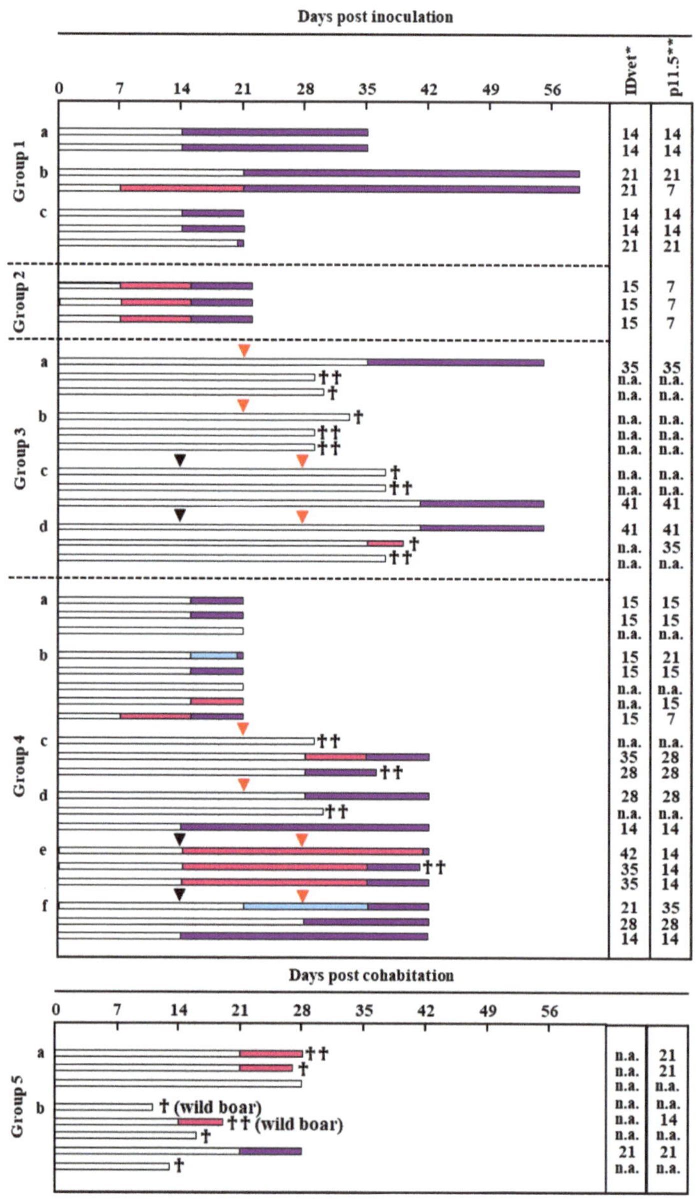

Figure 4. The time to the first detection of antibodies after ASFV inoculation in animals using a commercial ELISA kit or p11.5-indirect ELISA. Group 1, inoculated with OUR T88/3 (genotype I); Group 2, inoculated with Lisbon60V (genotype I); Group 3, inoculated with an attenuated AQS strain and wild-type AQS (genotype II); Group 4, inoculated with Arm07ΔMGF and wild-type AQS (genotype II); Group 5, cohabitation of a Kenya05/Tk-1 (genotype X)-infected pig and wild boar. Serum samples were collected from cohabitating animals. In the cohabitation study, ASFV-inoculated animals were excluded from this graph because they died at 9–13 dpi, and their serum samples were

not collected. The exact dates of virus transmission from the inoculated individuals to the uninfected animals were unknown, and thus, the time was based on the dates of inoculation. An overview of virus inoculation is presented in Supplementary Table S1. In the graph, each bar represents one animal. White bars, the period during which the serum antibodies were negative in both indirect ELISAs; light blue bars, the period during which the serum antibodies appeared positive only in IDvet indirect ELISA; pink bars, the period during which the serum antibodies appeared positive only in the p11.5-indirect ELISA; purple bars, the period during which the serum antibodies appeared positive in both indirect ELISAs. The time until anti-ASFV antibodies were first detected by IDvet indirect ELISA * and p11.5-indirect ELISA ** is presented. Black arrow, additional inoculation of attenuated virus; red arrow, additional inoculation of virulent parental virus; n.a., no antibody. For deceased animals, the time to death † or euthanasia because of human endpoints † † is indicated. All results represent pigs excluding the wild boar noted in brackets in Group 5b.

In the last experimental group (Group 5), we conducted cohabitation studies among wild boars inoculated with a genotype X ASFV (Kenya05/Tk-1) and naïve animals. In Group 5a, one wild boar that was injected 1×10^1 $TCID_{50}$ of Kenya05/Tk-1 cohabitated with three susceptible naïve pigs on the day of injection. Although the exact date of transmission was unclear, two of three pigs showed seroconversion after 21 days of cohabitation by p11.5-indirect ELISA but not by IDvet indirect ELISA. Similarly, in Group 5b, two Kenya05/Tk-1–inoculated wild boars (1×10^3 $TCID_{50}$/head) cohabitated with two naïve wild boars and three naïve pigs on the day of inoculation. Among the cohabitating animals, one wild boar and one pig became seropositive after 14 and 21 days of cohabitation, respectively, when examined by p11.5-indirect ELISA (wild boar) or both ELISAs (pig). In Groups 5a and 5b, we did not examine the serum samples of virus inoculated wild boars; hence, we omitted the data of those animals. Table 4 summarizes the results of all serological assays. For the detection of genotype I virus-specific antibodies in pigs (Groups 1 and 2), seroconversion was detected in all ASFV-inoculated pigs at later phases of infection, but p11.5-indirect ELISA allowed earlier antibody detection than IDvet indirect ELISA. Regarding the detection of genotype II virus-specific antibodies, although we found some inconsistent results in Arm07ΔMGF-inoculated pigs (Groups 4b and 4f), the time to detection was slightly faster for p11.5-indirect ELISA than for IDvet indirect ELISA. For the detection of genotype X virus-specific antibodies (Group 5), p11.5-indirect ELISA permitted more sensitive and earlier detection of humoral responses in ASFV-infected animals.

Table 4. Comparison of the performance of p11.5-indirect ELISA and commercial ELISA in the detection of specific antibodies in ASFV-infected animals.

	Genotype I		Genotype II		Genotype X
	Group 1	Group 2	Group 3	Group 4	Group 5
IDvet-indirect ELISA	7 (17)	3 (15)	3 (39)	15 (23.7)	1 (21)
p11.5-indirect ELISA	7 (15)	3 (7)	4 (38)	16 (19.1)	3 (21), 1 * (14)
Number of pigs used in the experiment.	7	3	12	20	6, 2 *

The number of animals that appeared positive for the antibodies by the end of the study is presented, and the number in parentheses indicates the mean time to a positive result after inoculation. When calculating the mean, samples that tested negative throughout the experimental period or until the death animals were excluded. *, a result obtained from a cohabitated wild boar.

4. Discussion

Currently, ASFV vaccines remain under development [21]. Therefore, disease prevention largely focuses on the implementation of effective biosecurity measures and the rapid detection and removal of affected animals. Antibody detection is particularly informative for monitoring virus circulation in the field, and it will possibly be useful for assessing immunological responsiveness in vaccinated animals when reliable vaccines become available for practical use [8]. For serological disease diagnosis, different types of serological ELISAs employing different target antigens such as indirect ELISAs using a mixture of ASFV p72, p62, and p32 (ID Screen African Swine Fever Indirect ELISA) or p30 (SVANOVIR ASFV-Ab, INDICAL, Uppsala, Sweden), blocking ELISA using p72 (INGEZIM PPA COMPAC, Ingenasa, Madrid, Spain), and competitive ELISA using p32 (ID Screen African Swine Fever Competition) are commercially available worldwide.

In our preliminary studies of ASFV-specific mAbs prepared from mice immunized with highly virulent viruses, we found that most of the established hybridoma clones produced antibodies against limited types of viral (mainly structural) proteins such as p72 and p11.5. Previous reports illustrated that p11.5 is the most abundantly expressed viral protein of ASFV during the replication cycle in vitro [16,17]. This suggests that p11.5 could likely be an appropriate target antigen for serological diagnosis.

p11.5 is a structural protein of ASFV encoded by the A137R gene. Gladue et al. recently reported that deletion of the gene in the highly virulent ASFV Georgia2010 (genotype II) significantly reduced its virulence in pigs, suggesting that "A137R-deficient Georgia2010" might be a vaccine candidate for ASFV [22]. Although the localization of p11.5 in a viral particle remains to be elucidated, it should be noted that both nucleotide and deduced amino acid sequences of the A137R gene are well conserved among isolates of various genotypes. Combined with its strong antigenicity, we speculate that ASFV p11.5 is an ideal target in serological investigations of various ASFVs of different genotypes. In this study, we examined the ability of p11.5-indirect ELISA to detect antibodies against genotype I, II, and X viruses and demonstrated the apparent applicability that this assay to the serological diagnosis of genetically various ASFV strains.

The competence of p11.5-indirect ELISA in early detection of ASFV-specific antibodies in affected animals was also evaluated, and with a few exceptions, the assay detected ASFV-specific antibodies in the samples at earlier times after infection than IDvet indirect ELISA regardless of the genotype tested. In addition, the assay detected antibody responses in some samples obtained from pigs inoculated with genotype II and genotype X viruses, whereas no antibody response was detected by IDvet indirect ELISA. Furthermore, in naïve animals cohabitated with wild boars infected with Kenya05/Tk-1, one susceptible wild boar tested positive by p11.5-indirect ELISA but remained negative throughout the experimental period by IDvet indirect ELISA. These results supported the superior sensitivity of p11.5-indirect ELISA compared to the commercial kit.

This novel indirect ELISA facilitated early detection of specific antibodies in ASFV-infected animals, permitting the practical identification of chronically infected pigs and precise monitoring of immunological responsiveness to ASFV raised by vaccines in future.

5. Conclusions

In this study, we developed a new indirect ELISA for serological diagnosis of ASF using the ASFV-derived p11.5 protein as the target antigen. The performance of the new ELISA was similar to or better than that of a currently available and globally used commercial product. The developed assay could permit early and reliable detection of ASF-affected animals, and it has the potential to diagnose disease caused by various strains of ASFV of different genotypes. Therefore, p11.5-indirect ELISA is promising as a tool for the serological diagnosis and field surveillance of ASF and the assessment of immunization efficacy when ASF vaccines become available in the future.

Supplementary Materials: The following supporting information can be downloaded at: https://www.mdpi.com/article/10.3390/pathogens12060774/s1, Table S1: Details of the ASFV inoculation protocol for pigs.

Author Contributions: M.W. and T.K. (Takehiro Kokuho) conceived and designed the experiments. M.W., T.K. (Tomoya Kitamura), M.I., K.-i.K. and K.M. performed the experiments. M.W. analyzed the data. K.N. provided the resources. M.W. and T.K. (Takehiro Kokuho) wrote the paper and designed the figure. All authors have read and agreed to the published version of the manuscript.

Funding: This study was conducted under the research project on "Regulatory research projects for food safety, animal health and plant protection (JPJ008617. 20319736)" funded by the Ministry of Agriculture, Forestry and Fisheries of Japan.

Institutional Review Board Statement: All the animal experiments performed in this study were approved by the animal care and use committee of the NIAH (approval numbers: 21-023, 21-052, 22-026, 22-058, 22-059).

Informed Consent Statement: Not applicable.

Data Availability Statement: Not applicable.

Acknowledgments: We thank Carrie Batten, The Pirbright Institute, Surrey, U.K. for providing ASFV OUR T88/3 isolate. We also thank Jose Manuel Sanchez-Vizcaino, Universidad Conplutense de Madrid, Madrid, Spain for providing ASFV Kenya05/Tk-1 isolate. We appreciate Ryoko Tsuji, Masayuki Shimizu, Ayaka Namikawa, Ken Ishii, Masayuki Kanda, Tatsuo Nakamura, Yuki Kimura and Yuki Takahashi for helping us collecting pig blood, processing serum samples, and caring for the pigs.

Conflicts of Interest: The authors declare no conflict of interest.

References

1. Blome, S.; Franzke, K.; Beer, M. African Swine Fever—A Review of Current Knowledge. *Virus Res.* **2020**, *287*, 198099. [CrossRef] [PubMed]
2. Galindo, I.; Alonso, C. African Swine Fever Virus: A Review. *Viruses* **2017**, *9*, 103. [CrossRef] [PubMed]
3. Eustace Montgomery, R. On a Form of Swine Fever Occurring in British East Africa (Kenya Colony). *J. Comp. Pathol. Ther.* **1921**, *34*, 159–191. [CrossRef]
4. Sánchez-Cordón, P.J.; Montoya, M.; Reis, A.L.; Dixon, L.K. African Swine Fever: A Re-Emerging Viral Disease Threatening the Global Pig Industry. *Vet. J.* **2018**, *233*, 41–48. [CrossRef]
5. Gogin, A.; Gerasimov, V.; Malogolovkin, A.; Kolbasov, D. African Swine Fever in the North Caucasus Region and the Russian Federation in Years 2007–2012. *Virus Res* **2013**, *173*, 198–203. [CrossRef]
6. Urbano, A.C.; Ferreira, F. African Swine Fever Control and Prevention: An Update on Vaccine Development. *Emerg. Microbes. Infect.* **2022**, *11*, 2021–2033. [CrossRef] [PubMed]
7. Zhang, H.; Zhao, S.; Zhang, H.; Qin, Z.; Shan, H.; Cai, X. Vaccines for African Swine Fever: An Update. *Front. Microbiol.* **2023**, *14*, 1139494. [CrossRef] [PubMed]
8. World Organisation for Animal Health. *Chapter 15.1. Infection with African Swine Fever Virus*; World Organisation for Animal Health: Paris, France, 2022.
9. Blome, S.; Gabriel, C.; Beer, M. Pathogenesis of African Swine Fever in Domestic Pigs and European Wild Boar. *Virus Res.* **2013**, *173*, 122–130. [CrossRef] [PubMed]
10. Salguero, F.J. Comparative Pathology and Pathogenesis of African Swine Fever Infection in Swine. *Front. Vet. Sci.* **2020**, *7*, 282. [CrossRef] [PubMed]
11. Sánchez-Cordón, P.J.; Nunez, A.; Neimanis, A.; Wikström-Lassa, E.; Montoya, M.; Crooke, H.; Gavier-Widén, D. African Swine Fever: Disease Dynamics in Wild Boar Experimentally Infected with ASFV Isolates Belonging to Genotype I and II. *Viruses* **2019**, *11*, 852. [CrossRef] [PubMed]
12. Kameyama, K.; Kitamura, T.; Okadera, K.; Ikezawa, M.; Masujin, K.; Kokuho, T. Usability of Immortalized Porcine Kidney Macrophage Cultures for the Isolation of ASFV without Affecting Virulence. *Viruses* **2022**, *14*, 1794. [CrossRef] [PubMed]
13. Kanazawa, I.; Karaki, H.; Washitani, I.; Hirohashi, S.; Asashima, M.; Osumi, N.; Taniguchi, M.; Nomoto, A.; Miyashita, Y.; Yano, H.; et al. *Guidelines for Proper Conduct of Animal Experiments*; Science Council of Japan: Tokyo, Japan, 2006.
14. Kilkenny, C.; Browne, W.J.; Cuthill, I.C.; Emerson, M.; Altman, D.G. Improving Bioscience Research Reporting: The ARRIVE Guidelines for Reporting Animal Research. *Osteoarthr. Cartil.* **2012**, *20*, 256–260. [CrossRef] [PubMed]
15. King, D.P.; Reid, S.M.; Hutchings, G.H.; Grierson, S.S.; Wilkinson, P.J.; Dixon, L.K.; Bastos, A.D.S.; Drew, T.W. Development of a TaqMan® PCR Assay with Internal Amplification Control for the Detection of African Swine Fever Virus. *J. Virol. Methods* **2003**, *107*, 53–61. [CrossRef] [PubMed]
16. Nishi, T.; Okadera, K.; Fukai, K.; Yoshizaki, M.; Nakasuji, A.; Yoneyama, S.; Kokuho, T. Establishment of a Direct PCR Assay for Simultaneous Differential Diagnosis of African Swine Fever and Classical Swine Fever Using Crude Tissue Samples. *Viruses* **2022**, *14*, 498. [CrossRef] [PubMed]
17. Kitamura, T.; Masujin, K.; Yamazoe, R.; Kameyama, K.; Watanabe, M.; Ikezawa, M.; Yamada, M.; Kokuho, T. A Spontaneously Occurring African Swine Fever Virus with 11 Gene Deletions Partially Protects Pigs Challenged with the Parental Strain. *Viruses* **2023**, *15*, 311. [CrossRef] [PubMed]

18. Boinas, F.S.; Hutchings, G.H.; Dixon, L.K.; Wilkinson, P.J. Characterization of Pathogenic and Non-Pathogenic African Swine Fever Virus Isolates from *Ornithodoros erraticus* Inhabiting Pig Premises in Portugal. *J. Gen. Virol.* **2004**, *85*, 2177–2187. [CrossRef] [PubMed]
19. Masujin, K.; Kitamura, T.; Kameyama, K.I.; Okadera, K.; Nishi, T.; Takenouchi, T.; Kitani, H.; Kokuho, T. An Immortalized Porcine Macrophage Cell Line Competent for the Isolation of African Swine Fever Virus. *Sci. Rep.* **2021**, *11*, 4759. [CrossRef] [PubMed]
20. Kanda, Y. Investigation of the Freely Available Easy-to-Use Software 'EZR' for Medical Statistics. *Bone Marrow Transpl.* **2013**, *48*, 452–458. [CrossRef] [PubMed]
21. Wang, T.; Sun, Y.; Huang, S.; Qiu, H.-J. Multifaceted Immune Responses to African Swine Fever Virus: Implications for Vaccine Development. *Vet. Microbiol.* **2020**, *249*, 108832. [CrossRef] [PubMed]
22. Gladue, D.P.; Ramirez-Medina, E.; Vuono, E.; Silva, E.; Rai, A.; Pruitt, S.; Espinoza, N.; Velazquez-Salinas, L.; Borca, M.V. Deletion of the A137R Gene from the Pandemic Strain of African Swine Fever Virus Attenuates the Strain and Offers Protection against the Virulent Pandemic Virus. *J. Virol.* **2021**, *95*, e01139-21. [CrossRef] [PubMed]

MDPI AG
Grosspeteranlage 5
4052 Basel
Switzerland
Tel.: +41 61 683 77 34

Pathogens Editorial Office
E-mail: pathogens@mdpi.com
www.mdpi.com/journal/pathogens

www.ingramcontent.com/pod-product-compliance
Lightning Source LLC
Chambersburg PA
CBHW041417210926
51850CB00013B/3301